MOBILE ROBOTS – CONTROL ARCHITECTURES, BIO-INTERFACING, NAVIGATION, MULTI ROBOT MOTION PLANNING AND OPERATOR TRAINING

Edited by **Janusz Będkowski**

Mobile Robots – Control Architectures, Bio-Interfacing, Navigation, Multi Robot Motion Planning and Operator Training
Edited by Janusz Będkowski

Published by InTech

Janeza Trdine 9, 51000 Rijeka, Croatia

Publishing Process Manager Bojana Zelenika

Technical Editor Teodora Smiljanic

Cover Designer InTech Design Team

Additional hard copies can be obtained from orders@intechopen.com

Mobile Robots – Control Architectures, Bio-Interfacing, Navigation, Multi Robot Motion Planning and Operator Training, Edited by Janusz Będkowski
p. cm.
ISBN 978-953-307-842-7

For EU product safety concerns:
IN TECH d.o.o., Prolaz Marije Krucifikse Kozulić 3, 51000 Rijeka, Croatia, info@intechopen.com or visit our website at intechopen.com.

Meet the editor

Janusz Bedkowski, Ph.D. in Automation and Robotics, adjunct in Institute of Automation and Robotics-Warsaw University of Technology. From 2006 to 2011, he collaborated with Industrial Research Institute for Automation and Measurements, Warsaw, Poland, concerning multi robotic inspection-intervention system development and integration. From 2009 to 2011, he collaborated with Institute of Mathematical Machines, Warsaw, Poland concerning mobile robot operator training tool development and semantic mapping. Dr. Bedkowski was granted a postdoctoral scholarship (CAS/19/POKL 15.05.2011-15.11.2011) in the Royal Military Academy (RMA), Brussels, Belgium, funded by Center for Advanced Studies, Warsaw University of Technology. His research interests are in inspection intervention robot systems, semantic mapping including 3D cloud of points processing, 6D SLAM, mobile robot operator training with AR techniques, and GPGPU computing applied for mobile robotics.

Contents

Preface

The objective of this book is to cover the advances of mobile robotic systems and related technologies applied especially for multi robot systems' design and development. The design of mobile robots control system is an important and complex issue, requiring the application of information technologies to link the robots into a single network. In recent years, a great number of studies of mobile robot applications have been proposed but still there is a need to provide software tools to integrate hardware and software of the robotic platforms. The robot control software consists of many interacting components and often possible accidents and problems arise from components interactions rather than the failure of individual ones. Autonomous robots may operate unattended and through an unsafe operation which can cause significant human, economic, or mission losses. Hence, to minimize the losses, software becomes crucial in robot control. This book discusses several robot control architectures, especially those considering safety and security.

Human robot interface becomes a demanding issue, especially when we try to use sophisticated methods for brain signal processing. Many researchers are interested in using neurophysiological recordings. The interaction between bioartificial brains and mobile robots is also an important issue, where new technologies for direct transfer of information between natural neuronal systems and artificial devices are investigated. The electrophysiological signals generated from the brain can be used to command different devices, such as cars, wheelchair or even video games. Devices that can interpret brain activity and use it to control artificial components are referred to as brain–machine interfaces and are rapidly developed for not only scientific but also commercial purposes.

During the last decade a number of developments in navigation and path planning including parallel programming improving the performance of classic approaches could be observed. Real applications often require the collaboration of mobile robots in order to perform required tasks. A cooperative path planning and a formation control of multi robotic agents will be discussed. Also, technical problems related to communication and distance measurement between agents will be shown. The design of a network architecture that allows mobile robots to cooperate efficiently, without negatively impacting the performance of each intra-robot communication is proposed.

Training of mobile robot operators is a very difficult task, not only because of the complexity of the mobile robot, but also because of several factors related to different task execution. The presented improvement is related to the problem of automatic environment model generation based on autonomous mobile robot observations. The approach related to semantic mapping is used and, in consequence, semantic simulation engine is implemented. The presented result is a new approach and can potentially improve the process of advanced mobile robot application design and professional training of the operators.

We have included seventeen reviewed chapters organized into five sections. We would like to thank all the authors for their contributions.

Janusz Będkowski, Ph.D.
Warsaw University of Technology
Poland

Introductory Chapter

Janusz Będkowski
Institute of Automation and Robotics, Warsaw University of Technology
Industrial Research Institute for Automation and Measurements
Poland

1. Introduction

This chapter is related to the recent advances in the mobile robotics for RISE (Risky Intervention and Surveillance Environment) applications. In RISE applications mobile robots are used for assisting human in risky tasks such as: victim search, victim transportation, hazardous materials disposal etc.

The State of The Art of the information systems Ahmad & Sumari (2008), artificial intelligence Ullman (2002) Herbert (1995), and mobile robotics Choset (2001) Braunl (2006) allows for development of increasingly sophisticated multi-robot systems Maslowski (2004) Baudoin et al. (2009) Arai et al. (2002). Available computing power of modern computers, and miniaturization has repeatedly increased the possibility of computational capabilities of new robots in the field of pattern recognition Cao et al. (2006), the construction of maps Yu & Hang (2006) Xiang & Zhou (2006) Thrun et al. (2001), communication Okada & Murakami (2007) in order to execute the mission. It is no longer a problem to execute complex calculations directly on the onboard PC of mobile robot, and as a result these machines become more autonomous. The following description of available technical solutions was limited to the representative examples of inspection - intervention robots.

The chapter is organized as follows: in section 2 State of the Art concerning inspection - intervention systems is shown. Section 3 concerns the examples of autonomous mobile robots and section 4 concerns the examples of remote-controlled mobile robots used in RISE applications. An example of command base stations is shown in section 5. Control systems of mobile robots are described in section 6. In section 7 an important aspect in RISE application SLAM is described and in section 8 the training systems for RISE are discussed. Section 9 concludes the chapter.

2. Inspection - intervention system

Mobile robot systems are used in inspection-intervention activities Maslowski (2005) Maslowski (2006). The basic element of the system is a mobile platform that is equipped with an advanced measurement system. The main task of the robot is to provide maximum amount of information concerning the environment directly to command center. The system of mobile inspection - intervention robots should enable the identification of hazards in the area and the possible intervention by remote-controlled robot equipped with a specialized arm. It is important to emphasize that mobile robots are usually a part of multi agent systems Leitao & Putnik (2001). It is possible to use different types of mobile units in an unknown terrain, thus there is the need of integration of incompatible interfaces in terms of systems from different manufacturers Alami et al. (1998). Integration is necessary to obtain a web connected mobile robot system. The illustration 1 shows a simplified diagram of the architecture of View Finder VF (2009) system. View-Finder system Baudoin et al. (2009) "Vision and Chemiresistor Equipped Web-connected Finding Robots" is a structure of web connected mobile robots, with the task of carrying out the inspection of area in which human life and health are highly threatened, cooperating with the services of the fire brigade during an inspection of chemically contaminated area. In the system, the autonomous robots are used for acquiring information from sensors installed on them, and remotely controlled robots performing intervention tasks. Robots first transmit measured data to a mobile base station, and then the data are available on CMIS system (Crisis Management Information System).

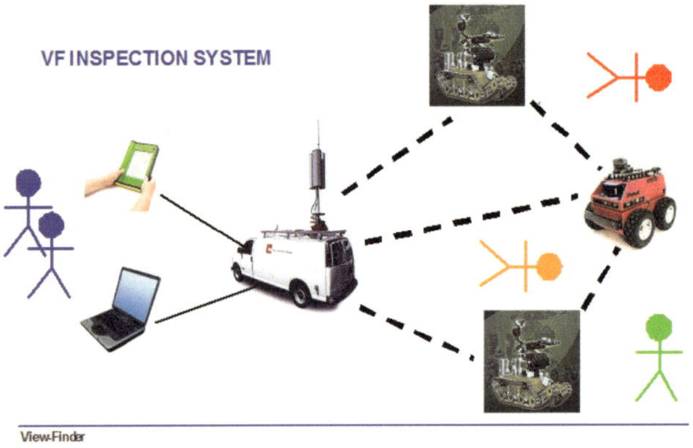

Fig. 1. Simplified diagram of the system architecture View-Finder VF (2009).

An example of an advanced mobile robot system is a border inspection system TALOS TALOS (2010). The system comprises three basic elements: unmanned ground vehicles and aircraft, whose job is to patrol the border, ground observation towers located on mobile platforms and command center to ensure the connectivity between components located on the ground, and the appropriate unit of the Border Guard. Another example of new ideas associated with inspection - intervention systems is the PROTEUS project PROTEUS (2011). A new element of the system is unmanned air vehicles supporting ground units.

In general multi-robot inspection-intervention system includes remote-controlled inspection-intervention robots with elements of autonomy, a fully autonomous mobile robots and mobile base station. All system components should be equipped with appropriate sensors to recognize the conditions in the environment and being able to share this information between system nodes.

3. Autonomous mobile robots

In this section a crucial examples from RISE applications point of view of autonomous mobile robots which are constructed for working in difficult circumstances Railhet et al. (2007), to search for victims, in the areas of high risk to human Murphy et al. (2000) Jacoff et al. (2002)

are shown. These robots are equipped with: a stereo microphone, accelerometer, camera Lee et al. (2005), chemical sensors Stetter et al. (2003), infrared sensors, infrared camera, sound sensors Valin et al. (2007), ultrasonic sensors and laser range finders.
CASTER robot (Figure 2) is an example robot capable of working under difficult conditions. The advantage over other robots is advanced construction, with an extensive module for victims detection, an advanced module for map building and modern human-machine interface. It uses sophisticated mapping technology based on 3D sensor-time-of-flight range camera Kadous et al. (2005).

Fig. 2. Robot Caster Kadous et al. (2005).

Figure 3 shows commercially available robots designed for scientific and research purposes with potential application in developing inspection - intervention systems. ATRVJr robot Lima et al. (2003) is equipped with basic sensors used for navigation and map construction. The construction of four-wheel differential drive is designed to overcome substantial obstacles, which makes the robot attractive in terms of test application. SeekurJr robot is successor of ATRVJr robot and it is designed also for scientific and research purposes. A distinguishing feature of this design is the integration of advanced laser sensors to the robot housing, so that achieved a greater degree of integrity of the mobile base. In addition use of waterproof materials in the robot chassis should be noted. This makes this design very attractive compared to another available solutions.

4. Remote-controlled mobile robots

The remote-controlled robots are developed in different sizes depending on the applications, which consequently affects their functionality. One can distinguish several classes of remotely controlled robots dedicated to the inspection-intervention tasks. They are:

- heavy inspection - intervention robots;
- medium inspection - intervention robots;
- lightweight inspection - intervention robots;

(a) ATRVJr (b) SeecurJr

Fig. 3. Autonomous mobile robots used in research.

- miniature inspection robots.

Inspection - intervention robots, depending on the application, differ not only in dimensions, but also equipment, including the type of sensors that are used. Figure 4 shows the heavy inspection - intervention Robot T-52 Enryu Hiyama (2004). This robot is distinguished by two arms mounted to move heavy objects of considerable size. This mobile unit is used, for example, for searching the rubble of disaster zones.

(a) Robot T-52 Enryu. (b) RISE task execution.

Fig. 4. Remote-controlled mobile robot T-52 Enryu.

Figure 5 shows a commercially available inspection - intervention INSPECTOR robot of middle class. This robot Maslowski (2001) is dedicated for supporting special forces in risky interventions Szynkarczyk (2005). INSPECTOR robot is built using modular construction, so that when damage occur to one of the modules as a result of a failure, the rest of the robot system retains its functionality, which has a positive impact on the completion of the mission task.

Fig. 5. Remotely controlled robot INSPECTOR.

In the area of direct rescue of human life, robots are used to transport victims of the crisis. Performing safe transport of humans is a difficult task from the perspective of remote control of mobile robot. Figure 6 shows an example of a robot equipped with two arms, performing the task of transporting accident victims.

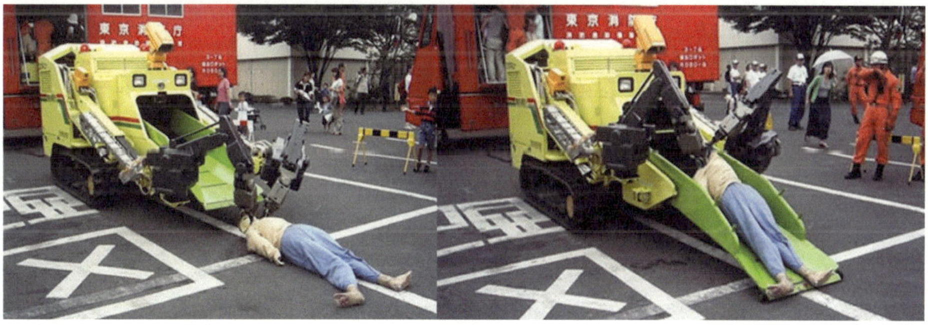

Fig. 6. Transporting accident victim.

Another group of inspection robots are small-size mobile robots. Figure 7 left shows the robot YUKO-SHITA developed by Sanyo. Figure 7 right shows a miniature robot TerminatorBot Voyles et al. (2004). It is able to effectively move in tight spaces with a pair of mechanical limbs. It is equipped with miniature video cameras, and is being used to search for victims in complex environment.

Figure 8 presents a hybrid technical solution dedicated to inspection - intervention system, unmanned ground/surface vehicle LEWIATAN Typiak (2008), which is able to overcome watery obstacles.

Based on these few examples given above it can be concluded that there are enough different types of mobile platforms to build a robotic inspection - intervention system. However, the major concern remains the development of multi - agent system structure, based on the robots of different manufacturers, designed to perform inspection - intervention tasks. Therefore the main effort is usually related to the integration technologies.

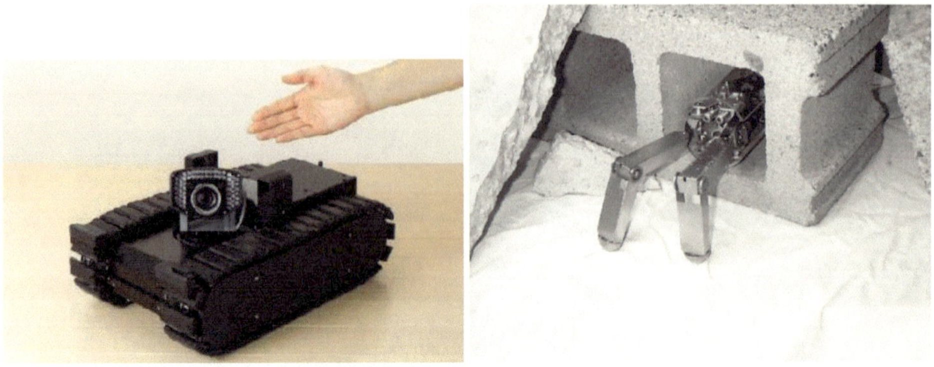

(a) Inspection robot YUKO-SHITA Sanyo. (b) Miniature robot TerminatorBot.

Fig. 7. Small-size inspection mobile robots.

Fig. 8. Unmanned ground/surface vehicle LEWIATAN.

5. Command base stations

An important subsystem of multi robot inspection - intervention system is the mobile command base station. Figure 9 a) shows the command base station of the robot Hiyama (2004) T-52 Enryu. The station has a collection of many displays of the view from the robot's cameras, and advanced ergonomic manipulator, to facilitate remote control of the robot's arm. Another example is the mobile command base station Typiak (2008) constructed in the Military University of Technology in Warsaw is shown in Figure 9 b). Mobile command base station (in this case operator's console) provides possibility to control the unmanned vehicle through the steering wheel and series of specialized switches. The construction of the control station limits the ergonomics and provides complex functionality in the form of numerous displays of robot state, the measurements from the sensors of the mobile platform and the view from on-board cameras. Figure 10 shows the functional architecture of the mobile base station developed in View Finder project Warlet et al. (2009). View Finder mobile base station is one of the two components of a mobile command center, a second component is a CMIS - Crisis Management Information System. The base station can distinguish the following components: BSC - the core of the base station which is a basic functionality of the station, MTE - mission editor, a tool for editing the mission, MPSEM - mission planning and execution monitoring module, MDRD - data recording during the execution of missions and sharing

them with other system components, HMI Clients - programs for human-machine interface, SDP - processing the data from sensors, interface for CMIS- crisis management information system, interface for robots and sensors.
To conclude this section it can be stated that modern mobile command base stations are using advanced information technologies in the field of: interactive computer graphics, distributed programming, databases, artificial intelligence, work with peripherals.

(a) Command base station of robot T-52 Enryu.

(b) Command base station of unmanned ground/surface vehicle LEWIATAN.

Fig. 9. Example of command base stations.

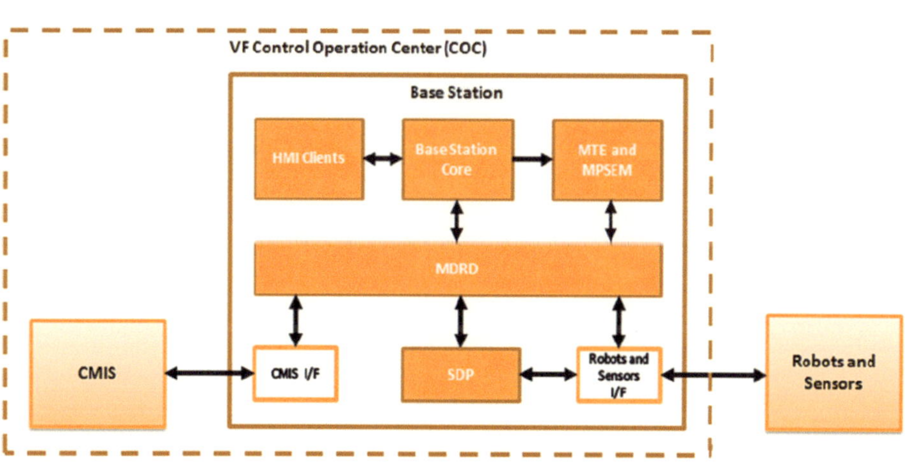

Fig. 10. Functional architecture of the mobile base station working in the View-Finder system Warlet et al. (2009).

6. Control systems of mobile robots

Control system of mobile robots is an important and complex issue, requiring the application of information technologies to link the robots into a single network. It is possible to use multi-agent system architecture Kramer & Scheutz (2007), a network of agents Tambe et al. (1995) and co-applicants Guilbert et al. (2003). Agents are characterized by autonomous action, mostly heterogeneous Baumann (1999) Hilaire et al. (2000). Such control system architecture is characterized by a variety of hardware components and software, often working under different operating systems. From the standpoint of software engineering it can be concluded, that several tools supporting the development of systems for mobile robots control had been developed. Below most common of them are described:

6.1 GenoM

It is an environment for supporting the design of control systems operating in real time Alami et al. (1998) Alami (2000). Environment helps in the process of software components integration available through numerous independent functional modules. The module can be remotely switched on, off, paused, and parameterized. Software modules are implemented in a form of standardized servers. Each module consists of two basic parts: a controller cooperating with the client program and the controller performing the current task.

6.2 DCA

Distributed Control Architecture DCA is designed to implement a robot arm control system Peterson et al. (2001). DCA architecture provides application messaging environment ACE (Adaptive Communication Environment), it can provide compatibility with other control systems. DCA architecture is characterized by a hierarchical structure, consisting of modules, supervisors and controllers.

6.3 MIRO

Architecture Miro (Middleware for Robots) Miro (2005) is object-oriented. The core of the architecture has been developed in C++ in Linux environment using CORBA and ACE technology. In connection with the use of CORBA, MIRO is independent of operating system or programming language. MIRO allows to control several robots, B21 and MobileRobots Pioneer. There have been many studies using MIRO, including maps construction algorithm and simultaneous localization SLAM Kraetzschmar et al. (2004), the algorithm of planning and navigation Kraetzschmar et al. (2000), a hierarchical behavior algorithm Utz et al. (2005) and multi - robot systems H. Utz (2004).

6.4 MARIE

Architecture MARIE Cote et al. (2004) (Mobile and Autonomous Robotics Integration Environment) supports designing cross-platform applications. MARIE is implemented in C++ using ACE. MARIE consists of following components: applications component, communication component, communication manager and manager applications. For the compatibility purpose the corresponding components for Player, CARMEN and Aria are delivered. There have been many systems using this architecture Lemay et al. (2004).

6.5 Player

The Player environment provides servers for configuration of multi - robot control system acting on the basis of defined interfaces Gerkey & Matari'c (2002) Gerkey et al. (2003) Gerkey

& Matari'c (2004) Gerkey et al. (2005). Therefore, the communication based on TCP / IP control system for mobile robots can be fully distributed and implemented in different programming languages (including C, C++, Tcl, Python, Java, Common Lisp). Player allows to connect multiple clients to one server, giving a degree of design freedom. It is important to emphasize that Player is common in research community.

6.6 MCA2

Architecture MCA2 (2009) (Modular Control Architecture) can be used for real time robot operating system design. Mainly dedicated to multi - robot applications. All features are implemented as modules that can be associated to groups. MCA2 distinctive feature is the usage of only floating point numbers as input and output between software components. Communication occurs through the use of low-level mechanisms that support SOCKETS API. MCA2 advantage is the ability to implement modules in real-time systems in the RT-Linux.

6.7 TeamBots

TeamBots is a collection of programs implemented using the Java language, dedicated to multi - agent systems Balch & Arkin (1999) Balch (2000) Balch (2004). The architecture supports drivers for robot Nomad 150. TeamBots developed appropriate modules implementing the control algorithms using the "hierarchical behavior" and goal oriented control. Communication between the modules is solved using SOCKETS and serial communication.

6.8 MissionLab

Software MissionLab Missionlab (2003) is dedicated to multi - robot systems including mission planning based on the model of military plans. Coexistence of agents, collaboration and coordination has been carried out in accordance with the idea of multi-agent systems. In MissionLab, a new agent description language CDL (Configuration Description Language) is proposed. The implementation of several drivers is available for the following robots: MobileRobots Pioneer, Robot ATRVJr, Urban, Evolution Robotics ER-1, Nomad 150/200. CDL language is compiled into the code CNL (Configuration Network Language) and as a result compiled into an executable program directly on the system. There have been many studies using MissionLab including high-level user assistance for robot mission specification Endo et al. (2004).

6.9 ARIA

It is basic software for MobileRobots' robots LaFary & Newton (2005). ARIA provides a set of C++ classes for the core functionality of the robot. ARIA architecture defines the structure of the robot consisting of sensors, effectors, and defines the physical elements of the low-level communication mechanisms. The software is fully commercial, offering a complete set of functionality needed to design multi - robot systems.

6.10 CARMEN

CARMEN Montemerlo et al. (2003) is written using C language to implement a uniform layer of communication for multi - robot systems. The architecture defines three main layers: the layer of interfaces for devices, data acquisition from sensors and control effectors, the second layer carrying out basic tasks for a robot: navigation, localization, tracking facilities, traffic planning, executing, and the third layer of applications designed by the user. CARMEN supports the following mobile platforms MobileRobotics family, ie: Nomadic Technologies, Scaut, XR4000, Segway, iRobot ATRV, ATRVJr, B21R. There have been many studies using CARMEN for example related to hierarchical models of activity Osentoski et al. (2004).

6.11 MRROC++

Program structure MRROC++ framework was based on theoretical considerations Winiarski & Zielinski (2005) on the concept of agent. MRROC++ consists of functional modules and patterns of use. With MRROC++ it is possible to construct a multi - robot system based on operating implementation of the relevant modules. The control system consists of the following processes: UI (User Interface Process) - the process responsible for communication with the operator, MP (Master Process) - the process of coordinating the work of all effectors, ECP (Effector Control Process) - the process responsible for the implementation of the tasks commissioned with effector, EDP (Effector Driver Process) - the process responsible for steering the effector, VSP (Virtual Sensor Process) - the process responsible for the aggregation of data from the sensors.

6.12 MRDS

Microsoft Robotics Developer Studio is a new project supporting the development of multi - robot systems. It is a powerful tool, through the use of modern communication technologies and simulation environment based on NVIDIA PhysX technology Buckhaults (2009). The MRDS supports building multi - robot systems. In addition, MRDS offers easy HMI design for web browsers.

6.13 CoRoBa

Architecture CoRoBa (Controlling Robots with CORBA) is organizing a mobile robot control system or a multi - robot system into three basic types of modules: SENSOR, PROCESSOR, ACTUATOR, where SENSOR is responsible for collecting data from the sensors, PROCESSOR makes logical operations, and the ACTUATOR controls the effectors. Additionally CoRoBa architecture defines mechanisms for sleep, wake and remote on / off data components. CORBA is independent from programming language or operating system Colon (2006).

6.14 ROS

ROS (Robot Operating System) ROS (2011) is a software framework for robot software development. It is providing operating system-like functionality for all mobile robot components. ROS provides standard operating system services such as hardware abstraction, low-level device control, implementation of commonly-used functionality, message-passing between processes, and package management, therefore it is used in modern robotic applications. Unfortunately the library is dedicated to a Unix-like systems.

7. SLAM in RISE

It seems that from RISE application point of view an interesting SLAM solution is so called 6DSLAM Nuchter et al. (2007) Sprickerhof et al. (2009). Alignment and merging of two 3D scans, which are obtained from different sensor coordinates, with respect to a reference coordinate system is called 3D registration Huber & Hebert (2003) Fitzgibbon (2001) Magnusson & Duckett (2005). Park Park et al. (2010) proposed a real-time approach for 3D registration using GPU, where the registration technique is based on the Iterative Projection Point (IPP) algorithm. The IPP technique is a combination of point-to-plane and

point-to-projection registration schemes Park & Subbarao (2003). Processing time for this approach is about 60*ms* for aligning 2 3D data sets of 76800 points during 30 iterations of the IPP algorithm. Unfortunately, the IPP algorithm has a problem concerning the scalability of the implementation. Fast searching algorithms such as the k-d tree algorithm are usually used to improve the performance of the closest point search Rusinkiewicz & Levoy (2001). GPU accelerated nearest neighbor search for 3D registration is proposed in work of Qiu Qiu et al. (2009).
A fast variant of the Iterative Closest Points (ICP) algorithm that registers the 3D scans in a common coordinate system and relocalizes the robot is shown in Surmann et al. (2004). Consistent 3D maps are generated using closing loop detection and a global relaxation. The loop closing algorithm detects a loop by registering the last acquired 3D scan with earlier acquired scans, e.g., the first scan. If a registration is possible, the computed error is in a first step divided by the number of 3D scans in the loop and distributed over all scans. The authors reported that the computation time of about 1 min per scan is acceptable, but that further improvement is needed. An algorithm for efficient loop closing and consistent scan alignment that avoids iterative scan matching over all scans is proposed in Sprickerhof et al. (2009). Detecting loops in the path is done by using the Euclidean distance between the current and all previous poses (distance threshold of 5 meters), or using GPS data if available. A threshold of minimal number of intermediate scans (e.g., 20) is used to circumvent continuous loop closing within consecutive scans.
Another scan registration approach using 3D-NDT (Normal Distribution Transform) is shown in Magnusson et al. (2007) and automatic appearance-based loop detection from 3D laser data using the Normal Distributions Transform is demonstrated in Magnusson et al. (2009).
Vision has also been used successfully for localization of a mobile robot Newman et al. (2006) Newman & Ho (2005) Cummins & Newman (2008). A comparison of loop closing techniques in monocular SLAM is show in Williams et al. (2009). Loop closure detection in SLAM by combining visual and spatial appearance was shown in Ho & Newman (2006). This approach relies upon matching distinctive signatures of individual local scenes to prompt loop closure. Another advantage is the possibility to enhance robustness of loop closure detection by incorporating heterogeneous sensory observations. The laser scan is divided into smaller but sizeable segments and the complexity of a segment is encoded via entropy. SIFT Lowe (2004) (Scale Invariant Feature Transform) descriptors are also used to match images. In Leong & Newman (2005) a methodology combining visual and spatial appearance is shown, whereas in Ho & Newman (2005), an approach based on multiple map intersection detection based on visual features appearance is shown. The authors in Ho & Newman (2007) encode the similarity between all possible pairings of scenes in a similarity matrix and then pose the loop closing problem as the task of extracting statistically significant sequences of similar scenes from this matrix. The analysis (introspection) and decomposition (remediation) of the similarity matrix allows for the reliable detection of loops despite the presence of repetitive and visually ambiguous scenes. Relaxing loop-closing errors in 3D maps based on planar surface patches were shown in Kaustubh Pathak & Birk (2009).

8. Training in RISE

A detailed description of computer based simulators for unmanned vehicles is shown in Craighead et al. (2007b). Also in Boeing & Bräunl (2007) the comparison of real-time physics simulation systems is given, where a qualitative evaluation of a number of free publicly available physics engines for simulation systems and game development is presented. Several

frameworks are mentioned such as USARSim which is very popular in research society Wang et al. (2003c) Wang et al. (2003a) Wang et al. (2003b) Balaguer et al. (2008) Greggio et al. (2007), Stage, Gazebo Rusu et al. (2007) Karimian et al. (2006), Webots Hohl et al. (2006) Michel (2004), and MRDS Buckhaults (2009) Sallé et al. (2007). Some researchers found that there are many available simulators that offer attractive functionality, therefore they proposed a new simulator classification system specific to mobile robots and autonomous vehicles Craighead et al. (2007a). A classification system for robot simulators will allow researchers to identify existing simulators which may be useful in conducting a wide variety of robotics research from testing low level or autonomous control to human robot interaction.
Training of the mobile robot operators is very difficult task not only because of the complexity of the mobile robot but also because of several factors related to different task execution. The study of the human-robot interactions (HRI) during a real rescue is presented in Casper & Murphy (2003) Bedkowski, Kowalski & Piszczek (2009). The research on simulation and training systems for mobile robots is shown in Xuewen et al. (2006). Another simulation engine - the Search and Rescue Game Environment (SARGE), which is a distributed multi-player robot operator training game, is described in Craighead, Burke & Murphy (2008) Craighead (2008) Craighead, Gutierrez, Burke & Murphy (2008). An advance computer techniques such augmented reality approach applied in training robotics is shown in Kowalski et al. (2008)Bravo & Alberto (2009). The simulation environment used as a testbed for simulation based approach for unmanned system command and control is demonstrated in Drewes (2006). The Modeling and simulation for the mobile robot operator training tool was presented in Bedkowski, Piszczek, Kowalski & Maslowski (2009).

9. Conclusion

This chapter is related to recent advances in mobile robotics for RISE (Risky Intervention and Surveillance Environment) applications. In RISE applications mobile robots are used for assisting human in risky tasks such as: victim search, victim transportation, hazardous materials disposal etc. From State of the Art discussed in this chapter it can be stated that it is no longer a problem to execute complex calculations directly on the onboard PC of mobile robot, and as a result these machines become more autonomous. Instead of this fact an effort has to be done to develop new software techniques to build secure and safety robotic platforms. Still an open problem is effective human robot interaction because of increased autonomy of robots. Another goal is an integration of State of the Art robotic technology to build sophisticated systems of collaborative autonomous robots.

10. References

Ahmad, A. S. & Sumari, A. D. W. (2008). *Multi-Agent Information Inferencing Fusion For Integrated Information System, Information Science and Computing Series*, School of Electrical Engineering and Informatics, Institut Teknologi Bandung, Bandung, Indonesia.

Alami, R. (2000). Around the lab in 40 days, *IEEE ICRA conference 2000*, San Francisco (USA), pp. 88–94.

Alami, R., Fleury, R., Fleury, S. & Ingand, F. (1998). An architecture for autosomy, *International Journal of Robotics Research, special issue on Integrated Architectures for Robot Control and Programming* .

Arai, T., Pagello, E. & Parker, L. E. (2002). Advances in multi-robot systems, *IEEE Transactions on robotics and automation* 18(5): 655–661.

Balaguer, B., Balakirsky, S., Carpin, S., Lewis, M. & Scrapper, C. (2008). Usarsim: a validated simulator for research in robotics and automation, *Workshop on Robot simulators: available software, scientific applications and future trends at IEEE/RSJ IROS*.

Balch, T. (2000). Hierarchic social entropy: An information theoretic measure of robot group diversity, *Autonomous Robots* 8(3): 209–238.

Balch, T. (2004). Teambots, http://www.teambots.org/.

Balch, T. & Arkin, R. (1999). Behavior-based formation control for multi-robot teams, *IEEE Transactions on Robotics and Automation* 20(5).

Baudoin, Y., Doroftei, D., Cubber, G. D., Berrabah, S. A., Pinzon, C., Penders, J., Maslowski, A. & Bedkowski, J. (2009). View-finder: Robotics assistance to fire-fighting services, *International Workshop On Robotics for risky interventions and Environmental Surveillance RISE09*, Brussels.

Baumann, J. (1999). Control algorithms for mobile agents, *Institut fur Parallele und Verteilte Hochstleistungrechner der Universitat Stuttgart* .

Bedkowski, J., Kowalski, P. & Piszczek, J. (2009). Hmi for multi robotic inspection system for risky intervention and environmental surveillance, *Mobile Robotics: Solutions and Challenges, Proceedings of the Twelfth International Conference on Climbing and Walking Robots and the Support Technologies for Mobile Machines*.

Bedkowski, J., Piszczek, J., Kowalski, P. & Maslowski, A. (2009). Modeling and simulation for the mobile robot operator training tool, *The 5th International Conference on Information and Communication Technology and Systems (ICTS), Surabaya, Indonesia*, pp. 229–236.

Boeing, A. & Bräunl, T. (2007). Evaluation of real-time physics simulation systems, *GRAPHITE '07: Proceedings of the 5th international conference on Computer graphics and interactive techniques in Australia and Southeast Asia*, ACM, New York, NY, USA, pp. 281–288.

Braunl, T. (2006). *Embedded Robotics Mobile Robot Design and Applications*, Springer Verlag.

Bravo, J. & Alberto, C. (2009). An augmented reality interface for training robotics through the web, *Proceedings of the 40th International Symposium on Robotics. Barcelona : AER-ATP, ISBN 978-84-920933-8-0*, pp. 189–194.

Buckhaults, C. (2009). Increasing computer science participation in the first robotics competition with robot simulation, *ACM-SE 47: Proceedings of the 47th Annual Southeast Regional Conference*, ACM, New York, NY, USA, pp. 1–4.

Cao, F., Tung, A. K. H. & Zhou, A. (2006). Scalable clustering using graphics processor, *Lecture Notes in Computer Science* 4016/2006: 372–384.

Casper, J. & Murphy, R. R. (2003). Human-robot interactions during the robot-assisted urban search and rescue response at the World Trade Center., *IEEE transactions on systems, man, and cybernetics. Part B, Cybernetics : a publication of the IEEE Systems, Man, and Cybernetics Society* 33(3): 367–385.
URL: *http://dx.doi.org/10.1109/TSMCB.2003.811794*

Choset, H. (2001). Coverage for robotics - a survey of recent result, *Annals of Mathematical and Artificial Intelligence* 31: 113–126.

Colon, E. (2006). Installation and configuration of CoRoBa (including MoRoS3D), technical report, royal military academy.

Cote, C., Letourneau, D., Michaud, F., Valin, J. M., Brosseau, Y., Raievsky, C., Lemay, M. & Tran, V. (2004). Code reusability tools for programming mobile robots, *Proc. IEEE/RSJ International Conference on Intelligent Robots and Systems (IROS)*.

Craighead, J. (2008). Distributed, game-based, intelligent tutoring systems - the next step in computer based training?, *Proceedings of the International Symposium on Collaborative Technologies and Systems (CTS 2008)*.

Craighead, J., Burke, J. & Murphy, R. (2008). Using the unity game engine to develop sarge: A case study, *Proceedings of the 2008 Simulation Workshop at the International Conference on Intelligent Robots and Systems (IROS 2008)*.

Craighead, J., Gutierrez, R., Burke, J. & Murphy, R. (2008). Validating the search and rescue game environment as a robot simulator by performing a simulated anomaly detection task, *Proceedings of the 2008 International Conference on Intelligent Robots and Systems (IROS 2008)*.

Craighead, J., Murphy, R., Burke, J. & Goldiez, B. (2007a). A robot simulator classification system for hri, *Proceedings of the 2007 International Symposium on Collaborative Technologies and Systems (CTS 2007)*, pp. 93–98.

Craighead, J., Murphy, R., Burke, J. & Goldiez, B. (2007b). A survey of commercial and open source unmanned vehicle simulators, *Proceedings of ICRA*.

Cummins, M. & Newman, P. (2008). Fab-map: Probabilistic localization and mapping in the space of appearance, *The International Journal of Robotics Research* 27(6): 647–665.

Drewes, P. (2006). Simulation based approach for unmanned system command and control, *Conference on Recent Advances in Robotics, FCRAR Miami, Florida*, pp. 189–194.

Endo, Y., MacKenzie, D. & Arkin, R. (2004). Usability evaluation of high-level user assistance for robot mission specification, *IEEE Transactions on Systems, Man, and Cybernetics* 34(2): 168–180.

Fitzgibbon, A. W. (2001). Robust registration of 2d and 3d point sets, *In British Machine Vision Conference*, pp. 411–420.

Gerkey, B., Howard, A. & Vaughan, R. (2005). Player/Stage, http://playerstage.sourceforge.net/.

Gerkey, B. & Matari'c, M. (2002). Sold!: Auction methods for multi-robot coordination, *IEEE Transactions on Robotics and Automation, Special Issue on Multi-Robot Systems* 18(5): 758–768.

Gerkey, B. & Matari'c, M. (2004). Are (explicit) multi-robot coordination and multi-agent coordination really so different?, *In Proceedings of the AAAI spring symposium on bridging the multi-agent and multi-robotic research gap*, pp. 1–3.

Gerkey, B., Vaughan, R. & Howard, A. (2003). The player/stage project: Tools for multi-robot and distributed sensor systems, *In Proceedings of the 11th international conference on advanced robotics*, Coimbra, Portugal, pp. 317–323.

Greggio, N., Silvestri, G., Menegatti, E. & Pagello, E. (2007). A realistic simulation of a humanoid robot in usarsim, *Proceeding of the 4th International Symposium on Mechatronics and its Applications (ISMA07), Sharjah, U.A.E.*

Guilbert, N., Beauregard, M., Michaud, F. & Lafontaine, J. (2003). Emulation of collaborative driving systems using mobile robots, *In Proceedings IEEE conference on systems, man, and cybernetics*, pp. 856–861.

H. Utz, F. Stulp, A. M. (2004). Sharing belief in teams of heterogeneous robots, *RoboCup-2004: The eighth RoboCup competitions and conferences*, Springer Verlag.

Herbert, S. (1995). *Artificial Intelligence: An Empirical Science*, Artificial Intelligence 77 (1): 95-127.

Hilaire, V., Koukam, A., Guer, P. & Muller, J. P. (2000). Formal specification and prototyping of multi-agent systems, *Proc. of the First International Workshop on Engineering Societies in the Agent World*, pp. 114–127.

Hiyama, Y. (2004). Rescue robot ENRYU, *SICE The 5thSI section academic lecture meeting*, pp. 999–1000.

Ho, K. L. & Newman, P. (2005). Multiple map intersection detection using visual appearance, *3rd International Conference on Computational Intelligence, Robotics and Autonomous Systems, Singapore*.

Ho, K. L. & Newman, P. (2007). Detecting loop closure with scene sequences, *Int. J. Comput. Vision* 74: 261–286.

Ho, K. L. & Newman, P. M. (2006). Loop closure detection in slam by combining visual and spatial appearance, *Robotics and Autonomous Systems* pp. 740–749.

Hohl, L., Tellez, R., Michel, O. & Ijspeert, A. J. (2006). Aibo and Webots: Simulation, Wireless Remote Control and Controller Transfer, *Robotics and Autonomous Systems* 54(6): 472–485.

Huber, D. & Hebert, M. (2003). Fully automatic registration of multiple 3d data sets, *Image and Vision Computing* 21(1): 637–650.

Jacoff, A., Messina, E. & Evans, J. (2002). Performance evaluation of autonomous mobile robots, *Industrial Robot An International Journal* 29(3): 259–267.

Kadous, M. W., Sheh, R. K.-M. & Sammut, C. (2005). Caster: A robot for urban search and resuce, *Technical report, University of New South Wales, Sydney, NSW 2052, Australia, 2005* .

Karimian, P., Vaughan, R. & Brown, S. (2006). Sounds good: Simulation and evaluation of audio communication for multi-robot exploration, *In Proceedings of the IEEE/RSJ International Conference on Intelligent Robots and Systems (IROS'06*, pp. 2711–2716.

Kaustubh Pathak, Max Pfingsthorn, N. V. & Birk, A. (2009). Relaxing loop-closing errors in 3d maps based on planar surface patches, *14th International Conference on Advanced Robotics (ICAR), IEEE Press*.

Kowalski, G., Bedkowski, J. & Maslowski, A. (2008). Virtual and augmented reality as a mobile robots inspections operators training aid, *Bulletin of the Transilvania University of Brasov. Proceedings of the international conference Robotics08, special issue vol. 15(50) series A, no. 1, vol. 2, Brasov, Romania*, pp. 571–576.

Kraetzschmar, G., Gassull, G. & Uhl, K. (2004). Probabilistic quadtrees for variable-resolution mapping of large environments, *M. I. Ribeiro and J. Santos Victor (Eds.), Proceedings of the 5th IFAC/EURON symposium on intelligent autonomous vehicles*.

Kraetzschmar, G., Sablatnog, S., Enderle, S., Utz, H., Simon, S. & Palm, G. (2000). Integration of multiple representations and navigation concepts on autonomous mobile robots, *Workshop SOAVE-2000: Selbstorganisation von adaptivem verhalten (Vol. 10/643)*, Ilmenau, Germany.

Kramer, J. & Scheutz, M. (2007). Development environments for autonomous mobile robots: a survey, *Autonomous Robots* 22(2): 101–132.

LaFary, M. & Newton, C. (2005). Aria html documentation.

Lee, S., Jung, B., Ryu, J., Oh, S., Oh, J. & Choi, C. (2005). Variable pulse mode driving ir source based 3d robotic camera, *MVA2005 IAPR Conference on Machine Vision Applications*, Tsukuba Science City, Japan.

Leitao, P. & Putnik, G. (2001). A multi-agent based cell controller, *Proc. Of the 8th IEEE International Conference on Emerging Technologies and Factory Automation*, Antibes, France, pp. 463–470.

Lemay, M., Michaud, F., L'etourneau, D. & Valin, J. (2004). Autonomous initialization of robot formations, *In IEEE international conference on robotics and automation*.

Leong, K. & Newman, P. (2005). Combining visual and spatial appearance for loop closure detection in slam, *2nd European Conference on Mobile Robots ECMR*.

Lima, P., Ribeiro, M. I., Custodio, L. & Santos-Victor, J. (2003). The rescue project - cooperative navigation for rescue robots, *Proc. of ASER'03 - 1st International Workshop on Advances in Service Robotics*, Bardolino, Italy.

Lowe, D. G. (2004). Distinctive image features from scale invariant keypoints, *Int. J. Comput. Vision* 60: 91–110.

Magnusson, M., Andreasson, H., Nüchter, A. & Lilienthal, A. J. (2009). Automatic appearance-based loop detection from 3D laser data using the normal distributions transform, *Journal of Field Robotics* 26(11–12): 892–914.

Magnusson, M. & Duckett, T. (2005). A comparison of 3d registration algorithms for autonomous underground mining vehicles, *In Proc. ECMR*, pp. 86–91.

Magnusson, M., Duckett, T. & Lilienthal, A. J. (2007). 3d scan registration for autonomous mining vehicles, *Journal of Field Robotics* 24(10): 803–827.

Maslowski, A. (2001). Surveillance mobile robot sr 11 inspector - from prototype to the real application, *Proc. of the Int. Conf. WESIC 2001*, Enschede.

Maslowski, A. (2004). Intelligent mobile robots supporting security and defense, *Proc. of the Int. Conference IMEKO TC-17 Measurement and Control in Robotics*, NASA Space Center, Huston, Texas, U.S.A.

Maslowski, A. (2005). Applied intelligent service and surveillance robotics, key-note paper, *Int. Symposium on Measurements and Control in Robotics, IMEKO TC-17 15th*, Brusseles, Belgium.

Maslowski, A. (2006). Intervention robotics - experience in poland, key-note paper, *Int. Workshop RISE 06*, Brusseles, Belgium.

MCA2 (2009). http://www.mca2.org/.

Michel, O. (2004). Webotstm: Professional mobile robot simulation, *CoRR* abs/cs/0412052.

Miro (2005). Miro - middleware for robots, Robotics Group, University of Ulm, http://smart.informatik.uni-ulm.de/MIRO/index.html.

Missionlab (2003). http://www.cc.gatech.edu/aimosaic/robot-lab/research/MissionLab/.

Montemerlo, M., Roy, N. & Thrun, S. (2003). Carmen, Carnegie Mellon Robot Navigation Toolkit, http://carmen.sourceforge.net/.

Murphy, R., Casper, J., Hyams, J., Micire, M. & Minten, B. (2000). Mobility and sensing demands in usar, *26th Annual Conference of Industrial Electronics Society, IECON 2000*, pp. 138 – 142.

Newman, P., Cole, D. & Ho, K. L. (2006). Outdoor slam using visual appearance and laser ranging, *Proceedings of the IEEE International Conference on Robotics and Automation (ICRA)*.

Newman, P. & Ho, K. L. (2005). Slam - loop closing with visually salient features, *IEEE International Conference on Robotics and Automation, 18-22 April*.

Nuchter, A., Lingemann, K. & Hertzberg, J. (2007). Cached k-d tree search for icp algorithms, *Proceedings of the Sixth International Conference on 3-D Digital Imaging and Modeling*, IEEE Computer Society, Washington, DC, USA, pp. 419–426.

Okada, M. & Murakami, K. (2007). Robot communication principal by motion synchronization rusing orbit attractor, *IEEE International Conference on Robotics and Automation*, Rome, Italy, pp. 2564–2569.

Osentoski, S., Manfredi, V. & Mahadevan, S. (2004). Learning hierarchical models of activity, *In IEEE/RSJ international conference on robots and systems (IROS 2004)*.

Park, S.-Y., Choi, S.-I., Kim, J. & Chae, J. (2010). Real-time 3d registration using gpu, *Machine Vision and Applications* pp. 1–14. 10.1007/s00138-010-0282-z. URL: *http://dx.doi.org/10.1007/s00138-010-0282-z*

Park, S.-Y. & Subbarao, M. (2003). An accurate and fast point-to-plane registration technique, *Pattern Recogn. Lett.* 24: 2967–2976.

Peterson, L., Austin, D. & Christensen, H. (2001). Dca: A distributed control architecture for robotics, *IROS 2001 - IEEE International Conference on Intelligent Robots and Systems*, Hawai, USA.

PROTEUS (2011). http://www.projektproteus.pl/.

Qiu, D., May, S. & Nuchter, A. (2009). Gpu-accelerated nearest neighbor search for 3d registration, *Proceedings of the 7th International Conference on Computer Vision Systems: Computer Vision Systems*, ICVS '09, Springer-Verlag, Berlin, Heidelberg, pp. 194–203.

Railhet, S., Wolf, J., Adra, A., Kabbara, R., Deshmukh, S., Garghouti, M., Nash, G., Belpaeme, T., Culverhouse, P., Robinson, P., White, P. & Bugmann, G. (2007). Sensor systems in a compliant geometry robot: Butlerbot, *Proceedings of Taros07*, Aberystwyth, UK, pp. 176–181.

ROS (2011). www.willowgarage.com/pages/software/ros platform.

Rusinkiewicz, S. & Levoy, M. (2001). Efficient variants of the ICP algorithm, *Third International Conference on 3D Digital Imaging and Modeling (3DIM)*.

Rusu, R. B., Maldonado, A., Beetz, M., Systems, I. A. & Munchen, T. U. (2007). Extending player/stage/gazebo towards cognitive robots acting in ubiquitous sensor-equipped environments, *in Accepted for the IEEE International Conference on Robotics and Automation (ICRA) Workshop for Network Robot System, 2007, April 14*.

Sallé, D., Traonmilin, M., Canou, J. & Dupourque, V. (2007). Using microsoft robotics studio for the design of generic robotics controllers: the robubox software, *ICRA 2007 Workshop Software Development and Integration in Robotics -"Understanding Robot Software Architectures"*.

Sprickerhof, J., Nuchter, A., Lingemann, K. & Hertzberg, J. (2009). An explicit loop closing technique for 6d slam, *4th European Conference on Mobile Robots, September 23-25, 2009, Mlini/Dubrovnik, Croatia, 229-234*.

Stetter, J. R., Penrose, W. R. & Yao, S. (2003). Sensors, chemical sensors, electrochemical sensors, *ECS Journal of The Electrochemical Society* 150(2).

Surmann, H., Nuchter, A., Lingemann, K. & Hertzberg, J. (2004). 6d slam - preliminary report on closing the loop in six dimensions, *Proceedings of the 5th IFAC Symposium on Intelligent Autonomous Vehicles (IAV '04)*.

Szynkarczyk, P. (2005). Neutralising and assisting robot smr-100 expert - design problematics, *Bulletin of the Polish Academy of Sciences Technical Sciences* 53(1): 87–92.

TALOS (2010). http://talos-border.eu/publications.html.

Tambe, M., Johnson, W. L., Jones, R. M., Koss, F., Laird, J. E., Rosenbloom, P. S. & Schwamb, K. B. (1995). Intelligent agents for interactive simulation environments, *AI Magazine* (16): 15–39.

Thrun, S., Hhnel, D. & Brugard, W. (2001). Learning compact 3d models of indoor and outdoor environments with a mobile robot, *IJCAI* .

Typiak, A. (2008). Using video camera for object's localization in surroundings of unmanned vehicle, *XII Conference on Automation - Innovations and Prospective*, Warsaw, pp. 451 – 458.

Ullman, E. (2002). *Programming the Post-Human: Computer science redefines life*, Harper's, Vol. 305, No. 1829: 60-70.

Utz, H., Kraetzschmar, G., Mayer, G. & Palm, G. (2005). Hierarchical behavior organization, *In Proceedings of IROS 2005*, Edmonton, Canada.

Valin, J. M., Michaud, F. & Rouat, J. (2007). Robust localization and tracking of simultaneous moving sound sources using beamforming and particle filtering, *Robotics and Autonomous Systems Journal (Elsevier)* 55(3): 216–228.

VF (2009). Vision and chemiresistor equipped web-connected finding robots, project STREP 6FP EU, contract no. 045541 2006-2009.

Voyles, R., Larson, A., Lapoint, M. & Bae, J. (2004). Core-bored search-and-rescue applications for an agile limbed robot, *Proceedings of the 2004 IEEE/RSJ International Conference on Intelligent Robots and Systems*, pp. 58–63.

Wang, J., Lewis, M. & Gennari, J. (2003a). A game engine based simulation of the nist urban search and rescue arenas, *Proceedings of the 35th conference on Winter simulation: driving innovation, December 07-10, New Orleans, Louisiana.*

Wang, J., Lewis, M. & Gennari, J. (2003b). Interactive simulation of the nist usar arenas, *IEEE International Conference on Systems, Man and Cybernetics*, Vol. 2, pp. 1327 – 1332.

Wang, J., Lewis, M. & Gennari, J. (2003c). Usar: A game-based simulation for teleoperation, *Proceedings of the 47th Annual Meeting of the Human Factors and Ergonomics Society, Denver, CO, Oct. 13-17.*

Warlet, F., Lancet, J., Motard, E., Pinzon, C., Mees, W. & Ilzkovitz, M. (2009). Crisis management supporting architecture in view-finder, *International Workshop On Robotics for risky interventions and Environmental Surveillance RISE'2009*, Brussels.

Williams, B., Cummins, M., Neira, J., Newman, P., Reid, I. & Tardos, J. (2009). A comparison of loop closing techniques in monocular slam, *Robotics and Autonomous Systems* .

Winiarski, T. & Zielinski, C. (2005). Implementation of position-force control in MRROC++, *Proceedings of the Fifth International Workshop on Robot Motion and Control RoMoCo '05.*, pp. 259–264.

Xiang, Z. & Zhou, W. (2006). 3d map building for mobile robots rusing a 3d laser range finder, *Lecture Notes In Control land Information Sciences, Volume 345/2006*, Springer Berlin/Heidelberg, pp. 785–790.

Xuewen, L., Cai, M., Jianhong, L. & Tianmiao, W. (2006). Research on simulation and training system for eod robots, *IEEE International Conference on Industrial Informatics*, pp. 810 – 814.

Yu, C. & Hang, D. (2006). A new 3d map reconstruction based mobile robot navigation, *8th International Conference on Signal Processing*.

Part 1

Mobile Robot Control Design and Development

1

Model-Driven Development of Intelligent Mobile Robot Using Systems Modeling Language (SysML)

Mohd Azizi Abdul Rahman[1], Katsuhiro Mayama[2], Takahiro Takasu[2], Akira Yasuda[2] and Makoto Mizukawa[2]
[1]*University Technology of Malaysia*[1]
[2]*Human-Robot-Interaction Lab, Shibaura Institute of Technology*
[1]*Malaysia*
[2]*Japan*

1. Introduction

In Japan alone the number of disabled and elderly people are growing rapidly and they usually require personal transportation vehicles such as cars and wheelchairs. However, people currently have limited capabilities or difficulties to operate these vehicles in such complex environments like shopping malls or tourist-attraction centers.

In recent years, a great number of studies of mobile robot applications have been proposed. However, only few of these studies have realized the model-based design approach in their entire development process. NEDO's Intelligent Robot Technology (RT) Software project (Okano et al.,2009) started to promote the robot technology as the basic knowledge and technology to solve various problems in daily life. In this context, an intelligent mobile robot has been developed for providing mobility to the elderly and physically unfortunate people. Besides that, the National Institute of Advanced Industrial Science and technology, also known as AIST has developed the RT-Middleware (RTM) in order to achieve efficiency in developing robot software components (Ando et al., 2005).

The main idea of this research is to develop robot software modules by looking from the system engineering analysis point of views. This is because most robotic systems are complex embedded systems. System engineering approaches focus on designing and constructing the complete system, and also on providing model reuse capabilities. Moreover, these approaches can enhance communications among the development teams, specification and design qualities and reuse of system specification and design artifacts. Modules' reusability is our main concern in this paper.

Past development efforts of robot software using Model Driven Architecture (OMG MDA, 2003) Model Driven Architecture® (OMG MDA, 2003) approach seem insufficient to support the demand of current industrial-to-domestic robot transitions. Developing intelligent robots in large scales is very demanding for experiment purposes. Thus, almost all robot systems have some common functions. However, much usable design information went to waste because of a serious lack of sharing and reusability. This motivates us to explore the

use of the MDA for the design and development of robot software modules in order to achieve module reusability.
This paper outlines the need for the MDA approach in developing reusable robot software modules that are expressed as models. Additionally, this approach is employed to encourage the use of model-driven engineering methodology (MDE, 2008), as it is less attractive to robotic software developers (Bruyninckx, 2008). More detail about MDA is presented in section 2.
The key issue is robot software module reusability and this paper concerns about the design process of robot software modules by using a model-based approach. This is shown by how existing RT-Components (RTCs) (Ando et al., 2008) can be mapped to the SysML models (see Figure 1).

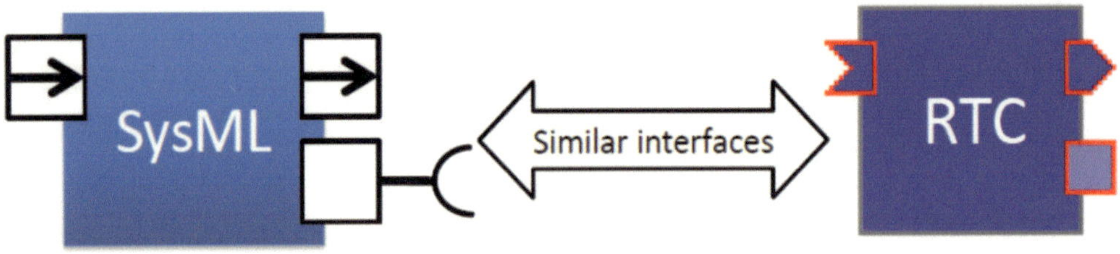

Fig. 1. SysML module to RTC block mapping concept

The scope of this study is to develop reusable intelligent RT modules that can be used to provide safe and convenient transportation service for disabled people in order to promote a barrier-free society. A mobile robot platform is currently developed for real implementation purpose at Shibaura Institute of Technology, Japan. In this study, we only concentrate on the initial phase of designing the software modules and the a mapping strategy of the SysML module into some existing RTCs (i.e. samples taken from other robotics project).

2. Model driven architecture, its need, and systems modeling language

In order to achieve the purpose of this study, we adopt an MDA approach to using models in robot software development. It prescribes certain kinds of models to be used, how those models are prepared, and the association of the relationships of the different kinds of models.

2.1 MDA approach

MDA (OMG MDA, 2003) is a software design approach for the development of software systems that describes the content of models by developing a language-neutral and separate from the implementation design software. MDA is a method known in model-driven engineering and the purpose of MDA is to change design information from implemented software into independent software by describing the coded content of a specific language-independent model. This allows the design information to be developed independently, making it possible to be changed without the use of implementing software. In MDA, the abstract model of information is called platform independent models(PIMs) and the model that corresponds to a specific programming language is called platform specific models (PSMs).

PIM describes higher abstraction level of a system design than a program does but does not describe anything regarding to the platform. Its content falls under the behavior and the internal structure of the system. The application information for the platform added into the PIM is called PSMs. Inside MDA, the PIM is converted into a coded programming language (PSM) using a tool. With the abstract representation model, the design information itself will become reusable and the production of the program that was reflected immediately from design information due to the semi-auto code generation is also improving.

The use of MDA allows the construction of design information that is no longer depending on the specified development techniques and technologies. With the design depending on which platform is being clearly specified, not just abstract design information, specific system information from a voluntary platform can also be reusable. Figure 2 shows MDA basic process and its usage in developing complex systems.

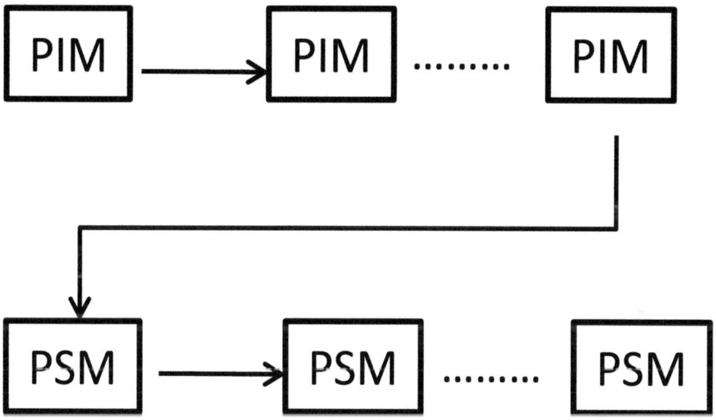

Fig. 2. MDA basic process of complex systems (OMG MDA, 2003)

2.2 The need of MDA in robotic software development

MDA is considered very useful in robotic software development. The OMGRobotics-Domain Task Force (OMG Robotics DTF, 2005) that aims to promote and integrate the OMG robot standards, has laid out one example of why MDA which was designed for robotic systems should be introduced and extended throughout the robotic field.

The MDA is considered very effective in handling functional decomposition problems in the intelligence module. Functional decomposition problems are problems that occur during the division of a robotic system due to different development teams having different ways of splitting functionality into components, making it difficult to use the independent developed modules mutually.

In order to solve this problem, it is important to standardize the division unit that will suit to robot functions. If intelligence modules are functionally decomposed with a mutual cooperation, not only that different module can be easily applied together, but also the exchangeability and continuity of the modules can be achieved. This can be compared to the benefit of implementing service-oriented architecture (SOA) that highly depends on the partition in services that is not a trivial problem. As we are concerned about reusability issues, about reusability issue, both MDA and SOA provide a concrete solution for that, but how we split functionality into components make an MDA approach is more preferable.

In this research, the focus is the output of the module and if the output is the same as the input, the implementation method used is considered correct. Therefore the major thing that has to be considered during this stage is how individual modules are divided. This is no other than the PIM in the MDA. By promoting an intelligence module with the same functional decomposition as well as combining different intelligence module with the same purpose will make it possible to realize a variety of application systems. If we apply MDA to the robotic planning stage and construct the PIM with sufficiently repeated discussion on the functional division, we will be able to develop a much proper robotic module development.

2.3 SysML

SysML is a general-purpose modeling language that is only associated with descriptive semantics. It was developed to support the specification, analysis, design, verification and validation of a wide range of complex systems (OMG SysML, 2010). The <<block>> is the basic unit of structure in SysML and can be used to represent the systems that may include hardware, software, information, processes, personnel, and facilities. The objective of SysML is to unify the diverse modeling languages currently used by system engineers. SysML reuses a subset of Unified Modeling Language (OMG UML, 2010) and provides additional extensions needed to address systems engineering aspects not covered by UML2. It includes nine diagrams (see Figure 3) that can be used to specify system requirements, behavior, structure and parametric relationships. Requirements diagram and parametric diagrams are the new diagram types proposed by SysML.

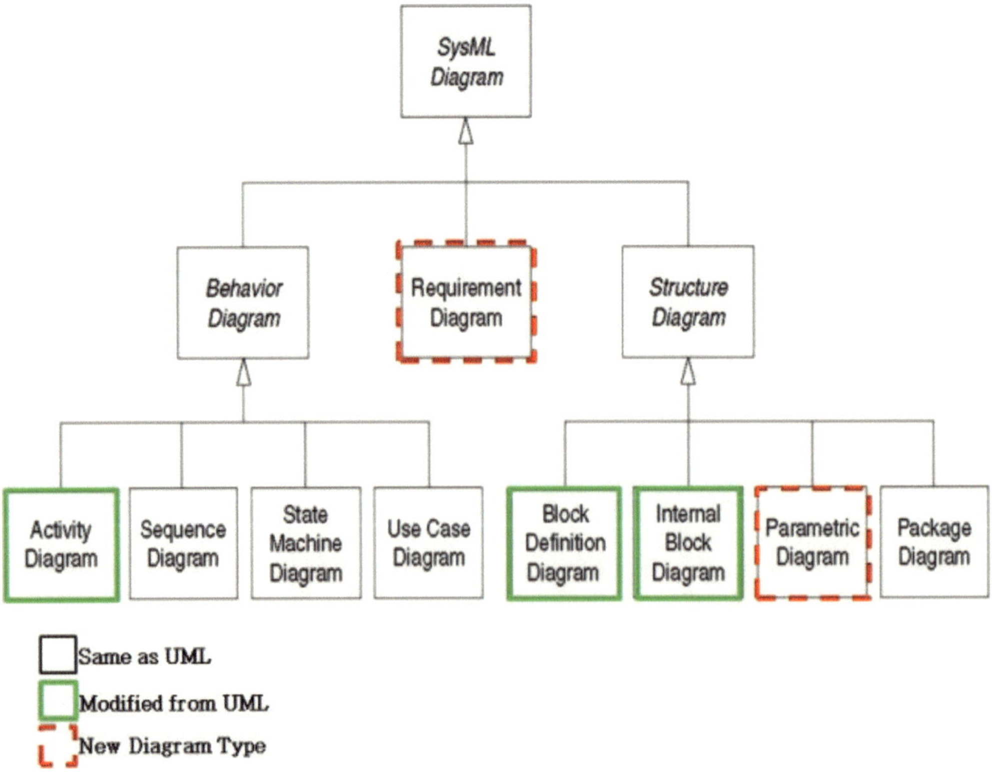

Fig. 3. SysML diagram taxonomy and its comparison with UML (OMG SysML, 2010)

SysML provides modeling constructs to represent text-based requirements and relate them to other modeling elements. The requirement diagram can be used to represent the relationships that exist between requirements and to visualize them. It provides a bridge between traditional requirements management tools and other SysML models. It can be used to depict the requirements in graphical, tabular, or tree structure format and highlight their relationships between requirements and other model elements that satisfy or verify them.

3. Modeling approach

3.1 Overview

In this paper, the progress of our model-based approach is presented by adopting the international standard modeling language SysML throughout our design process. All diagrams are made of SysML constructs.We study two problems for improving module reusability such as intelligent module division and system development based on the intelligent modules.

3.2 Purpose of this study

This paper describes our effort to adopt model-based approach to make robot models independent from any specific hardware and software platform by deriving reusable and versatile robot model. Thus, other developers will be able to refer to our designated models and customize it to meet their robot specifications. For making effortless transition from models to real system, we employ existing RT-Components developed by NEDO's project to reduce our burden on the development, as well as extending the reusability of RT modules. We have the following purposes to:

- make the model for each process to develop reusable models

In this research, we adopt model based approach to each design process from intelligent mobile robot basic functional analysis to final system analysis, depending on the implementation. This approach enables us to choose appropriate design decision expressed as models. For example, if the requirement is same, developers will reuse our requirement diagram. Otherwise, only design workflow should be reused.

- design PIM level and PSM level separately

Modeling process consists of platform independent model design and platform specified model design as previously described in Section 2.

- improve PSM reusability

We make use the RT-Middleware in our development as a software platform to improve software module and system reusability. Mappings between abstract blocks and RTC blocks were achieved.

3.3 Modeling process

Modeling activity conducted in this study is shown in Figure 4. The red box means "movement intelligence basic design" while blue box means "movement intelligence system design", and green box means "internal software system design".The paper-like diagrams are shown as the output of each analysis regarding to SysML diagrams. For brevity, we disregard some explanations about the design steps throughout the paper. However, the reader is advised to contact the authors for the full specifications. The following contents explain each of the steps:

1. Context analysis; this analysis reveals the supposed environment and is used for deriving requirement analysis and functional analysis. As a result, a context diagram is obtained in this analysis.
2. Requirement analysis; the functional and non-functional requirements based on the supposed environment and limitations are organized. In the process we make the requirements as well as making the connection between related requirements; and the solution for each requirement is shown.
3. Use case analysis; the required functions that based on the functional requirement are described. The use case diagram is used in this phase.
4. Hardware structure analysis enables us to organize abstracted hardware connection by using Block Definition Diagrams.
5. Software functional analysis; shows software including its relations. Each software block is abstracted from the necessary functions.
6. Software structure analysis defines the interface of each module using Internal Block Diagrams (IBDs).
7. Software specification analysis shows how its internal behavior is defined.
8. Behavioral analysis; Software system behaviors are designed using Sequence Diagram (SD) and State Machine (STM).
9. RTCs selection; RTC blocks are chosen to be mapped from the designed software modules.
10. Implementation of the system; Final step to test and evaluate some parts of the designed modules implemented on a real robot platform (not covered in this paper).

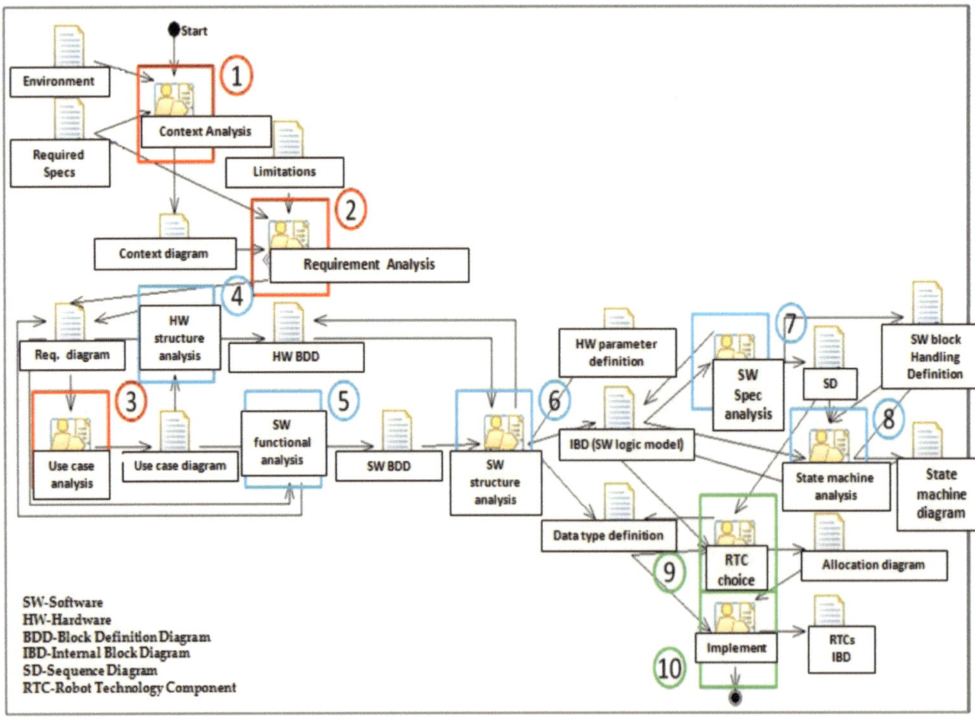

Fig. 4. Overall design process and modelling workflows

3.3.1 Context analysis

The PIM level modeling activity starts by abstracting information about the robot system in environments. Our mobile robot operates in an outdoor surrounding with pedestrians, bicycles, cars and various road conditions such as on the grass, pavement, slopes or ramps, and so on. The robot must be able to avoid the obstacles and make a correct path in various situations. Figure 5 shows an example of the context diagram that includes expecting objects in the surroundings and the disturbance elements in our mobile robot operating environments. Disturbance elements (e.g. light and building) are considered as the possible distraction sources that may affect the robot's sensing ability. The passenger is identified to be a user who interacts with our robot system. This categorization of object and environment is for deriving functional robot requirements.

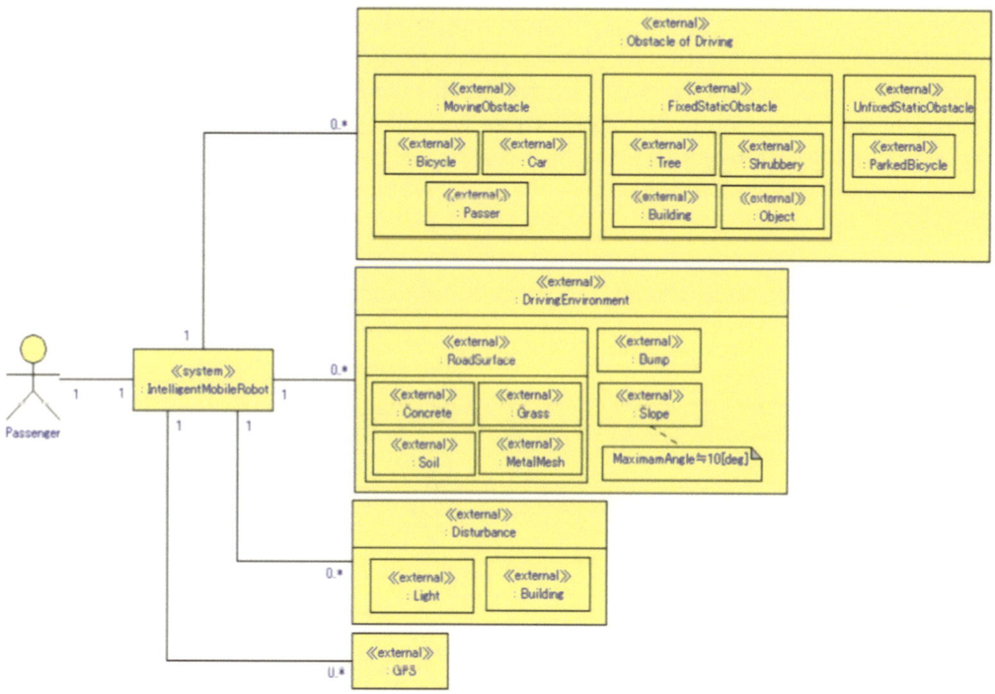

Fig. 5. Context diagram for operating environment analysis

3.3.2 Functional requirement analysis

A requirement diagram is a new diagram type in SysML that can be used to model the system requirements in more detail. Our mobile robot system requirements are described in the following SysML's requirement diagrams. They focus on the requirement mobility, the safety for operating in the environment and the extension of RT modules reusability.Figure 6 shows the basic requirements derived from the previous operating environment analysis. The top-level requirements are identified as "SafetyLocomotion" and "ExtendingRT-

ModuleReusability". The "SafetyLocomotion" is composed of other sub-requirements with respect to the locomotion strategies which are called "Controlled by Passenger", "AutonomousLocomotion", and "Enhanced Safety". Figure 7 illustrates the derived requirements and functions of robot locomotion for manual control mode (i.e. a joy-stick controller may be used as an input device to the system) and for autonomous locomotion mode. These requirements represent abstract levels of functional requirements to the concrete description with the specific devices or software modules. The autonomous locomotion is further derived into several specific components. The '<<deriveReqt>>' stereotype is associated for specifying a desired destination, planning a path, tracing trajectory, self-localization, and obstacle avoidance functionalities. In this paper, we omit the discussion on the "EnhancedSafety" requirement as a future work possibility by considering safety standardization imposed in Japan.

In Figure 7, the red rectangles represent the actual hardware of the robot system with the <<satisfy>> stereotypes associate to its upper layer blocks meaning that the chosen hardware should satisfy the designed requirements. As we can see, they are arranged somewhat hierarchically, with low-level derived requirements closer to the hardware. Constructing the system in a hierarchical manner like this helps us to maintain good design pattern. It also helps with traceability because we will have a clear understanding of the dependencies between requirements. These abstraction layers allow for easier model re-use and exchange.

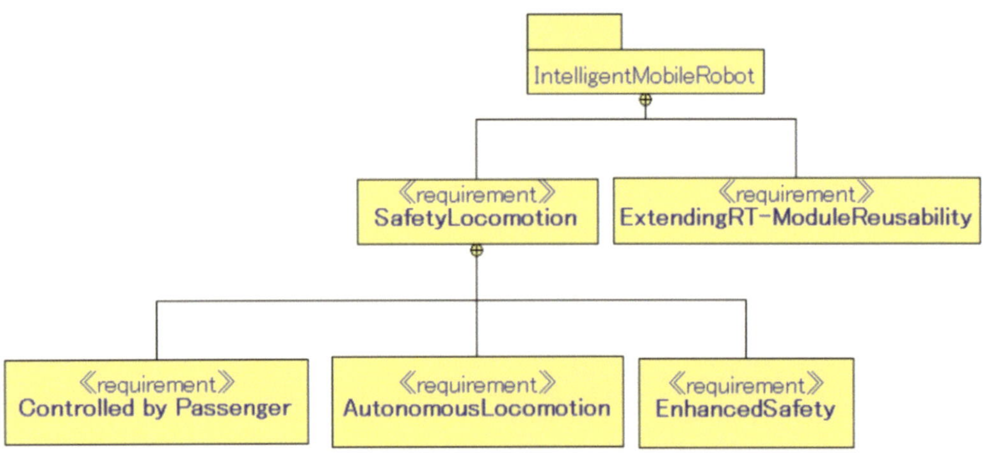

Fig. 6. Top-level requirements derived from operating environment analysis

For necessary functional analysis, we identified the following core functions for our mobile robot system to perform

- Specifying a destination point "InputGoal" function.
- Navigating to the specified destination by using path planning, path generating, trajectory generating, obstacles detecting, position localizing and error detecting modules.

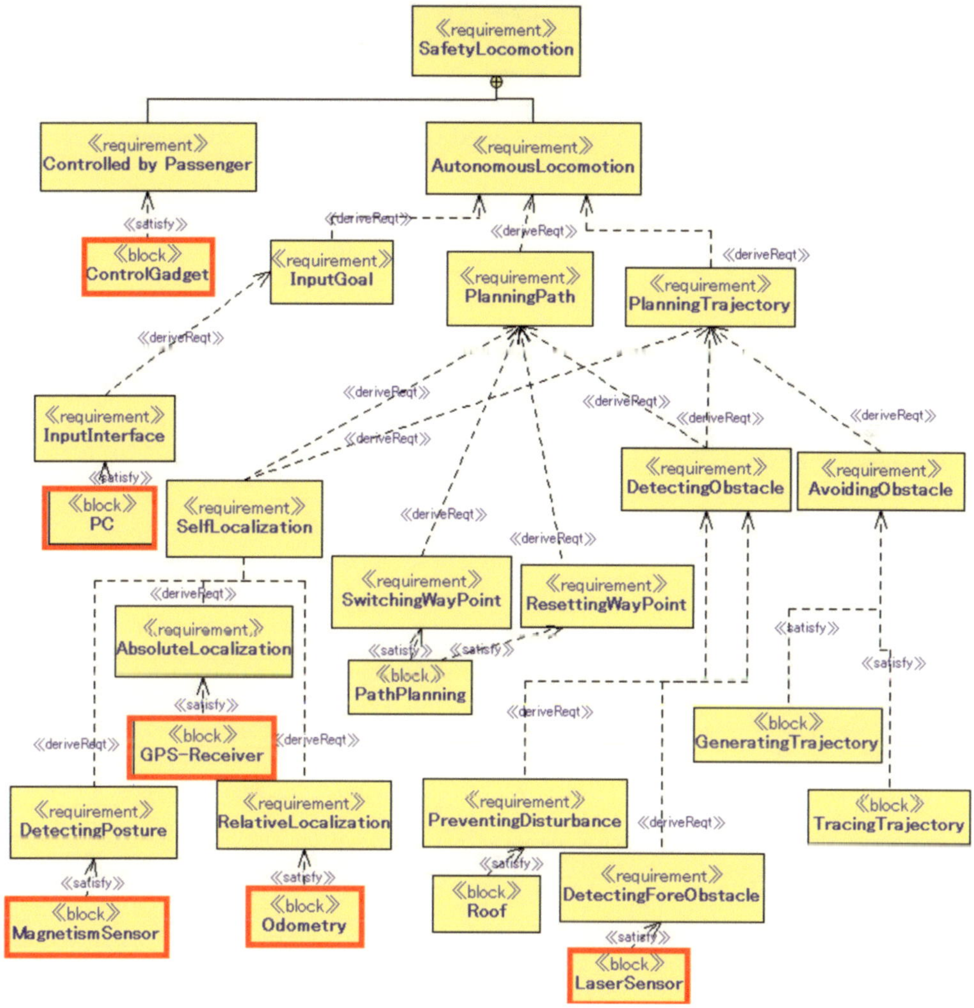

Fig. 7. Functional requirements and its derivation of "SafetyLocomotion" block.

3.3.3 Software and hardware composition

This section describes how we model the required software and hardware components, and decide the internal interfaces between these components. By implementing the common interfaces that were proposed by the working group of the NEDO project, our robot modules can easily adapt and apply the existing RTCs.

In addition, our designed models separate device specific components from components for the transporting functions. As a consequence, our models are truly independent from the specific devices especially for the wheeled platform, and easy to be reused in future. Figure 8 shows robot software composition diagram while Figure 9 shows its hardware profile's composition. The former diagram is made of SysML's internal block diagram (IBD) and the latter is made up of SysML block definition diagram (BDD). In Figure 8, we structure the software components as identified previously during the necessary functions analysis by connecting each software component through interface ports provided in SysML. To make the system complete, hardware configuration is shown in Figure 9 by profiling each hardware implementation using various sensors for robot localization, obstacle avoidance, battery malfunction, emergency switches and so on. Figure 10 displays our mobile robot platform, the Four-X developed at Shibaura Institute of Technology, Japan. This type of mobile platform is equipped with swivel-steered drive system. The following list specifies some parts of its hardware components:

1. Hemisphere A100 Global Positioning System (GPS) receiver.
2. On board PC (Panasonic CF-19).
3. Emergency Switch Button.
4. On board PC for data logging.
5. Notification lamp.
6. Bumper switch.
7. Motor controller (EPOS 70/10).
8. On board power supply (intelligent battery).
9. Laser range finder (UTM-30LX).

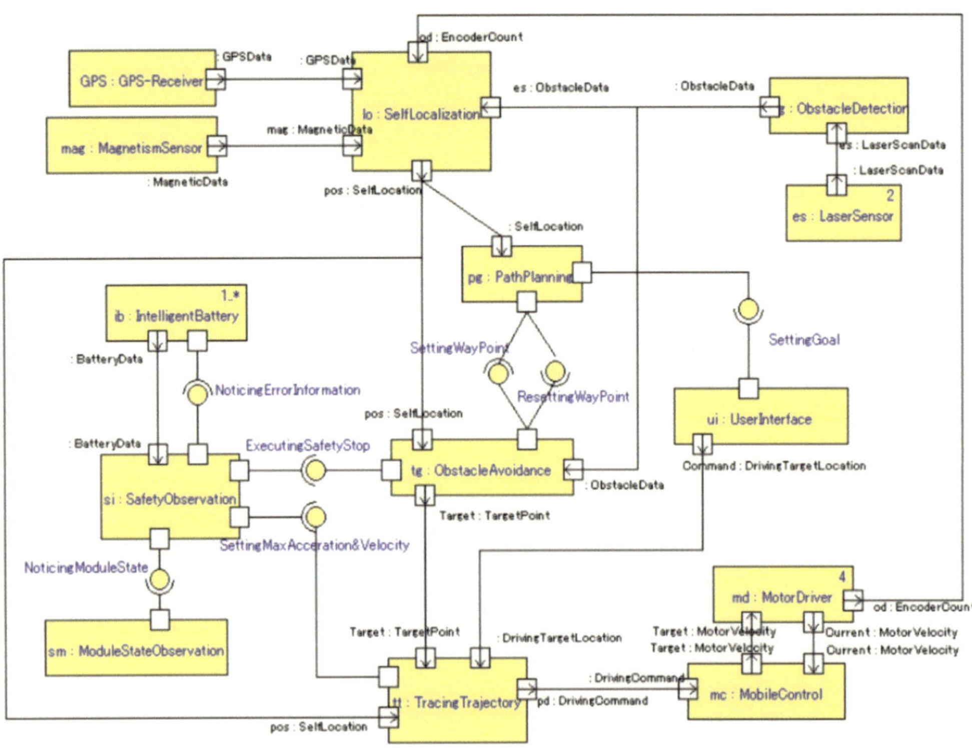

Fig. 8. Software modules composition and ports configuration modelled with the internal block diagram (IBD).

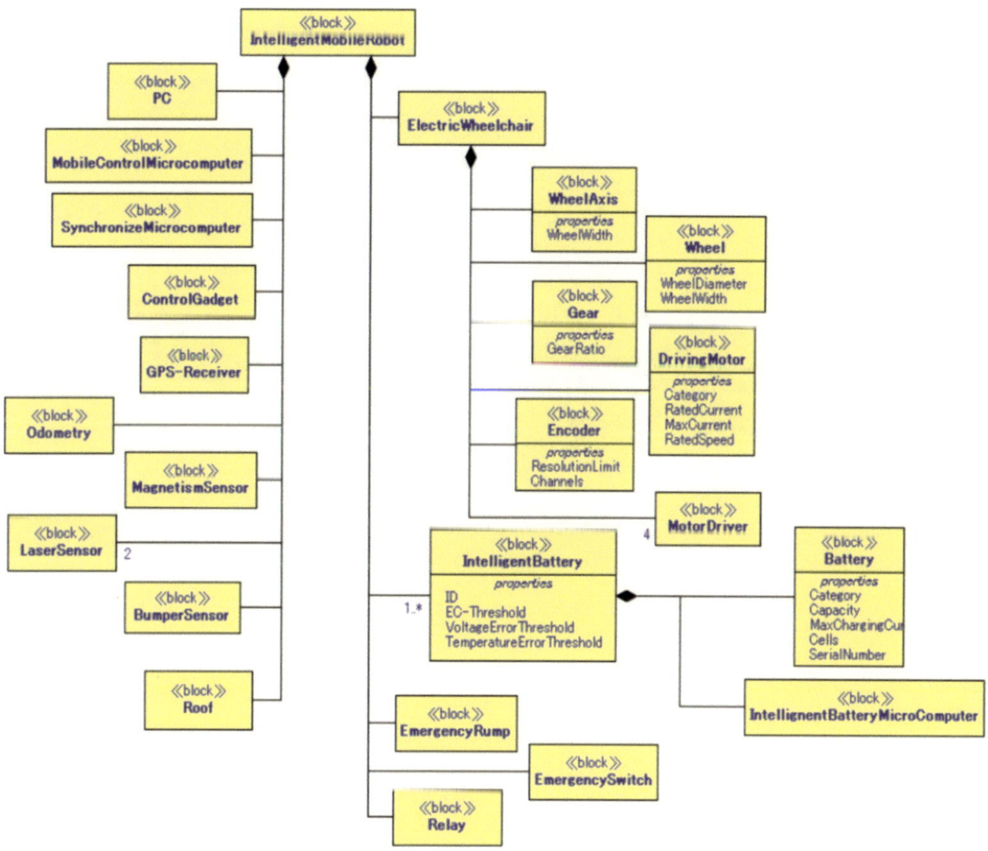

Fig. 9. Hardware modules composition using a block definition diagram (BDD).

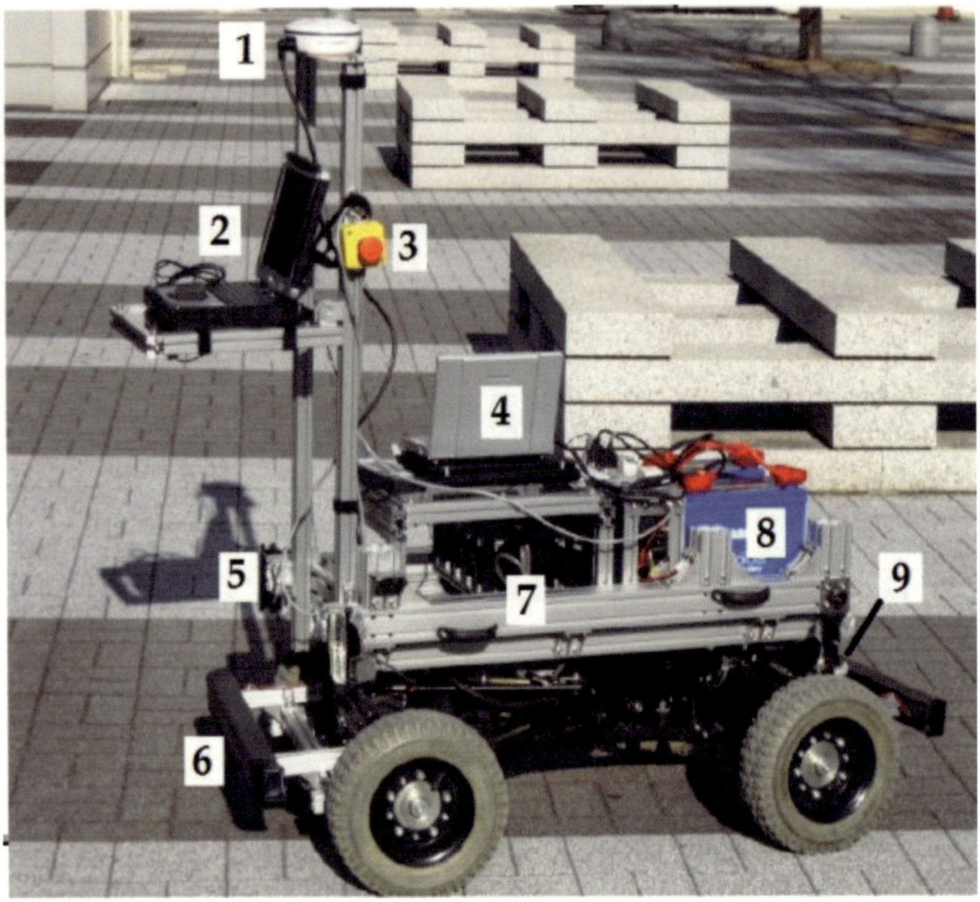

Fig. 10. The Four-X mobile robot platform

3.3.4 Mapping to the RT-components

The PSM modeling activity starts from here. We adopt a sample from Tohoku University's RT-Cs robot project to be reused in our software development. They have developed a versatile R-TC navigation system implemented on the Segway RMP 200 platform to complete the whole course in the Real World Robot Challenge 2009(Segway, 2009). Refer (Yuta et al., 2010) for information about the Tsukuba Challenge event. So, it is proven that their robot has a robust navigation system for using in outdoor environments. Outdoor navigation capability is one of the reasons why we adopt these RTCs beside its algorithm similarity. In addition, these RTCs also adopt a Global Positioning System (GPS) map, self-localization by using GPS and odometry, and obstacle avoidance by laser range finder (LRF). These similarities make our robot software development efforts slightly reduced because the concept of reusability is applied although with a different mobile robot platform.

In this phase, software modules are replaced with compatible RTCs blocks as we propose to select the suitable RTCs for the functionality of each SysML modules. As depicted in Figure 11, "SelfLocalization" module consists of four RTCs compatible modules, which are "CorrectorGPS", "CorrectorURG", "CurrentVelocity", and "PredictorOdometry". By making a wide variety of software elements, we can have more choices of compatible RTCs.

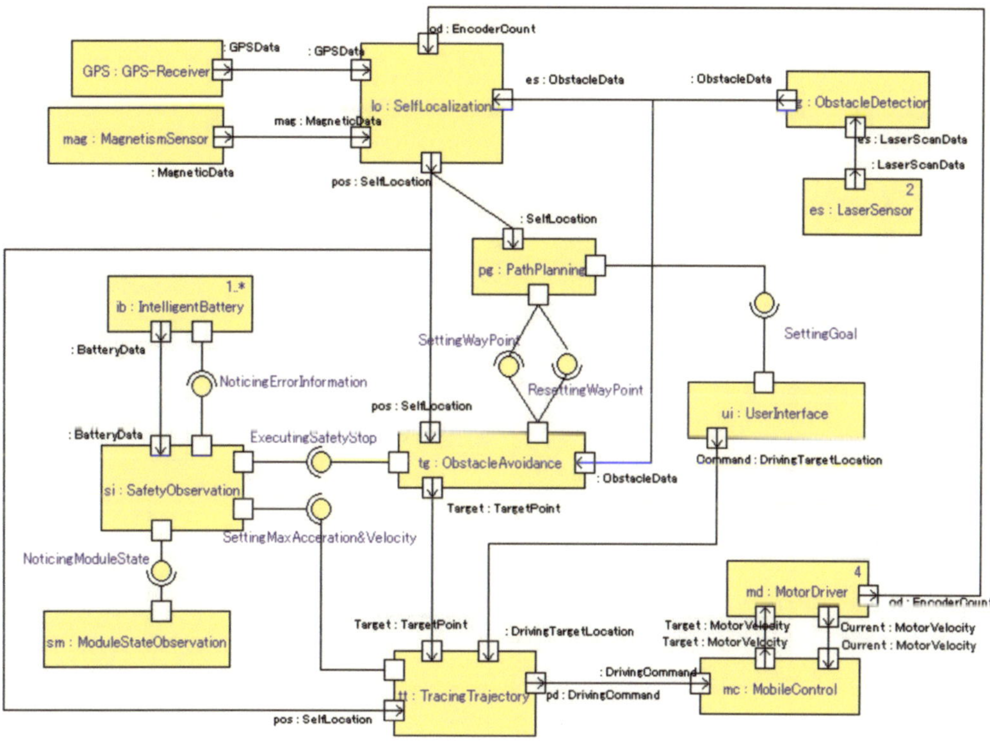

Fig. 11. Software modules (IBD) mapping technique

Thus, we set the suitable RTCs for functionality of software elements. To make a better understanding about this mapping technique, we provide a clearer mapping diagram by using <<allocation>> stereotypes (OMG SysML, 2010). Allocations are a set of notations (i.e. the new generic allocation dependencies) that allow different aspect of a SysML model to be related to each other. This is one of the SysML constructs to extend the UML capabilities. It is worth to notice that our SysML and RT-Middleware software platforms are based on C++ language.

As shown in Figure 12, the red box represents software element blocks (i.e. previously clarified in the analysis), and orange box determines real RTC system blocks. Therefore, these blocks have clear functionalities when each software block is allocated to other RTC system blocks with different specification or parameters. As a result, we model the full specification of software modules that have been replaced with RTCs modules and its connection. An example is given in Figure 13. This mapping technique has reduced our development effort as well as promoting module reusability.

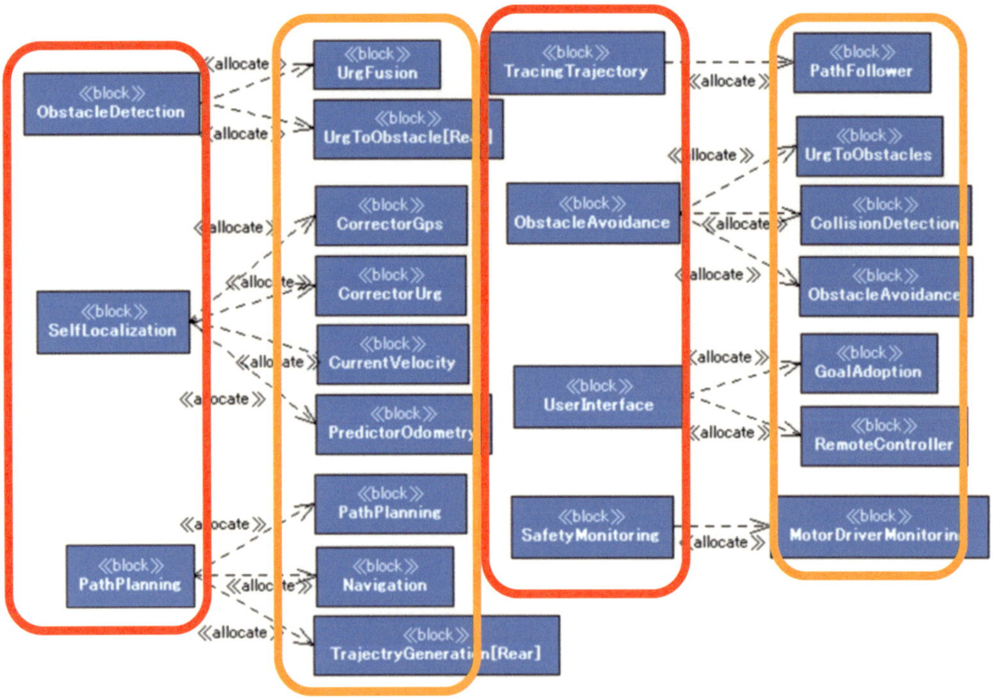

Fig. 12. Allocation diagram for mapping software component to compatible RTCs blocks.

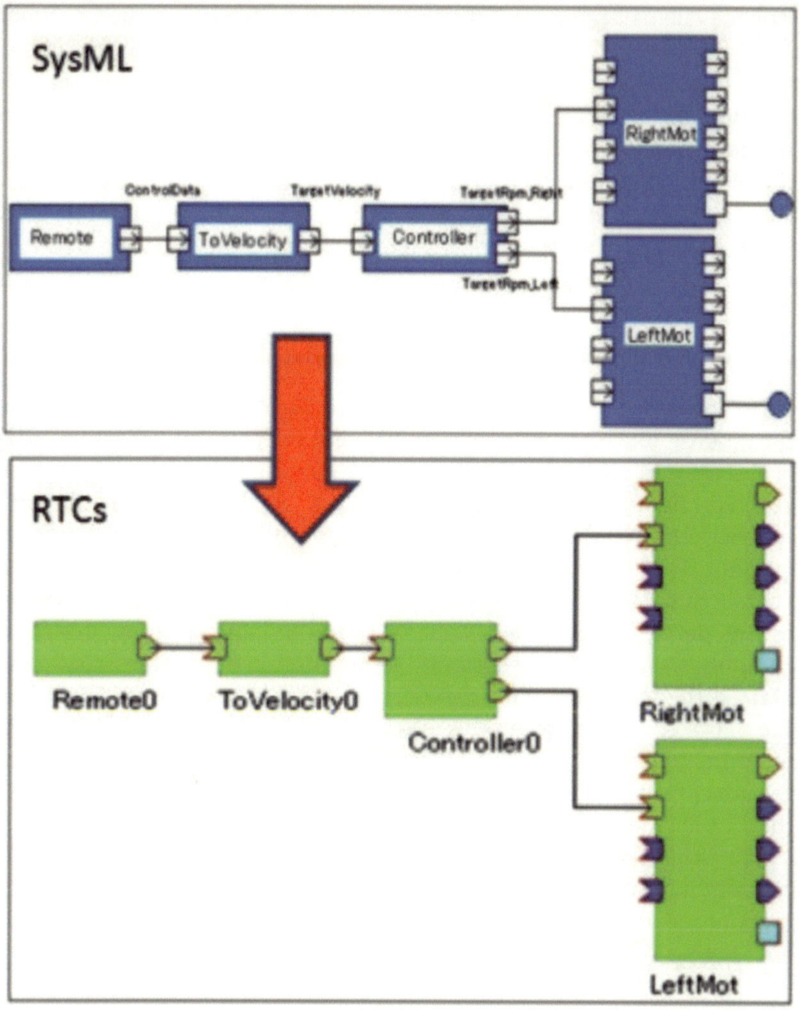

Fig. 13. The SysML to RT-Cs mapping.

4. Lessons learned

This section summarizes the lesson learned from the study where we successfully applied the model-based approach to developing mobile robot software modules for reusability purposes. As a result, we provide a modeling process to develop those modules.

4.1 The usage of SysML constructs for robotics

Through the study, we found that the OMG SysML standard was fit as a notation for specifying the requirements, modeling the system views (e.g. both structural and behavioral), decomposing into objects and/or blocks, and defining relationships between objects especially in a complex embedded system like a robotics system. Some diagrams and notations provided in the SysML were particularly important for abstracting information, analyzing the structure and behavior, designing the functionality, as well as modeling robot systems. In the case of context and requirement diagrams, functions (e.g. functional requirements) which robot system should perform can be identified and defined in terms of actors who are users of the robot system and use cases. Context and requirement diagrams define the information of some aspect of the robot system without knowing its internal structure.

Additionally, block definition diagram (BDD) notation is used to compose the structure model, which focuses on the static structure of a robot system (i.e. both software and hardware). The BDD also shows the blocks of objects in the system, their internal structure that include attributes, operations, parts, and their relationships to other blocks in terms of associations and generalization.

For internal block diagram (IBD), as its name, is used to show each part of the blocks that participate in software or hardware modules interface with each other by defining ports and flows in order to send and/or receive signals (see Fig.9 and Fig.11). Although SysML's behaviour diagrams, like sequence diagram (SD) and state machine diagram (STM) are not specifically described in this paper, we believe that its roles in modeling the behavior of our robot system are significantly important. SD plays its main roles in designing object interaction arranged in time sequence and also could be used to describe the task event sequencing logic. This is a common activity in designing real-time embedded systems especially in robotics. As for STM, it can be used to describe how state-dependent aspects of the system are defined by a finite state machine and can help in designing and in developing highly state-dependent systems. As depicted in Fig.12, the SysML specification (OMG SysML, 2010) defines an allocation notation using stereotypes to show how various elements are allocated to and from other elements. Such allocations may be used to show deployment or more generally to relate different parts of a model as the design progresses. Allocation models help us to map the SysML software modules to the compatible RTCs based on the C++ language.

In addition, by using the SysML notation for system engineering analysis, software and system engineers not only can communicate among themselves to develop and integrate specific components for providing various functions but also will encourage interest in model-driven engineering approach.

4.2 The future of robotics software development

To our best knowledge, not much effort has been done especially in robotics software and system development that use model-driven engineering (MDE) method by adopting a general modeling language like SysML. Both MDE and MDA are interconnected in terms of technology approaches to a broad range of systems and software engineering. The former tends to propel the development process while the latter is the separation of the operation of a system from the details of its platform in the development process.

According to (Bruyninckx, 2008), the robotics community is seriously neglecting the progress in MDE, hence discarding lots of opportunities to let software engineering and practice mature in the domain of robotics. Therefore, creating SysML profile for robotics is one of the solutions. For instance, the robotics literature includes bunch of articles about architecture which most of them use graphical models with boxes and arrows, but the meaning of these models has never been standardized, and the practical constraints on real-world implementations are implicit. In spite of that, standardization of SysML and its real-time and embedded specialization (MARTE) provides exceptional ways to start with reusable and semantically well-defined designs of complex software systems.

5. Conclusion and future work

In this paper, we have presented the basic specification of our mobile robot. The proposed mobile robot software system is fully based on model-driven engineering methodology to overcome the current difficulties of mobile robot development. The highlighted features of our approach are really straightforward, mobile platform independent, reusable model design, reduced development efforts, and also customizable system configuration.
Based on this specification, future work will focus on the detailed design for each module as well as extending SysML profiles that ideally suit to the robotics domain. Then the suitable RTCs will be selected from the RTC Centre of Intelligent RT Software Project. The reusability of the platform independent component, such as mobile control will be further extended by extracting the platform-independent parts.
As we believe that SysML can easily provide information models for capturing system aspects (i.e. both structure and behavior), we continue our effort to integrate some software tools for model-based design (e.g. Simulink and Modelica) in order to fully utilize SysML as a platform of model integration.

6. Acknowledgment

This work is supported by NEDO Japan (New Energy and Industrial Technology Development Organization) project on "Intelligent RT Software Project".

7. References

Ando, N.; Suehiro, T. ; Kitagaki, K. ; Kotoku, T., & Yoon, W.-K. (2005). RT-Middleware: Distributed Component Middleware for RT (Robot Technology), *Proceedingsof IEEE/RSJ 2005 International Conference on Robots and Intelligent Systems (IROS 2005)*, pp. 3555-3560, ISBN 0-7803-8192-3, Edmonton, AB, Canada, Aug 2-6, 2005

Ando, N.; Suehiro, T. & Kotoku, T. (2008). A Software Platform for Component Based RT-System Development : OpenRTM-Aist, *Proceedingsof the 1st International Conference on Simulation, Modeling and Programming for Autonomous Robots (SIMPAR 2008)*, pp. 87-98, ISBN 987-3-540-89075-1, Venice, Italy, Nov 3-7, 2008

Bruyninckx, H. (2008). Robotics Software: The Future Should Be Open. *IEEE Robotic & Automation Magazine*, Vol.5, No.1, (March 2008), pp. 9-11, ISSN 1070-9932

Model Driven Engineering, MDE (2008). Model-Driven Engineering. Available at http://www.theenterprisearchitect.au/archive/2008/model_driven_engineering/

Okano, K. & Yasukawa, K. (2009). NEDO: Overview of Intelligent RT Software Project, *Proceedingsof RSJ 27th Annual Conference on The Robotics Society of Japan*, RSJ2009 AC1D1-01 (in Japanese), Tokiwadai, Hodogaya, Yokohama, Sept 15, 2009

OMG MDA, Version 1.0.1 (2003). Model Driven Architecture Guide. Available from http://www.omg.org/cgi-bin/doc?omg/03-06-01

OMG Robotics-DTF, (2010). The OMG Robotics Domain Task Force. Available from http://www.robotics.omg.org/

OMG Systems Modeling Language, Version 1.2 (2010). SysML Specification. Available from http://www.omg.org/spec/SysML/2.3/

OMG Unified Modeling Language, Version 2.3(2010). UML Specification. Available from http://www.omg.org/spec/UML/2.3/

SEGWAY, (2009).Tohoku University team completes Tsukuba Challenge. Available from http://rmp.segway.com/2009/11/16/team-from-tohoku-university-completes-tsukuba-challenge-09/

Yuta, S.; Mizukawa, M.; Hashimoto, H.; Tashiro, H. & Okubo, T. (2010). Tsukuba Challenge 2009- Towards Robots Working in the Real World: Records in 2009. *Journal of Robotics and Mechatronics,* Vol.23, No.2, (April 2011), pp. 201-206, ISSN 1883-8049

2

Development of Safe and Secure Control Software for Autonomous Mobile Robots

Jerzy A. Barchanski
Brock University
Canada

1. Introduction

It is often said that autonomous robots should be *trustworthy* or *dependable*, meaning by this safe, secure, reliable, etc This terms are too general to be useful, so we prefer to limit the scope of this chapter to two of their inter-related components - safety and security. They must be built into a system from the start; it is difficult, if not impossible, to add them in an adequate and cost-effective manner later on.

We view autonomous robots as situated, real-time embedded systems endowed with enough intelligence to adapt to changing environment and learn from their experience. They may operate unattended and through an unsafe operation may cause significant human, economic or mission losses. The focus of this chapter is on safety and security of robot control software. This software allows unprecedented complexity of robotic systems, which goes beyond the ability of current engineering techniques for assuring acceptable risk.

Most of the publications on safety has a form of recommendations on providing safe environment for robot operators, like the Occupational Safety and Health Administration regulations or the more recent NASA recommendations for space robots. This approach is effective when accidents are primarily caused by hardware components failures.

As software becomes more and more important in robot control, we have to consider ways to prevent accidents caused by software.

Robot control software consists of many interacting components. Accidents arise in the interactions among the components rather than the failure of individual components.

The need for safety is obvious, but how to ensure it is less obvious. Autonomous robots may operate unattended and through an unsafe operation may cause significant human, economic, or mission losses. Similar problems were encountered early on in manufacturing automation; but autonomous mobile robots may change their behaviour and operate in much less controlled environments.

We will review at first the principal concepts of system safety like risk and hazard and some traditional approaches to dealing with them. We consider security as a subset of safety and we will present our point of view on this issue.

The present trend to make the robots more autonomous requires new approaches to deal with much more complex problems of their safety. After review of several robot control architectures from the viewpoint of their safety we present an approach based on systems theory. While the theory was developed long time ago it turns out very useful to ensure safety of complex software systems. We apply it in so called intention specification, which allows a designer to explain why a specific decision was made and to introduce safety constraints.

2. Principal concepts of system safety

Safety refers to the ability of a system to operate without causing an accident or an unacceptable loss (Leveson 1995). An accident is an undesired and unplanned (not necessarily unexpected) event that results in (at least) a specified level of loss. To prevent accidents, something must be known about their precursors, and these precursors must be under control of the system designer. To satisfy these requirements, system safety uses the concept of a hazard. There are many definitions of hazards. We will define a hazard as a state or set of conditions of a system that, together with other conditions in the environment of the system may lead to an accident (or loss). We define therefore a hazard with respect to the environment of the robot.

In case of physical systems, existence of a hazard depends on how the boundaries of the system have been drawn - they must include the object that is damaged plus all the conditions necessary for the loss.

This does not apply to software, since it is not a physical object, only an abstraction. Thus software by itself is not safe or unsafe, although it could theoretically become unsafe when executed on a computer. Thus we can talk only about the safety of software and its hazards in the context of a particular system design within which it is being used. Otherwise, the hazards associated with software do not exist. Due to this, the system safety engineers prefer to use the term software system safety instead of software safety.

A hazard has two important characteristics: severity and likelihood of occurrence. Hazard severity is defined as the worst possible accident that could result from the hazard given the environment in its most unfavourable state.

The hazard likelihood of occurrence can be specified either qualitatively or quantitatively. Unfortunately, when the system is being designed and hazards are being analysed and ranked as to which should be eliminated first, the information needed to evaluate the likelihood accurately is almost never available. It means, the best what can be done is to evaluate the likelihood qualitatively.

The combination of severity and likelihood of occurrence is often called the hazard level. Hazard level, along with two other factors related hazards, are used in the definition of risk. Risk is the hazard level combined with (1) the likelihood of the hazard leading to an accident and (2) hazard exposure or duration. Duration is a component of a risk, since the longer the hazardous state exists, the greatest the chance is that the other prerequisite conditions will occur.

The relationship between hardware and software hazards is shown on Fig.1, where:

Design dysfunction means any hazard inadvertently built-into the system design due to design and integration, e.g.: design error, design interface error/oversight, design integration error/oversight, tool errors;

Code error is any hazard inadvertently built-into the system design due to a coding error, e.g.: wrong sign (+/-), endless loop, language error;

Hardware induced software error - any hazard resulting from hardware failure causing safety critical software error, e.g. memory error, memory jump, bit error, change of value of a critical variable.

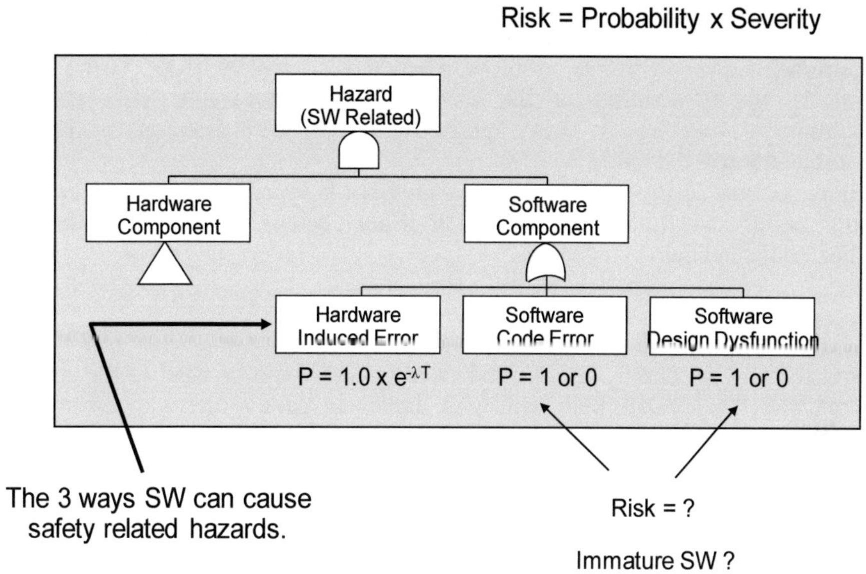

Fig. 1. Generic software hazard model

As an example of risk evaluation let's take the case when a computer controls movements of a robot.

The risk is then a function of the

- probability that the computer causes a spurious or unexpected robot movement,
- probability that a human is in the field of movements,
- probability that the human has no time to move or will fail to diagnose the robot failure,
- severity of worst-case consequences.

If the computer executes a robot control software monitoring the state of the system and including some safety function, then the risk is a function of the

- probability of a dangerous condition arising,
- probability of the computer not detecting it,
- probability of the computer not initiating its safety function,
- probability of the safety function not preventing the hazard,
- probability of conditions occurring that will cause the hazard to lead to an accident,
- worst-case severity of the accident.

In almost all cases, the correct way to combine the elements of the risk function are unknown, as are the values of the parameters of the function.

Traditional hazard analyses consist of identifying events that could lead the system to a hazardous state. These events are usually organized into causal chains or trees. Popular event-based hazard analysis techniques include Fault Tree Analysis (FTA) and Failure Modes and Effects Criticality Analysis (FMECA). Because of their reliance on discrete failure events, neither of these techniques adequately handles software or system accidents that result from dysfunctional interactions between system components.

Hazard analysis usually requires searching for potential sources of hazards through large system state spaces. To handle this complexity it is possible to use a State Machine Hazard Analysis (Leveson 1987). While any state machine modeling language can be used, it is more efficient to use a language providing higher-level abstractions, such as Statecharts (Harel 1987).

The tasks of the software development process that relates to software hazard analysis include:
- Tracing identified system hazards to the software-hardware interface and then into requirements and constraints on software behavior.
- Showing the consistency of the software safety constraints with the software requirements specification and demonstrating the completeness of the software requirements specification.

In addition, because software can do more than what is specified in the requirements (the problem of unintended functions), the code itself must be analyzed to ensure that it cannot exhibit hazardous behavior.

3. Security as a subset of safety

Safety and security are closely related, and their similarity can be used to the advantage of both (Barchanski J.A., 2004),. Both deal with threats or risks - one with threats to life or property, and the other with threats to privacy or security. Both often involve negative requirements or constraints that may conflict with some important system goals. Both involve protection against losses, although the type of losses may be different. Both involve global system properties that are difficult to deal with outside of the system context. Both involve requirements that are considered of supreme importance in deciding whether the system can and should be used - that is particularly high level of assurance may be needed, and testing alone is insufficient to establish those levels. In fact, a higher level of assurance that a system is safe or secure may be needed than that system performs its intended functions.

While we will consider mostly the issues of illegitimate access and usage of a robot there is another aspect less often encountered - denial of access. It is a threat in which the communication channel is made unusable by an attacker who transmits noise on purpose (jamming). While it is not possible to survive jamming when the attacker is all powerful and can make the entire spectrum unusable over the spatio-temporal range of interest, it may be possible to drive up the cost of jamming. Two widely used techniques for this are frequency hopping and direct sequence spread. In both techniques the receiver must know the pseudorandom sequence of frequencies used by the transmitter. An attacker wishing to jam the channel must either discover the sequence or jam a sufficient number of frequencies, therefore employing much more power (and money) than the transmitter.

It should be clear that this protection is conditional on transmitter and receiver having previously established some shared key (e.g. by imprinting as described in the following paragraph).

There are also important differences between safety and security. Security focuses on malicious actions, whereas safety is also concerned with well intended actions. In addition, the primary emphasis in security traditionally has been preventing unauthorized access to classified information, as opposed to preventing more general malicious activities. Note, however, that if an accident or loss event may be caused by illegitimate or malicious access or usage of a system, then security becomes a subset of safety.

4. Safety aspects of autonomous robots

The concept of autonomy plays an important role in robotics. It relates to an individual or collective ability to decide and act consistently without outside control or intervention. Autonomous robots hold the promise of new operation possibilities, easier design and development, and lower operating costs.

Achieving safety of autonomous robots is much more challenging than teleoperated robots.

To improve safety of autonomous systems, adjustable autonomy can be used, in which the robot is autonomous to some degree only so a human may still retain more or less control of its behavior. While verification of the safety of a fully autonomous robot designed for a critical mission requires extensive test and analysis, safety verification of a semi-autonomous robot designed for the same mission is less strict - it includes only design requirements analysis and testing.

Mobile robots control software can implement different robot control architectures.

The oldest architectures were of a hierarchical type, highly influenced by the AI research of its time. This meant a system having an elaborate model of the world, using sensors to update this model, and to draw conclusions based on the updated model. This is often called the sense-plan-act paradigm or deliberative architecture. The hierarchical architectures did not perform well partly because of the difficulty in modeling of the world, partly because of relying too much on inadequate sensors.

In 1987 Rodney Brooks revolutionized the field by presenting an architecture based on purely reactive behaviours with little or no knowledge of the world. The reactive architecture is inherently parallel, fast, operates on short time scales and is scalable. There are two general kinds of reactive architectures - subsumption and potential field (Arkin 1999). Reactive architectures eliminate planning and any functions that involve remembering or reasoning about the global state of the robot relative to its environment. That means that a robot cannot plan optimal trajectories, make maps, monitor its own performance or even select the best behaviours to use to accomplish a task (general planning).

The solutions to the drawbacks of reactive architectures appeared in hybrid architectures combining the reactive architectures with modified deliberative architectures. They combine different representations and time scales, combine closed-loop and open loop execution, may re-use plans and allow dynamic replanning [Murphy 2000].

There is a number of other robot architectures designed independently of the above classes (e.g., Alami et al, 1998). Especially interesting are architectures including learning or adaptive components. The ability to adapt is key to survival in dynamic environments. When robots are adaptive, the following questions need to be addressed:

1. Is learning applied to one or more robots?
2. If multiple robots are adaptive, are they competing or cooperating?
3. What element(s) of the robot(s) get adapted?
4. What techniques are used to adapt?

Learning may alter some components of robot strategy, for example, it may change the choice of actions to take in response to a stimulus. Learning may be motivated by observation, it could be initiated in response to success/failure, or be prompted by a general need for performance improvement. The motivation influences the choice of learning technique. Nearly every learning technique has been used for robots. The two most prevalent techniques are reinforcement learning (RL) and evolutionary algorithms (EA).

A very popular form of RL is Q-learning, where robots update their probabilities of taking actions in a given situation based on penalty/reward.

Introduction of learning makes behavior of robots significantly harder to predict. It is necessary therefore to develop efficient methods for determining whether the behavior of learning robots remains within the bounds of prespecified constraints (properties) after learning. This includes verifying that properties are preserved for single robots as well as verifying that properties are preserved for multirobot systems. We want the robots to be not only adaptive, but as well predictable and timely. Predictability can be achieved by formal verification while timeliness by streamlining reverification, using the knowledge of what learning was done. Fast reverification after learning must include proofs that certain useful learning operators (in case of EA) are a priori guaranteed to be "safe" with respect to some important classes of properties, i.e. if the property holds for the robot prior to learning, then it is guaranteed to still hold after learning (Gordon-Spears, 2001). If a robot uses these "safe" learning operators, it will be guaranteed to preserve the properties with no reverification required.

5. Hazard analysis of an exemplary case

As an example we will consider a mobile robot acting in a world where the three Laws of Robotics (Asimov 1950) should be satisfied. We will be concerned with the hazards resulting from the first two Laws, which are relevant for human-robot interactions :

First Law: "A robot may not injure a human being, or through inaction, may not allow a human being to come to harm".

From the first half of the Law: "A robot may not injure a human being" - the hazard is a harmful physical contact with a human.

It can be mitigated by a safety constraint inherent in the robot behavior **avoid _obstacles.**

The second part of the law :"A robot through inaction, may not allow a human being to come to harm" envisioned by Asimov as a kind of action requiring sacrifice of robot existence to protect its master, is left for future generations of roboticists. At present it may be implemented at most aposteriori by search and rescue robots.

The second law: "A robot must obey the orders, given to it by human beings except where such orders would conflict with the First Law."

This law has to do with tasks commanded by a human, like: **move-to-goal**, **grasp**, **collect-pucks**, etc. The robot must be able to communicate with a human operator to execute his orders. The examples of hazards violating this law are communication security hazards - e.g. receiving from an illegitimate operator false commands or requests which may cause wrong actions. To eliminate such hazards it must be possible to authenticate the human operator.

Most conventional authentication protocols (e.g. Kerberos) require usage of an online server. This is out of the question here. Another widely used authentication protocol which seems to be more suitable is one based on public key cryptography. It was in fact proposed for authentication of cooperating autonomous digger and dumper truck [Chandler et al 2001]. However the problem of online access to the server appears here as well. What we need is to be able to create a *secure transient association.* As well as being *secure,* the association must also be *transient.* When an operator changes, the robot has to obey the new operator. A solution for this dilemma is to use a metaphor of a duckling emerging from its egg - it will recognize as its mother the first moving object it sees that makes a sound, regardless of what it looks like: this phenomenon is called **imprinting.** Similarly, our robot will recognize as its owner the first entity that gives it a secret key. As soon as this imprinting key is received, the robot is no longer a newborn and will stay faithful to its owner for the rest of its life. The association is secure.

To make the association transient, the robot must be designed so it can die (loose its memory) and resurrect (be imprinted again). This security policy model, called "Resurrecting Duckling" (Stajano 2002) can be used not only in the communications between a human and a robot but between two robots as well, enabling robot-to-robot secure interaction.

We will focus in the following on the hazard involving collision with an obstacle, whether it is a human or not. To deal with this hazard, a robot should be able first to detect the obstacle, then recognize it and finally execute the avoidance manoeuvre. Obstacle recognition is useful only if the robot is supposed to cooperate with a human or another robot. Recognition of a human is quite a difficult task. It is necessary to use for this some special sensors (e.g. a suite of vision and thermal sensors). Recognition of another robot may be easier as it is possible to give the robot a special appearance (e.g. color). To reduce the risk of collision, robots must have some kind of a behavior to avoid obstacles. This behavior should be active all the time, concurrently with any other behavior active at the moment. We will discuss in the following how efficient are the obstacle avoidance behaviors in different robot control architectures.

6. Safety of different robot control architectures

6.1 Reactive architectures

As we have noticed above, the hierarchical architecture does not provide adequate functionality for obstacle avoidance. Much better equipped for this purpose are reactive architectures. The main reason for this is their direct connection of sensors to actuators. Even though the architecture does not have a memory-based world model, but as Brooks said "the world itself is its best model." Continuous sensing guarantees that it is always current, though not always correct, due to the sensors errors or failures. One of the characteristics of reactive architectures is their ability for graceful degradation of emergent behavior (Jones 2004). Let us consider a mobile robot moving to a goal and equipped with a number of different sensors. In the simplest case it will not discriminate between different kinds of obstacles. Failure of a sensor to detect an obstacle is called false negative. A false positive occurs when a sensor reports a condition that does not exists. From the viewpoint of safety a false negative implies a hazard to be dealt with. One way to mitigate this hazard is to use a set of different sensors invoking suitable behaviors. For example a robot may use for long distance journey a **sonar-avoid** behavior. While still some distance off, the robot will turn away from sonar detected obstacles. But sonar sensors are easily fooled. Smooth surfaces struck at shallow angles reflect sonar beams forward, not back toward the robot – thus no echo is ever detected, causing the sonar system to report a false negative. When this happens, the robot believes that the path is clear, even if it is not. In this case, the **sonar-avoid** behavior fails to trigger. But as the robot approaches the acoustically specular object, typically the infrared (IR) detectors will sense an obstacle and drive the robot away. But perhaps along with being too smooth and set at too oblique an angle, the obstacle's surface is also too dark to reflect IR radiation reliably back to the IR obstacle detector. Thus the IR detector may also fail to sense the object, generating a false negative of its own.

With no signal from the IR detector, the **IR-avoid** behavior does not trigger and the robot collides with the obstacle. The robot bumper now compresses and triggers the **Bumper-escape** behavior. Having failed to avoid the obstacle, the robot must now back up and try to turn away. Typically, the bump sensors can detect, at least crudely, where the collision has occurred, say, on the right, or the left of the robot. This knowledge can give the robot an indication of how to respond.

The performance of the robot may suffer, but the robot can still continue its mission. But suppose there is yet another difficulty, say, that the bumper is not a full-coverage bumper or that it has a dead spot at some point and it is exactly that point that the troublesome smooth, dark oblique object contacts the bumper. Now the bumper fails to report the collision and accordingly, the **Bumper-escape** behavior never triggers. We are left with our robot pressing against the object with all its might. Fortunately, an over-current sensor is available to detect this condition. When the drive motors are asked to work too hard, as can happen, when they force the robot against an object, motor current goes up. Too high current for too long a time is sensed and is used to trigger a **Stall-escape** behavior. Although there is no reliable way to tell where the blockage is located with respect to the robot, **Stall-escape** can at least command the robot to retreat and spin. So, the robot still can move towards its goal. We have assumed that all the problems are caused by a difficult-to-detect obstacle. But the same behavior would be produced by some inoperative sensors.

Note that the emergent desirable behavior does not require writing a special code that would explicitly instruct the robot to determine if a sensor is working properly. It is just the feature of behavior-based architecture.

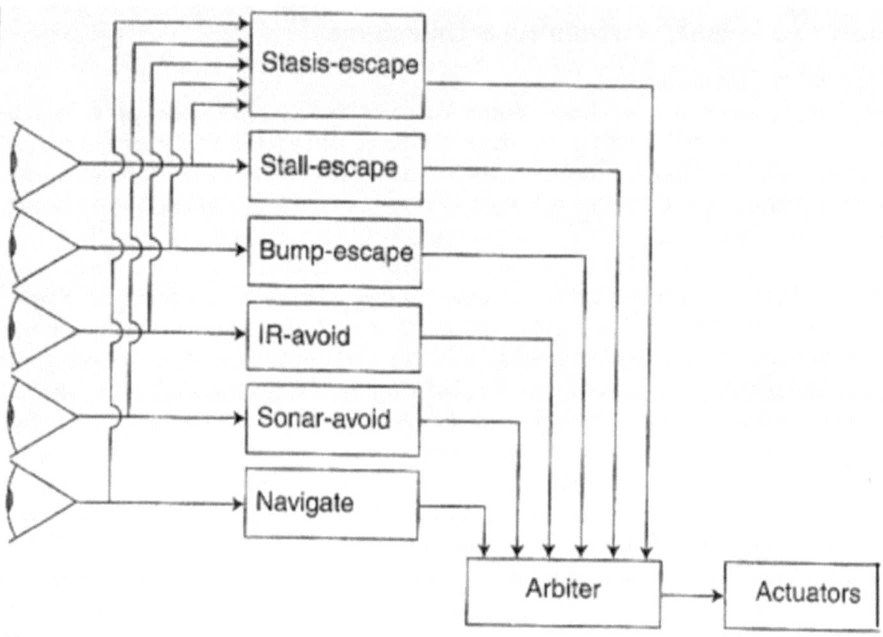

Fig. 2. Reactive architecture with graceful degradation.

The above examples show a crucial difference between the behavior-based approach of reactive architectures and the traditional deliberative approach. In reactive architectures we do not employ a single expensive sensor from which we must demand unattainable levels of precision and reliability. Rather we achieve superior results using a combination of relatively unreliable systems that work together to deliver safe behavior.

In the original subsumption architecture the speed was decided by the behavior subsuming all the other behaviors. Quite often the speed was fixed - the same for all the winning behaviors.

In case of reactive architecture using potential fields for coordination of behaviors (Arkin 1999) the emergent speed is the magnitude of the vector summed from all the active behaviors. It allows therefore to avoid obstacles while moving towards a goal - providing better overall behavior than subsumption architecture.

6.2 Hybrid architectures

The reactive architectures do not have an ability to monitor performance of the robot or to use e.g. an optimal speed and path to a destination, while avoiding obstacle. This can be mitigated by appropriately designed deliberative layer of a hybrid architecture. For example, in managerial architectures (Murphy 2000), the deliberative layer may include a Sensing Manager monitoring performance of the robot. If a behavior fails or a perceptual schema detects that sensor values are not consistent or reasonable, the Sensing Manager is alerted. It can then identify alternative perceptual schemas, or even behaviors, to replace the problematic behaviour immediately. Imagine a mobile robot in a convoy of robots hauling food to refugees. If the robot had a glitch in a sensor, it should not suddenly stop and think about a problem. Instead it should immediately switch to a back-up plan or even to smoothly slow down while it identifies a back-up plan. Otherwise, the whole convoy would stop, there might be wrecks, etc. The sensing manager working in the background mode, can attempt to diagnose the cause of the problem and correct it.

In some managerial architectures (e.g. SFX (Murphy 1996) speed control is considered a separate behavior. The safe speed of a robot depends on many influences. If the robot cannot turn in place, it will need to be operating at a slow speed to make the turn without overshooting. Likewise, it may go slower as it goes up or down hills. These influences are derived from sensors, and the action is a template (the robot always slows down on hills), so speed control is a legitimate behavior. But the other behaviors should have some influence on the speed as well. So, these other strategic behaviors contribute a strategic speed to the speed control behaviour. If the strategic speed is less than the safe speed computed from the tactical influences, then the output speed is the strategic speed. But if the tactical safe speed is lower, the output speed is the tactical speed. Tactical behaviors serve as filters on strategic commands to ensure that the robot acts in a safe manner in as close accordance with the strategic intent as possible.

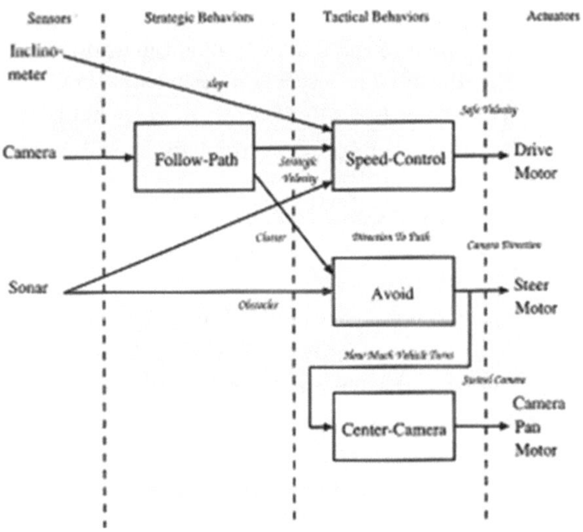

Fig. 3. Strategic and tactical behaviors for path following in SFX architecture.

An example of the Model-oriented architectures is the Task Control Architecture (TCA) (Simmons 1999). TCA has a flavor of an operating system. There are no behaviors per se, however many of low level tasks resemble behaviors. The reactive layer in this architecture called Obstacle Avoidance Layer takes the desired heading and adapts it to the obstacles extracted from the evidence grid virtual sensor. The layer uses a curvature-velocity method (CVM) to factor in not only obstacles but how to respond with a smooth trajectory for the robot's current velocity. Because this architecture does not have direct connection of sensors to actuators, it may have the same problems as the hierarchical architectures of the past - delayed response to events.

7. System-theoretic approach

To make a progress in dealing with safety in complex systems, we need new models and conceptions of how accidents occur, that more accurately and completely reflect the types of accidents we are experiencing today.
We use for safety analysis of robot control architectures a system-theoretical approach (Barchanski, J.A. 2006), which allows more complex relationships between events to be considered and also provides a way to look more deeply at why the events occurred. Whereas industrial safety models focus on unsafe acts or conditions and reliability engineering emphasizes failure events, a systems approach to safety takes a broader view by focusing on what was wrong with the system's design or operations that allowed the accident to take place.
System theory dates from the thirties and forties, and was a response to limitations to the classic analysis techniques in coping with the increasingly complex systems being built. Norbert Wiener applied this approach to control and communications engineering while Ludwig von Bertalanffy developed similar ideas for biology. It was Bertalanffy who suggested that the emerging ideas in various fields could be combined into a general theory of systems (Bertalanffy 1969) .
The systems approach focuses on systems taken as a whole, not on the parts examined separately. It assumes that some properties of systems can only be treated adequately in their entirety, taking into account all facets relating the social to technical aspects. These system properties derive from the relationships between the parts of systems: how the parts interact and fit together. Thus the system approach concentrates on the analysis and design of the system as a whole as distinct from the components or the parts. While components may be constructed in a modular fashion, the original analysis and decomposition must be performed top down.
The foundations of system theory rests on two pairs of ideas:
(1) emergence and hierarchy and (2) communication and control (Checkland 1981).

7.1 Emergence and hierarchy

A general model of complex systems can be expressed in terms of hierarchy of levels of organization, each more complex than the one below, where a level is characterized by having emergent properties.
Emergent properties do not exist at lower levels; they are meaningless in the language appropriate to those levels. Thus, the operation of the processes at the higher levels of the hierarchy results in a higher levels of complexity - that has emergent properties.
Safety is an emergent property of systems. Determining whether a plant is acceptable safe is not possible by examining a single valve in the plant.
In fact, statements about the "safety of the valve" without information about the context in which the valve is used, are meaningless. In a system-theoretic view of safety the emergent safety properties are controlled or enforced by a set of safety constraints related to the behavior of the system components.

A second basic part of system theory, the hierarchy theory, deals with the fundamental differences between levels of a hierarchy. Its ultimate aim is to explain the relationship between different levels, what generates the levels, what separates them, and what links them. Emergent properties associated with a set of components at one level in a hierarchy are related to constraints upon the degree of freedom of those components.
In a systems-theoretic view of safety the emergent safety properties are controlled or enforced by a set of safety constraints related to the behavior of the system components.
Safety constraints specify those relationships among system variables or components that constitute the non-hazardous or safe system states. Accidents result from interactions among system components that violate these constraints - in other words, from a lack of appropriate constraints on system behavior.

7.2 Communication and control

Regulatory or control action is the imposition of constraints upon the activity at one level of a hierarchy, which define the "laws of behavior" at that level yielding activity meaningful at a higher level.
Hierarchies are characterized by control processes operating at the interfaces between levels. Any description of a control process entails an upper level imposing constraints upon the lower. The upper level is a source of an alternative (simpler) description of the lower level in terms of specific functions that are emergent as a result of the imposition of constraints.
Control in open systems (those that have inputs and outputs from their environment) implies the need for communication. A system is not treated as a static design, but as a dynamic process that is continually adapting to achieve its ends and to react to changes in itself and its environment. To be safe, the original design must not only enforce appropriate constraints on behavior to ensure safe operation (the enforcement of the safety constraints), but it must continue to operate safely as changes and adaptations occur over time.

8. Intent specification

Conventional software engineering uses top-down and bottom-up hierarchies of two kinds:

- part-whole abstractions (where each level represents aggregation of components of the lower level),
- information-hiding abstraction (where each level is a refinement of the information at a higher level).

Higher level specifies the "what", lower level specifies the "how". There is no place to specify the "why" (intent, goal, purpose) of a level.
Systems-theoretical approach allows to do this in so called Intent Specification (Leveson 2000) which complements the conventional methodology - it implements the "means - ends" hierarchy, where each level provides the intent ("why") information about the level below.
The specification supports safety-driven development by tightly integrating the system safety process and the information resulting from it into the system engineering process and decision-making environment. The goal is to support design of safe systems rather than simply attempt to verify safety after-the-fact. Safety-related design decisions are linked to hazard analysis and design implementations so that assurance of safety is enhanced as well as any analysis required when changes are proposed.
The levels are used to specify user requirements, environmental assumptions and safety constraints.

Especially useful feature of the intent specifications is strong support of traceability. The structure of the intent specification is such that each level above provides rationale for *why* decisions at the lower levels were made the way they were. To be able to follow this reasoning easily, traceability links are encouraged throughout the document. For example, each hazard would link, within the same level, to the safety design constraints that mitigate the hazard. Each safety constraint would then link down from level one to level two where design decisions comply with and enforce the constraint. Those design decisions would link down to level three, the subcomponent requirements, where the subcomponent requirements adhere to the system level design decisions. These links go from the highest level goals of the system all the way down to the code and operator manuals. By following these links, one follows the reasoning behind the system's behavior and can evaluate the safety of changes to the system.
We will review in the following four levels of the intent specification

8.1 Design and safety constraints – level 1

Requirements document what things the system should do. Constraints document things that the system should *not* do. Constraints provide limits on the space of possible designs within which the system will behave as desired. Intent specifications frequently divide constraints into those that are related to safety and those that are not. Safety constraints are design constraints motivated by safety concerns. It is fairly easy to derive safety constraints from hazards. For example, if a hazard for an autonomous mobile robot is having its arm extended while it is in motion, then the safety constraint could be written that the robot must not move while its arm is extended. Safety constraints link to the hazards that generated them and to the design decisions that enforce them.

8.2 System design principles – level 2

The core of the level 2 of an intent specification is the set of design principles that specify how the design will satisfy the requirements documented in level 1 while not violating any design constraints. The functional design principles show the functional decomposition upon which the software logic is structured. It is useful to divide the robot control functionality into different operating modes (e.g. initialization, autonomy, operator, and safety). The functionality for each mode should be independent of the others. This feature allows the designers to change the internal logic of one mode without worrying about the effect the changes will have on the other modes.
The mode selection logic implements the mode transitions.

8.3 Black box specification – level 3

Level 3 is designed to provide the system designers with a complete set of tools with which to validate the specified requirements before implementation begins. Only black box (externally visible) behavior is included, i.e. the input/output function to be computed by the component. Black box models assist in the requirements review process by eliminating implementation details that do not affect external behavior and thus are not relevant in the validation of the requirements. For this purpose, a model of the robot control system is produced using formal specification language SpecTRM-RL (Leveson 2002).
Most model elements in SpecTRM-RL rely on AND/OR tables. The tables pictured below are part of the transitions for the "Distance" state in the obstacle avoidance behavior. AND/OR tables provide a very simple way to read the behavior of a component.

The first column of the table contains expressions that may be true or false. In the very first row of the table for the *Unknown* transition below, the expression is *System Start*. This expression is true at the instant the system is first started and false at all other times. The next row's expression is true when the *Reset* input has a value of *High* and false at other times.

When reading a table, the true or false value of this first column is matched against the values in subsequent columns. Subsequent columns must contain *T* indicating true, *F* for false, or * for don't care. If every row in the first column matches the values in some subsequent column, the table as a whole is true. If no column matches, then the table is not true. Another way of saying this is that rows are ANDed together and columns are ORed together

As an example, consider the *Unknown* transition below. If the table is true, then the system will transition to the state "distance to an obstacle" being unknown. This table can be read as saying that the distance transitions to unknown if the system is just starting, the system is being reset, or all the sensors are in the unknown state.

All the system state variables in a SpecTRM-RL model are required to have an "Unknown" value. A very common error found in requirements specifications and often associated with accidents is assuming that the computer always has an accurate (up-to-date) model of the controlled system. If input processing and feedback is disrupted for some reason (including temporary halting of the computer itself) however, the assumed controlled-system state may inaccurately reflect the real state. In SpecTRM-RL models, each state variable defaults to the unknown value at startup and returns to the unknown value after interruptions in processing or expected (and necessary) inputs are not received.

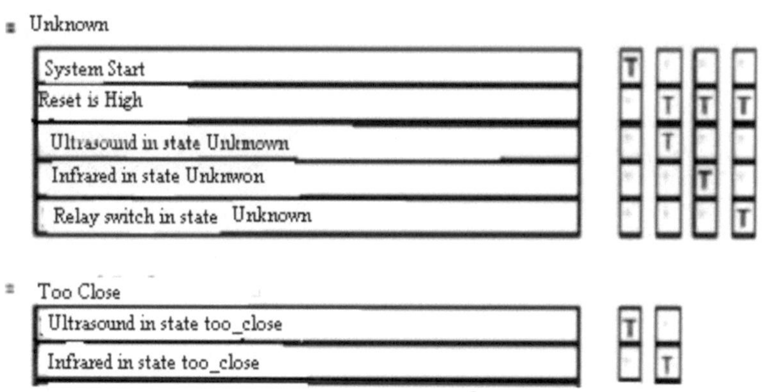

8.4 Physical and logical design – level 4

Level 4 describes the physical and logical designs of each subcomponent that allows them to meet the subcomponent requirements described at level 3. The component may be hardware, software, or some combination of the two.

Software Design Specifications can be represented in any design notation customarily used for software. Intent specifications can incorporate any desirable design notation, from formal predicate logic to UML diagrams.

Requirement specifications are mapped into design specifications of a selected robot control architecture. The mapping can be done in one of three ways:

Uncoupled: the mappings or functions from requirements to design principles are all one-to-one.

Loosely coupled: the mappings are one-to-many.
Tightly coupled: the mappings are many-to-many.
For any complex control system, a completely uncoupled designs, while allowing changes to requirements with minimal impact, is usually not practical.
A tightly-coupled system is highly interdependent. Each part is tightly linked to many other parts, so that a failure or unplanned behaviour in one, can rapidly affect the status of others. This is definitely not desirable.
The most appropriate and realistic is to design the system in such a way as to reduce the impact of requirements changes, i.e. to eliminate the rippling effects to other requirements, by designing the mappings to be one-to-many (loosely coupled). By this, a failure of a design element can be traced to incorrect or incomplete requirement.

8.5 Determinism analysis

A system is nondeterministic when it may display one of several behaviors for the same sequence of inputs. Nondeterministic systems are very difficult to analyze and test because of the element of chance in their behavior. It is strongly desirable for a safety-critical system to behave in a deterministic fashion.
SpecTRM-RL models are models of the requirements for system components. Nondeterminism in the behavior of a SpecTRM-RL model indicates inconsistency in the requirements. Additional information must be added to the specification to collapse the case of nondeterminism down to one deterministic choice. The following is an example of nondeterminism in a SpecTRM-RL model.

= Operational

Reset	T	*	*	*
Startup	*	T	T	T
Valid sonar	*	T	*	*
Valid infrared	*	*	T	*
Valid relay switch	*	*	*	T

= Internal Fault Detected

Internal fault detected	T	*
Startup	*	T
Time since last entered Startup > 3 seconds	*	T

This example is taken from the robot control mode. According to the first table, the robot goes to the Operational mode any time the Reset button is pushed. According to the second table's first column, the robot transitions to Internal Fault Detected whenever it was already in Internal Fault Detected. This specification is ambiguous as to whether a reset command should put the system in an operational state once a fault has been detected.

8.6 Completeness analysis

A system is complete when there is a specified response for every sequence of inputs that might come into the system. Incomplete systems have combinations of system states and inputs for which no response is specified. Incompletness occurs when there is no table in a state or mode definition that is true for some set of conditions. Additional information must be added to the specification to make one of the tables true.

9. Related methods

The closest to presented above approach to system safety is an industrial hazard analysis method called HAZard Operability analysis (HAZOP) developed for the British chemical industry in the 1950's. The goal of a HAZOP is to identify operational deviations from intended performance and study their impact on system safety (Soukas, 1988).
The HAZOP procedure is carried out by a HAZOP expert (the leader) and a team of system experts. The leader poses a battery of questions to the experts in an attempt to elicit potential system hazards. A HAZOP is potentially an exploratory analysis as neither potential faults nor hazards have been assumed beforehand. The HAZOP leader hypothesizes an abnormal condition and analysis proceeds in both directions, determining whether and how the condition is possible and what effects it has on the system.
The analysis is based on a systems theory model of accidents, in that it concentrates on the hazards that can result from components interaction, i.e. accidents are caused by deviations in component behavior.
HAZOP has several limitations. First, it is time- and labor –intensive, in large part due to its reliance on group discussions and manual analysis procedures. Second, HAZOP analyzes causes and effects with respect to deviations from expected behaviour, but it does not analyze whether the design, under normal operating conditions, yields expected behaviour or if the expected behaviour is what is desired.
A third limitations arises from the fact that HAZOP is a flow-based analysis. Deviations from within a component or processes are not expected directly; instead, deviation within a component (as well as human error or other environmental disturbance) is assumed to be manifested as a disturbed flow. A purely flow oriented approach may cause the analyst to neglect process-related malfunctions and hazards in favour of pipe-related causes and effects.
Because HAZOP concentrates on physical properties of the system (Souskas, 1988), it is not directly applicable to analyzing computer input and output.
There is a number of other manual methods for hazard analysis (e.g. Lear, 1993). All of these methods suffer from two weaknesses with respect to analyzing software. First, being manual techniques they depend on human understanding of the proposed software, which can be quite limited. Second, the manual techniques adhere to the HAZOP principle of identifying deviations in the connections, i.e., the computer inputs and outputs only. Accordingly, they do not provide guidance for following deviations into the control logic.

10. Conclusion

Knowledge of how to build safe and secure autonomous mobile robots is a prerequisite for their wide acceptance. In this chapter we have described relationship between robot autonomy and its safety and security. In contrast to traditional hazard analysis techniques used in electro-mechanical systems, we have introduced a technique based on system theory. We have outlined a methodology for building safe autonomous mobile robots based on hazard analysis and intent specification. The specification supports safety-driven development by tightly integrating the system safety process and the information resulting from it into the system engineering process and decision-making environment. The goal is to support design of safe systems rather than simply attempt to verify safety after-the-fact. Safety-related design decisions are linked to hazard analysis and design implementations so that assurance of safety is enhanced as well as any analysis required when changes are proposed.

11. References

Alami et al (1998), An Architecture for Autonomy, International Journal of Robotics Research, April 1998.

Arkin, R. (1999) Behaviour-based Robotics,The MIT Press.

Asimov I. (1950), I Robot, Doubleday, Arkin, R. (1999), Behavior-Based Robotics, The MIT Press

Barchanski J.A.(2004), Safety and Security of Autonomous Robots Software, IAESTED International Conference on Advances in Computer Science and Technology, St. Thomas, US Virgin Islands, November 2004.

Barchanski, J.A. (2006), System-Theoretic Approach to Safety of Robot Control Architectures, Canadian Conference on Electrical and Computer Engineering, 6 - 10 May 2006, Ottawa.

Bertalanffy, L., (1969) General Systems Theory: Foundations, Development and Applications, G. Braziller, New York,

Chandler, A. et al (2001) Digging into Concurrency, Technical Report, Computing Department, Lancaster University,

Checkland, P., (1981), Systems Thinking. System Practice, John Wiley &Sons, NY Harel, D.,(1987) Statecharts: A Visual formalism for complex systems, Science of Computer Programming, 8:231-274,

Gordon, D. (2001). APT Agents: Agents that are adaptive, predictable, and timely. In Lecture Notes in Artificial Intelligence, Volume 1871. Springer-Verlag.

Jones, J.L.,(2004), Robot Programming, A Practical Guide, McGraw-Hill

Murphy, R., (1996), Biological and Cognitive Foundations of Intelligent Sensor Fusion, IEEE Transactions on Systems, Man and Cybernetics, vol. 26, No 1.

Murphy, R., (2000), Introduction to AI Robotics, The MIT Press.

Lear, J.,(1993), Computer hazard and operability studies, In Sydney University Chemical Engineering Association Symposium: Safety and Reliability of Process Control Systems, October 1993.

Leveson N. G.(1987), et al, Safety Analysis Using Petri Nets, IEEE Transaction on Software Engineering, March 1987.

Leveson, N.G., (1995), Safeware: System Safety and Computers, Addison Wesley,.

Leveson, N. G. (2000), Intent Specification: An Approach to Building Human-Centered Specifications, IEEE Trans. on Software Engineering, January 2000.

Leveson, N. G., (2002), Safety-critical Requirements Specifications using SpecTRM, Trans. on Soft. Engineering

Simmons, R., et al, (1999), Xavier: An Autonomous Mobile Robot on the Web, Robotics and Automation, Magazine,

Souskas, J.,(1988), The role of safety analysis in accident prevention, Accident Analysis and Prevention,

Stajano, F., (2002), Security for Ubiquitous Computing,Wiley,

3

Control Strategies of Human Interactive Robot Under Uncertain Environments

Haiwei Dong and Zhiwei Luo
Japan Society of the Promotion of Science and Kobe University
Japan

1. Introduction

Actually, the research on human interactive robot (HIR) has been a topic of both science fiction and academic speculation for a long time. The origin of HIR as a discrete issue was stated by 20th century author Isaac Asimov in 1941, in his novel "I, Robot". He stated the Three Laws of Robotics[1] as, "

a. A robot may not injure a human being or, through inaction, allow a human being to come to harm.
b. A robot must obey any orders given to it by human beings, except where such orders would conflict with the First Law.
c. A robot must protect its own existence as long as such protection does not conflict with the First or Second Law. "

The three laws of robotics determine the idea of safe interaction which constitutes the basic rules of HIR. With the advances of artificial intelligence (AI), the HIR could eventually have more proactive behaviours, planning their motion in complex unknown environments. Nowadays, HIR are artificial agents with capacities of perception and action in the human's environment. Their use has been tended to be found in the most technologically advanced societies in critical domains as search and rescue, military battle, law enforcement, entertainment, hospital care, etc. These domains of applications imply a closer interaction with human. The concept of closeness is to be taken in its full meaning, HIR and humans not only share the workspace but also share goals in terms of task achievement. The HIR has to adapt itself to human's way of expressing desires and fulfill its task. Taking lifting up human in elder care for example, the human interactive robot RI-MAN, designed by the RIKEN Bio-Mimetic Control Research Center, communicates with human by listening and speaking, which makes it understand the human will (Onish, Luo et al. 2007). To fulfil the task, it also estimates the attitude of human body in real-time by tactile sense (Mukai, Onishi et al. 2008). This example contains two aspects of HIR, one is to understand the human mind and the other is to accomplish the manipulation. The former is based on AI techniques, like language comprehension, and the latter is relied on force control.

On the other hand, human's environments are much more complex. Thus, the HIR needs perceiving and understanding capacities to build dynamic models of its surroundings. It

---

[1] In science friction, the Three Laws of Robotics are a set of three rules written by Isaac Asimov, which almost all positronic robots appearing in his friction must obey.

needs to categorize objects, recognize and locate humans and further their emotions. Also in the case of human interactive robot RI-MAN (Mukai, Onishi et al. 2008), if he cannot navigate in the hospital and further more cannot locate the human, the understanding and good manipulation mentioned above make no sense. In our opinion, with the development of human society and robotics technology, HIR research becomes much important (Fig. 1).

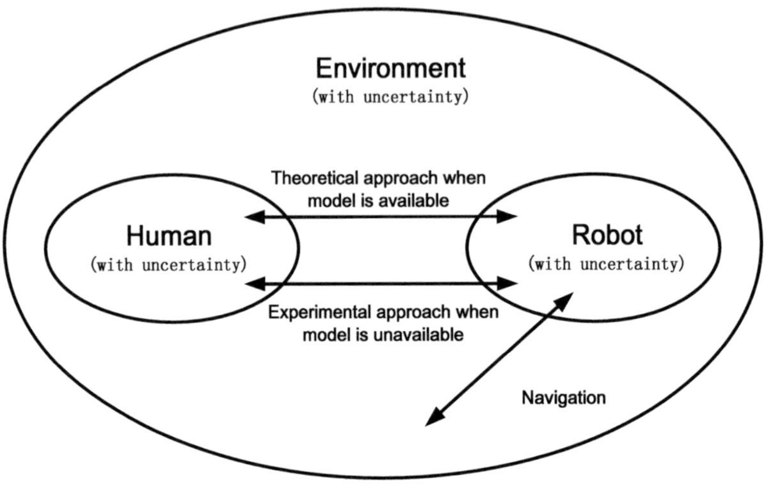

Fig. 1. Interaction relation between human, robot and environment.

As HIR interacts with human, sometimes we can model the process of human activity accurately and the robot-human interaction process can be simulated and further done by theoretical approach; sometimes human movement is too complex to model and in that case, the experimental approach is a good way. However, the above thing is to make the HIR capable to navigate in the human environment.

The chapter focuses on the issue of designing control strategies of human interactive robot where dealing with uncertainties is a critical issue. Actually, although there are many researches on HIR, these works do not concern too much on the uncertainty. The fact is that there are various uncertainties in the world which comes from robot, human, environment, etc. Developing human interactive robot, in some concept, is dealing with uncertainties. Actually, there are kinds of solving approaches for uncertainties according to circumstances: some uncertain can be assumed as Gaussian noise and based on the property of Gaussian noise, we can do estimation more accurately; some uncertainty coming from model reduction and this kind of uncertainty can be solved by control theory; some uncertainty is neither hard to model, nor difficult to determine the source. In that case, it is a better way to consider it as a black box and recognize it by system identification.

In our opinion, the issue of uncertainty has become a major stumbling block for the design of capable human interactive robot. To develop a human interactive robot, it is evitable to deal with uncertainties. Furthermore, managing uncertainty is possibly the most important step towards robust real-world HIR systems (Thrun, Burgard et al. 2005). The basic principle of this chapter is to design control for HIR by dealing with uncertainties. Specifically, there are a number of factors that contribute to the uncertainty of robot, human and environment:

Firstly, robot environment is inherently unpredictable. While the degree of uncertainty in well-structured environment such as assembly lines is small, environments such as highways and private homes are highly dynamic and in many ways highly unpredictable. The uncertainty is particularly high for robots to operating in the proximity of people.

Secondly, sensors are limited in what they can perceive. Limitations arise from several factors. The range and resolution of a sensor is subject to physical limitations. For example, cameras cannot see through walls, and the spatial resolution of a camera image is limited. Sensors are also subject to noise, which perturbs sensor measurements in unpredictable ways and hence limits the information that can be extracted. And finally, sensors can break. Detecting a faulty sensor can be extremely difficult.

Thirdly, robot actuation involves motors that are, at least to some extent, unpredictable. Uncertainty arises from the physical structure of actuator, i.e. actuator is impossible to achieve absolute precision. In addition, uncertainty also comes from control noise, wear-and-tear, mechanical failure, etc.

Fourthly, some uncertainty is caused by model approximation. Models are abstractions of the real world. As such, they only partially model the underlying physical process of the subject. Model errors are a source of uncertainty that has a great impact on the fulfillment of the robot task.

Fifthly, there are some uncertainties that are extremely difficult to model or are impossible to obtain. In that case, the uncertainty is very hard to deal with by analytic methods, like mathematical approaches. Also the evaluation of the uncertainty, both sources and magnitude are hard to be obtained.

In summary, the uncertainties caused by the five factors can be noted as environment uncertainty, sensor uncertainty, actuator uncertainty, model uncertainty, unmodeled uncertainty, respectively. The above five kinds of uncertainties involve nearly all the uncertainties human interactive robot may encounter. The control design of HIR in this chapter takes it a clue to deal with the uncertainties mentioned above. In other words, we correlate the basic HIR's technical problems with the five uncertainties. By dealing with the uncertainties, we design control for HIR and finally solve the basic technical problems of HIR. Specifically, the first basic problem is to make HIR capable to navigate in a large unknown environment which corresponds to dealing with environment uncertainty, sensor uncertainty and actuator uncertainty; the second basic problem is to propose approach for HIR to physically interact with human which has large degrees of freedom (DOF). This basic problem corresponds to dealing with the model uncertainty. The third basic problem is to design approach for HIR to mentally interact with human, i.e. extract the human's intention, which corresponds to dealing with unmodeled uncertainty. The detailed explanation on correlation between the basic problems and corresponding uncertainties is as follows:

The first basic problem concerns two issues as robot localization and environment cognition, i.e. to estimate the robot's position and the feature's position. Considering that robot's motion and observation are affected by Gaussian noise, its kinematics model and sensor model are added by uncertainty. Similarly, the position of feature (or landmark from the viewpoint of mapping) is also affected by Gaussian noise; herein we add Gaussian uncertainty in the model. With regards to these models with uncertainty, it is better to use a probabilistic approach to estimate robot's and landmark's positions, i.e. simultaneous localization and mapping (SLAM). By definition, SLAM is the process by which a mobile robot builds a map of an environment and at the same time uses this map to compute its own location. By SLAM, robot is able to navigate in an unknown environment freely. However, the problem is that the current SLAM approaches only fit the relative small environment because of the time-consuming computation. When solving the first basic problem, the main topic is to propose an efficient SLAM approach for large scale unknown environment.

The second basic problem contains two issues potentially: the interactive object is human dynamics with large degree of freedom; the interactive manipulation is physically done by force. As the first issue, human body has numerous bones and joints and the human model is a very complex one with large degrees of freedom. Exerting force on such a big model is very complex and such dynamic process is impossible to be calculated in real time. Here we use model reduction to decrease the DOF of human model whereas the reduction error (i.e. model uncertainty) comes out. For the second issue, external force is exerted on human. Unlike common object, human has passive moment in human joint and more complicated, sometimes moves at his or her will. Such force character provides a big challenge to control. To solve the basic problem two, the main topic is to propose an adaptive force control approach for HIR when physically contacting with human. More specifically, we take a typical case to research, i.e. how to lift human by HIR in nursing care.

The third basic problem focuses on obtaining the human's intention. Actually, the human's intention is very difficult to measure. The model for human mind is also extremely hard to build right now. For us, the human's intention is almost full of unmodeled uncertainty, i.e., we have little knowledge about it. In this case, treating it as a black box is a good way. We stimulate the black box (i.e. human dynamics system) and measure output. By choosing a suitable function to link the stimulus signal and output, we can obtain the intention model experimentally. The main topic here is to design an approach for extracting human's intention. Without loss of generality, in this chapter we consider the problem of estimating the human's intended walking speed.

To conclude, in this chapter, we design controls for human interactive robot by dealing with the environment uncertainty, sensor uncertainty, actuator uncertainty, model uncertainty and unmodeled uncertainty. Specifically, the typical problems of HIR we focus on are designing an efficient SLAM approach for large unknown environment; proposing an adaptive force control for lifting human up; estimating human's intended walking speed. The above three typical problems involve the basic problems when designing controls for HIR. The solution of them also provides a general solving framework for HIR, which is of great importance both in research and in application. The preceding researches relating with the above three typical problems are shown as follows.

2. Motion control for large environment navigation

2.1 Background

Extended Information Filter SLAM (EIF-SLAM) estimates the positions of robot and landmarks by updating information matrix and information vector. The total element numbers of information matrix and information vector are

$$\left(dof\left(landmark\right)\times n_{lm}+dof\left(robot\right)\right)^{2}$$

and

$$dof\left(landmark\right)\times n_{lm}+dof\left(robot\right)$$

where n_{lm} denotes the mapped landmarks. $dof(landmark)$ and $dof(robot)$ denote landmarks' degree of freedom (DOF) and robot's DOF, respectively. Actually, the dimension of the information matrix increases rapidly with the increase of the landmark number in the environment. For example, if a three-dimensional environment has 100 landmarks, the information matrix is a huge matrix with dimension of 303 by 303. Actually, the computational burden in EIF-SLAM is mainly due to the calculation of the information matrix.

Previous research has proven that information matrix is a naturally sparse matrix (Eustice, Singh et al. 2005). Hence, it is a good way to make use of this feature and change information matrix into a real sparse matrix for computation reduction. Until now, the successful research includes Thrun et al.'s work which solved a relatively large environment (i.e. Victoria Park) for the first time (Thrun, Koller et al. 2002); Eustice et al.'s work which constructed the map of Titanic ship in the dark ocean (Eustice, Singh et al. 2005). Among this work, the sparsification of information matrix was obtained by constructing a proper topological structure of Bayes network, which needs to classify the landmarks in advance. This is a time-consuming work and against the enhancement of efficiency. Actually, the intuitive idea is to set the near-zero elements to zero directly. Thrun tried this way (Thrun, Liu et al. 2004), however, Eustice proved that direct sparsification leads the algorithm diverge (Eustice, Walter et al. 2005), which indicates the improper sparsification process mentioned may cause the algorithm corrupt. If we want to pursue this kind of idea, we have to clarify the condition under which the algorithm maintains stable which constitutes the basis of this section. This section focuses on proposing an efficient stable sparsification SLAM approach for large scale environment.

2.2 Sparsing information matrix

2.2.1 Characters of information matrix

In fact, information matrix in EIF-SLAM has special structural characters (Dong, Luo et al. 2009) where the value denotes the numerical value of the element in information matrix located at $(row\ index,\ column\ index)$. It is convenient to divide the elements of information matrix into two parts: Part I denotes larger part where the elements have larger values. The elements along the main-diagonal and near the endpoints of sub diagonal line belong to this part. Part II is composed of other elements in the information matrix. Three main characters of information matrix can be stated as

a. Information matrix is symmetric along the main diagonal line;
b. Elements with huge values distribute in the neighborhood of the main diagonal line and the end points of the sub diagonal line (Part I);
c. For the elements of Part II, the value of the element decreases with the distance from the main diagonal line.

The reasons for character 1) to 3) can be explained as follows, respectively (Smith, Self et al. 1990; Liu and Thrun 2003; Eustice, Singh et al. 2005; Eustice, Singh et al. 2006).

a. Each element in the information matrix denotes the link strength between the corresponding landmark and robot. As we all know that the link strength has symmetry, i.e. if the link strength between A and B is π, then the link strength both from A to B and from B to A are π. Therefore, information matrix is a symmetric matrix;

b. From the viewpoint of Bayes network, the link strength shows the correlation. In SLAM problem, the largest correlation is the relation of robot itself and landmark itself. The correlation between the robot and landmark becomes weak with time. If the robot is currently observing a landmark, the correlation between the robot and the landmark is strong. The elements in diagonal line denote the link strength of robot versus itself, or landmark versus itself while the elements near the endpoint of information matrix show the link strength of robot versus current observing landmark. Hence, the elements with large value centralize in Part II;

c. The correlation becomes weak when time passes after observing it. When the robot observes the landmark again, the correlation increases again.

The most important characteristic of information matrix is that most elements of it are nearly zero. In other words, an information matrix is a nearly sparse matrix. The information matrix sparsification by different threshold $\chi_i \, (1 \le i \le 6)$ is shown in Fig. 2 where χ_i is

$$\begin{cases} \chi_{i+1} = 10^2 \times \chi_i \\ \chi_1 = 10^{-13} \end{cases} \quad (1 \le i \le 5) \tag{1}$$

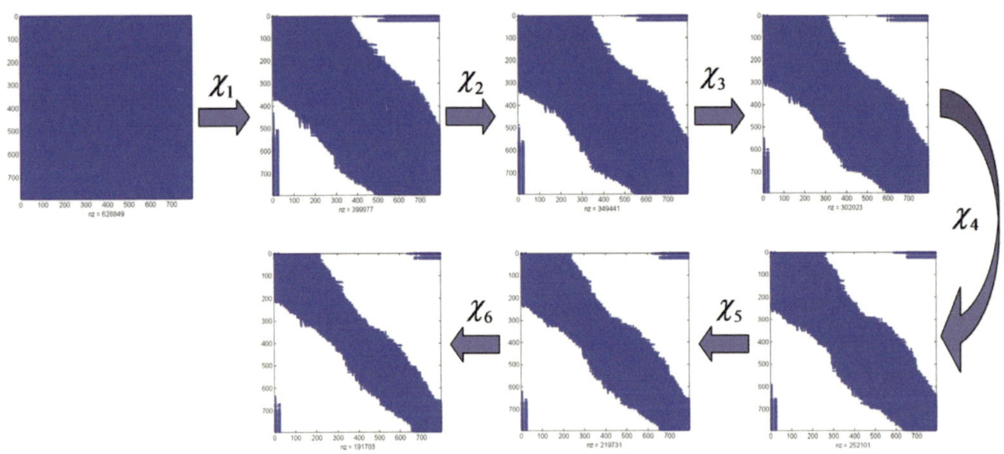

Fig. 2. Information matrix sparsification.

After computing the sparse ratios of the information matrices in Fig. 2, we can get the graph that illustrates how the sparse ratio changes with the threshold (Fig. 3) where the x-axis denotes the threshold and the y-axis denotes the sparse ratio. By using a curve fitting method, it shows that the distribution of elements in the information matrix satisfies a linear distribution in logarithmic coordinates. The sparse ratio increases with the threshold. However, sparsification error arises from the sparsification process. Furthermore, inappropriate sparsification may lead EIF-SLAM to diverge. Therefore, the key issue here is to find a condition under which the estimation results converge all the time.

From the illustrations above, the information matrix in EIF-SLAM algorithm is naturally sparse. Hence, the sparsification structure of the information matrix is very suitable to be sparsed. Moreover, because the computations are mainly on the calculation relating with information estimation variables, high efficiency is predicted by sparsing the information matrix. The following sparsification approach utilizes the characters of the information matrix to decrease the computational burden of EIF-SLAM.

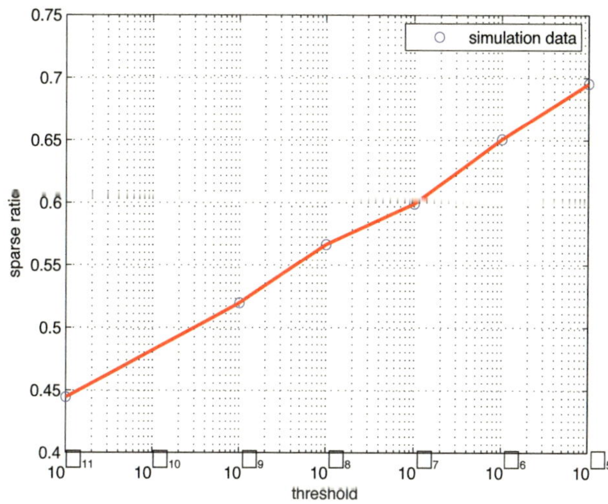

Fig. 3. Linear relation between sparse ratio and threshold.

From the illustrations above, we obtain that the information matrix in EIF-SLAM algorithm is naturally sparse. Hence, the sparsification structure of the information matrix is very suitable to be sparsed. Moreover, because the computations are mainly on the calculation relating with information estimation variables, high efficiency is predicted by sparsing the information matrix. The following sparsification approach utilizes the characters of the information matrix to decrease the computational burden of EIF-SLAM.

2.2.2 Sparsification approach

The mean vector μ and the covariance matrix Σ of the environment state vector ε are written in the form

$$\varepsilon = \begin{bmatrix} \varepsilon_1 \\ \varepsilon_2 \\ \vdots \\ \varepsilon_n \end{bmatrix}, \mu = \begin{bmatrix} \mu_1 \\ \mu_2 \\ \vdots \\ \mu_n \end{bmatrix}, \qquad \Sigma = \begin{bmatrix} \sigma_{11} & \sigma_{12} & \cdots & \sigma_{1n} \\ \sigma_{21} & \sigma_{22} & \cdots & \sigma_{2n} \\ \vdots & \vdots & \ddots & \vdots \\ \sigma_{n1} & \sigma_{n2} & \cdots & \sigma_{nn} \end{bmatrix} \tag{2}$$

Assume estimation error is a Gaussian noise, we have

$$\begin{aligned} &\mathrm{Prob}\left(\max_{1 \le i \le n}\{|\mu_i - \varepsilon_i| - \sigma_{ii}\} \le 0\right) \approx 0.68 \\ &\mathrm{Prob}\left(\max_{1 \le i \le n}\{|\mu_i - \varepsilon_i| - 2\sigma_{ii}\} \le 0\right) \approx 0.95 \\ &\mathrm{Prob}\left(\max_{1 \le i \le n}\{|\mu_i - \varepsilon_i| - 3\sigma_{ii}\} \le 0\right) \approx 0.997 \end{aligned} \tag{3}$$

By taking the relation between time estimation variables and information estimation variables

$$\begin{cases} \Lambda^{-1}\eta = \mu \\ \Lambda^{-1} = \Sigma \end{cases} \tag{4}$$

into Equation (3), we obtain

$$\begin{aligned} &\text{Prob}\left(\max\left\{\left|\Lambda^{-1}\eta - \varepsilon\right| - \text{Sqrt}\left(D(\Lambda^{-1})\right)\right\} \le 0\right) \approx 0.68 \\ &\text{Prob}\left(\max\left\{\left|\Lambda^{-1}\eta - \varepsilon\right| - 2\times\text{Sqrt}\left(D(\Lambda^{-1})\right)\right\} \le 0\right) \approx 0.95 \\ &\text{Prob}\left(\max\left\{\left|\Lambda^{-1}\eta - \varepsilon\right| - 3\times\text{Sqrt}\left(D(\Lambda^{-1})\right)\right\} \le 0\right) \approx 0.997 \end{aligned} \tag{5}$$

Here we propose Theorem 1 to clarify the conditions for sparsification for guaranteeing the convergence. In other words, from the viewpoint of mathematics, Theorem 1 gives the consistency condition; from the viewpoint of geometric, the mean remains in the range of the corresponding covariance under consistency condition.

Theorem 1

Assume the new information matrix after sparsification is written in the form of $\Lambda + E$, where E is the sparsification error matrix. If $\max\left\{\left(\Lambda^{-1}\eta - \varepsilon\right) \otimes H\eta\right\} \le 0$ is satisfied, we obtain

$$\max\left\{\left|(\Lambda + E)^{-1}\eta - \varepsilon\right| - \text{Sqrt}\left(D(\Lambda^{-1})\right)\right\} \le 0 \tag{6}$$

where

$$\begin{aligned} H &= -\overline{\Sigma}\left(I + \overline{\underline{E}}\underline{\Sigma}\right)^{-1}\overline{\underline{E}}\underline{\Sigma} \\ &= -\overline{\Sigma}\overline{\underline{E}}\left(I + \overline{\underline{\Sigma}}\overline{\underline{E}}\right)^{-1}\underline{\Sigma} \end{aligned} \tag{7}$$

$\overline{\underline{\Sigma}}$, $\overline{\Sigma}$, $\underline{\Sigma}$ and $\overline{\underline{E}}$ are defined as

$$\Sigma = \begin{bmatrix} \overline{\underline{\Sigma}} & \Sigma_2 \\ \Sigma_1 & \Sigma_3 \end{bmatrix}, \quad E = \begin{bmatrix} \overline{\underline{E}} & 0 \\ 0 & 0 \end{bmatrix}, \quad \overline{\Sigma} = \begin{bmatrix} \overline{\underline{\Sigma}} \\ \Sigma_1 \end{bmatrix}, \quad \underline{\Sigma} = \begin{bmatrix} \overline{\underline{\Sigma}} & \Sigma_2 \end{bmatrix} \tag{8}$$

where the element positions of $\overline{\underline{\Sigma}}$ are the transport element position of $\overline{\underline{E}}$. That is, the portioned matrices $\overline{\underline{E}}$, $\overline{\underline{\Sigma}}$, $\overline{\Sigma}$ and $\underline{\Sigma}$ are of order of $m_1 \times m_2$, $m_2 \times m_1$, $n \times m_2$ and $m_1 \times n$, respectively.

The proof of Theorem 1 is shown in (Dong, Tang et al. 2010). Consider Equation (6) with respect to Equation (5), it is shown that after sparsification, each estimated mean still inside the scope of corresponding covariance. That is to say, the estimated position of the robot or landmark is inside the estimation range, which indicates the estimation is reliable. In addition, the proof of Theorem 1 also gives a quite efficient way to get covariance matrix from information matrix without directly computing the inversion. According to Theorem 1, we can eliminate the information matrix under the condition of stabilization. In practice, loop-closure is used to compensate the estimation error which comes from sparsification. After sparsification, another theorem was proposed to evaluate sparsification error by deriving the upper bound of relative error ratios, including mean vector error ratio and covariance matrix error ratio (Dong, Tang et al. 2010).

2.3 Large complex environment simulation

The proposed sparsification SLAM approach was simulated in a large scale environment (about 400 landmarks). As the landmarks are very dense, the environment is a complicated one. Such environment can further verify the effectiveness of the proposed approach. In the simulation, the robot applied is a two-wheel robot with three degrees of freedom and the sensor arranged on the robot is range-bearing type sensor (such as laser, camera, etc.). The robot moves counterclockwise for two laps while utilizing sparsification approach to process SLAM.

2.3.1 Efficiency analysis

The main computational burden of EIF-SLAM is primarily from the three steps: motion update, features addition and observation update. The comparison of computational time between EIF-SLAM (denoted as stars) and sparsification method (denoted as crosses) is shown in Fig. 4. It shows that the sparsification approach is able to solve SLAM in the environment with 400 landmarks. Furthermore, according to the curve trend of computational time, we can predict that our sparsification method has an efficiency advantage for large environment.

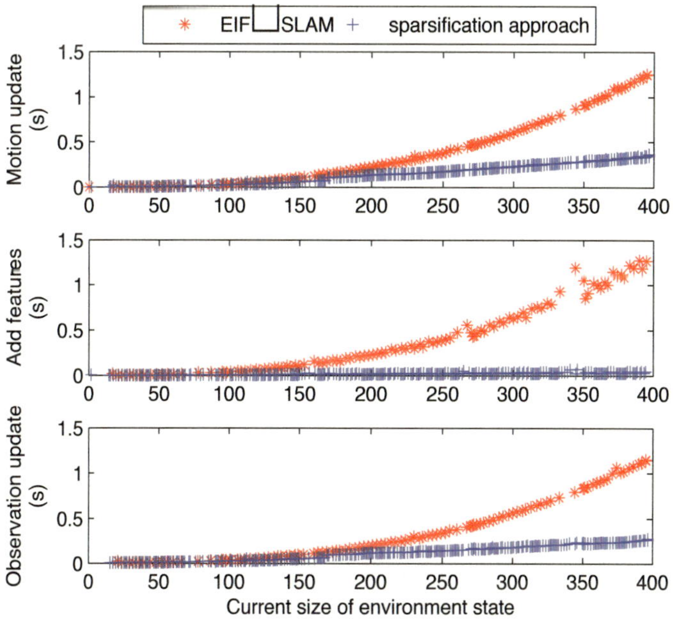

Fig. 4. Efficiency comparison between EIF-SLAM and sparsification SLAM.

2.3.2 Accuracy analysis

The comparison of error and covariance is shown in Fig. 5 where dash line and solid line denote estimation error and self-covariance of robot's position, respectively. Horizontal axis shows the time. Vertical axis denotes the estimation error and covariance where x and y denote the moving directions, respectively. It shows that estimation error and covariance decrease sharply at time 250, which indicates that there is a loop-closure at this moment.

In fact, during the first lap, the error and covariance by the sparsification method nearly have the same magnitude, (in Fig.5 (b), the solid line nearly coincides with the dash line), which indicates that the sparsification method eliminates considerable number of elements under the condition of consistency. In the second lap, the estimation error by EIF-SLAM (Fig. 5 (a)) and the sparsification approach (Fig. 5 (b)) seems the same. Thus, it can be concluded that by using loop-closure properly, the sparsification approach can obtain satisfying accuracy with high efficiency.

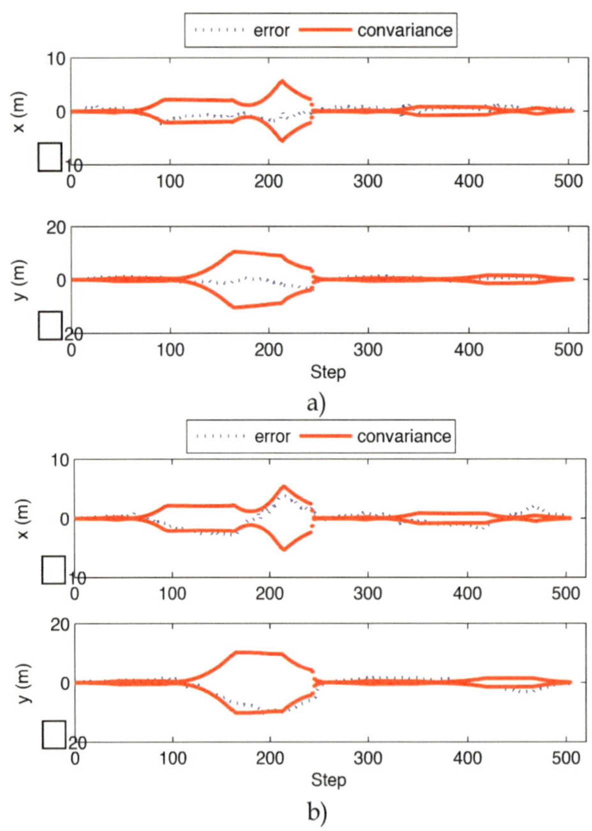

Fig. 5. Comparison of estimation error and covariance.

2.4 Summary

In this section, a sparsification EIF-SLAM approach was proposed to enable human interactive robot to navigate in the large unknown environment. According to the normal structural features of information matrix, we eliminate many of the near-zero elements in the information matrix. Such sparsification process is under the condition of consistency.

Hence the sparsification approach stays stable. In addition, the upper bound of estimation error is also given for evaluation. The large complex environment simulation indicates that the sparsification approach has the advantage of high efficiency and accuracy. The outdoor car-park experiment shows the ability to realize consistent estimation by the sparsification approach (Dong, Luo et al. 2009).

Compared to the previous researches on efficient navigation of human interactive robot, the sparsification approach proposed in this section gives a direct but effective way to obtain efficiency. Concerning parameters, we can compromise the efficiency and accuracy under the condition of stabilization. It is also noted that the consistency condition derived in this section also has great meaning for the future research on sparsification.

This section dealt with the environment uncertainty, sensor uncertainty, actuator uncertainty, and assumed them as Gaussian noises. The proposed sparsification SLAM approach eliminated these uncertainties and realized high efficiency of SLAM estimation.

3. Adaptive force control for lifting human

3.1 Background

It has been thought that the human body is composed of 206 bones and numerous human joints connecting adjacent bones. Based on the physiological structure of human joints, the human joints can be mainly divided into three types as the hinge (1 DOF), the pivot (1 DOF) and the ball socket (3 DOF). In dynamic equations, each DOF is expressed as one differential equation, which indicates that the overall set of equations of human body dynamics is a very huge one. The force interaction with such a big model needs considerable computation for human interactive robot.

As illustrated in Section 1, in the process of lifting human, the human body has to be considered as a free-floating multi-link rigid object with passive moments. Compared with one-end-fixed object, like manipulator, free-floating object is much more complex. The force acting on any part of the object would affect the attitude of the entire object. Moreover, although the multi-link object has been researched a lot, the one with large DOF and large redundancy has not fully been considered yet. Another problem is calculation; any computation on such a huge model needs much computational time whereas the safe lifting requires real-time computation.

In consideration of the difficulties mentioned above, our basic idea for lifting human comes from the daily experience. When we human lift a person, we do not care too much about the detailed dynamics, like the change of ankle angle, hand position and so on. What we do care about are the head position, the vertical deflection of upper limb and the hip angle. Here we call them "states of interest". From the viewpoint of system theorem, we treat the human body as a large redundant system whose dimensions are reduced by diverting the effects of other "joints of noninterest" to the "interested ones". The resulting body model is a reduced one with less DOF but unfortunately, has huge uncertainties (i.e. model uncertainty). Here we focus on proper model reduction and methods for dealing with the generating model uncertainty, which leads to a force control for lifting human.

3.2 Force interaction modeling

If we consider human body as a rigid multi-link object, each bone corresponds to a link and each human joint corresponds to a joint connecting adjacent links. Moreover, human joints have passive torques corresponding to the constriction forces and moments developed by ligaments, joint capsules and other soft tissues. Hence, we write the dynamics of human in the from

$$H(q)\ddot{q} + C(q,\dot{q})\dot{q} + G(q) = \tau_{pass} \tag{9}$$

where

$q_{n\times 1}$ Generalized body states, including the position of head and the angles of all the joints.

$H(q)_{n\times n}$ Inertia matrix, defined as a positive semi-definite symmetric matrix, containing information on the body's instantaneous mass distribution.

$C(q,\dot{q})_{n\times n}$ Centripetal and coriolis torques. $C(q,\dot{q})$ term contains products of angular speeds. When the degree of freedom is rotational, the terms of $C(q,\dot{q})$ represent the moments of centrifugal forces.

$G(q)_{n\times 1}$ Gravitational torques. Because $G(q)$ changes as the posture configuration of the human body model, its terms are functions of the generalized states.

$\tau_{pass\ n\times 1}$ Passive joint torque, including torques and moments arising from muscular activations and passive elastic structures surrounding the human joints.

Here the HIR's task is to lift human, i.e. to control the position and posture of the body to the desired states by external forces. Adding the force exertion in Equation (9), leads to:

$$H(q)\ddot{q} + C(q,\dot{q})\dot{q} + G(q) = \tau_{pass} + \tau_{rob} \tag{10}$$

where

$\tau_{rob\ n\times 1}$ The torques acted by the robot arms which is controllable.

Actually, the human interactive robot considered in this section has two manipulation arms, like RI-MAN (Mukai, Onishi et al. 2008). Force is applied to the fixed point at back and knees, shown as F_1 and F_2. Forces F_1 and F_2 have relations with τ_{rob} as

$$\tau_{rob} = J_1{}^T F_1 + J_2{}^T F_2 \tag{11}$$

where J_1 and J_2 are the Jacobian matrices of the human body model.

3.3 Human dynamics reduction

The basic idea in this section is to reduce the human body model into a small one with less degree of freedom, which includes the following three steps:

a. Choose "states of interest". These states include the fundamental performance indexes of the task. In other words, based on these states, we can determine whether the task is complete or not and furthermore, evaluate the performance is good or not. Let us define

$$\bar{q}_1 = [q_1, \cdots, q_m]^T{}_{m\times 1} \tag{12}$$

as the "states of interest" and

$$\bar{q}_2 = [q_{m+1}, \cdots, q_n]^T{}_{(n-m)\times 1} \tag{13}$$

as the "uninterested states" consisting of the other states. The overall state can be written as $q = [\bar{q}_1 \ \ \bar{q}_2]^T$.

b. Arrange the dynamic equation set. Based on the division of generalized states in (a), we change the element position of H, C, G, τ_{pass}, τ_{rob}. The expanded form of new dynamics equation set is

$$\begin{bmatrix} \bar{H}_{11} & \bar{H}_{12} \\ \bar{H}_{21} & \bar{H}_{22} \end{bmatrix} \begin{bmatrix} \ddot{\bar{q}}_1 \\ \ddot{\bar{q}}_2 \end{bmatrix} + \begin{bmatrix} \bar{C}_{11} & \bar{C}_{12} \\ \bar{C}_{21} & \bar{C}_{22} \end{bmatrix} \begin{bmatrix} \dot{\bar{q}}_1 \\ \dot{\bar{q}}_2 \end{bmatrix} + \begin{bmatrix} \bar{G}_1 \\ \bar{G}_2 \end{bmatrix} = \begin{bmatrix} \bar{\tau}_{pass,1} \\ \bar{\tau}_{pass,2} \end{bmatrix} + \begin{bmatrix} \bar{\tau}_{rob,1} \\ \bar{\tau}_{rob,2} \end{bmatrix}, \tag{14}$$

where the dimensions of sub block matrices of $\bar{H}_{11}$, $\bar{H}_{12}$, $\bar{H}_{21}$, $\bar{H}_{22}$ are $m \times m$, $m \times (n-m)$, $(n-m) \times m$, $(n-m) \times (n-m)$, respectively. The dimensions of the sub block matrices $\bar{C}_{11}$, $\bar{C}_{12}$, $\bar{C}_{21}$, $\bar{C}_{22}$ are $m \times m$, $m \times (n-m)$, $(n-m) \times m$, $(n-m) \times (n-m)$, respectively. The dimensions of vectors $\bar{G}_1$, $\bar{\tau}_{pass,1}$, $\bar{\tau}_{rob,1}$ are $m \times 1$, and $\bar{G}_2$, $\bar{\tau}_{pass,2}$, $\bar{\tau}_{rob,2}$ are $(n-m) \times 1$.

c. Construct the small dynamics equation set. First of all, we define the generalized states of reduced model as

$$q_s = \bar{q}_1 \tag{15}$$

Then extracting the parts of the human dynamics relating with q_s, we get

$$\begin{bmatrix} H_{11} & H_{12} \end{bmatrix} \begin{bmatrix} \ddot{\bar{q}}_1 \\ \ddot{\bar{q}}_2 \end{bmatrix} + \begin{bmatrix} \bar{C}_{11} & \bar{C}_{12} \end{bmatrix} \begin{bmatrix} \dot{\bar{q}}_1 \\ \dot{\bar{q}}_2 \end{bmatrix} + \bar{G}_1 = \bar{\tau}_{pass,1} + \bar{\tau}_{rob,1} \tag{16}$$

Considering that the dynamic model is time-varying, after arranging Equation (16), we obtain

$$\bar{H}_{11}(t)\ddot{\bar{q}}_1 + \bar{C}_{11}(t)\dot{\bar{q}}_1 + \left(\bar{G}_1(t) + \bar{H}_{12}(t)\ddot{\bar{q}}_2 + \bar{C}_{12}(t)\dot{\bar{q}}_2 - \bar{\tau}_{pass,1}(t)\right) = \bar{\tau}_{rob,1}(t) \tag{17}$$

By defining the inertia matrix, centripetal matrix, gravitational matrix and torque vector of the reduced human dynamics as

$$\begin{aligned} H_s(t) &= \bar{H}_{11}(t) \\ C_s(t) &= \bar{C}_{11}(t) \\ G_s(t) &= \bar{G}_1(t) + \bar{H}_{12}(t)\ddot{\bar{q}}_2 + \bar{C}_{12}(t)\dot{\bar{q}}_2 - \bar{\tau}_{pass,1}(t) \\ \tau_s(t) &= \bar{\tau}_{rob,1}(t) \end{aligned} \tag{18}$$

we obtain the general mechanical form of the reduced human dynamics

$$H_s(t)\ddot{q}_s + C_s(t)\dot{q}_s + G_s(t) = \tau_s(t) \tag{19}$$

where the subscript s denotes the small system. We consider the influences from "uninterested human joints" (in this case from state $\bar{q}_2$) as perturbations.

We change the attitude of reduced human model adaptively by estimating the parameters of H_s, C_s and G_s in real time. The detailed estimation meanings are as follows.

- Estimating H_s and C_s --- make the system adaptively adjust itself to various people with different weights.

- Estimating G_s --- eliminate the perturbations from other "uninterested joints".

Considering the basic principle above, the approach to be proposed in this section is to identify and control the reduced human dynamics at the same time. Assuming that the human parameters are totally unknown in advance, for the safety in the nursing activity, the identification process needs to be performed in real time. On the other hand, the weights and heights etc. of the human bodies are different between individuals. Hence, the strategy also has to be able to tolerate these individual difference.

3.4 Attitude control and human parameter identification

First of all, we assume that we do not have any priori knowledge before lifting human, i.e. the initial value of H_s, C_s, G_s are set to be zero matrices and zero vectors. The benefit of such assumption is that the proposed trajectory is much more robust and can be adaptive to various people with different heights and weights. Whereas, such assumption also leads to a problem, i.e. it generates much model uncertainties in the dynamics. To solve the above problem, we use robust controller to change the human attitude to the desired states. Moreover, online human parameter identification is also done so as to estimate the human body in real time.

For the convenience of mathematical derivation, we define the actual human parameter vector

$$P = \left[P_H^{\ T} \ \ P_C^{\ T} \ \ P_G^{\ T} \right]^T \tag{20}$$

where

$$P_H = \left[H_{s,11} \ H_{s,12} \ \cdots \ H_{s,1n} \ \cdots \ H_{s,n1} \ H_{s,n2} \ H_{s,nn} \right]^T \tag{21}$$

$$P_C = \left[C_{s,11} \ C_{s,12} \ \cdots \ C_{s,1n} \ \cdots \ C_{s,n1} \ C_{s,n2} \ C_{s,nn} \right]^T \tag{22}$$

$$P_G = \left[G_{s,1} \ G_{s,2} \ \cdots \ G_{s,n} \right]^T \tag{23}$$

and estimated human parameter vector as

$$\hat{P} = \left[\hat{P}_H^{\ T} \ \ \hat{P}_C^{\ T} \ \ \hat{P}_G^{\ T} \right]^T \tag{24}$$

where

$$\hat{P}_H = \left[\hat{H}_{s,11} \ \hat{H}_{s,12} \ \cdots \ \hat{H}_{s,1n} \ \cdots \ \hat{H}_{s,n1} \ \hat{H}_{s,n2} \ \hat{H}_{s,nn} \right]^T \tag{25}$$

$$\hat{P}_C = \left[\hat{C}_{s,11} \ \hat{C}_{s,12} \ \cdots \ \hat{C}_{s,1n} \ \cdots \ \hat{C}_{s,n1} \ \hat{C}_{s,n2} \ \hat{C}_{s,nn} \right]^T \tag{26}$$

$$\hat{P}_G = \left[\hat{G}_{s,1} \ \hat{G}_{s,2} \ \cdots \ \hat{G}_{s,n} \right]^T \tag{27}$$

Then the estimation error vector can be defined as

$$\tilde{P} = \hat{P} - P \tag{28}$$

In fact, not any combination of H, C and G corresponds to a physical system. Therefore, the first step is to prove that the reduced human model represents a real physical system. It is easy to verify that by proving that $\dot{H}_s - 2C_s$ is a skew-symmetric matrix. In other words, the reduced human model satisfies energy conservation where the detailed derivation is in (Dong, Luo et al. 2010).

We propose a theorem to change the "states of interest" of human body. It is composed of a human attitude control law and a human parameter identification law. In fact, the control process and identification process run at the same time. In the proof of Theorem 2, the global stability is shown by proving that the derivative of the constructed Lyapunov function candidate is less than zero.

Theorem 2

Consider a time-varying system with m-order

$$H_s(t)\ddot{q}_s + C_s(t)\dot{q}_s + G_s(t) = \tau_s(t) \tag{29}$$

without any pre-knowledge about H_s, C_s and G_s. The vector $q_{s,d}$ means the desired states. Define the sliding term s as

$$s = \dot{\tilde{q}}_s + \Lambda\tilde{q}_s = (q_s - q_{s,d})' + \Lambda(q_s - q_{s,d}) \tag{30}$$

where Λ is a positive diagonal matrix. Define the reference velocity $\dot{q}_{s,r}$ and reference acceleration $\ddot{q}_{s,r}$ as

$$\begin{aligned} \dot{q}_{s,r} &= \dot{q}_s - s \\ \ddot{q}_{s,r} &= \ddot{q}_s - \dot{s} \end{aligned} \tag{31}$$

If we choose the human attitude control law

$$\tau_s = \hat{H}_s(t)\ddot{q}_{s,r} + \hat{C}_s(t)\dot{q}_{s,r} + \hat{G}_s(t) - k \cdot \mathrm{sgn}(s) \tag{32}$$

and human parameter identification law

$$\dot{\hat{P}} = -\Gamma^{-1}\left[s_1\ddot{q}_{s,r}{}^T \cdots s_n\ddot{q}_{s,r}{}^T \; s_1\dot{q}_{s,r}{}^T \cdots s_n\dot{q}_{s,r}{}^T \; s_1 \cdots s_n \right] \tag{33}$$

under the assumption of

$$\left\| k \cdot s^T \mathrm{sgn}(s) \right\| > \left\| \tilde{P}^T \Gamma \dot{P} \right\| \tag{34}$$

the human dynamics tracks the desired state trajectory and the parameter estimation H_s, C_s and G_s converge to the actual human parameters asymptotically, where k and Γ are positive diagonal matrixes, $\mathrm{sgn}(\cdot)$ is a signal function.

The proof of Theorem 2 is shown in (Dong, Luo et al. 2010). It is noted that the signals required in the control law and identification law are s, $\dot{q}_{s,r}$, $\ddot{q}_{s,r}$ (Equation (32) and Equation (33)). According to the definitions of s, $\dot{q}_{s,r}$, $\ddot{q}_{s,r}$ (Equation (30) and Equation (31)), the basic signals required are q_s, $\dot{q}_s$, $\ddot{q}_s$. Actually, the "states of interest" q_s, $\dot{q}_s$ represent the basic attitude element of human, i.e. position, angle, linear velocity, angular velocity.

Hence, they are easy for measuring. The measurement can be achieved by binocular vision technology. However, $\ddot{q}_s$ is hard to measure. To avoid the acceleration signal, we use filtering technology. Specifically, let $w(t)$ be the impulse response of a stable, proper filter. For example, for the first-order filter $\varepsilon/(p+\varepsilon)$ where $\varepsilon>0$, $p=d/dt$, the impulse response is $e^{-\varepsilon t}$. Then using partial integration, $\ddot{q}_s$ can be integrated as

$$\begin{aligned}\int_0^t w(t-r)\ddot{q}_s dr &= w(t-r)\dot{q}_s\Big|_0^t - \int_0^t \frac{dw}{dr}\dot{q}_s dr \\ &= w(0)\dot{q}_s - w(t)\dot{q}_s(0) - \int_0^t \left[w(t-r)\dot{q}_s - \dot{w}(t-r)\dot{q}_s\right]dr\end{aligned} \tag{35}$$

which means $\ddot{q}_s = f(\dot{q}_s, w)$, i.e., the acceleration signal can be obtained from velocity signal.

3.5 Simulation

3.5.1 Simulation results

The simulation was implemented by coordination of three software packages, including AUTOLEV, MATLAB and VORTEX. The detailed cooperation relations are explained as follows. AUTOLEV is used to construct the human model (Kane and Levinson 1985). We choose MATLAB to process the main computation of solving ordinary differential equations; Although VORTEX is able to do physical simulation, the programming grammar is a bit complex. Here, we just use its stereoscopic presentation function to make animations.

In the simulation, we choose the head position (denoted as $P_{h,x}$, $P_{h,y}$, $P_{h,z}$), the angle drift off the horizontal line of lower-trunk (denoted as $\theta_{1,x}$, $\theta_{1,y}$, $\theta_{1,z}$), the angle of lower-trunk and upper-leg (denoted as $\theta_{2,x}$, $\theta_{2,y}$, $\theta_{2,z}$) to constitute the "states of interest", i.e.

$$q_s = \left[P_{h,x},\ P_{h,y},\ P_{h,z},\ \theta_{1,x},\ \theta_{1,y},\ \theta_{1,z},\ \theta_{2,x},\ \theta_{2,y},\ \theta_{2,z}\right]^T \tag{36}$$

The initial velocity and acceleration are set as $q_s(0) = \dot{q}_s(0) = 0$ and the desired states are set as

$$\begin{aligned}q_{s,d} &= [0.2,\ 0.8,\ 0.01,\ 0.01,\ 0.01,\ -0.7854,\ 0.01,\ 0.01,\ 1.5708]^T \\ \dot{q}_{s,d} &= \ddot{q}_{s,d} = [0,\ 0,\ 0,\ 0,\ 0,\ 0,\ 0,\ 0,\ 0]^T\end{aligned} \tag{37}$$

Applying the human attitude control law in Equation (32) and human parameter adaptation law in Equation (33), we obtain

$$\tau_s = \left[F_{h,x},\ F_{h,y},\ F_{h,z},\ \tau_{1,x},\ \tau_{1,y},\ \tau_{1,z},\ \tau_{2,x},\ \tau_{2,y},\ \tau_{2,z}\right]^T \tag{38}$$

Then we choose

$$\begin{aligned}F_1 &= \begin{bmatrix} F_{h,x} \\ F_{h,y} \\ F_{h,z} \end{bmatrix} \\ F_2 &\approx \begin{bmatrix} (\tau_{1,z}l_{1,z}\sin\theta_{1,z} + \tau_{2,z}l_{2,z}\sin\theta_{2,z}) + (\tau_{1,y}l_{1,y}\cos\theta_{1,y} + \tau_{2,y}l_{2,y}\cos\theta_{2,y}) \\ (\tau_{1,z}l_{1,z}\cos\theta_{1,z} + \tau_{2,z}l_{2,z}\cos\theta_{2,z}) + (\tau_{1,x}l_{1,x}\sin\theta_{1,x} + \tau_{2,x}l_{2,x}\sin\theta_{2,x}) \\ (\tau_{1,x}l_{1,x}\cos\theta_{1,x} + \tau_{2,x}l_{2,x}\cos\theta_{2,x}) + (\tau_{1,y}l_{1,y}\sin\theta_{1,y} + \tau_{2,y}l_{2,y}\sin\theta_{2,y}) \end{bmatrix}\end{aligned} \tag{39}$$

where $\left[l_{1,x},\ l_{1,y},\ l_{1,z}\right]^T$ denotes the distance between the head position and the buttock. $\left[l_{2,x},\ l_{2,y},\ l_{2,z}\right]^T$ denotes the distance between the buttock and the application point of F_2. By applying F_1 and F_2, human attitude control is achieved.

The energy, position and angle changes are shown in Fig. 6. It is seen that it takes about 1 second to accomplish the process of attitude change. There is a peak of kinematics energy at the time of about 0.2 second which means at that time, the attitude changes very quickly (Fig. 6 (a)). One reason is that we assume no pre-knowledge of the human body at the beginning of the simulation. The head position changed to the desired (0.2m, 0.8m, 0.01m) at about 1 second (Fig. 6 (b)). Compared with other joints rotating in x or y direction, the joints rotating in z direction change significantly. Thus, the angle changes of these joints affect the head position in x direction greater.

In (Dong, Luo et al. 2010), it is shown that $\dot{H}_s - 2C_s$ is a skew-symmetric matrix which indicates that parts of the states (or their linear combination) can be controlled as a whole. In the simulation, we constructed a new state which is the angle sum of head, chest, mid-trunk, and lower-trunk. The angle drift off the horizontal line of the new state changes to the desired -0.7854 rad (i.e. -45 degrees) at about 1 second as shown in Fig. 6 (c). The angle between lower-trunk and upper-leg changes to the desired 1.5708 rad (i.e. 90 degrees) at about 1 second (Fig. 6 (d)).

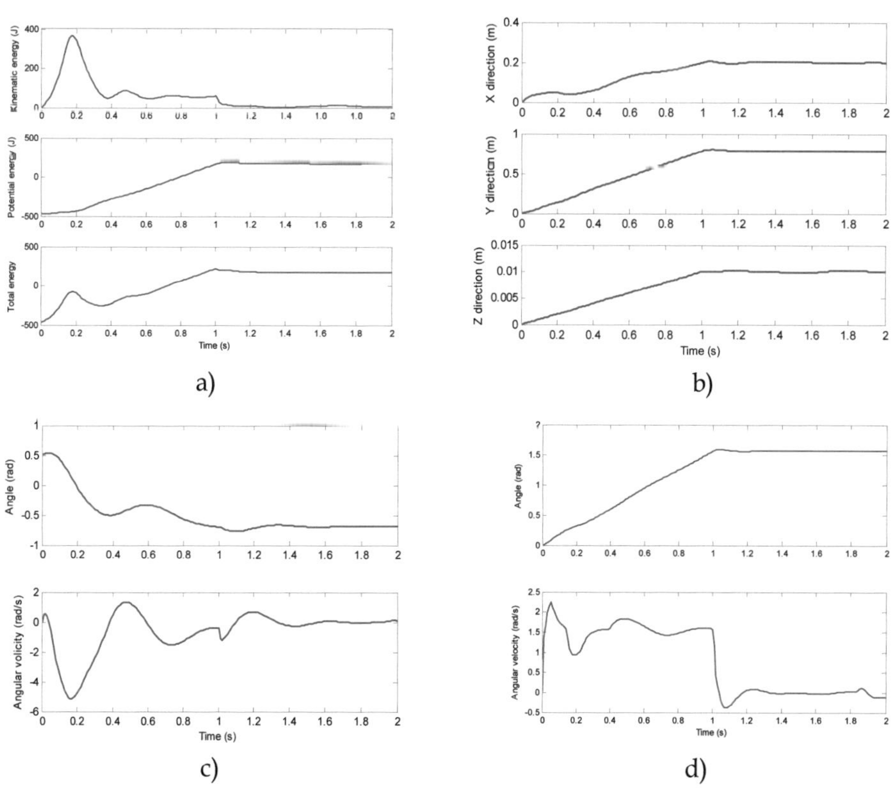

Fig. 6. Energy and angle change.

3.5.2 Animation

We imported the computed motion data into the VORTEX human model to make animation. Here the human model is the visible skeleton which was built in (Dong, Luo et al. 2010). Compared with the human model for simulation (16 links and 35 DOF), the human model for animation is a redundant one (54 links and 61 DOF). Here we just make the corresponding human joints rotate and maintain the redundant human joints fixed. As assumed in the simulation that the force contact type is contingence, we choose two cylindrical objects to represent robot arms.

The position of the head, angle drift off the horizontal line of lower-trunk, angle of lower-trunk and upper-leg are chosen as the "states of interest" (Fig. 7 (a)). The animation of lifting up human by the adaptive force control is shown in Fig. 7 (b) - (f). At the beginning of the simulation, we assume that we do not have any pre-knowledge about the human parameter. Hence, the initial values of $\hat{H}_s$, $\hat{C}_s$ and $\hat{G}_s$ are set to zero matrices (or zero vectors). As the identification of the human parameter goes on, $\hat{H}_s$, $\hat{C}_s$, $\hat{G}_s$ converge to the true values H_s, C_s, G_s.

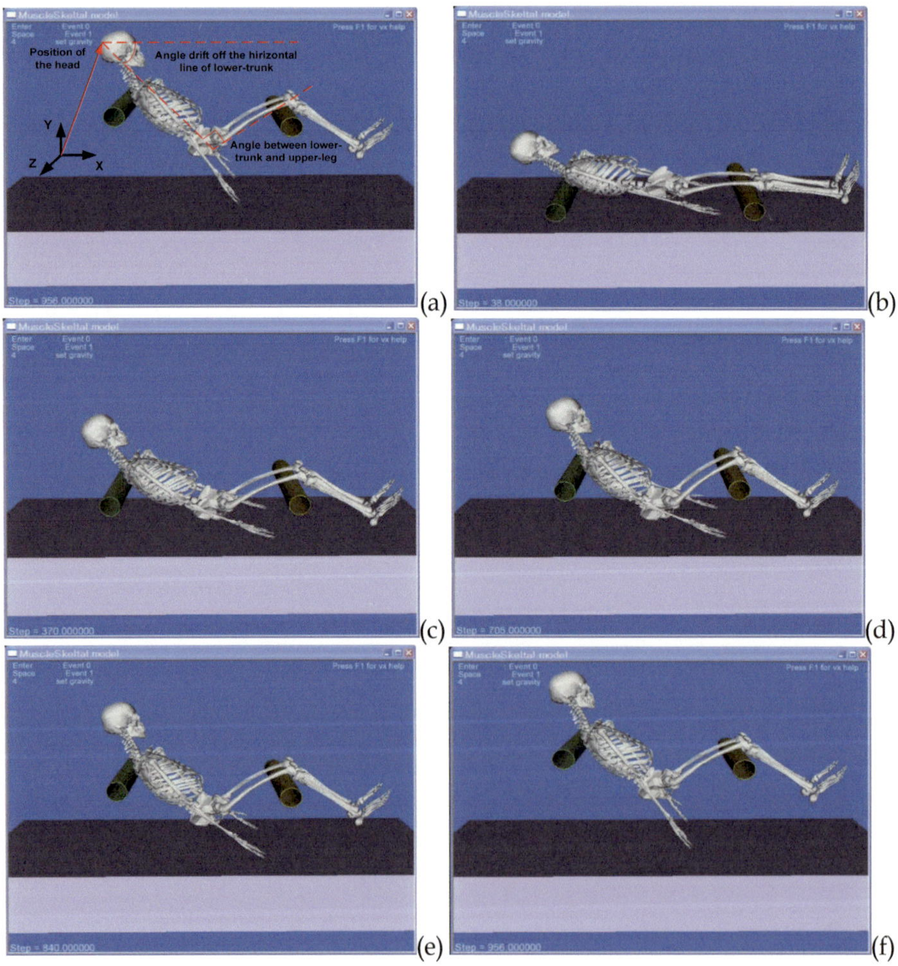

Fig. 7. "States of interest" in lifting human and Snapshots of the lifting human process (Dong, Luo et al. 2010).

3.6 Summary

In this section, an attitude control approach was proposed to lift human without regard to the individual differences, such as height, weight, and so on. We used robust adaptive control to eliminate the effects from the "uninterested joints" and identify the human parameters in real time. In addition, the convergence analysis, including tracking time and static tracking error, was also given. The approach was simulated by lifting a normal human body with two robot arms, which verifies the efficiency and effectiveness of the proposed strategy.

Compared with the previous researches, there are two novelties in the proposed approach. First is that it is not necessary to measure human, like height and weight, in advance because the approach can automatically identify the human parameter online. Second is that the attitude control law ensures the accuracy. Moreover, the robust controller which we proposed also has the ability to tolerate the model uncertainty of human.

Actually, lifting human is a typical problem when interacting with human by force. As human is such a big redundant model, a good solution is to reduce it and design controls. Here, there are two sources of model uncertainties: one is from the unmodeled dynamics which we cannot measure; the other is the generating model uncertainty which comes from model reduction. The effectiveness of the adaptive force control shows that by designing robust controller and online estimator, the two uncertainties mentioned can be solved.

4. Speed control by human's walking intention

4.1 Background

Human walking servo robot is a robot which adaptively adjusts its velocity to human's walking speed. In this section, we consider controlled treadmill as human walking servo robot. In the previous human walking servo robot applications, the subjects had to passively follow the given speed (Mcdermott, Ades et al. 2009; Powell, Stevens et al. 2009). However, in many application cases, the human walking servo robot control strategy motivated by human will is very desirable. However, human will is very complex and hard to be measured. Even if certain models are obtained by brain signal, like brain computer interface (BCI), the models are limited because of great unmodeled uncertainty.

Such a huge unmodeled uncertainty is very difficult to handle. Although the adaptive control with identification can tolerate some degree of unmodeled dynamics, it requires the unmodeled uncertainty is in some degree of limit. Once the model uncertainty is beyond the threshold, the control strategy corrupts. Considering this, we have to find another way which does not rely on model much. One constructive perspective is to consider the human process as a black box with multi-input and multi-output. Based on the experimental approach, we can evaluate the human process by the input stimulation (from u_1 to u_m) and output human response (from x_1 to x_n). The objective is to find a relation f_i to satisfy $x_i = f_i(u_1,\cdots,u_m)$ $1 \le i \le n$ where f_i can be linear or nonlinear function. By determing f_i $(1 \le i \le n)$, we can evaluate the human process. This section focuses on designing control for human walking servo robot by human's walking intention. As we consider the human process as a multi-input-muilti-output (MIMO) black box, the main topic is to identify the MIMO human process.

4.2 Human walking intention extraction

4.2.1 Characteristic index

The coordinate definition is shown in Fig. 8. Here based on the real product of force plate, we assume that the true origin of the strain gauge force plate is not at the geometric center of the plate surface. There is a pad on the force plate. After a series of calibrations of the true origin, the true origin O' is at $(0,0,h)$.

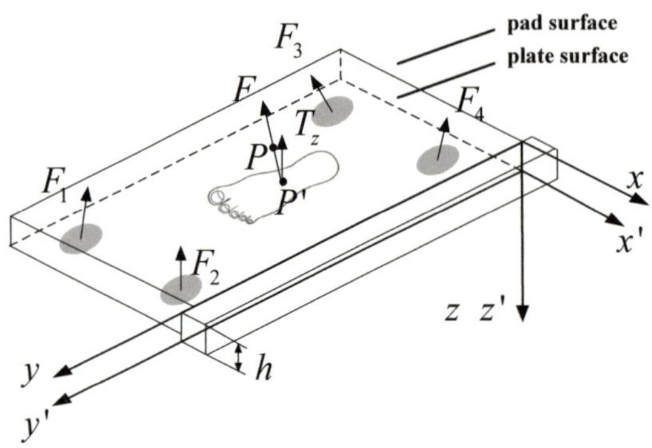

Fig. 8. Coordinate system.

According to the coordinate, the reaction force of ground to feet was simulated where the human model is built by OPENSIM in Section 4.4. It is shown that in one dynamic circle, the force F_z is a bell-shape signal; F_y is a sine-shape signal (Fig. 9 (a)).

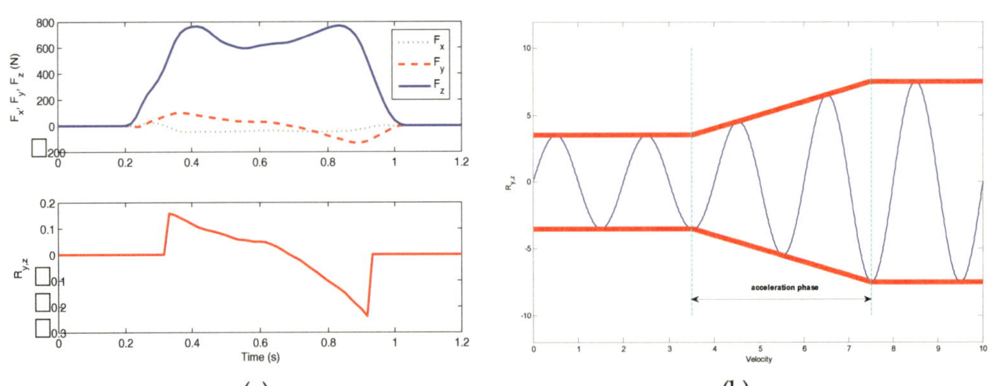

(a) (b)

Fig. 9. Envelope of R_{yz} in acceleration.

We can explain the reason of such curve shape as follows. When foot gets in touch with the surface of force plate, F_z increases very rapidly. At the same time, foot has to make a break to adjust its speed to the velocity of the force plate by fiction. After break process, foot applies a force with inverse direction to drive the leg to take a step, i.e. make a preparation for higher speed of leg in the next moment. It is noted that, compared with the previous break process, F_y changes its direction at this time. Until now, the foot is on the force plate. Hence, F_z maintains large (for a normal person with 70 kg weight, F_z is about 700 N).

Finally, the subject alternates the other foot to support body and F_z decreases rapidly. Considering the fact that when F_z is large enough, one foot is firmly on the human walking servo robot. We define a ratio index $R_{y,z}$ as

$$R_{y,z} = \begin{cases} \dfrac{F_y}{F_z} & F_z \ge \xi \\ 0 & others \end{cases} \tag{40}$$

where ξ is a threshold. In normal cases, ξ is set as $\max\{F_z\} \times 80\%$. According to the curve shapes of F_y and F_z, the curve shape of $R_{y,z}$ is a composite signal of connection of sine-shape signals and zero signals.

To prove the above simulation result and corresponding analysis, we use Bertec treadmill TM07-B to complete a verification experiment. It is noted that under the treadmill, there are two force plates individually measuring the interaction force and moment. In the experiment, the reaction data are measured when the treadmill velocity varies from 1.0 m/s to 1.6 m/s.

Without loss of generality, the zero signals are ignored in analysis, which leads to that $R_{y,z}$ changes to a connection of various sine-shape signals with different magnitude. It can be inferred that $R_{y,z}$ is a characteristic index for the walking intention: when the subject intends to speed up, $\left\|R_{y,z}\right\|_\infty$ becomes large, i.e. both the magnitudes of peak value and valley value become large. After acceleration, $R_{y,z}$ returns to a new equilibrium state (Fig. 9 (b)). In other words, the envelope curve of $R_{y,z}$ determines the intended walking speed. The deceleration case can be explained in a similar way.

4.2.2 Walking intention modeling

To model the characteristic index $\left\|R_{y,z}\right\|_\infty$, we use least-squares regression method to identify it. Specifically, first of all define the local peak value $R_{y,z}^+$ and local valley value $R_{y,z}^-$ of $R_{y,z}$ as

$$\begin{aligned} R_{y,z}^+ &= \max_{0 \le t \le T}\{R_{y,z}\} \\ R_{y,z}^- &= \min_{0 \le t \le T}\{R_{y,z}\} \end{aligned} \tag{41}$$

where T is the period of $R_{y,z}$. We can get a sequence of measurements

$$(V_{\text{intend},1}, R_{y,z,1}^-), (V_{\text{intend},2}, R_{y,z,2}^-), \cdots, (V_{\text{intend},n}, R_{y,z,n}^-) \tag{42}$$

Assuming that $R_{y,z}^-$ is predicted as a function of the intended walking speed V_{intend}, then one can model this situation by

$$R_{y,z}^- = f_w(V_{\text{intend}}, \lambda_w) + \varepsilon_w \tag{43}$$

where λ_w is a parameter vector. The random variable ε_w is independent of V_{intend} and on average it is equal to zero, i.e. $E(\varepsilon_w) = 0$. We want to find f_w that fits the measurement data best and we define the loss function to measure the quality of the fit as

$$L = \left\| R_{y,z}^{-} - f_w(V_{\text{intend}}, \lambda_w) - \varepsilon_w \right\|_2 \tag{44}$$

We minimize it over all choices of parameter vector λ_w. To find the approximation function, we write

$$\frac{\partial L\left(R_{y,z}^{-}, f_w(V_{\text{intend}}, \lambda_w), \varepsilon_w\right)}{\partial \lambda_w} = 0 \tag{45}$$

In this section, we choose the other three model candidates: quadratic, cubic, 4th degree polynomial. In the experiment, we obtain $R_{y,z}$ ten times when the treadmill velocity varies from 0.1 m/s to 1.0 m/s. The observation samples are shown in Table 1.

Velocity (m/s)	$R_{y,z}^{+}$	$R_{y,z}^{-}$	Velocity (m/s)	$R_{y,z}^{+}$	$R_{y,z}^{-}$
0.1	0.0080	-0.0213	0.6	0.0872	-0.1172
0.2	0.0310	-0.0353	0.7	0.1078	-0.0834
0.3	0.0716	-0.0542	0.8	0.1126	-0.1224
0.4	0.0622	-0.0695	0.9	0.1153	-0.1187
0.5	0.0957	-0.0884	1.0	0.1841	-0.1421

Table 1. Observation samples.

For the purpose of evaluating the candidate models, we define residual norm as the sum of square of deviations

$$S_r = \sum_{i=1}^{n} \left\| R_{y,z,i}^{-} - \hat{R}_{y,z}^{-} \right\|_2 = \sum_{i=1}^{n} \left\| R_{y,z,i}^{-} - f_w\left(V_{\text{intend}}, \hat{\lambda}_w\right) \right\|_2 \tag{46}$$

where $\hat{R}_{y,z}^{-}$ is the estimation of $R_{y,z}^{-}$ based on the particular model. Take linear model for example, $\hat{R}_{y,z}^{-} = \hat{\lambda}_{w,0} + \hat{\lambda}_{w,1} V_{\text{intend}}$. The regression results are shown in Table 2.

Regression Model	Results	Norm of Residuals
linear	$R_{y,z}^{-} = -0.13V_{\text{intend}} - 0.011$	0.039887
quadratic	$R_{y,z}^{-} = 0.067V_{\text{intend}}^{2} - 0.2V_{\text{intend}} - 0.00072$	0.034701
cubic	$R_{y,z}^{-} = -0.08V_{\text{intend}}^{3} + 0.19V_{\text{intend}}^{2} - 0.25V_{\text{intend}} + 0.0022$	0.034094
4th degree polynomial	$R_{y,z}^{-} = -0.6V_{\text{intend}}^{4} + 1.1V_{\text{intend}}^{3} - 0.56V_{\text{intend}}^{2} - 0.097V_{\text{intend}} - 0.0021$	0.031844

Table 2. Regression results.

Because the residual norms of the four types do not have big difference, any model of the four is able to describe the human's intended walking speed. Therefore, for simplicity we choose the linear model as follows.

$$R_{y,z}^{-} = -0.13 \cdot V_{\text{intend}} - 0.011 \tag{47}$$

To summarize, the human walking intention is extracted in four steps as follows. The first step is to collect a sequence of measurements $(V_{\text{intend},1}, R_{y,z,1}^{-}), \cdots, (V_{\text{intend},n}, R_{y,z,n}^{-})$. The second step is to select a model type for intended walking speed based on Equation (43). Take linear model for example, $R_{y,z}^{-} = \lambda_{w,0} + \lambda_{w,1} V_{\text{intend}}$. The third step is to define the loss function (Equation (44)) and compute the estimation of unknown coefficients. In the case of linear model, $\hat{\lambda}_{w,0}$ and $\hat{\lambda}_{w,1}$ are derived in (Dong, Luo et al. 2010). The fourth step is to compute the residual norms of the candidate models (Equation (46)). Considering model complexity and accuracy, choose a proper model.

4.3 Adaptive speed control

Nowadays human walking servo robot's application has become more and more popular in many fields like athletic exercise, rehabilitation training. In the following part, a speed control by human's walking intention is proposed by which the subject control the speed of human walking servo robot at his (or her) will.

4.3.1 System construction

The human walking servo robot used in the application is Bertec treadmill TM07-B. When the user walks on the treadmill, the dual force plates (placed under the human walking servo robot) measure the force and moment signals in x, y, z directions as F_x, F_y, F_z, M_x, M_y, M_z and output them as analog signals. After amplifying them and doing analog-to-digital (A/D) conversion, the digital signals are transferred to PC. Based on the force signals, human's walking intention is identified by the least-squares method illustrated in Section 4.2.2. Then the human's intended walking speed is used to drive the motors corresponding with left belt speed control and right belt speed control, respectively (Fig. 10). Here from the viewpoint of control, the control plant is Bertec treadmill TM07-B which is controlled by the inner speed controller in the inner loop feedback (Dong, Oshiumi et al. 2010).

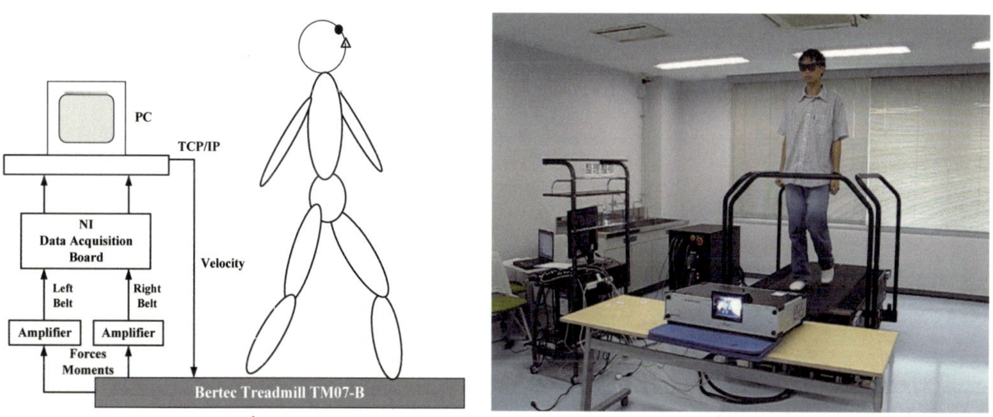

a) b)

Fig. 10. System construction of human walking servo robot.

When the subject walks on the treadmill, there is some kind of perturbation adding to the control signal. It is noted that our control objective is to control the treadmill by human's intended walking speed. Hence, we have to establish an outer loop feedback between the subject's intended speed V_{intend} and the treadmill's inner speed controller.

4.3.2 Experiment results

To calculate the transfer function $G(s)$, we use the characteristic index of intended walking speed in Equation (47). Hence, the intended walking speed is calculated as follows.

$$V_{\text{intend}}(k)=\begin{cases} V_{\text{intend}}(k-1) & R_{y,z}^{-}(k)=0 \\ -\left(R_{y,z}^{-}(k)+0.011\right)/\,0.13 & others \end{cases} \tag{48}$$

There are many reasons, like sensor noise, causing computed intended walking speed not smooth. Here, we choose a delay-line filter to solve the problem. Finally, the intended walking speed driving the treadmill's inner controller is

$$V_{\text{treadmill}}(k)=\frac{1}{4}\sum_{i=1}^{4}V_{\text{intend}}(k-i) \tag{49}$$

Good performance can be obtained by using the above filter for many times. Fig. 11 shows a dynamic walking process with acceleration motion, deceleration motion and uniform motion (constant speed motion) in the time interval $[28,34]$, $[55,65]$, $[15,24]$, respectively.

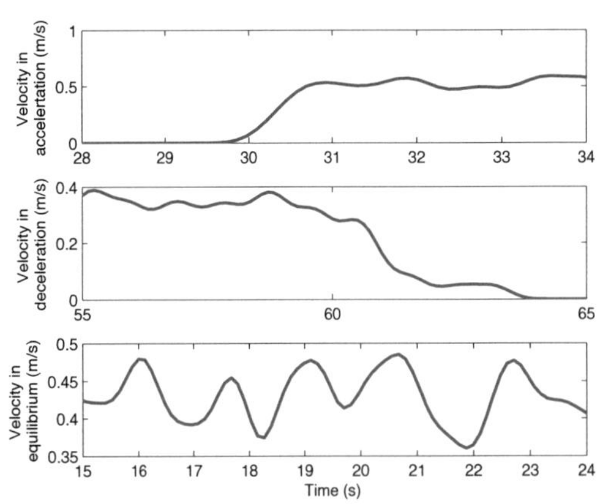

Fig. 11. Speed control result.

4.4 Summary

This section proposed a method for extracting human's walking intention and based on it, a speed control was proposed for adaptively driving human walking servo robot by human's will. The ground reaction force is considered as the indicator of the human's intended walking speed. By analyzing the walking simulation, we found a characteristic index which has significant relevance with the intended walking speed. After processing least-squares regression, four kinds of candidate models were obtained. For simplicity, we chose the linear model. The extracted human intention was used to control the human walking servo robot. The control performance shows the effectiveness of the proposed method for extracting human walking intention.

Compared with previous researches, there are two novelties to be illustrated. First is that the proposed method for human walking intention extraction simply uses the ground reaction forces to do estimation. The method is so easy-to-use that it even can be used on single-chip. The second is that in the previous human walking servo robot control, the subject had to follow the speed of human walking servo robot passively. While the proposed speed control can adaptively adjust the velocity of human walking servo robot to the subject's, which shows great potential both in research and real-world applications.

Actually, when HIR interacts with human by human's intention, it is extremely hard to model or identity the unmodeled uncertainty because we know so little about human's intention. In this case, it is a good way to analyze it just from the relation between inputs and outputs. The proposed speed control based on human's walking intention shows that by identifying the relation mentioned, we can deal with the unmodeled uncertainty from human's intention.

5. Conclusion

In this chapter, we designed kinds of control strategies of HIR and solved three typical problems for HIR. A sparse extended information filter SLAM approach was proposed for HIR's navigation in a large unknown environment, which ensures HIR can fulfil its task in human surrounding environments. An adaptive force control was designed for HIR To lift human solving the physical interaction under an uncertainty of human dynamics of dealing with the complex human dynamics during interaction. A speed control by human's intended walking speed was designed, which gives a solution to extracting human's intention. Compared with other HIR approach, this chapter takes fully into account of uncertainties HIR encounters, including environment uncertainty, sensor uncertainty, actuator uncertainty, model uncertainty and unmodeled uncertainty. The contribution of this chapter is not only to give specific HIR controls for specific cases, but also to provide a good solving framework for HIR problem.

6. References

Dong, H., Z. Luo, et al. (2009). Sparsing of information matrix for practical application of a robot's SLAM. IEEE/RSJ International Conference on Robotics and Automaion.

Dong, H., Z. Luo, et al. (2009). Sparsing of information matrix for practical application of a robot's SLAM. IEEE/RSJ International Conference on Robotics and Automation.

Dong, H., Z. Luo, et al. (2010). "Adaptive attitude control for redundant time-varying complex model of human body in the nursing activity." Journal of Robotics and Mechatronics 22(4): 418-429.

Dong, H., Z. Luo, et al. (2010). "Adaptive attitude control for redundant time-varying complex model of human body in the nursing activity." Japanese Journal of Biomechanics in Sports and Exercise 14(1): 52-60.

Dong, H., Z. Luo, et al. (2010). Adaptive treadmill control by human will. 13th International Conference on Climbing and Walking Robots and the Support Technologies for Mobile Machines.

Dong, H., Z. Luo, et al. (2010). Reduced model adaptive control for carrying human beings with uncertain body dynamics in nursing care. IEEE/ASME International Conference on Advanced Intelligent Mechatronics.

Dong, H., T. Oshiumi, et al. (2010). Development of a 3D interactive virtual market system with adaptive treadmill control. IEEE/RSJ International Conference on Intelligent Robots and Systems.

Dong, H., J. Tang, et al. (2010). "A novel information matrix sparsification approach for practical implementation of SLAM." Advanced Robotics 24(5-6): 819-838.

Eustice, R., H. Singh, et al. (2006). "Exactly sparse delayed-state filters for view based SLAM." IEEE Transactions on Robotics 22: 1100-1114.

Eustice, R., M. Walter, et al. (2005). Sparse extended information filters: insights into sparsification. IEEE International Conference on Intelligent Robots and Systems.

Eustice, R. M., H. Singh, et al. (2005). Exactly sparse delayed-state filters. IEEE International Conference on Robotics and Automation.

Kane, T. R. and D. A. Levinson (1985). Dynamics: Theory and Applications, McGraw-Hill.

Liu, Y. F. and S. Thrun (2003). Results for outdoor-SLAM using sparse extended information filters. IEEE International Conference on Robotics and Automation.

Mcdermott, M. M., P. Ades, et al. (2009). "Treadmill exercise and resistance training in patients with peripheral arterial disease with and without intermittent claudication: a randomized control trail." Journal of American Medical Association 301: 165-174.

Mukai, T., M. Onishi, et al. (2008). "Development of the tactile sensor system of a human-interactive robot "RI-MAN"." IEEE Transactions on Robotics 24: 505-512.

Onish, M., Z. W. Luo, et al. (2007). Generation of human care behaviors by human-interactive robot RI-MAN. IEEE International Conference on Robotics and Automation.

Powell, W., B. Stevens, et al. (2009). "Treadmill interface for virtual reality vs. overground walking: a comparison of gait in individuals with and without pain." Cyberpsychology and Behavior 12.

Smith, R., M. Self, et al. (1990). Estimating Uncertain Spatial Relationships in Robotics, Springer-Verlag.

Thrun, S., W. Burgard, et al. (2005). Probabilistic Robotics, The MIT Press.

Thrun, S., D. Koller, et al. (2002). Simultaneous Mapping and Localization with Sparse Extended Information Filters: Theory and Initial Results, Carnegie Mellon University.

Thrun, S., Y. Liu, et al. (2004). "Simultaneous localization and mapping with sparse extended information filters." International Journal of Robotics Research 23: 693-716.

4

Research on Building Mechanism of System for Intelligent Service Mobile Robot

Xie Wei, Ma Jiachen and Yang Mingli
Harbin Institute of Technology
China

1. Introduction

With the development of computer, communication, automation and system integration technologies, advanced research in service mobile robot has made rapid process. The modern intelligent service mobile robot (ISMR), viewed as an every day life "partner", has entered a critical period of industrialization. It not only takes place of people to do dirty, dull and dangerous (3D) work, but also becomes people's friend and chummy assistant in the future. But how to design robot system scientifically and efficiently is a hot issue in robotics because the design of robotic system is often so complex that it is difficult to seek underlying problem-solving approach. The lack of integrated methods with reuse approaches leads robotic developers to reinvent the wheel each time a new project starts(Chella et al., 2010) .Until today, the methods for designing intelligent robot may be concluded two approaches which are "Trial and error "and "Repeated optimization" (Nehmzow, 2003). So many researches have begun to study in this area from both robotic science and engineering. To sum up, there are mainly two research ideas which are "seek for systematic and scientific methodology" and "seek for general system architecture" for striving to achieve "methods reuse" and "architecture reuse".

As for seeking for systematic and scientific methodology, there are many arguments in robotics. In the same time, there are two research thoughts for designing intelligent robot (Joseph et al., 2004), one is "top-down" which is based on symbolic processing; and the other is "bottom-up" which is based on principles from biology. Currently, the two research methods are coexisting and have comprehensive integration trend. Pfeier provided the synthetic methodology of "understanding by building"(Pfeier, 2005). They mainly include two views of "constructing a model-computer or robot" and "abstract general principles from the model". He had concluded some principles including emergence, diversity-compliance, time perspectives, frame-of-reference, three constituents, complete agent, parallel loosely coupled process, sensory-motor coordination, cheap design, redundancy, ecological balance and value for guiding people to design intelligent system. Yavuz had proposed a new conceptual approach to the design of hybrid control architecture for autonomous mobile robots (Yavuz, 2002). It took the advantages of various control structure and then integrated them using comprehensive methods. And then, he proposed an integrated approach to the conceptual design and development of an intelligent autonomous mobile robot (Yavuz, 2007). A clearer view of the design approach based on function-oriented interdisciplinary were shown and the main functions of intelligent autonomous mobile robot were divided into Mobility, Navigation and Autonomy. Similarily, corresponding to design needs, five non-functional requirements including simple overall structure, cost effective, robust structure, intelligent and adaptive behaviour, and reliable operation were provided. Finally, those requirements were decomposed into smaller functions which could be performed by mechanical, electronic and software system through FD tree. In short, with the increased complexity of the system, the relationship between functional modules become more and more complex, so the overall system analysis and design still face many difficulties.

As for seeking for general architecture, many scholars tend to explore a common system reference model to reduce the duplication of design process. Architecture is the backbone of building complex intelligent systems (Ève coste-Manière, 2000). It can describe clearly that how the system is down into subsystem and how the subsystem achieves the overall function through the interaction of organic. Appropriate architecture make the robot system analysis become simple, so general architecture research has made some progress. For example, NASA and NBS's NASREM level reference model (Albus et al. 1989), Zeebo open architecture(Bogdan, 2006), ROACS(Real-time open-architecture control system) (Gu. J.s, 2004) and Daqing Wang's MRR (Modular and re-configurable robot) (Daqing wang, 2006), but they sill lack of effective methods of that how to analyze and design robotic system based on architecture systematically.

On the whole, in order to solve the difficult problems, we need to consider it from two aspects. One is the research of the scientific, systematic analysis methods, which is related to robotic development; and the other is the research of architecture, which is related to system model.

In essence, the robotic system is a complex information process system. But it is different from the computer, and its design must be considered together with the hardware and software. The systematic and comprehensive analysis of the intelligent service robot is still rare and not unified modelling methods and tools so far. Therefore, it is imperative to explore new methodological guidance for building mechanism. Our proposal tries to merge the strong points of the study methods mentioned above while minimizing their weak points, provide an open architecture for ISMR, and then explore the building mechanism of system for intelligent service mobile robot from metasynthesis's point of view.

The chapter is organized as follows. The features of analysis and design for ISMR are introduced in section 2. The metasynthesis of open architecture for ISMR is discussed in section 3. And the development model based on architecture is proposed in section 4. the system model based on Agent is provided in section 5. Finally, take a specific example to illustrate the efficiency and rationality of building mechanism in section 6. In addition, conclude and discuss this new methodological guidance of system analysis for ISMR and further research directions.

2. The features of system analysis and design for ISMR

2.1 The difference between robot and computer

Some people may ask, Robot system is ultimately embedded computer systems, we can use computer analysis methods to guide design robots, and compared to computer system, are there other special aspects in robotic? Robot system and logical relationship between the flow of information structure is calculated on the basis of computer platform, but the robot design and computer design is different.Because of the design objectives and the special constraints, the robot design is more complex than the computer . To successfully design the robot system, we must first understand the difference between the robot and the computer(as shown in Table.1), and then absorb and inherit of the computer system scientific analysis methods.

Point of difference	Robot	Computer
Information stype of input	Changing stimuli information are provided by sensors	Discrete information are providesd by key or mouse
Environment identification	Robots themselves	Computer users
Run way	Produce intelligent behavior accoring to chance from environment	Compute the need results from each operation of program
Design verification	Have difference between simulation model and the actual	Good repeatability of Simulation model
Intelligent forms	Deliberative reasoning and behaviourism	Symbolic reasoning
Environmental adaptability	Adape dynamic and unkown environment	Adape kown environment only
Real-time	Require to response changes of environment quickly	Relatively weak

Table 1. Comparative analysis between robot and computer

Overall, the robot is similar to organisms, which has a large number of complex sensors, and provide information or stimuli by them to understand and identify the environment, after that produce intelligent behavior through intelligent information processing to form a comprehensive closed-loop system. So the robot system design, not only need computer knowledge to support, but also need expertise and experience in robotics. Therefore, The combination of scientific methodology to build a computer system, robot designed common design theory, principle or a combination of domain modeling mechanism, are important condition for effectively building robotic systems.

2.2 The features of ISMR

For the robot, it's applications and the environment are very complex and there are still no unified theory and tools, so compared with the computer system, the robot can not be isolated only from the system of symbolic reasoning mechanisms, and it should be integrated and designed form the impact of three factors including body, environment and the task. In fact, the process of analysis and design is an integrated system, which turn these three system domain model to information mechanism.

In the intelligent service robot system, the robot body, the environment and the task work together, determine the robot's intelligent behavior. Therefore, when we conduct system analysis and design, the main line through the design of information systems includes not

only general information processing mechanisms, but also intelligent simulation or generation mechanism.Different Intelligent generation mechanism corresponding to different intelligent information processing method.
To sum up, intelligent robot with the characteristics of the following seven aspects(Zhu miaoliang, 2001).
F1: "survival" in a real the objective environment
F2: has some intelligence (intelligent generation mechanism)
F3: autonomous (independent of the problem-solving)
F4: has some ability to learn
F5: complex structure, Full range of levels
F6: belong to hybrid intelligent systems
Compared with general robot, ISMR is developed with the aim to perform operations in special no-industrial tasks, so it has some new features as follows excepting above(Xie wei, 2010).
F7: intelligent demanding
F8:module versatility
F9:friendly human-robot interaction
F10:secure high-performance
As can be seen from the above characteristics, building intelligent mobile robot is a complex process of comprehensive integration. With a variety of subjects (control theory, operations research, developmental psychology and systems theory, etc.) of the theories and methods to the penetration of artificial intelligence, comprehensive and integrated approach is an effective way to analyze the design of intelligent mobile robot system. Summarizing the design of robotic systems in recent years, according to Mr. Qian Xuesen's comprehensive integrated methodology(Cao Longbin, Dai Ruwei, 2008), comprehensive and integrated strategy of the robot can be summarized as the top-down and bottom-up, system abstraction and simulation complementary, more angle comprehensive modeling approach, multi-disciplinary, human (qualitative) and robots (quantitative) combination, complementary non-functional and functional division and from qualitative analysis to quantitative modeling and integration analysis.

3. The open reference architecture of ISMR

If we want to build an intelligent robotic system, the first step is to determine the appropriate architecture. The architecture includes information processing and control system, which ensures reasonable coordination, openness, and scalability of robotic system. The same as the brain, the architecture is the platform "body" for realizing intelligent behaviours, so new developments in artificial intelligence can provide some new ideological for designing ISMR. Actually, robotic system design involves two aspects of engineering and science. If there is no systematic analysis and design, any design methods based on "repeated optimization" and "trial and error" are difficult to design the overall behaviour offline. In this section, the typical reference architecture in the past 30 years are classified and summed up systematically from system analysis perspective, and then some common principles for design architecture are concluded on the basis of previous classified research. Finally, the open reference architecture of ISMR is provided.

3.1 The classification of intelligent robotic architecture

People had provided much model architecture for intelligent robot system from different angles since early 1980's. The fact recognized by everyone is that three primitives including S (sensor), P (plan) and A (action) are the foundation of constructing architecture. S represents the ability to percept the internal and external state changes; P represents the ability to make decisions autonomously based on conditions, status and constraints; A represents the basic actions or behaviors of robot. According to this relationship between the three primitives, the structure of intelligent robotic system is classified three paradigms correspondingly including deliberative ($S \rightarrow P \rightarrow A$), reactive ($S \rightarrow A$) and hybrid (deliberative/reactive). Recently, the development of distributed artificial intelligence (DAI) had created new conditions for realizing hybrid paradigm. Parallelism, distribution, open and scalability will become new characteristics. Therefore, we divide previous reference architecture into categories which are traditional architecture and agent-based distributed architecture.

3.1.1 Traditional architecture

The traditional architecture is divided into three categories on the basis of three research paradigm (deliberative, reactive and hybrid). We sum up their characteristics from design thought, analysis methods, intelligent generation mechanism, and typical examples as shown in Table 2.

Category	Deliberitive	Reactive	Hybrid
Design thought	Simulate human thinking such as introspection and deliberative	Simulate organisms perception-action response mechanism	Intergration of deliberative, reactive advantages
Analysis methods	Top-down	Bottom -up	Metasynthesis
Intelligent generation mechanism	Symbol docytrine	Behaviorism	Comprehensive integration and connectionist
Typical examples	Shakey, NASREM, 4-D/RCS	Subsumption and reactive based on motor schema	3T, AuRA, TCA and Sahira

Table 2. The comparison of three research paradigm

3.1.2 Agent-based distributed hybrid architecture

Recently, more and more researches have introduced multi-agent system (MAS) to construct mobile robot system. Based on this theory, robot's frame looks like a network that is made of agent node. Agents are independent and of self-awareness, so the large-scale problems were expected to solve by this way. Therefore, in the module design, there are many researchers to replace modules with agent and the multi-agent system (MAS) mechanism based distributed control has provided a new train of though for designing ISMR. Now, more and more researchers use agent paradigm to solve complex problems. A multi-agent architecture for mobile robot control was provided by Kolp (Kolp 2006), he compared it to other classical structures such as control-loop, layered and task-trees and the results had shown that the structure-in-5 and joint-venture based on MAS had obvious advantages. Grelle (Grelle 2007) used the agent paradigm to design very complex control strategy through multi-objective, fuzzy c-means, and genetic algorithms optimization. Under the hybrid architecture, the traditional control model was extended in agent-based distributed system. The classical architecture including the integration architecture of MAS and collaborative (Bianca, 2008), SC-agent architecture (Posadas, 2009) . Because the agent can decouple hardware and software, some methods of soft computer may be used to address cooperation and competition between agents. The approach of neural network based on evolutionary and reinforcement to design agent were presented in (Strnislav, 2010), and the results had shown Q learning algorithm was more adaptative for agent.

3.2 Discussion of common principles

Through classification and summary of past typical architecture, we not only can absorb previous experience and lessons, but also induct some common design principles for guiding follow-up study. Inspired by Pfeifer's new ideas of embodied cognition (Pfeifer, 2005), we can understand intelligence system by building. Historical experience have also proved that some new ideas or thought emerge from building a real physical system. Therefore, the general methodology on the basis of previous experiences and views are expected to get. And then, borrowing ideas from Pfeifer, we have concluded six common design principles for constructing architecture of intelligent robot system (Xie wei 2010).

- Principles of modular division based on S, P and A
- Hierarchical principle of architecture
- High cohesion and low coupling
- Design principle of redundancy
- Coordination principle between deliberative and reactive
- Principle of open-hardware based computer platform

Although the robot intelligence is so mysterious that people understand it only from a different side, we still can conclude some common design principles which are guidelines for designing efficiently robot system.

3.3 The open reference architecture of ISMR

From the development of architecture in last 30 years, we find that hybrid paradigm is effective design methods for constructing intelligent robot system. The computing models are the basis of robotic analysis and design, which have changed from process-oriented, objected-oriented and component-oriented to agent-oriented. And agent-oriented distributed hybrid architecture has more obvious advantages such as modularity, integration capability, scalability and openness than traditional architecture. But no matter which type of computing model is selected, it must have open reference architecture, only in this way, the design of ISMR is expected to change from code reuse, module reuse, component reuse to architecture reuse.

Combining six common principles above to metasynthesis, and referencing to common core mechanism of "information-Knowledge-intelligence converter", an open and space-oriented reference architecture of ISMR with learning mechanism is presented in this paper as shown in Fig.1.

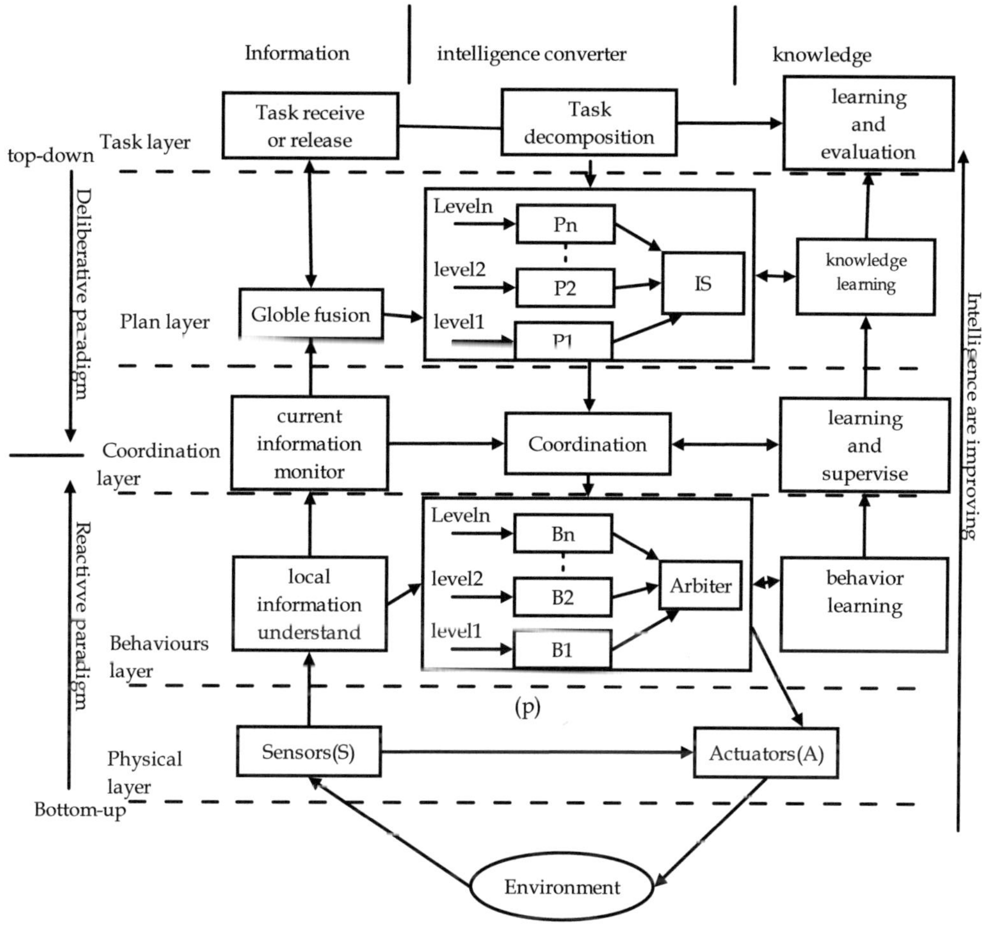

Fig. 1. The open and space-oriented reference architecture of ISMR with learning mechanism

As can be seen from Fig.1, in the space of the layered architecture, both vertical and horizontal dimensions are used to build the analytical framework. It is from the lower physical layer, behaviours layer, coordination layer, plan layer to task layer in vertical dimension, and from the left to right, it is divided into information, intelligence and knowledge conversion mechanism. The physical layer mainly includes sensors and actuators. And the openness is embodied in IS (intelligent strategy) and reactive intelligence based on successive hierarchical mechanism. Under this architecture, it is easy to form a comprehensive integrated system of division and interaction. From low to high, it shows the intelligent space hierarchy and intelligence are increased gradually. However, from the left to right, it reflects the intelligence produced by the common core mechanism. Therefore, this architecture has reflected the full range of intelligent frame structure of system content.

4. The development model of ISMR based on architecture

Each robot has its own architecture, like everyone has the body. Good architecture not only makes the problem easier to solve, but also is more conducive to the integration of complex subsystems. This is the purpose of our study architecture. The traditional design of robotic systems' development life cycle is divided into problem areas and the determines of the objective function, system analysis, system design, system implementation and system testing five stages(Zhu Miaoliang 2001). As the complexity of robotic systems are increased, it is difficult in the requirements analysis phase of an overall understanding of the system, so the actual development requires repeated changes and additions, more coupled subsystems, more experience components, lower development efficiency. Once the changes in demand or changes in the field of environment, much places needed to be modified or redesigned. The architecture as the backbone of the system, can help us build from the system point of view of system modules and intelligent interaction between this organization. Therefore, the section integrates with complex systems integration strategy, based on the architecture of the software engineering analysis method (Zhang Youshen, 2004 Hejian, 2004), and proposes development model of intelligent robot system based on architecture as shown in Fig.2

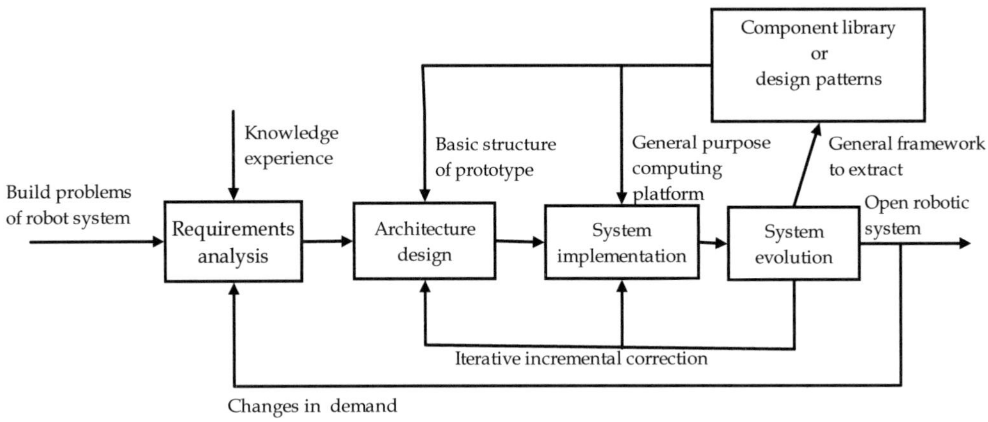

Fig. 2. Development model of intelligent robot system based on architecture

In Fig.2, the core is the architecture design. Architecture likes a bridge, which connect the robot system problem solving, needs analysis and information system. Inspired by the software architecture research ideas, the development model of Fig. 1 is divided into four design space, which are functional design (FD)space based on demand analysis, architecture design (AD) space based on architecture, system design (SD) space based on system implementation and system evolution (SE) space based on evolution(including testing, optimization and iterative modification), which are functional design (FD)space based on demand analysis, architecture design (AD) space based on architecture, system design (SD) space based on system implementation and system evolution (SE) space based on evolution(including testing, optimization and iterative modification), and finally the open robotic system are formed and some design models or components are built, then the development life cycle is completed. Compared to traditional system development model, it not only reflects the comprehensive and integrated methodology from the qualitative to quantitative, but also makes the system design from code reuse, reuse components to reusable design patterns, and improve the system development efficiency.

4.1 Requirements analysis

Robotic system requirements analysis include three aspects which are the problem description and classification (solve problems), determine the scope (constraints) and design goals (performance indicators). Has been proved(Hejian, 2004), 70% of software errors are that the result of requirements analysis is not clear, so pre- requirements analysis for the problem-solving and building system are critical. Description of the problem refers to analyze problems needed to slove clearly. Define the scope of the real refers to determine the system's objects and the environment. Only on the basis of two points of rigorous analysis in front, we determine the appropriate performance indicators. This is the same as the concept of three elements of the principle of "niche" (Pfeifer, 2005) . That is to say, on the basis of the expected behavior of the robot, we define the living environment of the robot, or applications. Such as service robots and industrial robots, 'niche' have different requirements, service robots are mainly used in homes, hotels and other more services social environment, but industrial robots are used in specific occasions such as factory environment which only need relatively simple interactions.

It can be seen from Fig.2, for demand analysis, there are two important external factors, including qualitative experience and knowledge or some common principle which are integrated into the design requirements analysis and combined qualitative design and specific areas of application requirements analysis together to solve human-computer collaborative problem. In the problem solving process, how to divide the problem is the key problem. Function-oriented classification method are easy to decouple hardware and software, and easy to analogy organisms, so it is generally recognized by every one. This method considers in conjunction with hardware and software to simplify system integration difficulties. But in the practice, the design of building robot system also face a number of non-functional requirements such as cost, robustness, operability and openness. When the three important aspects of requirements analysis is clear, we can use the existing experience or knowledge to identify problem-solving approach. When the results of the requirements analysis are converted into a functional description (the main function of the robot needs to be done) and non-functional description (constraints and performance indicators), we can see that requirements analysis is multidimensional. In the process of requirements analysis, regardless of how complex the problem to be solved, the problem can be driven in the framework of a causal relationship between the division. From different angles, this section provides three sets of functional design space for robot system (FD) to visually describe the needs analysis results.

$$FD = \{Fr, NFr, Ds\} \tag{1}$$

In which: $Fr = \{F1, F2, ..., Fn\}$ Functional requirements set

n is a natural number

$NFr = \{NF1, NF2, ..., NFn\}$ Collection of non-functional requirements

$DS = \{Ds1, Ds2, ..., Dsn\}$ Set of design specifications

For example, in this analysis mode, Hakan Yavuz's intelligent autonomous mobile robot system functional requirements are expressed as a set:

$$Fr = \{Mobility, Navigation, Autonomy\}.$$

Of course, the requirements analysis is not static, on the one hand, before the next system design it should assessed and changed repeatedly, and then be formed to documented description normative; on the hand other, after systems construction is built completely and fulfilled, changes are caused by increasing or reducing part of the functional and non-functional requirements. So the analysis needs to consider the dynamic update mechanism. These two external factors form requirements analysis process of comprehensive and integrated robotic system.

4.2 The design of architecture

According to IEEE610.12-1990 for the definition of software architecture: "Architecture is a system basic structure which contents component, the relationship among the components, the relationship between component and environment and a principle of guiding the design and evolution of the content above. " It can be seen that, components, connectors, environment and principle are of its constituent elements. At present there are not formed accepted general definition, but "Software Architecture = {components, connectors, constraints}" is generally reach a consensus. In the robot system, reference to the definition of software architecture studied by Bass, Shaw, Boehm and so on, the architecture can be understood as an abstract system specification, mainly used to describe interconnect and the set of various functional components, also includes guiding design principle. In the system development model based on robot architecture, the main task of designing architecture is to design the content based on function, to make out the content or basic structure and achieve these content and the relationship between them clearly, to make out system analysis model, to form the structure of the system design space (AD).

4.2.1 The abstract of system component

Robot academic generally accepted that, functions of robot are divided into three categories: sense (S), plan (P) and achievement (A).From the existing literature, it has not find more universal significance than these three functional classification, meanwhile S, P and A was also seen as the three primitives of building a robot system. Although the learning (L) to build intelligent systems as a primitive is still debated, in this section, we consider the learning paradigm as the fourth one apply to the design of the system, because learning is the most important manifestation of intelligence and integrated in the first three primitives. According to "By constructing to understand intelligent systems" and the relationship between these types of research paradigms, the robot basic system, combining to intelligent organization, is abstracted as three species: deliberate, reaction and hybrid (deliberate/reaction), corresponding to the three basic architecture. The characteristics of these three basic architectures are shown in Table 2. According to these research ideas, the components can be replaced by these four primitives. Intelligent organization form and architecture design of system are united and then to the component model of building intelligent systems are achieved.

Throughout development of research on intelligent abstraction mechanisms recent years, it experienced from the symbolic intelligence (the physical symbol hypothesis as the representative), connected intelligence (the artificial neurons as the representative), on-site intelligence (perception - action mechanism as the representative) to the community smart (Agent, represented in the social interaction, organization and emergence, etc.). As can be seen from Table 2, the basic architecture design and intelligent abstraction mechanism are integrated to consider.

4.2.2 Basic architecture analysis model

System modeling is the core of the architecture, the environment and the objects intelligent mobile robot systems to deal with are very complex, there is no unified theory and tools.

Building abstract and appropriate model help us to analyze the integrated mechanism and intelligence generation mechanism of architecture. It's easy to do dynamic optimized analysis, and also to predict the results of running in the environment. Recent cognitive structure, neurological psychological structure and architecture based on learning and emotion all can be regarded as evolutions of these three basic architectures. Therefore, if we consider the basic architecture as the study prototype, the system model is easy to discussed.

From a macro point of view, the level of robot intelligent will be reflected by intelligent behavior finally. While the most important factor of affecting intelligent behaviour is the sense of the environment (S), so if we take IB (intelligent behavior) as output, the environment as input, then the basic architecture of a simplified model is expressed as:

$$f : E \rightarrow IB \tag{2}$$

Among them, f means mapping mechanism from environment to intelligent behaviour. In the intelligent service robot system, the environmental status information are divided into two categories: external and internal environment. External environment includes robot running ecological environment and task information released by human or other robot; internal environment includes information about robot's understanding of environment (global and local) and introspection information (dynamic cognitive knowledge, learning, evaluation, memory information, etc.). Intelligence mainly is showed as intelligent strategy selection and the generation of correct behavior. In different environment, the mapping mechanism f is different. So based on functional design space requirements and the basic architecture, intelligent robotic systems analysis model will be built. System architecture design space (AD) can be expressed as a set:

$$AD = \{A_1,\ A_2,\ A_3, \ldots, A_n\} \tag{3}$$

A_i means the basic structure of subsystem

According to this conversion from function design space to structure space, how to make mapping mechanism? This mapping mechanism must be met the real-time requirements by relatively simple linear transformation (including general and special) or fuzzy query table. Linear transformation method describes model f by a linear mathematical model or ARMAX(Nehmzow, 2006)system identification method. Method of fuzzy query table $Q(s, a)$ achieves nonlinear f mapping mechanism through off-line fuzzy logic, and complex behavior will be real-time and dynamic completed, looking up table online. This analysis based on bottom-up and interaction principle based on perception-reaction have laid the foundation for the realization of reactive intelligent behavior. It is noteworthy that, in the lowest level of response mode, closed-loop control system as chief, such as motion control system, has the strongest instantaneity, can also be attributed to the special analysis model: $S \rightarrow A$. The mapping mechanism is designed with control system design theory.

Deliberative style basic architecture ($S \rightarrow P \rightarrow A$), is came true through intelligent information processing mechanism (mapping or policy). Planning includes task decomposition, scheduling decisions, global and local path planning. Online and off-line planning are coexistence. On-line planning includes environmental awareness, sensor fusion, task decomposition and scheduling, cooperative control and real-time path planning;

off-line planning includes building a library of knowledge and experience, the identification of demonstration representation and trajectory planning. The generation of intelligent strategies come from planning. Driven by goals or tasks, the sensor information are fused and processed into knowledge, intelligent strategies are self-generated and then intelligent behaviours are expected. The traditional deliberative design focuses on logical reasoning-search. In recent years, some soft calculation methods (such as fuzzy logic, neural networks, evolutionary computation, etc.) are applied to the deliberative reasoning design which reflect the comprehensive integration trend. For example, ANN hybrid architecture (Xu Yu-ru, 2007) used by XU Yu-ru combined with symbolic reasoning and artificial neural network, fully embodied the strengths of complementary of fusion.
Of course, if you reuse the basic architecture, then the corresponding mapping mechanism can be reused. This gives convenience to system analysis and design. For example, obstacle avoidance and along the wall of the differential behavior robot studied by Joseph (Jones, J.L. 2004), the same type of obstacle avoidance sensors are equipped on robot's left and right sides. He took mapping mechanism as common linear transformation model A, provided us with general ideas based on behavior analysis model. Simply select the appropriate transformation matrix A, reactive intelligent behavior can be achieved.
Intelligent behavior IB matrix expressed as:

$$IB = [Tr\ R\ \ Tri] \tag{4}$$

Tr : Translation speed; R : Rotation speed; Tri : Trigger flag
Assuming the sensor input El and Er, The matrix is expressed as:

$$E = [El\ \ Er\ \ 1] \tag{5}$$

Status signal: 1 indicates that this behavior is triggered, 0 indicates that this behavior is not triggered
Then there exists a common linear transformation matrix T .Such that $IB = ET$.
In its light-seeking robot design, if $Tr = V$, V means speed navigation; $R = K(El - Er)$, K for the normalization factor, the corresponding linear transformation matrix is:

$$T = \begin{bmatrix} 0 & K & 0 \\ 0 & -K & 0 \\ V & 0 & 1 \end{bmatrix} \tag{6}$$

4.2.3 Architecture of the spatial distribution

Currently, the concept that intelligent system should be layered has been widely recognized by everyone. The four-model computer systems and the seven-model computer network architecture, fully reflects the level of modeling ideas. Intelligent robot system with multi-level, three-dimensional and flexible, dynamic organizational structure and other features, are the same as cognitive scientists believe that animals have different levels of spatial capacity (reflection, integration, learning and cognitive). Therefore, the simulated biological process of intelligent systems also run through this philosophy. For example, the hierarchical structure of Saridis, level from low to high and intelligent increasing, the control precision is decreasing. The subsumption structure of Brooks, with that the high-level tolerant and include the low-level behavior and have the characteristics of stacked layers, this hierarchy has provided a convenient for building robotic system. In this framework, or the distribution of each layer, the component form of organization may be divided into centralized, distributed and hybrid (centralized/ distributed). The intelligence are increasing with the level of compatibility, from low to high. The comparison of three kinds of organization for designing system in Fig.1 are described in Table 3.

Category	Centralized	Distributed	Hybrid
Describe Ci is component	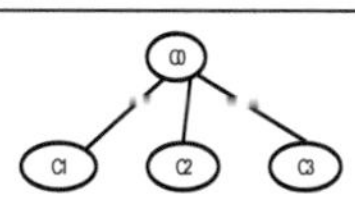	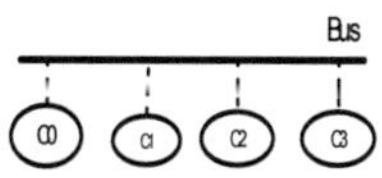	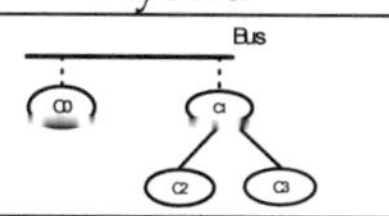
Scheduling	C0 plans and schedulings global information	Every component has local information	Balance the first two
Advantages	Facilitate unified planning and management	Distributed management and Heterogeneous components can be integrated	Metasynthesis
Disadvatages	Single information flow and weak rebustness	Difficult to deal with global information and need collaboration and competition	Design complexity and diffcult to integrate

Table 3. Comparison of three organized manner based on component

4.2.4 The comprehensive integration of structure space

In the process of function design space are converted into the structure designing space, the functional set is converted into the basic architecture by the primitive. Then the non-functional requirements and performance indicators are reflected in the architecture of the constraints. If we use these components to achieve the function of Fr set, the connection between components and the algorithm can reflect the constraints of Ds and NFr set. Thus, according to functional and non functional requirements, we determine the basic structure of the analysis model, and select appropriate mapping mechanism to realize, and achieve a function designing space completely transforms into a subset of structure designing space (basic structure). Then the problems are transformed into that how to use these form the basic structure of the whole system model.

From the view of system theory, complex systems have a Character of "hierarchy" and typically have a structure of modular, so we use the method of hierarchical - aggregation and block-integrate modeling to solve this problem. Nearly 30 years analysis research of robot architecture shows that hybrid architecture is still dominant when the basic model and its organization are identified, we use the comprehensive integrated spatial structure of intelligent service as Fig.1 to build system gradually.

The robotic system building process are descried simply as following after the AD set of structure design space is determined, we can use the method of combination modeling to integrate $AD = \{A_1, A_2, A_3, ..., A_n\}$ as the spatial structure under the frame in Fig.1. Each A_i represents a basic architecture, including components, connectors and constraints, which are used to form the overall system model. In this modelling process, according to the constraints, used the basic modular architecture, based on deliberative and reactive paradigms and different three-dimensional spatial structure of the 5-layer classification method, the association model is established , and then the various subsystems are comprehensive integrated organically. Finally, the overall model are formed. Since the independence of each layer, it enables the level of the surface to be extended independently. Combined with bottom-up, coordination ensures the deliberation and reaction layer to perform in parallel, thus, the openness and real-time have greatly improved.

4.2.5 Architecture assessment

The specific evaluation of architecture mainly comes from the constraints of FD. After FD are transformed into AD, these constraints were implicated in the design of architecture. Architecture evaluation requires a certain standard. Kolp(Kolp, 2006), based on research in Akin, derived from four qualitative evaluation criterias : support for modular, well-targeted features, easy to transplant to other areas and robustness. Anders Oreback (Anders, 2003) comprehensively evaluates the architecture of the mobile robot from aspects of hardware abstraction, scalability, uptime, software features, the implementing agency control modeling, tools, methods, documentation and others. Kolp put forward four qualitative evaluation standards from the perspective of organization theory: coherence, predictability, fault tolerance and adaptability, emphasis on evaluation of software systems . Drawing on these assessment criteria, particularly on the base of Anders Oreback, and the architecture design and analysis phase, we propose eight quality attributes to qualitatively assess the architecture design's reasonabilityas shown in Table 4. The architecture trade off analysis method(ATAM) is used to assess the architecture overall.

No	Quality attributes	Describe
1	Niche targetability	Functional integrity of the design space and consistency to architecture mapping
2	Modularity	Support modular, tight cohesion and loose coupling
3	Openess	Scalability, portability and reconfigurability
4	Adapt to the environment	Have adapt to the unkown and dynamic environment
5	robustness	Anti-interference ability and measures
6	Real-time response capacity	Real-time response
7	Security	Meet the three principles of Asimov
8	Document	Well detailed decumants

Table 4. Quality attributes of architecture structure assessment

4.3 System implementation

When the structure of the architecture design space is established, we form the specific system design space (SD), according to the whole comprehensive integration of assessed architecture. On this basis, we choose a suitable computing platform designed to achieve specific functions, and then to form the model for engineering system design. System

designing model includes the selections of computing platform and computational model. Therefore, the system design space is expressed as a collection of
SD = {computing platform, computing mode}.
Robot system is an information processing system in essence, whether deliberative plan or the acts reactive of synthesis, ultimately it comes down to a collection of computer-based hardware and software platform. Therefore, at this stage, we can use computer software engineering approach to modeling and analysis.
The difference between system design and architecture model is that the step of the former faces up with implementation and its main indicators are achievability and operability.We must firstly analyze quantitatively design specifications according to AD space including functional requirements and non-functional requirements from FD. Depending on the selected computer platform, we identified to achieve the basic architecture structure of the software modules (algorithms) and hardware modules (sensors, actuators and computing platforms), scope and interaction protocol between modules. In the design of architecture, when hardware module division has been very clear, we can determine the technology solutions of the hardware, according to the algorithm complexity of basic architecture model and the spatial and temporal distribution of the whole architecture. Next, the solution of the core issue is to choose the appropriate mode of computing, and use the appropriate tools to analyze and design a specific algorithm, so as to achieve comprehensive integrated modelling from different modular granularity. According to this idea, the design process of robotic system is converted into the next software engineering design problems for specific applications.
To sum up, the robot software design, there are four key application computing models: process-oriented (Process-oriented) approach, object-oriented (Object-oriented) approach for the component (Component-oriented) methods and agent-oriented (Agent-oriented) approach. In order to select the appropriate computing model, we conduct a comparative analysis from six areas, as shown in Table 5.

Category	Process-oriented	Object-oriented	Component-oriented	Agent-oriented
Basic unit	Sub program	object	component	agent
Computer process	Function call	Message, response	Message, response	Message, response
Abstrct element	Function	Object, class, modular	Component	Include organization, actor, target etc. excepting object and component
Granularity	small	smller	thicker	thick
Model tools	Flow chart	UML, ROSE, Usecase Diagram etc.	COM/DCOM Java Bean CORBA	AML, AUML, Gaia etc.
Typical applications	MCU, DSP program, etc	PC, ARM program and ROS design	Design of ditributed system based on WEB	Construction of complex intelligent system

Table 5. Comparison of computing model

On the view of the application of the robot engineering, these four computing modes are co-existence in the system design. We select the appropriate calculation patterns according to the algorithm, organizational structure identified in the system model and spatial distribution of the system. Then design the sub-module and the sub-system and determine the software development environment, following the corresponding modeling tools. So it is easy to build robot system software architecture. In addition, when we have completed the transition from system architecture space mode to system design space mode, we can test and evolve the algorithm of structure designing space through the simulation space or combination of hardware. modify the architecture design model by iterative incremental ways, until it meets system-related design specifications, finially establish a table or graph form the detailed system design documentation.

4.4 System evolution

Robot systems analysis and design is a dynamic iterative process. In the life cycle of systems analysis and design, the need of users and application areas of the environment may all change. Therefore, in architecture design, we need to consider the openness and scalability of architecture, in order to create conditions for the evolution of the system. Following successful tests, the system will extract some of the common framework to form the evolution of the design space (SE). If the system needs change, we must first update the functional design space, then search the evolution space (SE) that if there are ready component, and select add or reuse an existing structure, to achieve this function to find the basic architecture, in order to build a system design model. Accordingly, if the system tests have problems, we can date back to FD design space via the elements of SD design space. In the systems analysis and design of the architecture-based design and development model, system evolution is divided into two main lines. The first one is done to the structure design space from the functional design space, and done to system design space from the design space. Another is the change in demand, the use of the basic architecture and components extracted from homogeneous to update (add or delete) AD and SD elements. The dynamic evolution of the model shown in Fig. 3, this figure showing the relationship between the four design space.

When we decompose the robot system development process model into FD, AD, SD, SE four design space, we can use the collection analysis methods of robotic systems to get the qualitative and quantitative information. Here, we introduce a collection of characteristic functions, in order to complete the conversion of the four design space and structural modeling.

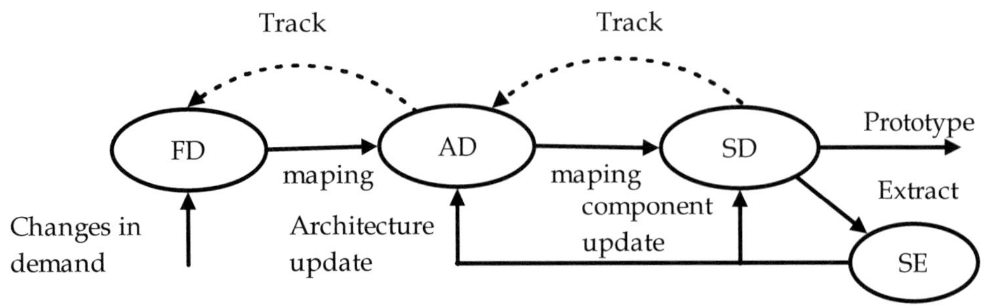

Fig. 3. Models the dynamic evolution of the system

Based on the U of A is a subset of the function:

$$\mu_A(x)=\begin{cases}1 & x\in A\\ 0 & x\notin A\end{cases} \tag{7}$$

Set A is called the characteristic function. According to the nature of the characteristic function, the development model can be represented by equation (8) to. For any $x_i \in FD$, $y_i \in AD$, $z_i \in SD$ and, $k_i \in SE$, $i=1,2,3,\ldots,n$, present $f^{'}$, $f^{'}: FD \to AD \to SD$, satisfies the equation:

$$\begin{cases}\mu_{FD}(x_i)\mu_{AD}(y_i)\mu_{SD}(z_i)=1\\ \mu_{SE}(k_j)\le \mu_{SD}(z_i)\end{cases} \tag{8}$$

5. Case study

Let us take a specific example to illustrate theory above. The intelligent gas-check service mobile robot (IGSMR) is used to check toxic gas autonomously through roaming or default route in chemical industry zone. And it can cooperate with human and help people detect gas leakage. Under the open and space-oriented reference architecture of ISMR with learning mechanism, according to the development model of ISMR based on architecture, the building process of IGSMR is divided into the design of FD, AD, SD and SE. And the design steps can be described as following.

1. Design functional space according to requirement analysis and ouput FD set.
2. Design the basic aechitecture analysis model accoring to FD set, and then integrate comprehensively the architecture space, output system reference architecture and AD set.
3. Select approprite computing model according AD set and Table 3, and then form system design model, finally output SD set and cooresponding physical implementation model.
4. Evolve (interative incremental correction) in the light of Fig.3 and test system, if it does not meet the requirements, return 1), otherwise, extract common components to SE.
5. Output detailed and standard documents after finishing the first four steps.

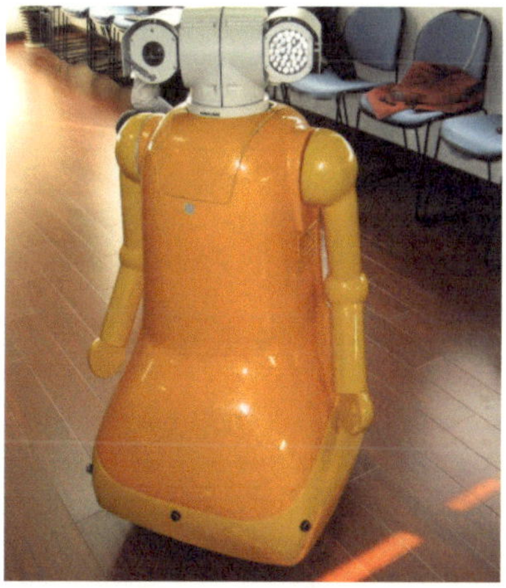

Fig. 4. Mulan No.1

Fig. 5. Mulan No.2

Subset of AD	Basic architecture	Environment sensor(E)	Intelligent Behaviors(IB)	Mapping Mechanism(f)
A_1	Reactive	Code/decode	Differential drive	PID
A_2	Reactive	Infrared	Obstacle avoidance	Generalized linear transformation
A_3	Deliberative	Magnetic sensor	Navigation	Fuzzy logic
A_4	Reactive	Ultrasound	Own roaming	Fuzzy logic and Reinforcement learning
A_5	Deliberative	Network signal	Runing state control	Generalized linear transformation
A_6	Deliberative	Power check	Independent charge	Fuzzy logic
A_7	Reactive	State detection within environment	Status indication	linear transformation
A_8	Deliberative	Sound	Sound identification	Fuzzy logic and leaining on-line
A_9	Deliberative	Photosensitive	Image processing	Soft computing
A_{10}	Deliberative	Gas-check sensors	Alarm or pre-process	Generalized linear transformation

Table 6. Architecture design space of IGSMR

From this design process, we can see that the design steps above are easy realized by computer. Because of space restrictions of this article, we manily focus on the architecture design of space. The architecture design is the core of buidling IGSMR, and the core of basic architecture is the mapping mechanism. If we grasp the core mechanism, we build robot ic system easily according to Fig.1. The physicals of IGSMR with navigation(Mulan No.1) and own roaming (Mulan No.2) are shown such as Fig.4 and Fig.5.

6. Conclusion

In this article, we have presented a new building mechanism of system for intelligent service mobile robot employed to address the robot system analysis and design problems. On the basis of previous studies, we provide firstly an open and space-oriented reference architecture of ISMR with learning mechanism, and then learning from software research methods, we present the development model of intelligent service robot system based on architecture which provide some engineering principles and theory foundation for building robot system effectively.

The design processes of ISMR are divided into four dynamic design spaces (FD, AD, SD and SE) which are contributed to integrate comprehensively. In addition, the three-dimensional spatial structure and how to design it, and formal theoretical model based on set analysis is provided. Compared to methods of conceptual design and integration (Yavuz, 2007), the methods in this paper have some obvious merits including intuitive, level clear, detailed size classification and easy computer-aided analysis, etc. From the case study of IGSMR, we conclude that this strategy not only have clear organization of intelligence, but also have high scalability and efficiency compared with traditional design methods. During the process of developing IGSMR, we use this method based on architecture to make overall efficiency increased by 40%. Moreover, the methods of analysis and design for ISMR have provided a new thought for architecture optimization and reuse of intelligent robotic system.

7. Acknowledgment

The authors wish to thank the helpful comments and suggestions from our teachers and colleagues in intelligent detection and control lab of HIT at Weihai. And thank the Shanghai Chemical industry park public pipe gallery Co.,Ltd.This work is supported by the study fund of HIT at Weihai (No.IMVQ02020003 and IMJQ21080002).

8. References

Albus.J.S, McCain, R. Lumia(1989), NASA/NBS standard reference model for telerobot control system architecture. *National instutute of standards and technology,* technical Report.

Anders Orebäck and Henrik I.Christensen(2003). Evaluation of architecture for mobile robotics. *Autonomous robots*, Vol.13, pp.33-49

Bogdan.R, Radu.R, Robotin, et al. (2006). Towards open architecture for mobile robots :Zee Ro, *Proceedings of IEEE international conference on automation, quality and testing, robotics,* pp. 260-265.

Cao longbin, Dai Ruwei (2008), *Fundamentals, conceptss, analysis and design and implementation,* Post and Telecom press, China

Chelloa. A. ; Cossentino.M. ; et al. (2010). Agent-oritented software patterns for rapid and aforable robot programing.*The Journal of system and software,* Vol.83, pp. 557-573.

Daqing Wang, (2006), *Developmentn of hybrid architecture of the control system for modular and reconfigurable robot(MRR) manipulators,* University of Toronto

Ève coste-Manière, Reid simmons (2000). The backbone of robotic system, *Proceedings of 2000 IEEE international conference on robotics & Automation,* pp. 67-72.

Grelle.C, et al. (2006) Agent-based architecture for designing hybrid control systems. *Information science,* Vol.176, No. 9, pp. 1103-1130

Gu.J.S, C.W .de Silva(2004). Development and implementation of a real-time open-architecture control system for industrial robot systems. *Engineering Applicationa of artifical intelligence,* Vol.17, No.5, pp. 469-483

Hejian, Jia xiaolin, et.al (2004). Model of software analysis and design process based on architecture. *Journal of Xi,an Jiaotong university,* Vol.38. No.6, pp.591-594.

Inniceni.B, et al.(2007) A multi-agent architecture with cooperative fuzzy control for a mobile robot. *Robotics and autonomous system,* Vol.55, No.12, pp. 881-891

Jones, J.L(2004), *Robot programing –A-practical guide to behavior-based Robotics,* McGraw-Hill Companies.inc

Manuel Kolp, P.G.J.M (2006), Muti-agent architecture as organizational structures. *Auton Agent Multi-Agent sys,* Vol.13, No. 3, pp. 3-25

Nehmzow.U. (2006)., *Scientific methods in mobile robotics-quantitative analysis of agent behaviour,* Spring-Verlag london

Pfeifer.R, Fumiya Iida, Josh Bongard (2005). New robotics : Design principles for intelligent system. *Artifical life,* Vol.11, No.2, pp. 99-120

Posadas, J.L.. et al.(2008) Agent-based distributed architecture for mobile robot control *Engineering applications of artificial intelligence,* Vol.21, No. 6, pp. 805-823

Slusn, S., R.Neruda and P.Vidnerov(2010) Comparison of behaviour-based and planning techniques on the small robot maze exploration problem. *Neural networks,* Vol.23, No. 4, pp. 560-567

Tan zhen, hejian, et al.(2002). *Software architecture,* Xi'an Jiaotong university press, China

Xie Wei, Ma jiachen, Yang mingli (2010). Agent-oriented architecture for intelligent service robot. *Proceedings of the 8th world congress on intelligent control and automation,* pp. 5964-5967

Xie Wei, Ma jiachen, Yang mingli (2010). Design of hybrid architecture for intelligent service mobile robot. *Proceedings of IEEE international conference on electrical and control engineering,* pp. 740-743

Xu Yu-ru, Xiao kun(2007). Technology development of autonomous ocean vehicle. *Acta automatica sinica,* Vol.33, No.5, pp.518-521

Yavuz.H (2002). A new conceptual approach to the design of hybrid control architecture for autonomous mobile robots. *Journal of intelligent and robotic system,* Vol.34, pp. 1-26

Yavuz.H (2007). An integrated approach the conceptual design and development of an intelligent autonomous mobile robot. *Robotics and autonomous systems,* Vol.55, pp. 498-512

Zhang yousheng(2004). Architecture-based software development model. *Compute engineering and application,* Vol.34, pp. 29-32, China

Zhu Miaoliang (2001), *Autonomous intelligent system,* Zhejiang university press, China

5

A Robotic Wheelchair Component-Based Software Development

Dayang N. A. Jawawi et al.*
Universiti Teknologi Malaysia
Malaysia

1. Introduction

A robotic wheelchair system provides mobility for handicapped and elderly people who are unable to operate classical wheelchair system. Software development for such system is challenged by requirement for multi-disciplines expert knowledge which includes embedded systems, real-time software issues, control theories and artificial intelligence aspects. Software reuse is an approach to provide a way to reuse expertise that can be used across domains in software engineering. Software reuse can be a mechanism to support the attempts to transfer technology from other engineering fields to rehabilitation engineering. For example, (Bonail et al., 2009) and (Cheein et al., 2009) have attempted to transfer software platform and algorithms from robotic technologies to rehabilitation engineering software development. The technologies transfer requires a methodological support to enable a systematic software reuse of the multi-disciplines expert knowledge.

Software reuse is one of the promising approaches to increase software productivity and improve its quality, as well as to decrease the costs of software development. This is because of software reuse uses existing software either in the form of component or knowledge to construct new software. Yet, applying software reuse in Embedded Real-Time (ERT) domain, such as robotic wheelchair sets major challenges to the software development process due to the resource-constrained and real-time requirements of the system.

In order to overcome multi-constraints and multi-disciplinary knowledge in ERT software development problems, Component-Based Development (CBD) method becomes a promising approach for ERT software development (Bunse & Gross, 2006; Carlson et al., 2006). Existing industrial component technologies such as OMG's CORBA Component Model (CCM), Microsoft's (D) COM/COM++, .NET, SUN Microsystems' JavaBeans and enterprise JavaBeans, are not suitable to develop ERT systems because they do not address the non-functional properties in ERT systems.

With the purpose to meet the requirements of ERT systems, a number of component technologies such as Koala (Ommering, 2000), PECOS (Nierstrasz, et al., 2002) and KobrA (Atkinson, et al., 2002) have emerged. However, these component technologies still have some weaknesses. Koala and PECOS cannot support multi-disciplinary knowledge, but they can

---

* Suzila Sabil, Rosbi Mamat, Mohd Zulkifli Mohd Zaki, Mahmood Aghajani Siroos Talab, Radziah Mohamad, Norazian M. Hamdan and Khadijah Kamal
Universiti Teknologi Malaysia, Malaysia

support different multi-constraint extra-functionality requirement. On the contrary, Koala supports resource constraint and PECOS supports timing problems. KobrA does not support ERT system development, but it has an extension that calls MARMOT (Bunse & Gross, 2006) to support multi-disciplinary knowledge in ERT system development. The only limitation of MARMOT is minimum support for multi-constraint extra-functionality requirement.

Therefore, PECOS is suitable to support multi-constraint in extra-functionality requirements whereby, MARMOT is good to support multi-disciplinary knowledge. An integration of PECOS and MARMOT can be a promising strategy to support the two issues. Implying a methodological support to enable a systematic CBD can be an important approach with consideration of the two issues. Currently, there are many Object-Oriented Analysis and Design (OOAD) methodology available, but for Component-Oriented Analysis and Design (COAD) method, the focus is still on PC based domain such as the web application (Lee and Shirani, 2004) and simulation systems (Gong et al., 2010). Component technologies also improved along with engineering practices, but they lack of a methodology that uses components within such a paradigm (Dogru & Tanik, 2003).

Motivated by these challenges, the focus of this chapter is to propose a method for developing robotic wheelchair software using a set of reusable software components obtained from mobile robot software. The method was adapted from general ERT component technologies and was applied to a robotic wheelchair CBD. The method is a combination of MARMOT and PECOS technologies aiming to support CBD methodological with purpose to solve multi-constraint extra-functionality requirement and multi-disciplinary knowledge that are required in the robotic wheelchair software development. The proposed systematic CBD process model is based on the Component-Based Software Engineering (CBSE) that is defined by Wang and Qian (2005). CBSE is a combination of Component-Oriented Analysis (COA), Component-Oriented Design (COD) and Component-Oriented Programming (COP) and Component-Oriented Management (COM). This chapter focuses on COA, COD and COP process of development, and this chapter defines the COA, COD and COP modelling and deployment activities of the method in a form of process model representation. The process model depicts understandable integration between MARMOT and PECOS.

This chapter documented the applicability of the process model in a wheelchair software development and implementation. This implementation showed how the process model helped the wheelchair hardware and software engineer to identify the possible software reused component in the early stage of the system and software development. The amount of software component reused in the wheelchair CBD from a mobile robot system was discussed and compared with a reused case from a mobile robot to another mobile robot CBD. The objective of the comparison was to identify the differences between the software reuse to support technologies transfer from robotic technologies to rehabilitation engineering with software reuse within robotics systems. Software reuse in robotics domain is one of the focuses area in current robotic research, example of the study are by Nesnas et al. (2006), Jang et al (2010) and Mallet et al. (2007).

The layout of this chapter is as follows: Section 2 discusses the robotic wheelchair design considerations and the hardware description. In section 3 the strategy to define the process model to support the CBD of the wheelchair software are described. The process model to support the CBD method is described in Section 4. Section 5 illustrates the process model in a CBD of the robotic wheelchair software. The comparison result to compare the process model with other models will be discussed in detail in Section 6. The Section 7 concludes the chapter.

2. The robotic wheelchair system

A prototype of robotic wheelchair was developed to support our researches in ERT software engineering and rehabilitation robotics. The main considerations when developing this prototype were to have a low cost and 'open' system such that it enables different aspects of hardware and software experimentations to be performed.

Due to these considerations, rather than basing the prototype on a standard power wheelchair such as the smart wheelchair system developed by Simpson et al. (2004), a commercially available manual wheelchair was used as the base for the robotic wheelchair prototype. An easily detachable add-on unit was developed to be attached to the manual wheelchair without significant modifications to the manual wheelchair. This add on unit consists of two geared direct current (DC) motors, two 12V batteries and a control box with embedded processor and associated electronics. Fig. 1 shows the prototype of robotic wheelchair which consists of a standard manual wheelchair and add on unit. The DC motors together with two small wheels provide the motorized wheels for mobility. The DC motors are powered by the 24V 14 Ampere obtained from the two 12V batteries. The voltage regulators in the control box provide 5V supplies for the sensors and the embedded processor from the two batteries.

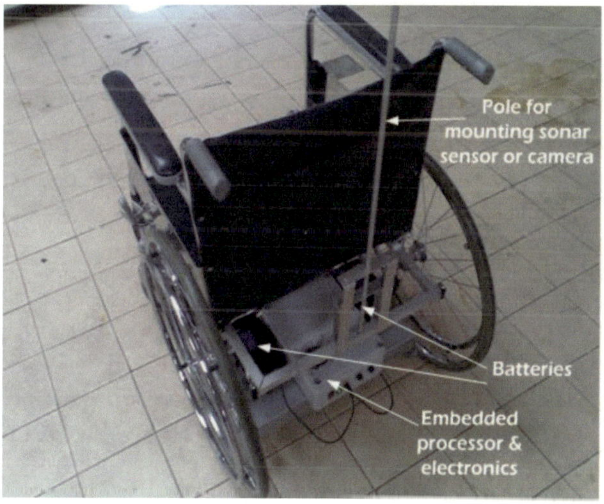

Fig. 1. The prototype of robotic wheelchair. The motorized wheels are under the add on unit

To provide the capability for sensing the environment, sets of sensors are attached to the robotic wheelchair. Infra red (IR) distance sensors are used to detect obstacles in the direction of the robotic wheelchair movement. Four IR sensors are mounted at the front and two IR sensors are mounted at the back. Sonar sensor is used to map room environment or detect far obstacles. It is mounted on the pole above the user head.

Currently, the robotic wheelchair supports two ways of controlling. The first way is through the usual joystick or keypad control. The second way is through the movements of head. Head movements control is particularly useful for severely-handicapped people who have spinal cord injury or quadriplegia which cannot use their hands to control the wheelchair. In the developed prototype, the head movement control of the robotic wheelchair is achieved with the help of accelerometer or tilt sensor. The accelerometer senses head movements and based on the predetermined direction of movements, the robotic wheelchair can be controlled accordingly. Fig. 2 shows the side view of the robotic wheelchair with the locations of sensors.

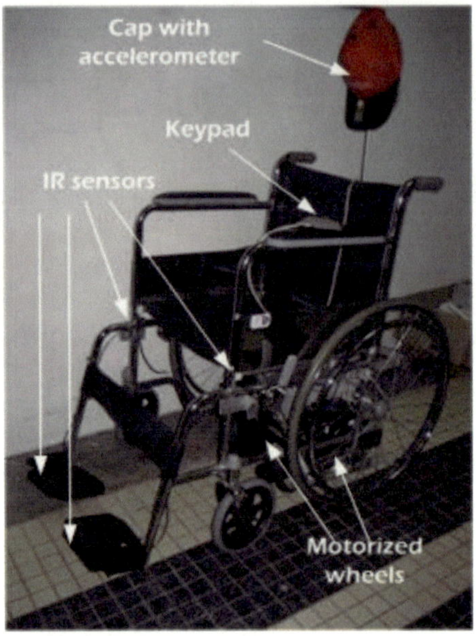

Fig. 2. The side view of robotic wheelchair prototype with sensors locations.

The embedded processor provides the intelligent decision making and motor control. All the sensors outputs are processed by the embedded processor and control the two DC motors as desired by the user. Automatic reactions such as stop or avoid when obstacles are detected are also handled by the embedded processor.

In the robotic wheelchair prototype an ATMEL ATMega32 8-bit single-chip microcontroller is used as the embedded processor. The ATMega32 has 32 Kbytes of flash program memory, and 2 Kbytes internal Random Access Memory (RAM). The ATMega32 also includes an 8-channels 10-bit analogue-to-digital converter (ADC), three timers, parallel input-output ports and several serial communication interfaces including Serial Peripheral Interface (SPI), Inter-Integrated Circuit (I2C) and Universal Synchronous Asynchronous Receiver Transmitter (USART). Fig. 3 shows the interfaces between the embedded processor, sensors and actuators in the robotic wheelchair.

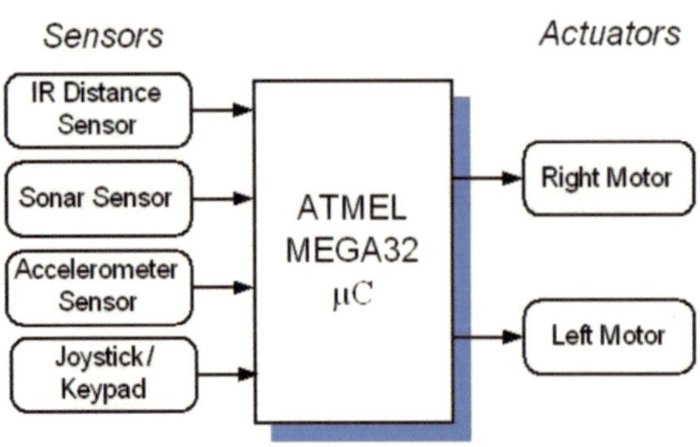

Fig. 3. The sensors and actuators in the robotic wheelchair.

3. The strategy to define the component-based development phases

A strategy is needed to allow the mapping of MARMOT and PECOS methods in order to identify overlapping phases in between the two methods. The mapping of these two methods is required in order to propose a new component-oriented developed method based on the two selected component models. Fig. 4 shows the models and phases of the methods. It comprises of three development process which are COA, COD and COP. The development process is then divided into five phases, which are analysis, early design, detailed design, composition and implementation. COA level involves the analysis phase; COD level involves two phases, which are early design and detailed design; and COP level involves composition and implementation phase.

The analysis phase includes two strategies, which are preliminary information and system architecture planning. Preliminary information consists of the basic requirements of the use case diagram, use case description and interaction model. The system architecture planning transforms the circuit into Unified Modelling Language (UML) representation and in turn produces a preliminary containment hierarchy of the whole system.

The early design phase contains the finer-grain component of software that further divides into two parts, which are specification and realization. In specification of components, three models are produced, which are functional model, behavioural model, and structural model. The structural model is also produced in realization of components, as well as the activity model and interaction model.

LEVEL	CBD PROCESS	MODEL DIAGRAM		CM
COA	Analysis Phase	Preliminary Information		MARMOT Method
		- Use Case Diagram - Use Case Description - Interaction model		
		System Architecture Planning		
		- Hardware UML representation - Preliminary Containment Hierarchy		
COD	Early Design Phase	Finer-Grain Component		
		Specification - Functional Model - Behavioural Model - Structural Model	Realization - Structural Model - Activity Model - Interaction Model	
	Detailed Design Phase	Details Information		← Integration Point
		Regular Containment Hierarchy		
COP	Composition Phase	Integration Process		PECOS Component Model
		Composition Diagram		
	Implementation Phase	Generation Code		
		Code Template		

Fig. 4. Models and phases of the method

In order to elaborate upon the preliminary hierarchy containment based on the finer-grain component, regular containment hierarchy is produced at the detailed design phase. Thus at implementation phase, the integration process is performed, and it is represented by the component composition diagram. Based on the component composition diagram, a code template is generated, which includes single component code template and component composition code template.

The models and phases in this research are derived from MARMOT and PECOS models to create a new method; the MARMOT method is applied at analysis, early design and detailed design phase while PECOS is applied at detailed design, composition and implementation phase. The integration point between MARMOT method and PECOS component model is at the detailed design phase, as shown in Fig. 4.

The central notion in COA and COD is the component. COP supports constructing software systems by composing independent components into software architecture. Software component is grouped with its component infrastructure. Component infrastructures include three elements, which are component model, component connection and component deployment. The analysis phase supports COA method, whereas early design and detailed design phases support COD. COP method is supported by two phases, which are composition and implementation phase. The next section provides detailed discussion on the process model of the integrated MARMOT and PECOS method. The process model clarifies the integration point of the two methods.

4. A component-based development process model for embedded real time software

Defining the integrated process of MARMOT and PECOS methods in a systematic form is important to enable and support the development of CASE tool for ERT system. Here, the process is represented using Software Process Engineering Meta-model (SPEM) (Schuppenies & Steinhauer, 2002). SPEM is a meta-model that is used to describe a concrete software development process or family of related software development process. The Fig. 5 below shows the SPEM icons that are used in this project:

a. Activity: is the main subclass of Work Definition, it describes a piece of work performed by one Process Role.
b. Document: a stereotype of work product.
c. Process Role: the performer of Activities and responsible for a set of Work Products.
d. Phase: a specialization of Work Definition such that its precondition defines the phase entry criteria and its goal (often called a "milestone") defines the phase exit criteria.
e. Work Definition: kind of Operation that describes the work performed in the process.
f. Work Product: an artifact is anything produced, consumed, or modified by a process.

There are five phases in the process model; analysis phase, early design phase, detail design phase, composition phase and implementation phase. Each of these phases produces a work product for the related activities and has its own responsibilities in designing one or more artefacts. Fig. 6 illustrates the phases in COA, COD and COP process using SPEM. The details for each phase are discussed in the following subsection.

Fig. 7 illustrates the use case diagram for analysis phase, whereas Fig. 8 represents the use case for early design and detailed design phase. Meanwhile, Fig. 9 illustrates the use case diagram for composition and implementation phase. These use case diagrams show the

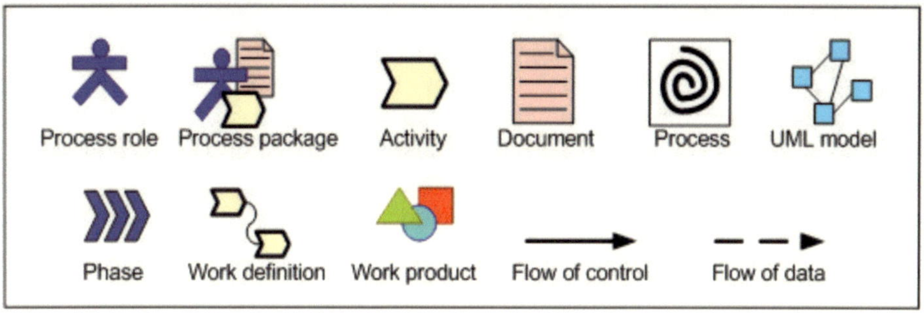

Fig. 5. SPEM Icons

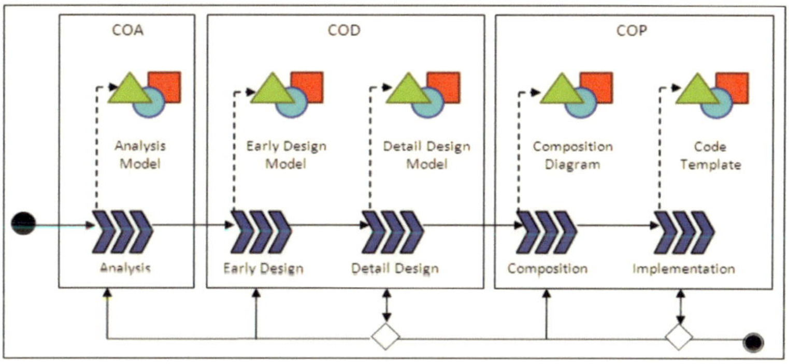

Fig. 6. Process model phases for the method

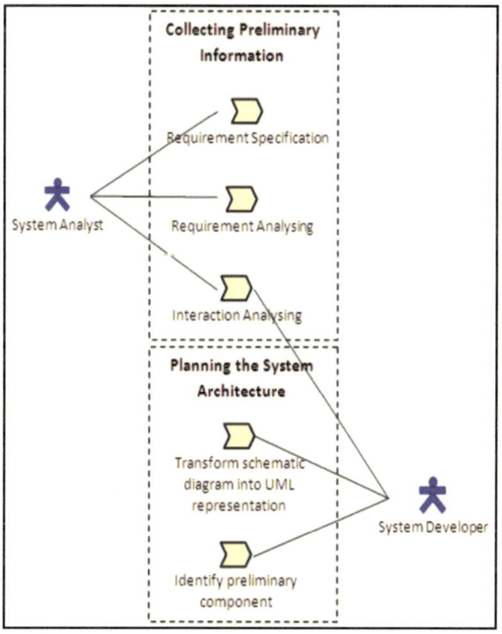

Fig. 7. Use case diagram for analysis phase

relationships between the process and the main activities in the software process based on the three level developments, which are COA, COD and COP.

In the COA level, analysis phase is divided into two main activities. The first activity is collecting preliminary information which consists of three processes; requirement specification, requirement analyzing and interaction analyzing. The second activity is planning the system architecture which includes two processes; manually transforming electrical schematic diagram into UML representation and identifying the preliminary component.

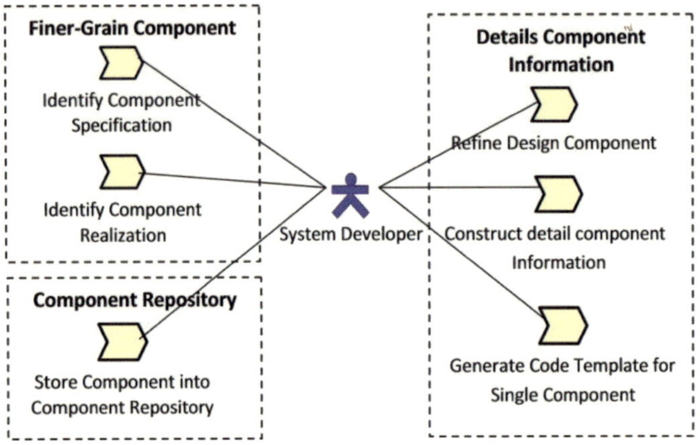

Fig. 8. Use case diagram for early design and detailed design phase

The COD level includes two phases, which are early design and detailed design phase. In the early design phase, the finer-grain component includes two activities. The first activity is identifying component specification that will produce a functional model, behavioural model and structural model. The second activity identifies component realization that

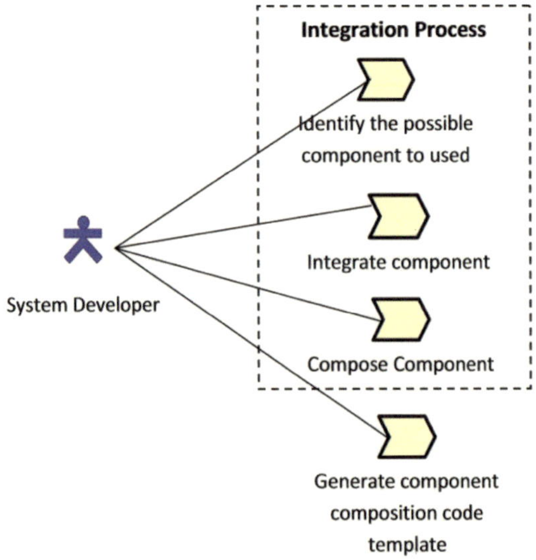

Fig. 9. Use case diagram for composition and implementation phase

produces a structural model, activity model and interaction model. 1 Meanwhile, in detailed design the component information is transformed into a graphical representation. The activities are refining design component, constructing detail component information and generating the code template for single component.
The COP level of CBD process consists of composition and implementation phase. Process integration starts at the composition phase and includes three activities, which are identifying the possible component, component integration and component composition. The implementation phase is to generate component composition code template.

4.1 Analysis phase

The purpose of analysis phase is to analyze the ERT system requirement. As was previously mentioned, ERT system requirement involves multi-disciplinary knowledge. Therefore, analyzing the ERT system requirement does not focus on software only but also on the hardware. In this analysis phase, the MARMOT method is used to support the multi-disciplinary knowledge required for ERT system. It is divided into two parts; the first part is collecting the preliminary information and the second part is planning the system architecture. Collection of the preliminary information can be further divided into three activities, which are problem analysis, requirement description and interaction analysis. System architecture planning includes two activities, which are transforming electrical schematic diagram and identifying a preliminary component based on transformation electrical schematic diagram into UML representation.
The analysis phase starts with analyzing the requirement, by identifying the problems and by looking at the application functionalities. It also identifies the user who is interacting with the application. This process produces the use case model of the system. The next activity involved is describing the requirement description thoroughly so as to produce the use case description table. This description table describes the multi-disciplinary knowledge of ERT system requirements. After that, interaction analysis is carried out to briefly draft the interaction between each state in order to produce the interaction model. Fig. 10 shows the flow of the activity in analysis phase, which is described in terms of work definition and work product as input.
The next step involves in obtaining the information directly from the electrical schematic diagram in order to identify the possible preliminary component software. The electrical schematic diagram is manually transformed into UML and all hardware parts are removed to obtain a list consisting of only software parts. The software parts can be a component with a stereotype <>. The list of the possible software components is represented by the containment hierarchy diagram. Fig. 10 shows the work definition and work product of analysis phase.

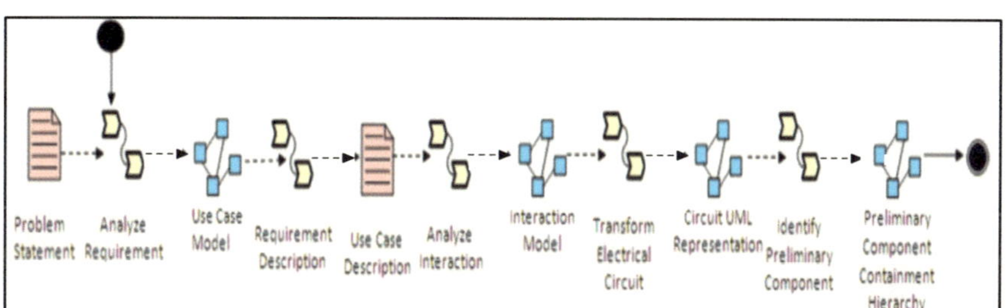

Fig. 10. The analysis phase description in terms of work definition and work product

4.2 Early design and detailed design phase

The purpose of an early design phase is to fine-grain the preliminary component containment hierarchy that is identified from the analysis phase. The preliminary component containment hierarchy at analysis phase is used as input for early design phase.

The early design phase has two main activities, which are identifying component specification and component realization for each software component. These two activities produce UML diagram artefacts. Component specification produces three models, which are structural, behaviour and functional model whereas component realization also produces three models, which are structural, activity and interaction model. In this stage, multi-constraint extra-functionality requirement which focuses on timing is considered as model for the software component specification and realization, and is represented by timing diagram.

At the detailed design phase, the integration point between the MARMOT and PECOS concept is produced. The detailed design phase includes three activities, which are refining the design component, constructing detail component information, generating the code template for single component and storing component into the component repository. Refinement of the design component involves some additional component to model the software behaviour. As a result, it produced a new version of preliminary component containment hierarchy. The process repeats until the component containment hierarchy fulfils the multi-disciplinary requirement. After the preliminary component containment hierarchy matures, the detailed component information is constructed.

In this phase, PECOS modification is integrated whereby each component includes not only the functional requirement but also multi-constraint extra-functionality requirement. In this research, the focus is on timing of extra-functionality requirement. Therefore, regular component containment hierarchy is produced where the component includes timing, priority, outport and inport. Based on the detailed information of each component, the code template produces and stores the component into the component repository. It then produces component lists of the entire application. Fig. 11 shows the steps at early design and detail design phase.

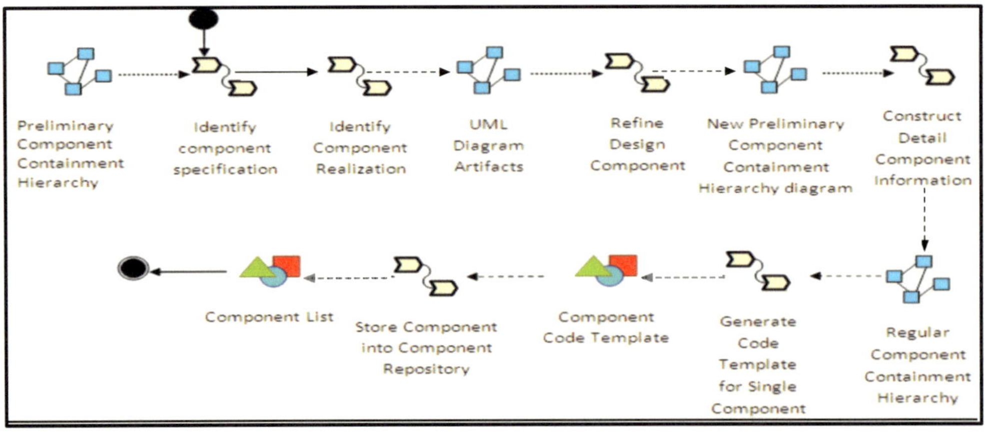

Fig. 11. Design phase description in terms of work definition and work product

4.3 Composition and implementation phase

The purpose of this phase is to integrate the components and generate the composition component code template for the entire application. The integration process integrates more than one component. In this stage, a component can have a sub-component, and it produces the composition component diagrams for the application.

Once the composition is completed, the next step is to allocate property bundles value such as period and priority for active components using real-time scheduling theory. Meanwhile, the code template for the composition diagram is generated in the implementation phase. Fig. 12 illustrates each step at composition and implementation phase.

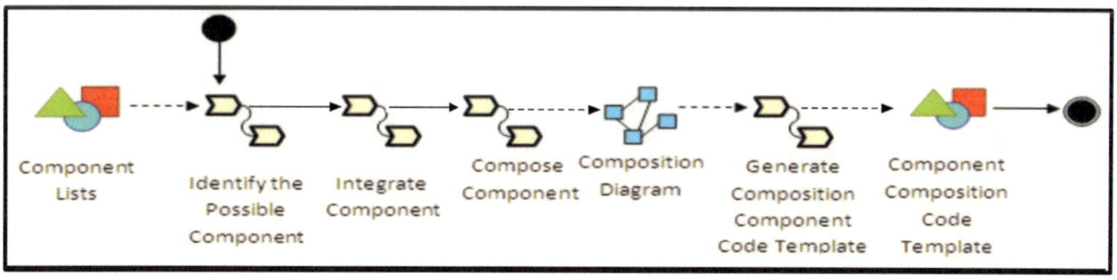

Fig. 12. Composition and implementation phase description in terms of work definition and work product

5. The wheelchair software development

The Intelligent Wheelchair (I-Wheelchair) case study, described in Section 2, was implemented to apply the integrated process model and show how the reuse activities from mobile robot systems to the I-Wheelchair were performed. The I-Wheelchair case study is an embedded system and it is relatable with the resource constraint of real-time and multi-disciplinary requirements.

I-Wheelchair software development as represented by the process model includes five different phases, which are analysis, early design, detailed design, composition and implementation phase. The following sub-section further elaborates each phase.

5.1 Analysis phase

As mentioned before, MARMOT method is implemented at the analysis phase. In this case, this phase involves processes like defining diagrams and textual specifications from the context realization of the I-Wheelchair system. It is divided into two parts, which are preliminary information and system architecture planning. Preliminary information produces use case diagrams, use case description and interaction model; and system architecture planning produces hardware UML diagram and preliminary containment hierarchy. The following sub-sections will be based on the I-Wheelchair case study.

5.1.1 Preliminary information

Preliminary information includes three activities which are requirement specification, analysis requirement and analysis interaction. The requirement specification of I-Wheelchair system is represented by the use case diagram, in which it consists of a textual and a graphical representation as shown in Fig. 13.

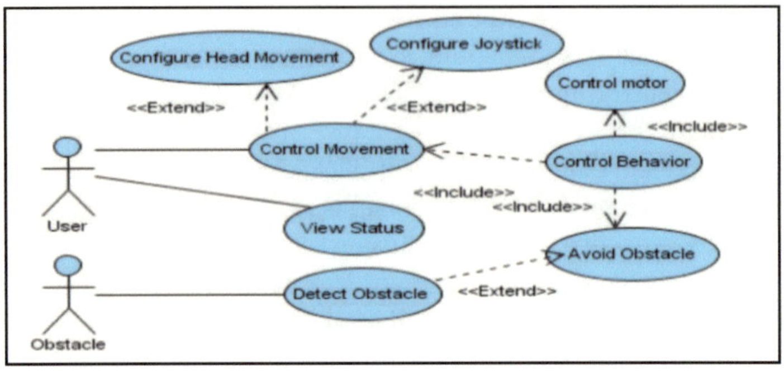

Fig. 13. I-Wheelchair use case diagram

The actor User initiates the task of controlling the I-Wheelchair movement whereby the actor Obstacle initiates the task of avoiding an obstacle around its environment. Control Movement and View Status use cases is controlled by User actor. Meanwhile, Detect Obstacle use case is controlled by Obstacle actor and extended by Avoid Obstacle use case. The use case description table is used to provide more information with regards to the I-Wheelchair use case diagram. An example of a detailed use case description table for Detect Obstacle use case is as shown in Table 1. The description table provided detailed information of Detect Obstacle use case which includes the name of use case, responsible actors, use case goals, use case descriptions, exceptions, rules, quality requirements, input/output, pre and post-conditions of the use case.

Name	Detect Obstacle
Actor	Obstacle
Goal	To detect any obstacle at front or back using IR sensor
Description	Get input data from sensor (HMC) or signal (joystick) to detect obstacle at the front or back in range distance min=30 cm and max=80 cm.
Exception	N/A
Rules	N/A
Quality Requirement	N/A
Input/Output	Input: IR sensor back IR sensor front Output: Motor
Pre-Conditions	Detect any objects that defend the movement whether at the front or back
Post-Condition	Move follow current direction

Table 1. Detect Obstacle Use Case Description

To analyze the interaction of the I-Wheelchair system, the interaction model diagram is used as shown in Fig. 14. The purpose of an interaction model is to represent the general flow of the I-Wheelchair control movement and obstacles avoidance by the two actors; User and Obstacle. It also illustrates several alternative actions that can be performed, and represents typical interaction of the operation for the overall system.

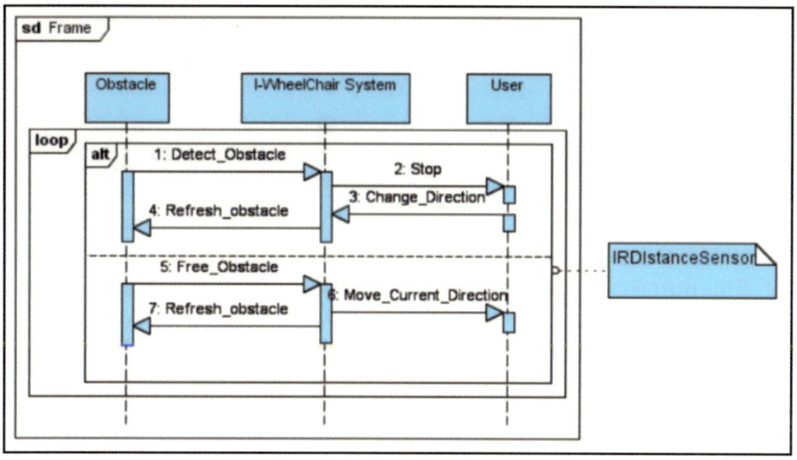

Fig. 14. Obstacle interaction model for avoiding obstacles

Fig. 14 shows the interaction model for Obstacle use case. This model represents behaviour of the Obstacle use case, in which includes two behaviour: if the I-Wheelchair system detects an obstacle, it will stop moving and change the direction of its movement, and after that it will start to detect other obstacles; if no obstacle is detected, the system will move according to the current direction that is given by the user and it will start to detect other obstacles. A semaphore used in the implementation of the Obstacle interaction model to avoid deadlock situation.

The infrared distance sensors at the back and front of the I-Wheelchair system are used to detect obstacles that labelled as input to the system as mentioned in Table 1. Therefore, the use case diagram, use case description table and interaction model can be useful for the ERT system developer to translate the hardware requirement into software requirement at the analysis phase.

5.1.2 System architecture planning

In this stage, the system architecture is planned by identifying the preliminary components. The identification of the component hardware is done based on the electrical microcontroller schematic diagram. As it requires software to calculate the control signal to be sent to motors, the motor is considered one of the initial hardware components. The microcontroller schematic diagram will be transformed into the UML presentation manually. The UML representation diagram represents the hardware component that may include software component.

Therefore, the five hardware components identified are Sensor, Controller, Joystick, Motor and Human Computer/Robot Interaction (HCI or HRI). These components are represented by a component tree called preliminary containment hierarchy as shown in Fig. 15. The I-Wheelchair is composed of both hardware and software components. The Controller hardware component consists of Driver and Application, with the Driver acting as the communicator between the software and the hardware components or called "wrapper hardware" in MARMOT. Hence, software component should be under the Driver component and an Application component refers to the behaviour of the software component.

5.2 Early design phase

MARMOT was implemented based on the software requirement documentation of the I-Wheelchair. The discussion on the modelling for the I-Wheelchair was divided into two separate descriptions; specification and realization. As mentioned before, specification produced three types of model; functional, behavioural and structural model, represented by the operation specification table, UML state diagram and class diagram respectively. Realization produced the activity and structural model as well as the interaction model which are represented by the activity diagram, class diagram and interaction diagram respectively. The difference between the specification and the realization class diagrams are the details of the information provided where the specification class diagram only included the basic information while the realization class diagrams provided more detailed information such as the operation and the attribute. Fig. 16 and Fig. 17 show the examples of realization models, i.e. the Sensor class diagram and the Sensor interaction diagram.

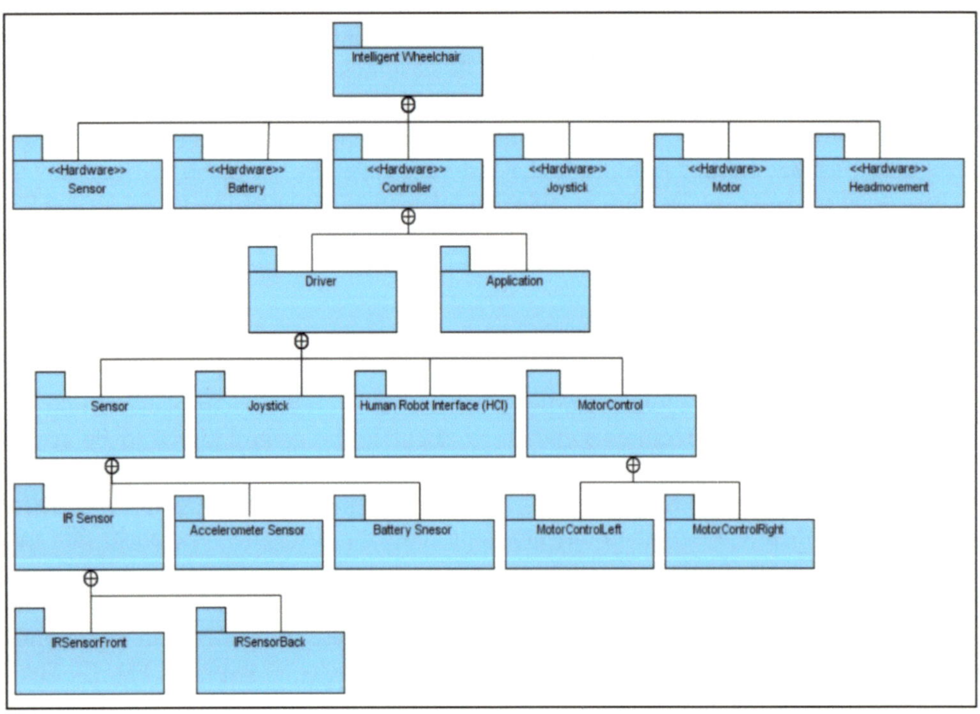

Fig. 15. Preliminary Containment Hierarchy

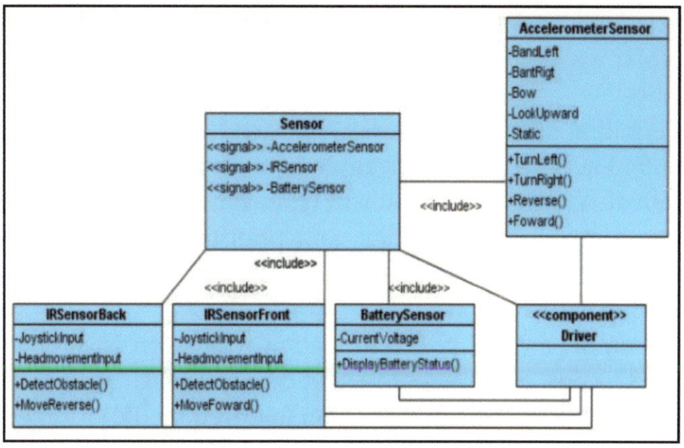

Fig. 16. Structural Model (Class Diagram)-Realization (Sensor)

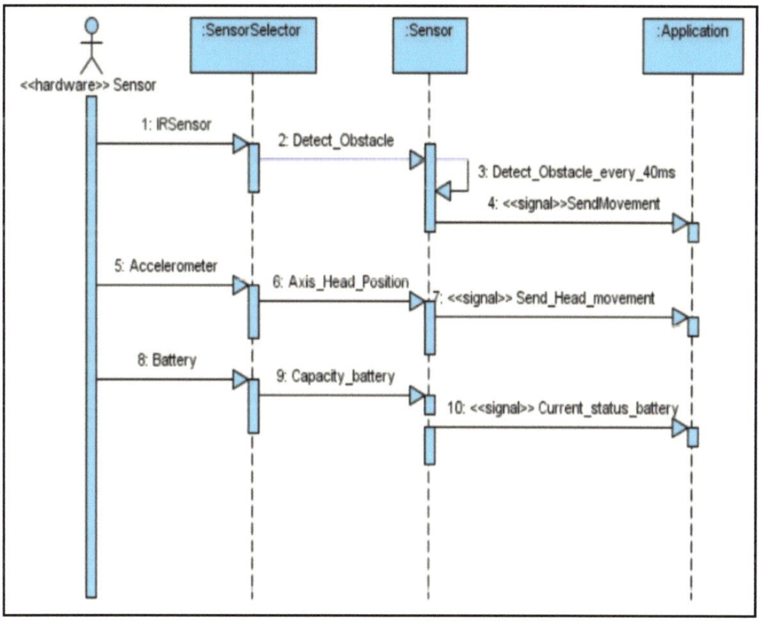

Fig. 17. Interaction Model-Realization (Sensor)

5.3 Detailed design phase

The integration of MARMOT and PECOS was implemented in this phase by extending the MARMOT Regular Containment Hierarchy diagram. The hierarchy diagram of the I-Wheelchair system in this phase is derived and extended from the Preliminary Containment Hierarchy diagram from Fig. 15, which included the component name and also has the detailed description of the component diagram information. The regular containment hierarchy of the I-Wheelchair shown in Fig. 18 includes the hardware and software components. The hardware component is represented by the <component> stereotype with six hardware components such as sensor, LED, controller, joystick, motor and accelerometer and was initially derived from the schematic diagram. The Controller component interacts with the I-Wheelchair application using the device driver to support software component structural modelling.

The component definition used in this work was extended from the original PECOS model (Jawawi et al., 2006). Fig. 19 shows an example of a composite component definition. The component diagram also contained component types such as active and passive. In order to represent an active component, initial time and priority are set in the right side of the component. If there is no set value for timing and priority in that component, it means that the component is a passive component.

Furthermore, each component has port information which consisted of two types of port; inport and outport. The port name is represented by a code such as IP00, signifying that the component has an inport port at the first position followed by IP01 for the inport port at the second position as illustrated in Fig. 19. The same rules were applied for the outport port. The components without shadow represent a leaf component (without any sub-components inside) while the components with shadow indicate that it has sub-components called composite components as illustrated in Fig. 18.

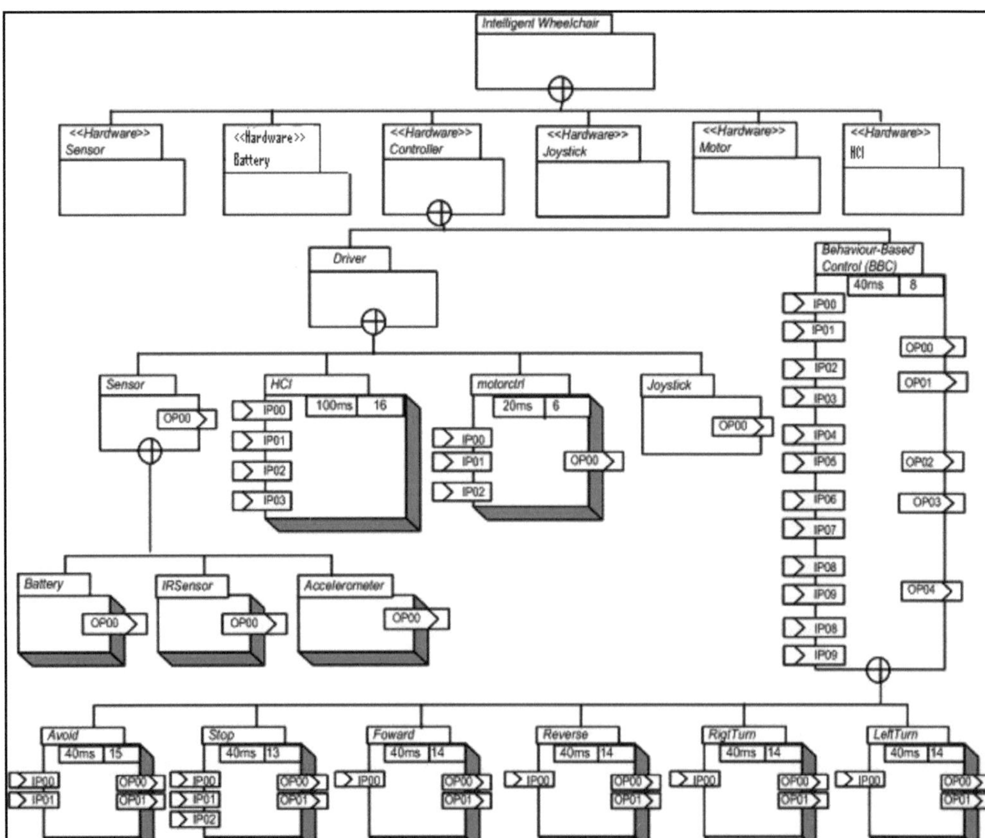

Fig. 18. Regular Containment Hierarchy

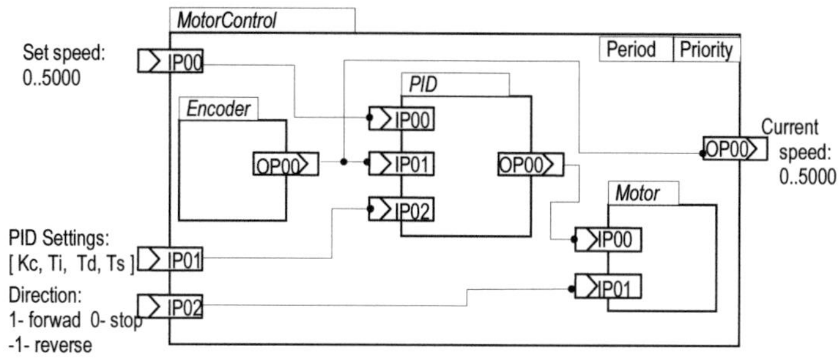

Fig. 19. MotorControl composite component

The Driver component is made up of four software components which are Sensor, HCI, Motorctrl and Joystick as shown in Fig. 18. These four software components were initially derived from hardware components where some modifications have been made based on the software design; for example, the Battery hardware component was not utilized as a Battery software component because Battery is one of the sensors.

The purpose of the Application software component is to support behaviour modelling of the I-Wheelchair application. Behaviour Based Contol (BBC) was used in the I-Wheelchair application to support behaviour modelling where BBC controls the behaviour of the I-Wheelchair with its six different behaviours including AvoidObstacle, Stop, Forward, Reverse, TurnRight and TurnLeft. BBC software components require 12 input ports to receive data from six behaviours: Avoid, Stop, Forward, Reverse, TurnRight and TurnLeft software components behaviour. Two output ports from the BBC software component were sent to the motorctrl_left software component while two other output ports were sent to motorctrl_right software component and one output port was sent to the HCI component.

5.4 Composition and implementation phase

Fig. 20 illustrates the I-Wheelchair component composition which includes 11 active and seven passive components, consisting of eight leaf components (without sub-component) and nine composite components (with sub-component) as shown by the blocks with shadow. Every single component provides the ports and connection lines to show the overall composition of the intelligent wheelchair.

From this composition diagram, code template was generated based a Component-Oriented Programming (COP) framework as proposed in Jawawi et. al (2007). The code consists of three block codes, which are the synchronization part, execution part, initialization part. The synchronization part synchronizes the connection port, the execution part chooses a behaviour based on the current command and the initialization part gives a default value.

The COP is supported by an experimental component composition tool. Fig. 21 shows an interface of component composition using the COP tool and all the new and reusable components from other application are listed on the right hand site of the composition. The COP tool is still at development stage, current version of our tool supports component integration, composition and code generation of the structure composition. The designer needs to manually code the functional behaviour of each component. Once the composition is completed, the next step is to analyse the property bundles value such as period and priority for active components using real-time scheduling theory. The scheduling analysis of the composition can be done since the timing required for the analysis were documented with the designed composition.

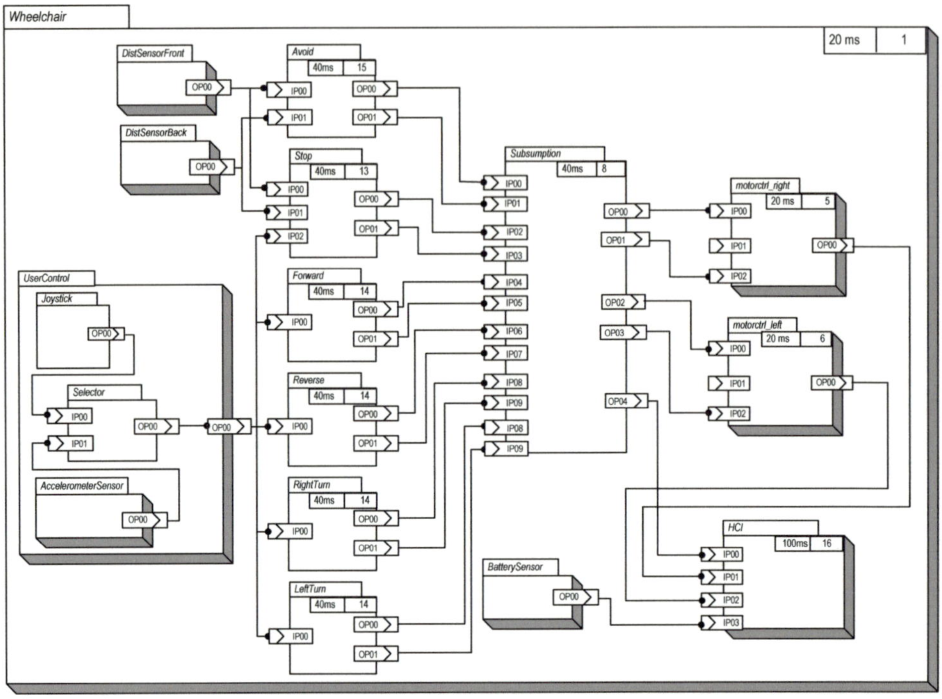

Fig. 20. I-Wheelchair component composition

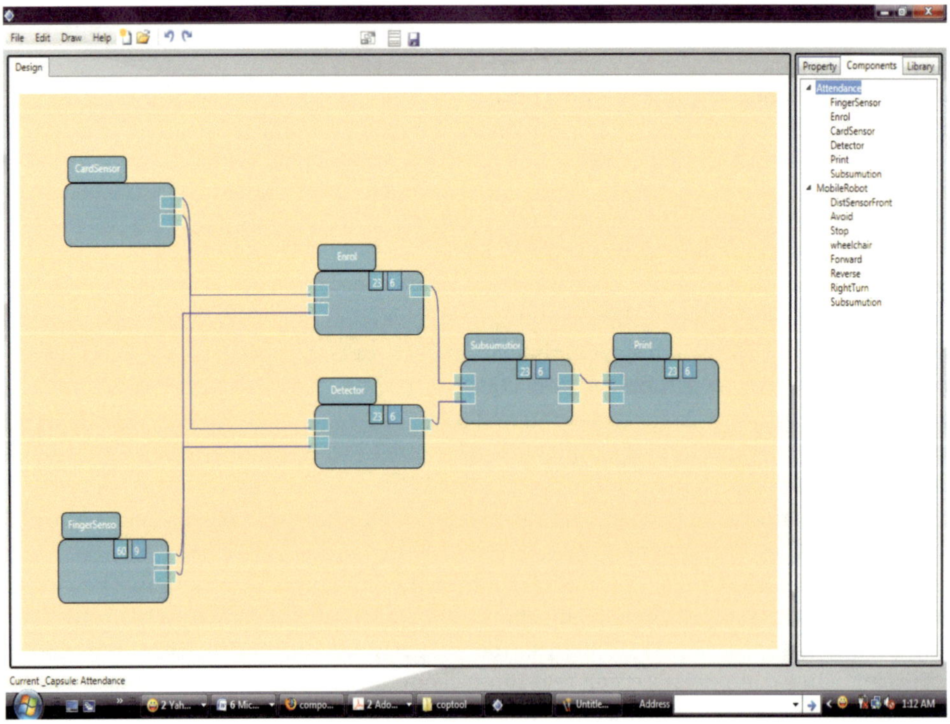

Fig. 21. COP tool's interface to support component composition and code generation

5.5 The software reuse results

The case study application showed how the process model can be used to support the CBD activities of the wheelchair software. The result of implementing these activities on the I-Wheelchair was software with 3583 lines of C codes which generates firmware with code size of about 8.6 Kb with RAM usage of about 1.6 Kb. The software was implemented on an ATMEGA32 8-bit microcontroller with 2 Kb RAM, which illustrated an implementation on a self-contained I-Wheelchair system. This size proves a light-weight solutions integrated in the developed process model is suitable for resource-constrained ERT systems. The real-time performance of the system was predicted during components composition. This prediction was verified in performance testing during implementation phase where the wheelchair showed a reliable behaviour especially on hard real-time tasks.

The design and implementation of the wheelchair software design components were derived from our previous studies and development of mobile robot software systems. One of the studies was the component engineering process of the reusable design components in the Autonomous Mobile Robot (AMR) software development documented in Jawawi et. al (2007). Each component in the reused component repository was modeled using the COP framework. The reusing process of the software components from an AMR system to the I-Wheelchair aims to analyze the reuse level resulting from the implementation of the process model activities. Amount of reused components were used to assess a reuse improvement effort by tracking percentages of reused components with two types of component reuse: reuse as is and instantiated reuse.

The main concern in this amount of reused components is to measure the reuse of the software components on different platforms and different physical sizes of the AMR and the I-Wheelchair systems. The strategy adopted in analyzing the amount of reused components in this work is to compare the percentage of component reused in two cases:

1. Case 1 Reuse: MobileRobot1 to MobileRobot2 software reuse as reported in Jawawi et. al (2007) and
2. Case 2 Reuse: MobileRobot1 to I-Wheelchair software reuse as shown in this section.

The differences between the three systems are tabulated in Table 2 and Table 3. Table 2 is to differentiate the platform and size of the systems and Table 3 is to differentiate the number of hardware used in the systems. The code used in the table is as: Mtr – Motor, Fan, Enc – Encoder, DS - Distance sensor, PS - Proximity sensor, TS - Temperature sensor, LS - Light sensor, AS - Accelerometer sensor and KP – Keypad.

Case-studies	Processor Type	Size (cm)	Shape	EPROM (Kilobytes)	RAM (Kilobyte)
MobileRobot1	AMD188ES (16 bits)	40	round	64	128
MobileRobot2	ATMEL AVR MEGA32 (8 bits)	16	octagonal	32	2
I-Wheelchair	ATMEL AVR MEGA32 (8 bits)	Standard size	Standard manual	32	2

Table 2. The AMR and wheelchair systems processor platform and size

From Table 3, two hardware components exist in both systems are motor and distance sensor. The types of distance sensor used in three case-studies are the same but the types of motor used in the three cases studies are different.

Case-studies	Mtr	Fan	Enc	DS	PS	TS	LS	AS	KP
MobileRobot1	2	0	2	2	2	0	0	0	0
MobileRobot2	2	2	0	1	0	1	1	0	0
I-Wheelchair	2	0	0	6	0	0	0	1	1

Table 3. The AMR and wheelchair systems sensors and actuators

To measure the software reuse rate, the number of reused component from the components repository will be analyzed. The parameters identified to analyze the amount of reused components in the two reused cases are: reused 100% from design components repository without changes (reuse as is), new components developed (new) and reused components by instantiating the components from MobileRobot1 for the MobileRobot2 or the I-Wheelchair software development (instantiated reuse). Table 4 shows the number of reused components in the two reused cases and the components are grouped into five groups according to our components repository groups.

Component Groups	MobileRobot1 to MobileRobot2			MobileRobot1 to I-Wheelchair		
	Reused as is	**Instantiated reuse**	**New**	**Reused as is**	**Instantiated reuse**	**New**
Sensors	2	1	1	2	1	1
Actuators and Motor Control	0	5	0	0	4	0
Behavior and Subsumption	3	1	1	2	2	3
Human-Robot Interfaces	3	0	0	2	0	3
Input-Output	0	3	0	0	4	0
Total components	**8**	**10**	**2**	**6**	**11**	**7**

Table 4. The number of reused and new components in the reuse cases

Table 4 shows higher number of components reused without changes in robot to robot reused case as compared to robot to wheelchair reuse case. This was due to the differences in the components in HRI group and Behavior and Subsumption group. The wheelchair system required different HRI devices and the behavior of wheelchair is not fully autonomous, which still require human control to navigate the wheelchair system. It led to more number of new behavior components or changes to the existing behavior components. Apart from HRI group and Behavior and Subsumption group, other groups showed the same reused pattern.

The analysis work was to identify the rate of components reused from the design repository integrated in the proposed process model. The component reused percentage is the calculation of percentage of reused software components over total software components in the composition. The summary of the component reused percentage for both reused cases are shown in Fig. 22. The figure shows up to 90% component reused rate achieved in designing MobileRobot2 software from the MobileRobot1 design components and 71% components reuse achieved in designing I-Wheelchair software from MobileRobot1 components. It showed that the reuse as is percentage for Case 1 is higher than Case 2. The instantiated reuse rate in both reused cases were about the same but the reused as is rate was higher in Case 1. This was due to high number of new components required in the wheelchair system as shown in Table 4.

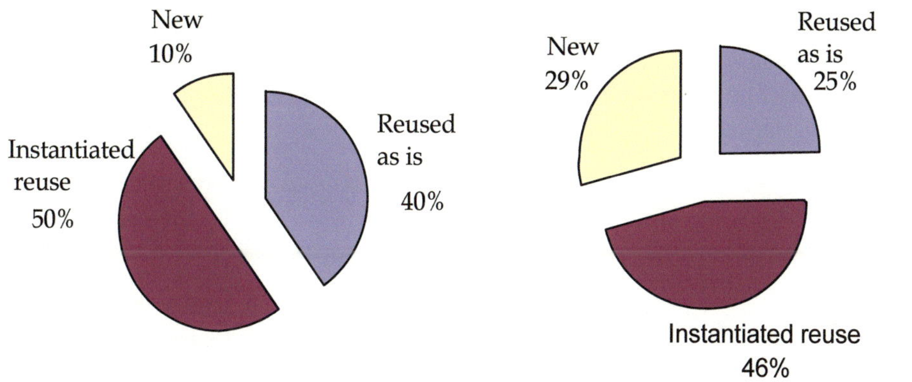

(a) Case 1: MobileRobot1 to MobileRobot2 (b) Case 2: MobileRobot1 to I-Wheelchair

Fig. 22. Component reused percentage in the two reused cases

The amount of component reuse analyzed in this implementation aims to study the possibility to achieve software reuse from mobile robot software to robotic wheelchair software. The component reused results showed a high rate of component reused was shown in mobile robot to wheelchair reused case, but this rate is lower than mobile robot to mobile robot reused case. This high rate was possible in this implementation due to the same behavior-based control paradigm (Brooks, 1986) and sensors' types used in both wheelchair and mobile robot systems. The systematic process model used in this wheelchair implementation help the wheelchair developers to identify the possible reused component in at the early stage of the software development.

6. Comparison results of component-based development process models

This section discusses the results comparison between the integrated MARMOT and PECOS method with other method, based on the criteria as shown in Table 5.

No.	Groups	Criteria
1	Evaluate the domain application of CBD method	Application domain
2	Evaluate the representation of component in the respective phase of the development process	Component representation in analysis Component representation in design Component representation in implementation
3	Evaluate which process development is supported in the CBD method	Component support in a development process Reusability
4	Evaluate the supportive tool, modelling techniques used and supportive ERT requirements of the CBD methods	Modelling techniques use Tool support Multidisciplinary support Multiple constrain support

Table 5. Summary of the evaluation criteria

The comparisons were made on two existing CBD methods; COMponent-based design of software for Distributed Embedded Systems - version II (COMDES-II) (Ke et al., 2007) and ACCORD/UML (Gerard et al., 2002). ACCORD/UML's targeted audience includes engineers who are not software experts and COMDES-II is a component based software framework intended for efficient development of distributed embedded control systems with hard real-time requirements. Table 6 shows the evaluation summary of CBD methods based on the evaluation criteria.

Criteria/CBD Methods	**COMDES-II**	**ACCORD**	**Integrated MARMOT & PECOS**
Application Domain	Hard real time requirement	Real time application engineering for engineer (not software expert)	Multi-disciplinary and multi-constraint development
Component representation in analysis	Actor Model	DAM (detailed Analysis Modelling) model	Preliminary containment hierarchy
Component representation in design	Not applicable	Not applicable	Block diagram
Component representation in implementation	Function Block	DAM (detailed Analysis Modelling) model	Graphical component composition and code template
Modelling techniques use	Actor model and function block model	UML with extension	UML 2.0 and COP composition
Tools support	Workbench	UML based tool	Any tool that supports UML 2.0, and COP tool
Component support in a development process	Design and implementation	Analysis and prototyping	Analysis, design and implementation
Reusability	Function block model	State-machine inheritance	Detailed design component block and code template
Multidisciplinary support	Software	Software	Hardware and software
Multiple constraints support	Concurrency, time multitasking	High level concurrency	Timing Analysis

Table 6. Comparison of CBD methods

Based on the general criteria evaluation results, it showed all three methods support component at analysis phase and implementation phase. Only the proposed integrated method represent components at design phase using block diagram introduced by MARMOT to be included as part of it containment hierarchy diagram and represented in component composition diagram at implementation phase.

COMDES-II and ACCORD methods combine the design phase component representation with the analysis phase. The integrated method defines the component representation in each phase, this enable all multi-disciplines experts for a system like the robotic wheelchair systems to define and specify their own components. Both the integrated method and ACCORD modelling techniques use UML except COMDES-II method that using actor model and function block model. The UML modelling enables multi-disciplines engineers to communicate and share their software components.

The integration of MARMOT and PECOS is shown to support both multi-disciplinary and multi-constraint real-time requirement. ERT CBD methods criteria showed that only the integrated MARMOT and PECOS method supported multi-disciplinary including software and hardware and multi-constraint ERT requirement. COMDES-II and ACCORD only support software disciplinary element. All three CBD methods support different level of real-time constraints; for example ACCORD supports high level concurrency, the integrated method supports timing analysis and COMDES-II supports concurrency and time multitasking. This is due to different nature of ERT systems targeted by different methods and different real-time requirements supported such soft real-time, hard real-time, time-triggered or event triggered.

The integrated MARMOT and PECOS aims to produce a systematic CBD process model of ERT system where the development process model can support multi-disciplinary knowledge and multi-constraint extra-functionality requirement. The component representation systematically supported all phases: the analysis, design and implementation phases, this can provide more reusability facilities in all phases.

The systematic CBD process model allows the multi-disciplines experts in the robotics wheelchair CBD development to develop their reusable components and integrates their components with other. Among the software components developed by different discipline reused in the I-Wheelchair system development are such as BBC component, motors component, sensors and input-output components.

7. Conclusion

This chapter proposed a method to enable reuse of mobile robot software into a robotic wheelchair (I-Wheelchair) software system. The robotic wheelchair system prototype was developed to test the implementation result of the proposed method. The component-based development (CBD) of the robotic wheelchair software using a set of reusable components, and the software composition of the wheelchair software were illustrated. The integration of PECOS into MARMOT method would produces a systematic CBD process model in terms of multi-disciplinary knowledge and multi-constraint extra-functionality requirement. The application of the CBD method in the I-Wheelchair system illustrated how the MARMOT and PECOS were integrated to produce a systematic development method for CBD. The method guides developers to model and deploy the robotic wheelchair software using the

CBD activities and the phases are defined by the process model. With the strengths of MARMOT and PECOS integrated into the method, it enabled the functional and non-functional requirements, especially the timing properties, to be modelled explicitly in all phases. Furthermore, the method also considered multi-disciplinary requirements in its models which help software engineers to focus on their problem-solving of the robotic wheelchair system development. The activities to reuse software from existing robotics systems were explicitly supported in the proposed process model.

The implementation of the I-Wheelchair case study, demonstrated that the proposed method has helped to guide the CBD of the software system from the analysis of the product to the implementation phase through the defined process model. The component reused rate achieved in the wheelchair implementation from a mobile robot design repository was high up to 71% and this implementation was compared with components reuse from the same mobile robot design repository to another mobile robot system to identify the pattern of reuse in both reused cases. No direct relation between the process model proposed and the reused rate studied in this chapter but the process model helped developer to identify and implement the component in all CBD phases. For future works, the existing component-based development tools will be refined further to support COD and COP methods and to integrate the tools with the UML models produced in the COA method.

8. Acknowledgment

Special thanks to the Universiti Teknologi Malaysia for financing and funding this research through Research University Grant and also to our Embedded & Real-Time Software Engineering Laboratory (EReTSEL) members for their continuous support.

9. References

Atkinson, C.; Bayer, J.; Bunse, C.; Kamsties, E.; Laitenberger, O.; Laqua, R.; Muthig, D.; Paech, B.; Wust, J. & Zettel J. (2002). *Component-Based Product Line Engineering with UML*, Addison Wesley Professional, ISBN 978-0201737912, Boston, USA.

Bonail, B.; Abascal, J. & Gardeazaba,l L. (2009). Wheelchair-based Open Robotic Platform and its Performance within the AmbienNet Project, *Proceedings of the 2nd International Conference on PErvasive Technologies Related to Assistive Environments (PETRA '09)*, ISBN 978-1-60558-409-6, Corfu, Greece, June 09-13, 2009.

Brooks R.A. (1986). A Robust Layered Control System for a Mobile Robot. IEEE Journal of Robotics and Automation. RA-2(1): 14-23.

Bunse, C. & Gross, H. G. (2006), Unifying Hardware and Software Components for Embedded System Development, In: *Architecting Systems with Trustworthy Components*, Reussner R. H., Stafford J. A. and Szyperski C. A., pp. 120-136, Springer-Verlag, ISBN 978-3540358008, Berlin, Germany.

Carlsona, J.; Håkanssonb, J. & Pettersson, P. (2006). Save CCM: An Analysable Component Model for Real-Time Systems, *Proceedings of the International Workshop on Formal Aspects of Component Software (FACS 2005)*, vol. 160 , Macao, October 24-25, pp.127-140.

Cheein, F. A. A.; Cruz, C.; Bastos, T. F. & Carelli, R. (2009). SLAM-based Cross-a-Door Solution Approach for a Robotic Wheelchair. *International Journal of Advanced Robotic Systems*, Vol. 6, No. 3. pp. 239-248, ISSN 1729-8806.

Dogru, A. H. & Tanik M. M. (2003). A Process Model for Component-Oriented Software Engineering. *IEEE Software*, Vol. 20 No. 2, (March 2003), pp. 34-41.

Gerard, S; Terrier, F. & Tanguy, Y. (2002). Using the Model Paradigm for Real-Time Systems Development ACCORD/UML. *Proceedings of the Workshops on Advances in Object-Oriented Information Systems (OOIS '02)*, Montpellier, France, September 2, pg. 260-269.

Gong, J.; Peng, Y.; Hao, J.; Huang J., & Huang, K. (2010), Research on Component-Oriented Methodology for Constructing Simulation Systems. *Journal of System Simulation*, pp. 1-10.

Jang, C; Lee, S-I; Jung, S-W; Song, B; Kim, R; Kim, S & Lee, C-H. (2010). OPRoS: A New Component-Based Robot Software Platform. *ETRI Journal*, Vol. 32, No. 5, (October 2010), pp 646-656.

Jawawi, D.N.A; Deris, S. & Mamat, R. (2006). "Enhancements of PECOS Embedded Real-Time Component Model for Autonomous Mobile Robot Application". *Proceeding of The 4th ACS/IEEE International Conference on Computer Systems and Applications*, pp. 882 - 889. Dubai/Sharjah, March 8-11 2006.

Jawawi, D. N. A.; Mamat, R. & Deris, S. (2007). A Component-Oriented Programming for Embedded Mobile Robot Software. *International Journal of Advanced Robotic Systems*, Vol. 4, No. 2, (September 2007), pp. 371-380. ISSN 1729-8806.

Ke, X.; Sierszecki, K. & Angelov, C., (2007). COMDES-II: A Component -Based Framework for Generative Development of Distributed Real-Time Control Systems, *Proceeding of 13th IEEE International Conference on Embedded Real Time Computing System and Applications (RTCSA 2007)*, pp. 199 - 208, ISBN 978-0-7695-2975-2, Daegu, Korea, *August 21-23*, 2007.

Lee, S. C. & Shirani A. I. (2004). A Component Based Methodology for Web Application Development. *Journal of Systems and Software*, Vol.71, No. 1-2, (April 2004), pp.177-187.

Mallet A.; Kanehiro F.; Fleury S. & Herrb M. (2007). Reusable Robotics Software Collection. *Proceedings of Second Int. Workshop on Software Development and Integration in Robotics (SDIR)*. Roma, Italy, April 2007.

Nesnas I. A.D.; Simmons R.; Gaines D.; Kunz C.; Diaz-Calderon A.; Estlin T.; Madison R.; Guineau J.; McHenry M.; Shu I-H. & Apfelbaum D. (2006). CLARAty: Challenges and Steps Toward Reusable Robotic Software. *International Journal of Advanced Robotic Systems*, Vol. 3, No. 1, (2006), pp. 23-30. ISSN 1729-8806.

Nierstrasz, O.; Arévalo, G.; Ducasse, S.; Wuyts, R.; Black, A.; Müller, P.; Zeidler, C.; Genssler, T. & van den Born, R. (2002). A Component Model for Field Devices. *Proceedings First International IFIP/ACM Working Conference on Component Deployment*, pg. 200-209, June 20-21, ISBN 3-540-43847-5. Berlin, Germany, June 20-21, 2002.

Ommering, R.; Linden, F.; Kramer, J. & Magee, J. (2000). The Koala Component Model for Consumer Electronics Software. *IEEE Computer*, Vol. 33, No. 3, (Mar 2000). pp. 78 - 85, ISSN 0018-9162.

Schuppenies, R. & Steinhauer, S. (2001). Software Process Engineering Metamodel, OMG group, November 2002.

Simpson, R.; Lopresti, A. E.; Hayashi, S.; Nourbakhsh, I. & Miller, D. (2004). The Smart Wheelchair Component System. *Journal of Rehabilitation Research & Development.* Vol. 41, No. 3B, (May/June 2004), pp 429–442.

Wang, A. J. A. & Qian, K. (2005). *Component-Oriented Programming*. Wiley-Interscience, ISBN 0471644463, Danvers, USA.

Part 2

Brain-Machine Interfacing

6

EEG Based Brain-Machine Interfacing: Navigation of Mobile Robotic Device

Mufti Mahmud[1], Alessandra Bertoldo[2] and Stefano Vassanelli[3]
[1]*NeuroChip Laboratory, Department of Human Anatomy and physiology, University of Padova*
[2]*Department of Information Engineering, University of Padova*
[3]*NeuroChip Laboratory, Department of Human Anatomy and physiology, University of Padova*
Italy

1. Introduction

During the last decade, rapid development of sophisticated methods for brain signal recordings along with availability of various efficient computational resources and the improving knowledge about brain dysfunctions have turned many researchers' interest in using large scale neurophysiological recordings for therapeutic and replacement strategies (Mason & Birch, 2003; Millán et al., 2003). Many patients with physiological disorders such as Amyotrophic Lateral Sclerosis (ALS) or injuries such as high-level spinal cord injury suffer from disruption of the communication path between the brain and the body. People with severe motor disabilities may lose much of their voluntary muscle control. The disabled people with the above mentioned problems are forced to accept a reduced quality of life, resulting in dependence on caretakers and escalating social costs (Vaughan et al., 2003). Most of the existing assistive technology devices for these patients are not usable because these devices are dependent on motor activities from specific parts of the body. Alternative control paradigms for these individuals are thus desirable (Fatourechi, 2008).
The electrophysiological signals generated from the brain can be used to command different devices, provided that the person who will control the device should also be able to control the generation of these signals. Studies showed that with sufficient training, people can control the generation of certain brain signals (Ohno et al., 2006). Having generated these signals, they can be conditioned and processed to perform the specific work for which they are generated. In other words, the interface can be made able to adapt and understand the meaning of these signals and work accordingly. If this type of Brain-Machine Interface (BMI) is successfully implemented, they can be used in developing sophisticated assistive devices (such as, a robotic wheelchair) to carry the people with motor dysfunction (Ferreira et al., 2008).
Previous works in development of BMI show that the signal acquisition and processing are getting complicated with the growing availability of more sophisticated recording devices (Cheein & Postigo, 2005; Ferreira et al., 2008; Moon et al., 2005; Mourino, 2003; Rani & Sarkar, 2005). To overcome these complexities rather simple method is required to couple easily recordable neuronal signals with the robotic device (Mahmud et al., 2010; 2009). This chapter illustrates a simple BMI system using EEG signals recorded through conventional EEG

acquisition devices and use these signals in commanding a robotic device's navigation. The figure 1 shows the schematic diagram of the proposed BMI.

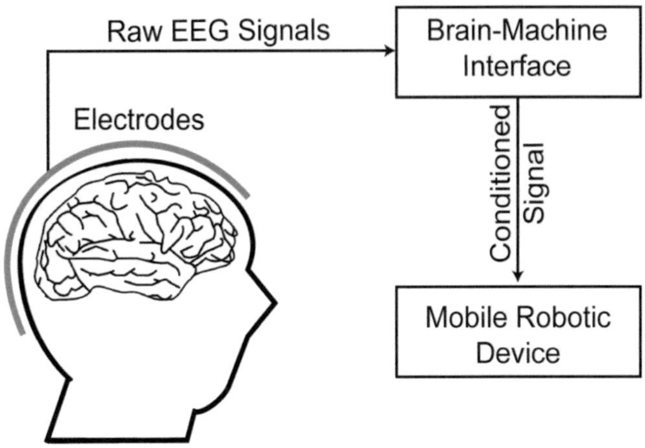

Fig. 1. The schematic diagram of the proposed Brain-Machine interface system where the devices can be controlled only with conditioned brain signals.

The BMI model was implemented using a two-layer approach. The first layer (also referred as upper layer) dealt with the signal acquisition and generation of control signals from the acquired EEG signals. The second layer (also referred as lower layer) contained the commanding and controlling modules of the robotic device, where the calculated binary control signals (BCS) were fed into the robotic device and its navigation was controlled. The two-layered approach separated the signal acquisition and processing phase from the robotic device commanding and controlling phase. This separation allowed a higher level of abstraction in programming sophisticated signal processing and analysis algorithms to calculate the BCS. Also, being separated from the signal processing and analysis layer, the robotic device interfacing was further simplified just to respond to the BCS to command and control the robotic device, thus enhancing reusability of the method. The figure 2 shows the schematic flowchart outlining the major steps of the BMI method. The first and second boxes from the top belong to the upper layer performing signal acquisition and processing and the third and fourth boxes belong to the lower layer doing the communication with the robotic device for commanding and controlling.

Considering the two-layered design of the BMI, the model was successfully tested with EEG signals generated by two distinctive actions and these two approaches will be elaborated in this chapter. The two distinctive event information extracted from the EEG in generating the BCS were:

- ERD (Event Related De-synchronization) and ERS (Event Related Synchronization) of the EEG signal (Pfurtscheller & daSilva, 1999).
- The event related evoked response (as in case of saccadic eye movement) is used as a second approach (Ohno et al., 2006).

The EEG signals generated by the above mentioned phenomena were used in calculating the BCS. Once the BCS was calculated, it was transmitted into another computer to command and control the mobile robot accordingly.

In the upper layer, the EEG signal acquisition was performed using a framework built in Matlab Simulink (`http://www.mathworks.com`) and the signal processing was performed

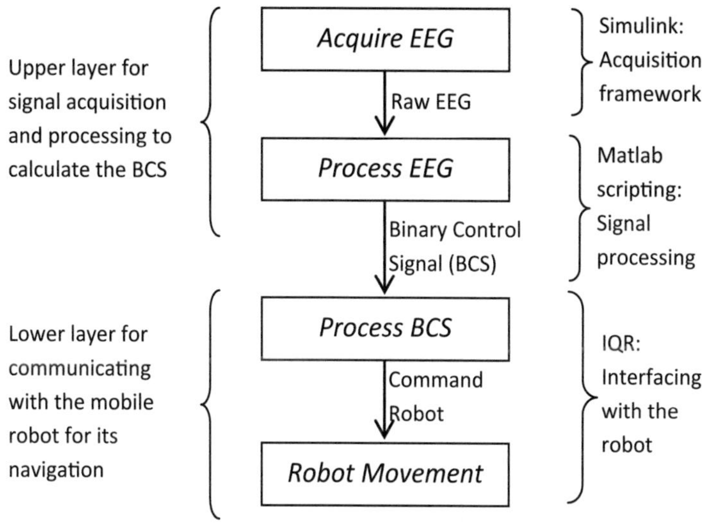

Fig. 2. Flowchart outlining the major steps of the BMI system, and their inputs and outputs. The curly braces on the right side categorize the steps based on the tools used in implementing those steps for the interfacing system. The curly braces on the left side show the two-layers and their components.

using Matlab scripting. In the lower layer, the interfacing with the robotic device was done using an open-source program called 'IQR' (Bernardet et al., 2002; Bernardet & Verschure, 2010) capable of mimicking neuronal network behavior based on the BCS.
After successful completion of signal acquisition, processing, and interfacing, the mobile robot was able to navigate through a predefined path showing the bright possibility of this method is designing assistive devices for the disabled.

2. The brain-machine interface system

2.1 The Electroencephalogram (EEG)

The Electroencephalogram (EEG) signals are widely used in the medical field for diagnosing diseases and studying brain functions. They are generated by firing of neuronal populations in the brain that propagates through the cortex. These propagated signals are recorded along the scalp using standard Ag-AgCl electrodes. A mapping of these electrode positions according to the 10-20 international system is shown in the figure 3. In general, EEG refers to the recording of the brain's spontaneous electrical activity over a short period of time, usually 20-40 minutes. Through the EEG technique evoked potentials (EP) and event-related potentials (ERPs) are recorded. The EP involves averaging the EEG activity time-locked to the presentation of a stimulus of some sort (visual, somatosensory, or auditory). Whereas, ERPs refer to averaged EEG responses that are time-locked to more complex processing of stimuli and used in cognitive science, cognitive psychology, and psychophysiological research. Necessary information related to a part of the brain can be obtained from the signals recorded by the corresponding electrodes placed in that part. For example, the visual cortex is located at the caudal portion of the brain (occipital lobe), thus, the signals recorded by the electrodes placed at the occipital lobe will provide information regarding the brain activity of the occipital lobe caused by visual stimuli.

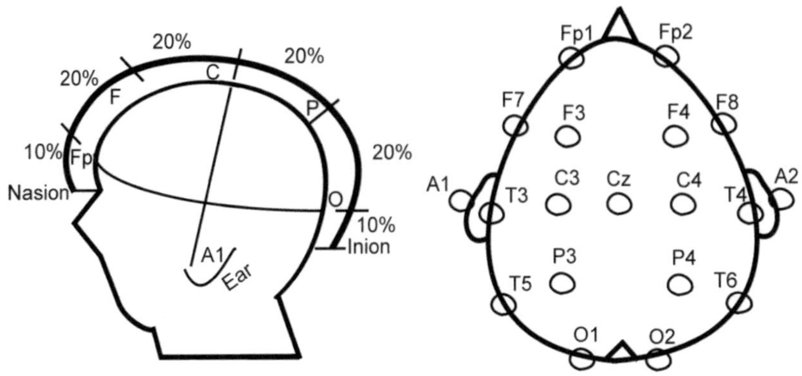

Fig. 3. Electrode mapping as per the 10-20 international system for EEG recording.

The spontaneous activity of the brain produces the signal with amplitude usually under 100 μV and a frequency ranging from little above the DC voltage up to 100 Hz. The EEG signal is recorded from the scalp and is extracellular in nature consisting of a few individual signals with different frequency bands, namely - delta (δ) with a frequency range of 0-3 Hz, theta (θ) ranges from 4-7 Hz, alpha (α) varies from 8-12 Hz, beta (β) is from 12-30 Hz, and gamma (γ) has a band of 34-100 Hz. These different signals have their own clinical implications in disease diagnosis (Niedermeyer & Lopes da Silva, 2005).

2.2 Event information selection from the EEG

As mentioned earlier, in this work we used two approaches to extract two different signal events from the recorded EEG signals to generate the BCS:

- In the first approach, the EEG signals were processed to extract the ERD and ERS, which appear in the α band (8 to 12 Hz) of the EEG spectrum. These ERS and ERD are event-related phenomena corresponding to a decrease (ERD) or an increase (ERS) in the signal power. These phenomena can be easily visualized by observing the alpha band power of the signal. The ERD and ERS patterns are usually associated to a decrease or to an increase, respectively, in the level of synchrony of the underlying neuronal populations. During the ERD and the ERS, the EEG signal changes sharply which can be detected in the α band of the EEG spectrum.
- In the second approach, the EEG signals were processed to extract the information related to the saccadic eye movement in the signals recorded from the occipital region of the scalp (indicated by 'O1' and 'O2' in the surface view of the electrode mapping, figure 3). Based on the work done by K. Ohno (Ohno et al., 2006), it is evident that the saccadic eye movement is well represented in the EEG during the memory tasks prior to the actual eye movement. These phenomena of changing EEG can be visualized by observing the signals recorded from the 'O1' and 'O2' electrodes (see figure 11 for the acquired signal). Depending on the direction of the saccade, the recording from the contralateral electrode changes sharply in the EEG which can be detected easily in the EEG spectrum.

These sharp changes were transformed into BCS and were sent to the robotic device for its navigation. In case of the saccade evoked EEG, refresh rate and discrimination between left and right decisions were important, but due to our work's scope (proof-of-principle) these aspects were not considered.
The point to be noted here is that, the approaches discussed here (the ERS and ERD and / or the EEG evoked by saccadic eye movement) are proof-of-principles of a simple BMI

model. The choice of using these type of signals was an arbitrary decision to show the model's workability and thus do not impose a restriction for this BMI to use EEG signals generated by any other phenomena that can be transformed or conditioned using sophisticated signal processing tools to a BCS. Also, it is possible to model imaginary movement in the EEG by means of Hidden Markov Models and coherence (Souza et al., 2010) which can also be used in this BMI.

2.3 The e-puk mobile robot

In the BMI system, we used the e-puck (Mondada et al., 2009) mobile robot to demonstrate the system's workability. The e-puck is an educational desktop mobile robot developed at the Ecole Polytechnique Fédérale de Lausanne (EPFL) for a broad exploitation in teaching activities. The basic configuration of an e-puk contains:

- A microcontroller
- A set of sensors and actuators:
  - Eight infrared (IR) proximity sensors
  - A 3D accelerometer
  - Three microphones
  - A color CMOS camera
  - Two stepper motors
  - A speaker
  - Eight red light emitting diodes (LED)
  - A set of green LEDs
  - A red front LED placed beside the camera
- A user interface containing:
  - Two LEDs show the status of the battery
  - An interfacing connector to an in-circuit debugger
  - An infrared remote control receiver
  - A classic RS232 serial interface
  - A Bluetooth radio link
  - A reset button
- Mechanics

The specification provided above is the basic configuration of the e-puk. There are possibilities to extend the e-puk to perform many other tasks. Detailed information about e-puk's design and devices can be found at `http://www.e-puck.org/index.php`.
Figure 4 shows the e-puk robot's electronic design outline and the figure 5 depicts the mechanical architecture.

2.4 Overview of the BMI system

The figure 6 shows the schematic cartoon of the BMI system with each numbered element representing an interface in the process of communicating to the mobile robotic device. The number 1 in the diagram represents the EEG recording cap with the electrodes (figure 7, (a)); the signals acquired by the electrodes were sent to number 2, which represents the electrode connector and/or multiplexer (figure 7, (b)). The multiplexed signal from each channel was then sent to the preamplifier, (number 3 in the diagram, figure 7, (c)) where the signals were amplified with a predefined gain and transferred to the computer for digitization at 256

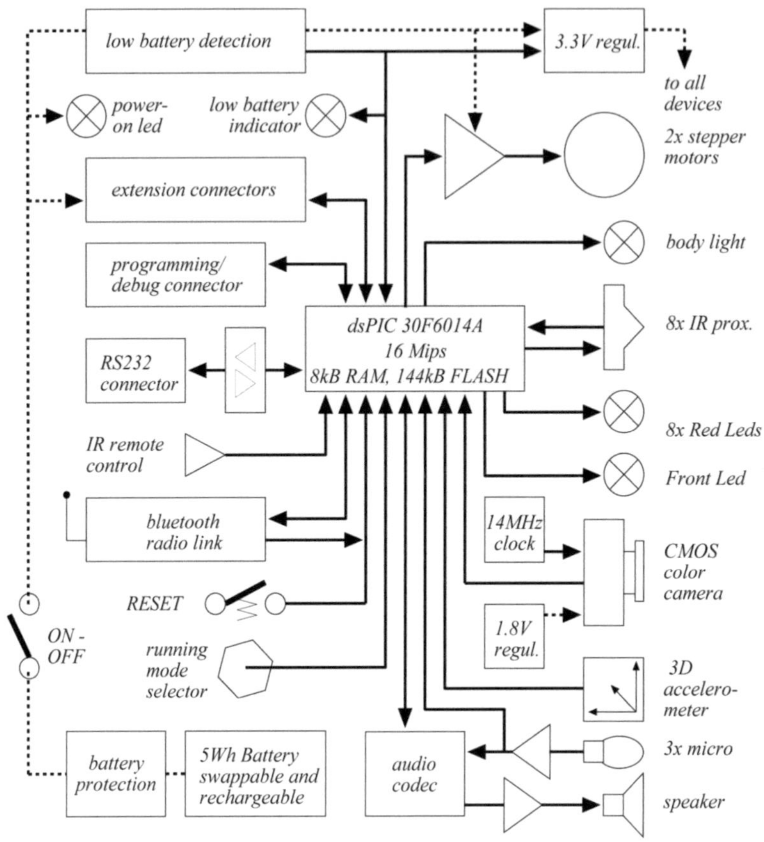

Fig. 4. The electronic design outline of the e-puk robot as described in Mondada et al. (2009).

Hz (number 4 in the diagram) and further process them to extract the relevant information (either ERS/ERD or events caused by saccadic eye movement) from the raw EEG signals. The number 5 shows the user-diagram protocol (UDP) transfer of the conditioned signal to another computer (represented by number 6) running on Linux for generating the BCS to be sent to the mobile robotic device. Finally these command signals were sent to the robotic device using Bluetooth (number 7). The numbers 1 to 5 shows the processing interfaces at the upper layer and 6 to 8 are the interfaces at the lower layer (see figure 2).

2.5 Data acquisition

The EEG signals were acquired using a four channel commercial EEG recording device, g®.MOBIlab, manufactured by the g.tec medical engineering GmbH, Austria (`http://www.gtec.at/`) as seen in figure 7. Two different configurations of channels were used in selecting the two different signal features as listed below:

- In case of ERS/ERD: two of the four channels were used in recording simultaneous signals from two healthy subjects skulls. One of the electrodes was placed in the 'O1' to acquire the EEG activity with reference to the other one placed at the 'F' or 'F_P'. The acquired EEG signals were then processed to extract the ERS/ERD complex.
- In case of saccade related evoked potential: three out of the four channels were used during the recording. Two of them were used in recording simultaneous signals from 'O1' and

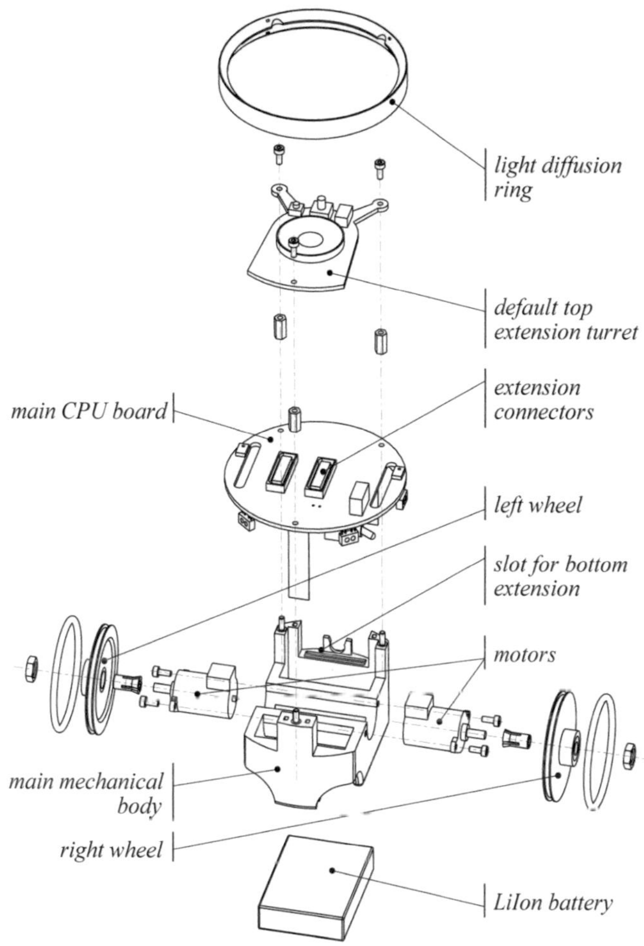

Fig. 5. The mechanical architecture of the e-puk robot as described in Mondada et al. (2009).

'O2' and the third electrode was used as a reference, placed at 'F' or 'F_P' position. EEG signals recorded using the 'O1' and 'O2' electrodes were processed to extract the saccade related event information.

The manufacturing body provided a module and a framework to acquire and further process the EEG signals to extract feature information. This module was designed using Matlab Simulink with a possibility to incorporate S-functions written in Matlab Script for signal processing. The module worked as an interface between the recording device and the computer used for EEG recording. It also provided a framework which was extended to suit the necessity of on-line signal processing based on the applications. Therefore, the module was reengineered to analyze and process the EEG signals acquired from the occipital region of the subject. After having read the data, they were amplified with a certain gain and sent on-the-fly to the computer for on-line processing. The figure 8 shows the schematic diagram of the signal acquisition and processing module. The signals detected by the recording electrodes were fed into a preamplifier, where the signals from different channels were separated and then amplified by a predefined gain (shown as right faced triangles). These amplified signals

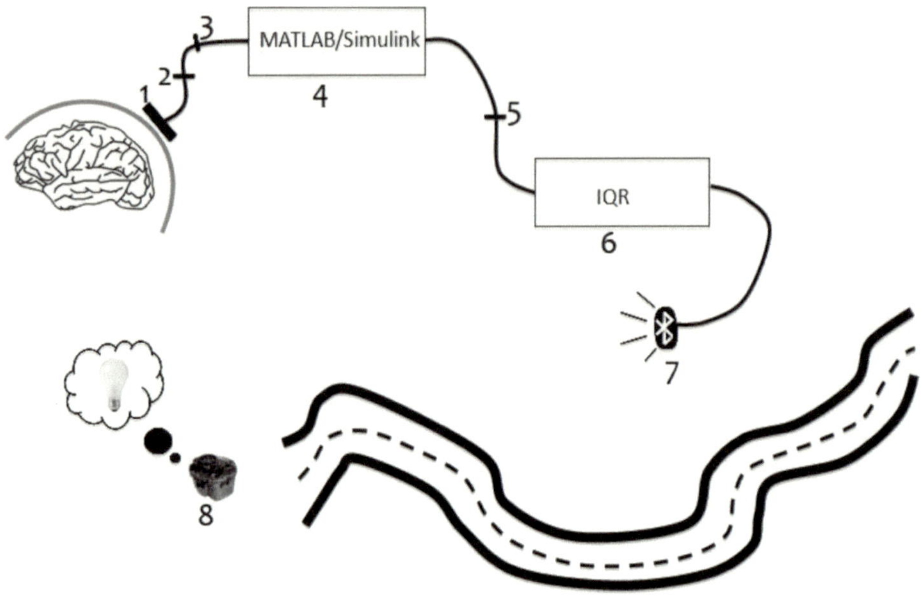

Fig. 6. The schematic cartoon of the Brain-machine Interface system.

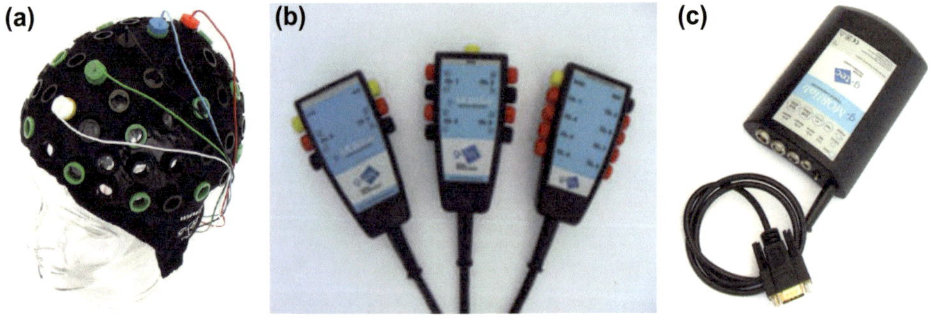

Fig. 7. EEG acquisition devices: (a) cap with electrodes, (b) connector and/or multiplexer, and (c) preamplifier.

from both the channels were then fed into the computer at a sampling rate of 256 Hz through a universal serial bus (USB) port for further processing.

2.6 Processing of EEG signals and BCS generation at the upper layer

The processing of raw EEG signals acquired by the electrodes for the generation of BCS was performed in the upper layer of the BMI flow (see figure 2). The following subsections describe the pre-, post-processing, and extraction of event related information from the EEG.

2.6.1 Pre- and post-processing

After the detection of the EEG signals by the electrodes and before extracting the signal events (ERS/ERD or event caused by saccadic eye movements) to generate the BCS, artifact removal was performed on-the-fly to remove artifacts related to eye blink, cardiac rhythms, and body movements from the signals using processes proposed in Haas et al. (2003); Rohalova et al.

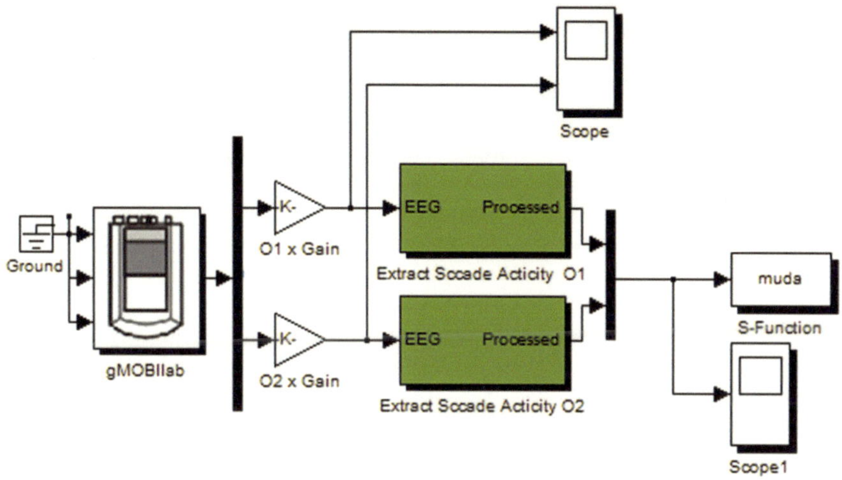

Fig. 8. Schematic of EEG signal acquisition and conditioning process.

(2001). The power line noise (50-60 Hz components) was removed using a stop-band filter of 50 and 60 Hz.

As our targeted events changed sharply from the baseline, we implemented a dynamic threshold based on the EEG signal's standard deviation to detect the occurrence of an event. A detected event denoted a high (i.e., 1) signal in the BCS, and the absence of an event was represented by a low (i.e., 0) signal. In figure 8, the green rectangular boxes at the center represent the signal processing steps mentioned above. These processed individual channels were then multiplexed and sent to another computer running on Linux through an UDP port using a crossover ethernet cable for interfacing with the robot to control it.

2.6.2 Processing for ERS/ERD

Generally, presence of a visual stimuli, i.e., when the eyes are open, the signal power in the alpha band decreases, characterizing an ERD, and when the eyes are closed (providing with a few or even no visual stimuli), the visual cortex is relaxed, the signal power increases denoting an ERS. This type of potential is believed to arise from the oscillation of postsynaptic potentials in the neocortex. Functionally, alpha wave has been interpreted as an idling rhythm that diminishes when eyes are opened or during mental activity. It is not clear whether alpha rhythms are pure noise or the product of chaotic processes, nor whether there are distinct generators of alpha activity (Laufs et al., 2003). The detected and artifact removed EEG signals were continuously scanned for occurrence of ERD/ERS to isolate them with a high signal-to-noise-ratio (SNR). An important issue in the EEG based signal processing was that, the levels of EEG change very often, therefore, a calibration step was required to detect the basic ERS/ERD level or baseline before sending the signals to the robotic device. To extract only the alpha wave, the recorded EEG signals were filtered through an on-line band-pass filter with pass-band and stop-band cutoff frequencies as 8 Hz and 12 Hz respectively. Through this filtering all kind of artifacts were removed from the EEG leaving only the clean alpha wave containing the ERD and the ERS. Finally, using the dynamic threshold the BCS was generated.

2.6.3 Processing for saccadic events

To extract the events from the EEG signals related to the saccadic activity, both the channels signals were filtered using a band-pass filter with the low-pass and high-pass cut-off frequencies as 15 Hz and 100 Hz, respectively. This filtering eliminated the possibility of recording the alpha wave generated by the occipital region of the brain. Then the signals were scanned for the occurrence of the sharp change in their amplitude using the dynamic threshold mentioned earlier. The system was trained before the actual experiment due to the signal variability from person to person. This threshold got renewed every 100 ms by calculating the ratio between the usual electrical activity and the existing threshold.

2.7 Interfacing with the robotic device at the lower layer

At the lower layer of the BMI system, interfacing with the robotic device was performed through processing of the BCS (see figure 2). To process the BCS and to command the robotic device an open source software named 'IQR' (Bernardet et al., 2002; Bernardet & Verschure, 2010) was used. The IQR provides a flexible platform for simulating large scale neural networks and developing robust applications to interface with robotic devices. Using this software it is possible to design large scale neuronal populations, interconnect these populations, and control their activation or inactivation by providing excitatory or inhibitory synapses.

To achieve our goal of steering wheels of the robotic device (number 8 in schematic diagram of figure 6) using the BCS (i.e., the extracted information from the EEG, ERS/ERD or saccadic events), three modules were developed in the IQR as shown in figure 9 (a), (b), and (c). As seen in figure 9 (a), two main processes were designed to receive the BCS coming from the UDP port ('simulink IN' process) and sending a command to the mobile robot ('Robot' process). These processes (figure 9 (b) and (c)) were carefully designed, each with a number of neuronal populations creating a neuronal network to perform the desired task . The 'Simulink IN' process received the binary control signals from each EEG channel in the upper layer (indicated by the 'M' and an inward arrow to the square boxes in figure 9 (c)); on the other hand, the 'Robot' process was designed to control the movement of the robot's wheels (the communication with the mobile robotic device is indicated by the outward arrow and 'M' at the bottom square box of figure 9 (b)). Due to the binary control signals' binary nature they were used to act as synapse, either excitatory or inhibitory, to make a population of neurons active or inactive.

The 'Simulink IN' process contained two neuronal populations: 'Channel 1' and 'Channel 2'. 'Channel 1' neuronal population received the input from one EEG channel and 'Channel 2' from the other. The main purpose of this process was to continuously scan the BCS and send the relevant synaptic signal (high in case of a 1 and no synapse in case of a 0) to the 'Robot' process to communicate with the mobile robotic device.

The 'Robot' process was designed to communicate and command the robotic device. This was done by five neuronal populations (as seen in figure 9 (b)). These neuronal populations were connected, as required, using excitatory and inhibitory connections to form a network capable of driving the robotic device. The five populations were to perform their predefined tasks based on the input they received from the 'Simulink IN' process. The name and purpose of the neuronal populations are:

- **Motor**: This neuronal population represented and communicated with the motor of the robotic device. The robotic device was mainly commanded using this neuronal population.
- **Left Wheel**: This neuronal population received input from the 'Channel 1' and was connected to three other populations. It was mainly responsible to make the left wheel to move forward.

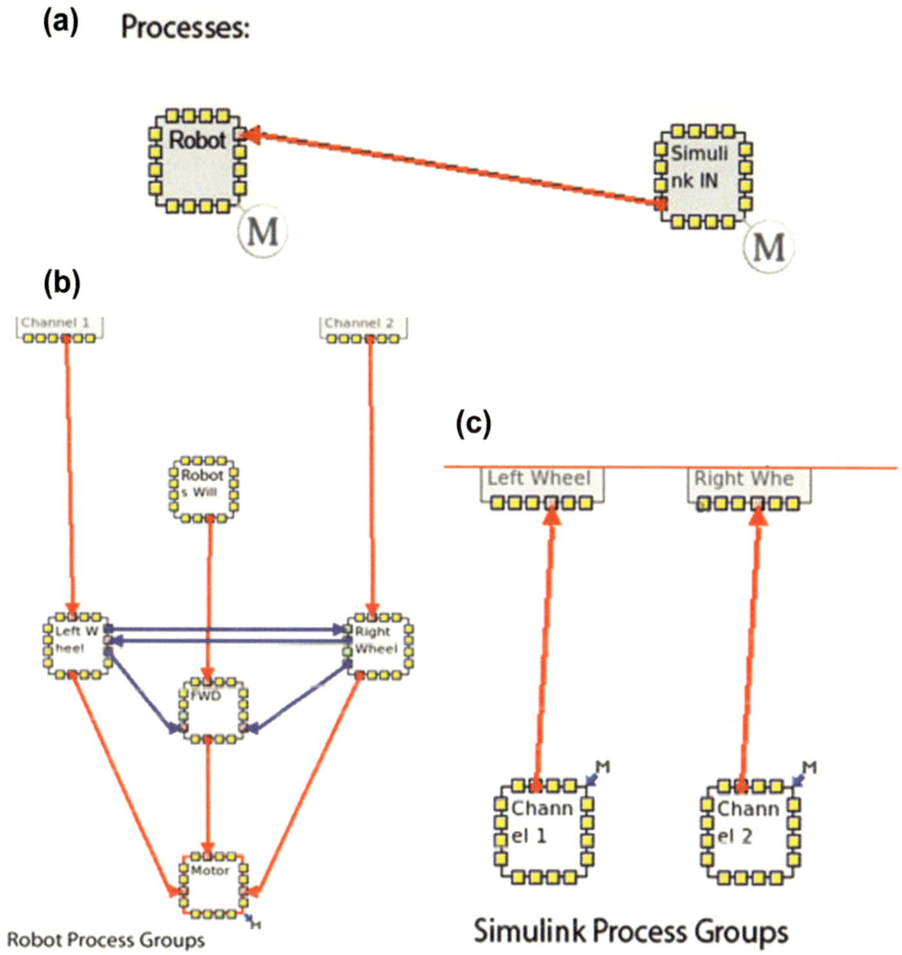

Fig. 9. IQR modules for generating and commanding the robotic device. The square boxes represent neuronal populations and the lines represent connectivity between two populations. The excitatory and inhibitory connections are shown using red and blue lines, respectively.

- **Right Wheel**: Contrarily, this neuronal population received input from the 'Channel 2' and was mainly responsible for the right wheel's forward movement. This was also connected to three other neuronal populations.
- **FWD**: This neuronal population represented the combined movement of the left and the right wheels. When a synapse was made to the 'motor' neuronal population, both the wheels would move forward.
- **Robot's Will**: This neuronal population represented the robot's will to move, thus, always provided high synapses to the population it was connected to (i.e., 'FWD').

At rest (when there was no input), the 'Robot's will' neuronal population constantly generated excitatory synapse to the 'Motor' through the 'FWD' neuronal population that kept the robotic device moving forward. However, to enable turning of the robotic device, the input control signals from the channels were fed to the neuronal populations representing the wheels. These two populations were connected to the 'Motor' which controlled the movement of the individual wheels. Each time a neuronal population corresponding to a wheel fired: an excitatory synapse was sent to the 'Motor', an inhibitory synapse to the other wheel inactivating that wheel's activity, and another inhibitory synapse to the 'FWD' neuronal population making the physical stepper motor to stop working for the previous command and to get ready to process the new command.

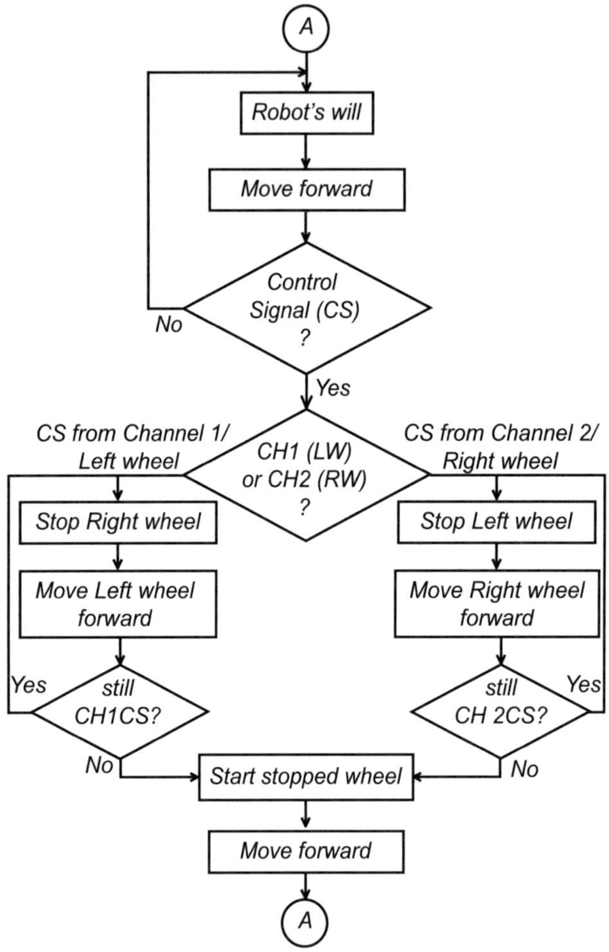

Fig. 10. Flowchart of the IQR modules' communication for the robotic device's navigation.

The Figure 10 depicts the flowchart of the IQR modules' communication strategy for the robotic device's navigation. As explained earlier, at the beginning the robotic device kept on moving in the forward direction and waited for a command. Once a neuronal population corresponding to a wheel received a synaptic input, based on the channel from where it was generated, the wheel corresponding to the other channel was stopped and the forward driving motor was initiated for the wheel corresponding to the control signal generating channel. This operation caused the robot to take a turn (right or left) based on the received control signal. During the turn the robot constantly scanned for continuation of high control signal (i.e., a stream of continuous 1s) and once the control signal was low (i.e., 0), the stopped wheel was restarted thus continuing the forward movement.

For instance, let us assume that the 'Left wheel' neuronal population received a synaptic input. This inactivated the 'Right wheel' and the 'FWD' neuronal populations through inhibitory synapses, and activated the 'Motor' neuronal population by an excitatory synapse; as a result, the right wheel was stopped, but the left wheel kept on rotating allowing the robotic device to turn right. Therefore, a synapse from a wheel was required to stop the other wheel to steer the robotic device to follow a predefined course by taking a left and a right turn.

3. Discussion

The BMI model was thoroughly tested with EEG signals containing both the event evoked responses (ERD/ERS and events caused by saccadic eye movement). As an example of the BMI system's workability, here we demonstrate it using the event information caused by the saccadic eye movements. The ERD/ERS based interfacing can be found in a previous publication (Mahmud et al., 2009).

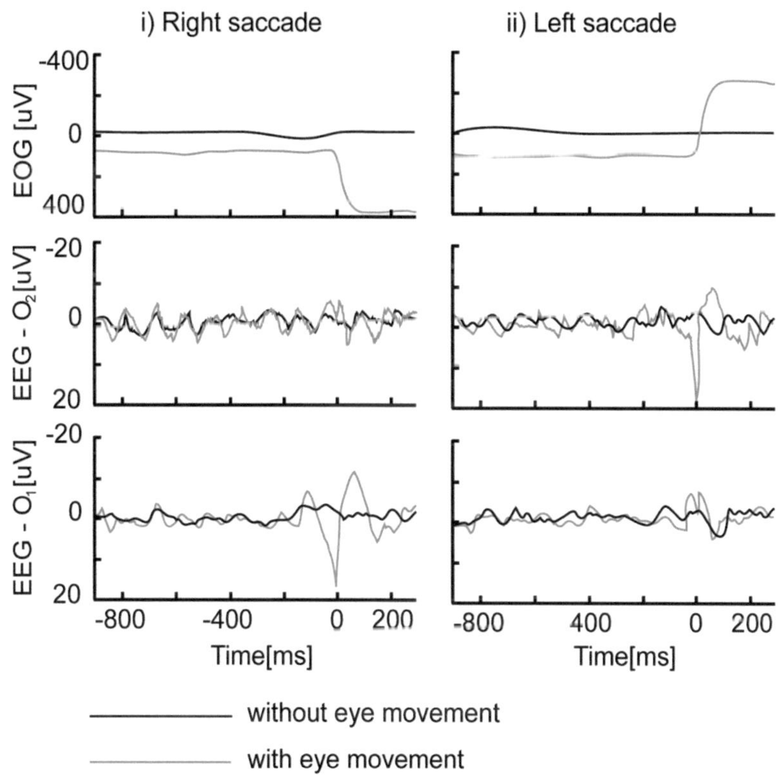

Fig. 11. Signals recorded by EOG and 'O1', 'O2' electrodes of the EEG from one subject while performing the saccadic movement during an experiment.

Healthy subjects were plugged in with the necessary equipments of EEG signal detection (see figure 7). The recorded and artifact removed EEG signals related to the saccadic eye movement is shown in figure 11. The saccadic movement direction of the subject was clearly reflected through sharp changes of amplitudes in the recorded signals from the contralateral electrodes. The dynamic threshold detected this sharp change and generated the controls

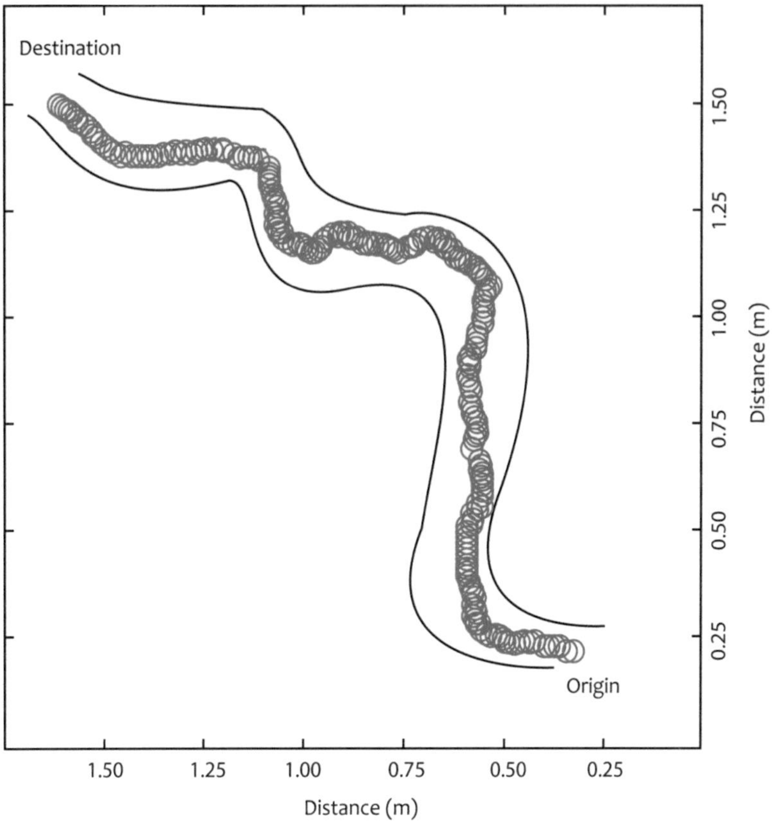

Fig. 12. Navigation result of the robotic device during an experiment to follow a predefined course.

signals which in turn triggered the activation / inactivation of particular neuronal populations by providing an excitatory / inhibitory synapses for taking a turn. The excitatory synapse from the forward movement neuronal population kept the robot moving forward; to turn the robotic device left, the left wheel was stopped through an inhibitory synapse generated by the right wheel neuronal population, and vice-versa for a right turn. A combination of these saccadic movements can guide the robot to follow a predefined course, as seen in the figure 12.

The time course analysis of the robot's movement was not performed during the experiments. However, the robot could move at a speed of 13.816 cm/s, and in an ideal situation it could travel a 1 m straight path in around 8 s [1].

4. Conclusion

This Brain-machine interface based on the EEG demonstrated here is a proof-of-principle to reduce the complexity that is gradually prevailing upon the very potential field of rehabilitation. By applying this technique it is possible to provide mobility to the people with motor dysfunction. The system was tested thoroughly using ERD/ERS events of the α and

[1] Each wheel of the e-puk had a diameter of 41 mm and the stepper motor could rotate 1000 steps/s denoting a single revolution of a wheel (`http://www.e-puck.org/index.php`).

the events caused by the saccadic eye movement. However, the usage of saccade evoked EEG signals in this work is just to demonstrate the proposed model's workability. It is very much possible to adapt and extend the BMI model to use any other type of signal (voluntary or involuntary) through proper signal processing techniques capable of transforming the input signal to a binary decision signal. The work in progress is to extend this technique to control assistive robotic devices (e.g., robotic wheelchair) for the disabled.

5. References

Bernardet, U., Blanchard, M. & Verschure, P. (2002). Iqr: a distributed system for real-time real-world neuronal simulation, *Neurocomputing* 44(46): 1043–1048.

Bernardet, U. & Verschure, P. (2010). iqr: A tool for the construction of multi-level simulations of brain and behaviour, *Neurocomputing* 8(2): 113–134.

Cheein, F. & Postigo, J. (2005). A fast finite state machine design for a brain computer interface, *Proceedings of the XI Reunion de Trabajo en Processamiento de la Informacion y Control*, Argentina.

Fatourechi, M. (2008). *Design of a self-paced brain computer interface system using features extracted from three neurological phenomena*, Ph.D. Dissertation, The University of British Colombia, Canada.

Ferreira, A., Celeste, W., Cheein, F., Bastos-Filho, T., Sarcinelli-Filho, M. & Carelli, R. (2008). Human-machine interfaces based on emg and eeg applied to robotic systems, *Journal of NeuroEngineering and Rehabilitation* 5(1): 10.

Haas, S., Frei, M., Osorio, I., Pasik-Duncan, B. & Radel, J. (2003). Eeg ocular artifact removal through armax model system identification using extended least squares, *Communications in Information and Systems* 3: 19–40.

Laufs, H., Kleinschmidta, A., Beyerlea, A., Egera, E., Salek-Haddadib, A., C., P. & Krakow, K. (2003). Eeg-correlated fmri of human alpha activity, *NeuroImage* 19(4): 1463–1476.

Mahmud, M., Hawellek, D. & Bertoldo, A. (2010). Eeg based brain-machine interface for navigation of robotic device, *Proceedings of the 3rd IEEE/RAS-EMBS International Conference on Biomedical Robotics and Biomechatronics (BioRob2010)*, Tokyo, Japan, pp. 168–172.

Mahmud, M., Hawellek, D. & Valjamae, A. (2009). A brain-machine interface based on eeg: extracted alpha waved applied to mobile robot, *Proceedings of the 2009 ECSIS Symposium on Advanced Technologies for Enhanced Quality of Life (AT-EQUAL 2009)*, Iasi, Romania, pp. 28–31.

Mason, S. & Birch, G. (2003). A general framework for brain-computer interface design, *IEEE Trans. Neural Syst. Rehabil. Eng.* 11(1): 70–85.

Millán, J., Renkens, F., Mouriño, J. & W., G. (2003). Non-invasive brain-actuated control of a mobile robot, *Proceedings of the 18th International Joint Conference on Artificial Intelligence*, Acapulco, Mexico, pp. 1121–1126.

Mondada, F., Bonani, M., Raemy, X., Pugh, J., Cianci, C., Klaptocz, A., Magnenat, S., Zufferey, J., Floreano, D. & Martinoli, A. (2009). The e-puck, a robot designed for education in engineering, *Proceedings of the 9th Conference on Autonomous Robot Systems and Competitions*, Castelo Branco, Portugal, pp. 59–65.

Moon, I., Lee, M., Chu, J. & M., M. (2005). Wearable emg-based hci for electric-powered wheelchair users with motor disabilities, *Proceedings of 2005 IEEE International Conference on Robotics and Automation (ICRA 2005)*, Barcelona, Spain, pp. 2649–2654.

Mourino, J. (2003). *EEG-based analysis for the design of adaptive brain interfaces*, Ph.D. Dissertation, Universitat Politecnica de Catalunya, Barcelona, Spain.

Niedermeyer, E. & Lopes da Silva, F. (2005). *Electroencephalography: Basic Principles, Clinical Applications, and Related Fields*, Lippincott Williams & Wilkins, Philadelphia, USA.

Ohno, K., Funase, A., Cichocki, A. & Takumi, I. (2006). Analysis of eeg signals in memory guided saccade tasks, *IFMBE Proceedings of World Congress on Medical Physics and Biomedical Engineering*, COEX Seoul, Korea, pp. 2664–2667.

Pfurtscheller, G. & daSilva, F. (1999). Event - related eeg / meg synchronization and desynchronization: basic principles, *Clinical Neurophysiology* 110(11): 1842–1857.

Rani, P. & Sarkar, M. (2005). Emg-based high level human-robot interaction system for people with disability, *Proceedings of 2005 IEEE International Workshop on Robot and Human Interactive Communication (ROMAN 2005)*, Nashville, Tennessee, USA, pp. 280–285.

Rohalova, M., Sykacek, P., Koska, M. & Dorffner, G. (2001). Detection of the eeg artifacts by the means of the (extended) kalman filter, *Measurement Science Review* 1: 59–62.

Souza, A., Santos-Filho, S., Xavier, P., Felix, L., Maia, C. & Tierra-Crioulo, C. (2010). Modeling movement and imaginary movement eeg by means of hidden markov models and coherence, *Proceedings of the ISSNIP Biosignals and Biorobotics Conference (BRC2010)*, Vitoria, Brazil, pp. 86–91.

Vaughan, T., Heetderks, W., Trejo, L., Rymer, W., Weinrich, M., Moore, M., Kübler, A., Dobkin, B., Birbaumer, N., Donchin, E., Wolpaw, E. & Wolpaw, J. (2003). Brain-computer interface technology: a review of the second international meeting, *IEEE Trans. Neural Syst. Rehabil. Eng.* 11(2): 94–109.

7

Bioartificial Brains and Mobile Robots

Antonio Novellino[1], Michela Chiappalone[2], Jacopo Tessadori[2], Paolo D'Angelo[1], Enrico Defranchi[1] and Sergio Martinoia[2,3]
[1]*ETT S.r.l.*,
[2]*Italian Institute of Technology*,
[3]*Dept. Electronics and Biophysical Engineering – University of Genova, Italy*

1. Introduction

The growth in neuroscience discoveries continues to be explosive, with new frontiers being reached every year in the understanding of new principles of the central and peripheral nervous system (CNS and PNS), in the interplay between structure and function at different scales (from molecules to behavior), and with the introduction of new technologies for direct transfer of information between natural neuronal systems and artificial devices.
Rapid advances in biomedical engineering and computer science are producing the methodologies required for predictive models of neural function that can interact with the brain in real time. The continuous achievements in microelectronics that allow ever-greater circuitry miniaturization together with increased speed and computational capacity are providing the next-generation hardware platforms for neuroprostheses and Brain Computer Interfaces (BCIs) or Brain Machine Interfaces (BMIs). On the other hand, in the last ten years, demonstrations of direct, real-time interfaces between living brain tissues and artificial devices, such as computer cursors, robots and mechanical prostheses, have opened new avenues for experimental and clinical investigations (Nicolelis and Lebedev, 2009). Interest in these BMIs has been kindled by the contribution that they may make to the treatment or rehabilitation of patients suffering from severe motor disabilities (Daly and Wolpaw, 2008; Hatsopoulos and Donoghue, 2009). When motor pathways fail, BMIs offer a physical bridge for movement intention to reach the external world (Donoghue, 2008). The first experimental demonstration that ensemble of cortical neurons could directly control a robotic manipulator was given in 1999 by the group of M. Nicolelis at the Duke University (USA) (Chapin et al., 1999). Since then, a continuous stream of research papers in the BMI field came out. Many groups (Nicolelis and Chapin, 2002; Taylor et al., 2002; Andersen et al., 2004; Schwartz, 2004; Hochberg et al., 2006; Velliste et al., 2008) have successfully shown the possibility of using cortical signals, recorded from different subjects, like rats, monkeys or humans, to move an artificial effector and, in these systems, the feedback was constituted by vision, tactile information and proprioception. BMIs have also been recently used as a tool for studying neural processing of information (Nicolelis and Lebedev, 2009). Nevertheless the limitations arising from the reduced possibility of a full control and observation of the system do not allow a systematic study on how information is processed and transmitted within and among cell-assemblies.

On the contrary, cultured cells or tissue allow for a proper control and observation of the phenomena taking place at cellular and cell-assembly levels but lack of an actual physiological situation in which the brain has a body and the body can interact with a specific environment. Since behavior emerges from complicate interactions of the nervous system with the body and the environment (Chiel and Beer, 1997), "embodiment" and "situatedness" are likely to be crucial in studying the mechanisms of sensory information processing, control and adaptation in living systems. For these reasons, closed-loop systems interfacing nerve cells to external devices have been used as an electrophysiological tool for studying adaptation and coding properties (Reger et al., 2000; DeMarse et al., 2001). More precisely, this framework allows to study interactions within cells and cell assemblies and to investigate the cellular and population mechanisms underlying learning and memory.

These mechanisms can be studied at different complexity levels. Networks of cultured neurons coupled to Micro Electrode Arrays (MEA) represent a kind of intermediate level where the control over the system is high and where it is possible to study universals of learning and memory, regardless of specific realizations (Marom and Shahaf, 2002). It should be noted that this neural preparation is an extremely simplified model of the brain: first of all it has a bi-dimensional structure, in spite of tri-dimensional brain organization; furthermore neural cells are dissociated and they re-organize autonomously even if they preserve functional synaptic interconnections. Nonetheless, the general morphological and physiological properties of the cell populations in culture correspond closely to the features of the original tissue (Shahaf and Marom, 2001).

Following the "embodied neurophysiology" approach, we interfaced a mobile robot with a population of neurons, extracted from rat embryos and cultured over a MEA (Novellino et al., 2007). The proposed paradigm represents an innovative, simplified and controllable bi-directional closed loop system where it is possible to investigate the dynamic and adaptive properties of a neural population interacting with an external environment by means of an artificial body (i.e., the mobile robot). Similar closed-loop approaches have been used to quantify the complexity of neural dynamics (Kositsky et al., 2009), to investigate the transition among different electrophysiological regimes (Wagenaar et al., 2005) and to investigate basic mechanisms of learning (Shahaf and Marom, 2001; le Feber et al., 2010). Closed-loop experiments are also relevant to the technology of neural interfaces (Mussa-Ivaldi and Miller, 2003; Nicolelis, 2003).

Along this line, Mussa-Ivaldi and coworkers (Reger et al., 2000) at Northwestern University proposed the first hybrid neuro-robotic system, an innovative experimental paradigm aimed at studying learning (in particular, sensorimotor adaptation) and synaptic plasticity in the nervous system (Fig. 1A). They bi-directionally connected the brain of a lamprey, maintained alive in vitro, to a small mobile robot. The robot had the function of an artificial body (physical embodiment) that provided sensory feedback to the brain and was controlled by the recorded electrophysiological signal converted in motor commands. In particular, the brain of a sea lamprey larvae was extracted and placed in a recording chamber where it was maintained at a constant temperature in a Ringer's solution. The activity was recorded from the visually identified neurons of the left and right Posterior Rhomb-encephalic Reticular Nuclei (PRRN) of the lamprey brain, and a simple interface decoder converted the spiking activities into driving signals for the corresponding wheels of the small robot. In parallel, an electrical stimulus, whose frequency was modulated by the light intensity perceived by the sensors housed in the robotic body, was delivered near the region in which the axons of the Intermediate and Posterior Octavo-motor nuclei (nOMI and nOMP) cross.

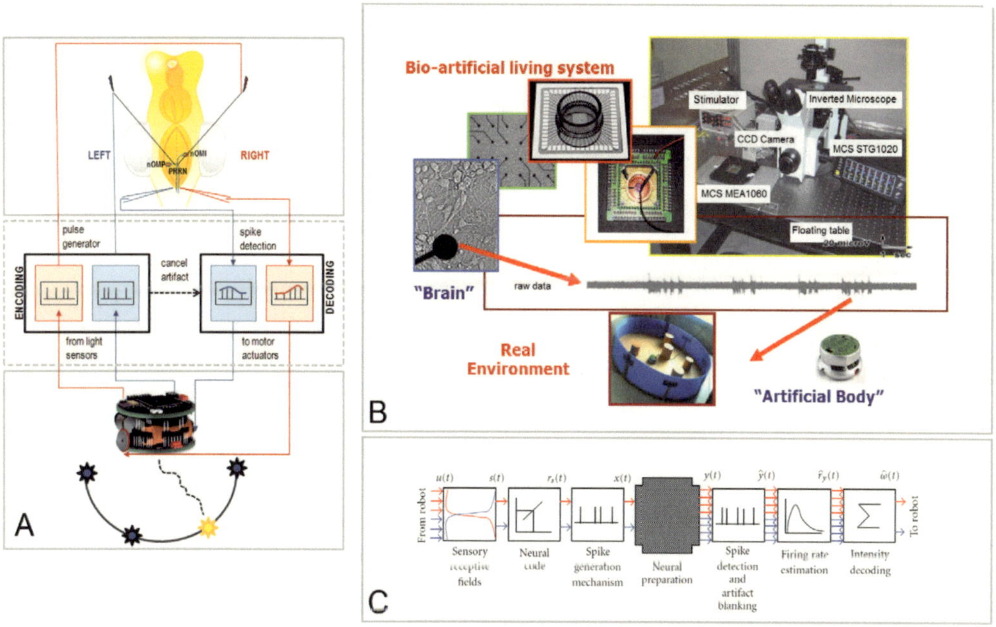

Fig. 1. A) Implementation of the first BMI realized at Northwestern University: a hybrid neuro-robotic system connecting a lamprey's brainstem to a small mobile robot. Signals from the optical sensors of the robot (bottom) are encoded by the communication interface into electrical stimuli, whose frequency depends linearly upon the light intensity. Stimuli are delivered by tungsten microelectrodes to the right and left vestibular pathways (top. nOMI and nOMP: intermediate and posterior octavomotor nuclei). Glass microelectrodes record extracellular responses to the stimuli from the posterior rhombencephalic reticular nuclei (PRRN). Recorded signals from right and left PRRNs are decoded by the interface, which generates the commands to the robot's wheels. These commands are set to be proportional to the average firing rate calculated from the corresponding side of the lamprey's brainstem. The robot is placed in a circular arena with light sources on the periphery. The neural system between stimulation and recording electrodes determines the motions in response to each light source. Modified from Mussa-Ivaldi et al., 2010, with permission. B) Implementation of the closed-loop system involving a bioartificial brain and a mobile robot. Dissociated cortical cells from a rat brain are grown onto a MEA. Once the in vitro neuronal network is formed, it is inspected under a microscope for visual compliance with internal standards. Furthermore, after three weeks of culturing, it is tested for activity, which should be synchronous in the whole network. If such criteria are met, the network is connected to the amplifier and its activity is recorded from selected electrodes to be decoded into motor commands for the artificial body, thus enabling interaction with the physical environment. C) General computational architecture of the closed-loop system: the signals coming from the infrared sensors (IR) of the robot are translated into patterns of stimuli that are delivered to the neural preparation through a set of selected stimulating electrodes. Then the activity recorded by two groups of electrodes is evaluated in terms of firing rate (i.e., mean number of detected spikes/s) and used as driving speed for each of the robot's wheel.

One of the main achievements provided by these experiments was that the alteration of the sensory input caused short- and long-term adaptive changes in the robot behaviors. This suggests that this system can be used as an experimental paradigm for investigating the properties of synaptic plasticity in the context of sensorimotor adaptation. However, one of the major limitations of this experimental set-up was the small number of electrodes used and the simple neuronal circuitry involved in such studies.

An alternative approach for investigating how neuronal information is processed was then proposed by Steve Potter's group at the GeorgiaTech where a new system, to interface an in vitro neuronal population to a computer-simulated environment, was developed. In particular, they employed Micro Electrode Arrays (MEA) dishes to record the neural activity from a population of neurons and use that activity to control a simulated animal, called 'Neurally Controlled Animat' (DeMarse et al., 2001). According to the developers, this approach makes possible a correlation between neural morphology, connectivity, and distributed activity, not feasible with in vivo neural interfaces (Nicolelis et al., 1998; Wessberg et al., 2000). In the preliminary experiment (DeMarse et al., 2001) they created a virtual environment where a simulated body was moved by the spatio-temporal patterns of activity across the living neuronal network coupled to a MEA. The motor commands for the Animat were computed from spatio-temporal patterns that were detected in real time from neural activity. A feedback signal was provided for each movement resulting in a collision with walls or barriers within the virtual environment, by means of electrical pulses. The main achievement of this work relies in the possibility to investigate a system where a living neuronal network communicates in real-time with an external agent which provides feedback to the network, thus generating electrophysiological patterns that can be associated with different behaviors. The concept was then extended to further hybrid devices that Potter and collaborators called "Hybrots", built by interfacing neuronal networks to other artificial devices (Bakkum et al., 2004; Bakkum et al., 2007). The preliminary results indicated that modifications in neural behaviors may occur as a consequence of the inner characteristics (timing) of the sensory feedback stimulation.

The proposed paradigm was not that far from having a simplified tool for studying "learning", seen as an "experience dependent" process (i.e. adaptive process) where the wiring process of the brain needs to be led by experience and by incoming sensory information.

With these premises, following what have been specifically proposed by our group, in the next sections we present the architecture and the developed closed-loop neuro-robotic system (Fig. 1B), the achieved results, limitations of the proposed paradigm and future prospects. Finally we introduce and discuss some open issues and future visions in the field of neuro-hybrid systems and BMIs.

2. A bioartificial brain bi-directionally connected to a mobile robot

Embodiment has been suggested to be an essential condition for the emergence of 'intelligent' behaviors (Chiel and Beer, 1997). This has been shown by pioneer publications (DeMarse et al., 2001; Shahaf and Marom, 2001; Martinoia et al., 2004), in which experiments have been performed with populations of neurons, micro-electrode arrays (MEAs) and then further extended to an actual neuro-robotic platform (Bakkum et al., 2007; Novellino et al., 2007) up to a neural inspired control in a biomimetic robot (von Twickel et al., 2011) and to a "ratcar" in which a vehicle is moved following electrophysiological signals extracted by the cortical motor area of a rat (Fukayama et al., 2010). Among the possible approaches related to the choice of the robotic platform and of the control system (artificial, bioartificial or biological), there are common issues that can be explored by means of this experimental paradigm, namely: (i) how to convey information to the brain and (ii) how to extract information from the brain. In this framework, we designed an experimental paradigm in which the neuronal cortical culture coupled to a MEA is used as a central processing unit, i.e. the "bioartificial" brain, and it is then connected to an artificial body, i.e. the mobile robot (Fig. 1B). The proposed system acts as a BCI with the general purpose of enabling embodied in vitro experiments on sensorimotor adaptation and learning.

2.1 The bioartificial brain and the electrophysiology recording system

The brain model is an in vitro neuronal network of neurons and glia. In particular, the primary cultures of cortical neurons were prepared from fetal day 18 Wistar SPF rats according to previously described procedures (Chiappalone et al., 2006). Briefly, the cortex was dissociated through enzymatic and mechanical dissociation (0.125% Trypsin - DNAse I 0,025mg/ml - BSA 0.3% solution in HBSS without calcium and magnesium). The cells were seeded on standard and ad-hoc (Berdondini et al., 2005) 60-electrode TiN-SiN MEA chips with internal reference (Multi Channel Systems, Reutlingen, Germany) pre-coated with poly-D-lysine and laminin (0.1 mg/ml diluted in sterile MilliQ water) as 50 μl droplets (1500-2000 cells/mm^2), with subsequent addition of 1 ml of medium after the cells were attached (approximately 2 hours). Cultures were maintained in neurobasal medium (NB) supplemented with 2% B27 and 1% Glutamax-I, and half volume of the medium was exchanged once a week. Cells were maintained at 37°C in a humidified atmosphere of 5% CO_2 until their use.

In these conditions, neuronal cells spontaneously grow by extending their dendritic and axonal arborisations re-establishing synapses. Within a few days of in vitro culture, the neurons already form a functionally connected network reaching the mature state in three weeks (Novellino and Zaldivar, 2010; Hogberg et al., 2011). The neuronal networks coupled to MEA forms a long-term non-destructive two-way interface with the cultured neuronal tissue; and under healthy conditions it is possible to maintain such cultures for months (Potter and DeMarse, 2001).

Experiments were performed in the range 18–42 DIVs, when the neuronal network reaches its "mature" state (Fig. 2), consisting of synchronized clustered activity with minute-to-minute fluctuations in the probability of firing (Marom and Shahaf, 2002).

If a network is electrically stimulated, the neuronal dynamics can deeply change (Novellino et al., 2003); however when weak stimuli are applied, the network shows dynamics whose spatio-temporal features are reproducible to a good degree (Jimbo et al., 2000; Vajda et al., 2008). For this reason, a very critical issue in neuroscience is the definition of a procedure able to induce pathway-specific synaptic changes at the network level either in the direction of potentiation or depression. In literature there are only a few examples of plasticity on cultured networks of cortical neurons (Jimbo et al., 1999; Tateno and Jimbo, 1999; Shahaf and Marom, 2001). More recently, our group demonstrated that a 'network potentiation' could be achieved by using a novel experimental protocol of electrical stimulation (Chiappalone et al., 2008). A high frequency sequence of bursts (internal frequency 20Hz) called 'tetanus' is delivered from one site of the MEA, coupled to a low frequency stimulation, either in-phase or out phase with respect of the tetanic burst, delivered from another electrode called 'co-activation' site.

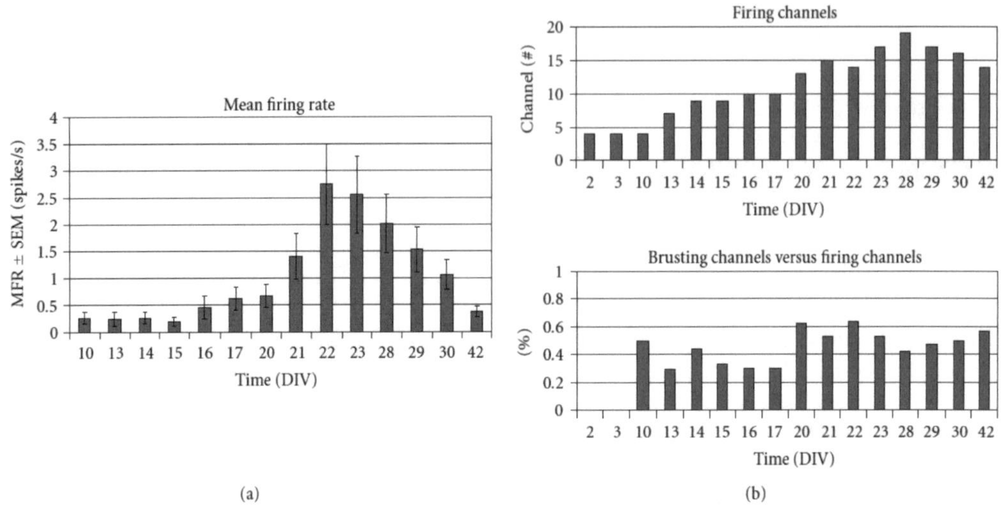

Fig. 2. The mean firing rate (MFR) reports the average spike number during the investigation period (i.e., 10 minutes) and it is an indicator of the spontaneous neuronal activity. Random spiking activity can be recorded at a very early stage of the network development, that is, DIV 2 or 3. The network is still immature and spiking is slow, sporadic, and sparse (i.e., active channels are likely to be far and unsynchronized). After one week of in vitro culture (e.g., DIV10) it is possible to record more frequent activity. The network reaches its maturity at the third week in vitro when activity reaches its peak, and the maximum number of active channels is recorded (top right), and then it starts decreasing. Burst behavior appears during the second week of culture (bottom right). We monitored the network behaviour up to DIV 42.

Experiments regarding the neuronal network electrical stimulation provide a lot of information about the possible functional states in which the network can fall in response to a well-defined stimulation. To mimic this natural morphological condition a specifically designed chip for studying in vitro neuronal network dynamic in interconnected sub-populations of neurons has been used. In particular the device presents physical barriers that induce a minimal constraint of the network in sub-populations and these clusters are interconnected via integrated microchannels (Berdondini et al., 2005). The described devices were demonstrated to guarantee a spontaneous synchronized global behavior and at the same time the sub-population of the stimulated cluster presents a higher probability of evoked response without preventing the stimulus propagation and activation of other sub-populations (Berdondini et al., 2005).

In order to record activity, the neuronal network has been coupled to the MEA60 system supplied by Multi Channel Systems (MCS, Reutlingen, Germany). Briefly the experimental setup is constituted by the following elements (Figure 1.B): a neuronal preparation cultured over an MEA, a mounting support with a 60-channel amplifier (gain 1200x), a homemade 8-channel stimulus generator, to deliver both current and voltage desired stimulating signals, an inverted optical microscope, connected to a CCD camera (Hamamatsu, Japan), to visually monitor the cells during the experiment, an antivibration table and a Faraday cage (for further details see Novellino et al. 2007).

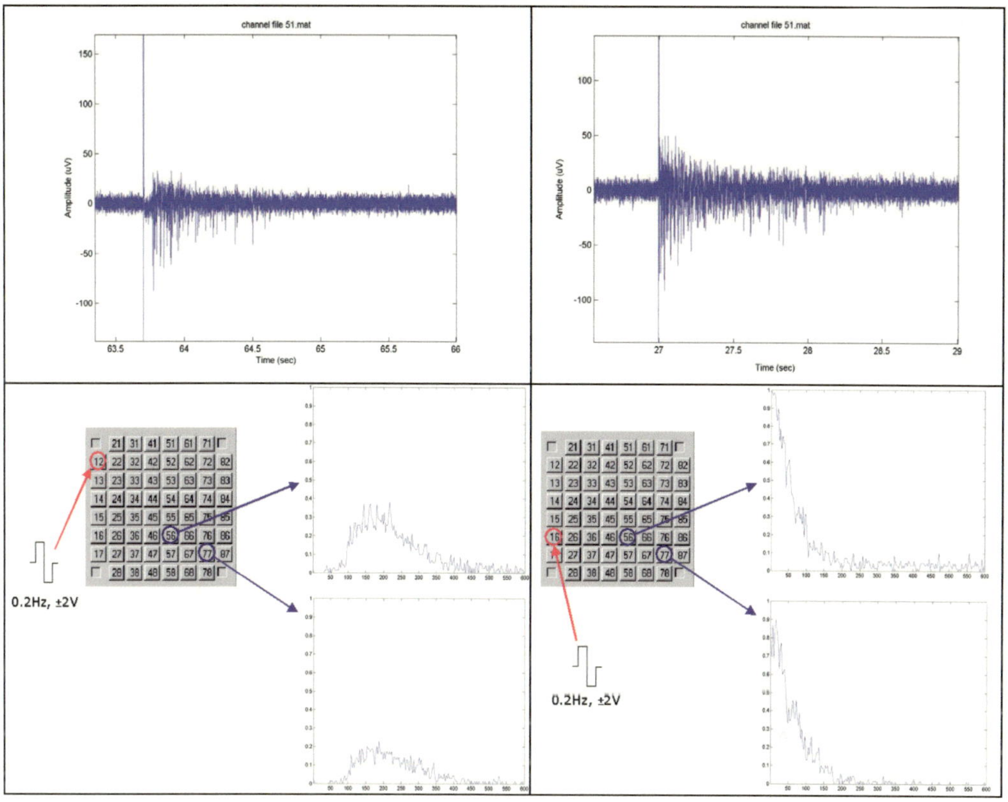

Fig. 3. Example of the neuronal activity recorded while applying an electrical stimulation to cell preparation. Top left, Zoom of the response after a stimulus: a "delayed" burst on channel 51 is evoked by the stimulus from channel 12. Top right, An early burst on channel 51 evoked by the stimulus from channel 72. Bottom left, Comparison between the PSTHs built by means of the activity arising from channels 56 and 77 after stimulation from electrode 12. The shape of the PSTH is the same on the two considered channels ("delayed response"). Bottom right, Comparison between the PSTHs built by means of the activity arising from channels 56 and 77 after stimulation from electrode 16. The shape of the PSTH is the same on the two considered channels ("early response").

The artificial body is a small mobile robot (Khepera II, K-team, http://www.k-team.com) equipped with eight infrared (IR) proximity sensors that provide information about the distance of the robot from obstacles and two wheels whose speeds are determined by neurons activity. These allow movements of the robot inside a circular arena (80 cm diameter), and interaction with several cylindrical obstacles.

2.2 Computational architecture, coding and decoding of information

To specifically implement a demonstration and to carry out experiments in the direction of adaptation in a closed-loop scenario, we developed an experimental set-up in which a bioartificial brain (the neuronal network) and its body (the mobile robot) were challenged to learn by experience a simple 'intelligent' adaptive behavior: to move in an unknown environment avoiding obstacles.

A model of adaptive behavior is represented by an agent who is motivated in trying to survive in a defined environment, without any external (i.e., human) help. The agent may generate its actions exclusively from the available sensory information, or may use some kind of previous "experience." The former type of agent is generally referred as "reactive" (Maes, 1994; Brooks, 2002).

To establish a bidirectional communication between the neuronal preparation and a mobile robot, the electrophysiological signals need to be translated into motor commands for the robot (decoding of neural activity), and at the same time the sensory signal from the robot need to be translated into a pattern of electrical stimulation (coding of sensory information). Figure 1C presents the general computational architecture of the proposed closed-loop system that can be summarized in the following three main logical blocks (i.e., from left to right in Figure 1.C).

1. Coding (i.e. the representation of the external sensory input in terms of electrical stimulation from the robot to the neural network): while the robot freely moves into the playground, its IR sensors see whether or not an obstacle is in the proximity and where it is (left or right side). The IR signals are weighted according to a sensory receptive field law and the two resulting stimulation signals, relative to the right and left eye of the robot, are then coded into a feedback stimulation according to a "binary" information coding (i.e. 0 or 1 Hz). The electrical stimulus is delivered when the sensory feedback overcomes a threshold, corresponding approximately to the presence of an obstacle at 5 cm distance.
2. Processing of electrophysiological signals: the spontaneous or evoked electrophysiological activity is sampled at 10 kHz and processed by means of an event detection algorithm (i.e. spike detection, see further), which also removes stimulation artifacts.
3. Decoding (from the neural preparation to the robot): the processed electrophysiological signals are translated into motor commands for the robot. In particular the instantaneous firing rate is low pass filtered to compute the "neural activity" that is then decoded into motor commands (wheels speed) according to a winner-takes-all (WTA) algorithm (see below).

The motor commands ω(t), that is, the angular speeds of the wheels, are obtained by implementing the following winner-takes-all (WTA) mechanism:

$$\begin{cases} \omega_L(t) = \begin{cases} \left(\omega_0 - \sum_{i=1}^{N} C_i * \left[\hat{r}_i(t)\right]_R\right) & \omega_L \geq \omega_R \\ -\omega_b & \omega_L < \omega_R \end{cases} \\ \omega_R(t) = \begin{cases} \left(\omega_0 - \sum_{i=1}^{N} C_i * \left[\hat{r}_i(t)\right]_L\right) & \omega_R \geq \omega_L \\ -\omega_b & \omega_R < \omega_L \end{cases} \end{cases}$$

where ω_b is a constant angular speed (up to 2 rad/s), ω_0 is the maximum angular speed (i.e., 5 rad/s); $\hat{r}_i(t)$ is the instantaneous firing rate of the recording site i, C_i is a normalization coefficient and represent the inverse of the estimation of the maximum value that can be reached by the instantaneous firing rate on each group (left versus right) of electrodes. L and R indicate signals pertaining respectively to the left and the right wheel.

The control law implements inhibitory crossed connections between inputs and outputs: in absence of neural activity, the robot moves according to the maximum angular speed of both the wheels, whereas an increase in the activity results in a fixed decrease of the speed of the opposite wheel. WTA helps the robot turning away from the obstacle by increasing its rotational speed, on equal evoked response. This allows to better achieve a reactive control task, overcoming the problems caused by the short range of the IR sensors of the robot.

2.3 Data processing and statistics

Spike detection. The electrophysiological signals $y(t)$ acquired from MEA electrodes must be pre-processed in order to remove the stimulus artifact and to isolate spikes from noise. The spike detection algorithm uses a differential peak-to-peak threshold to follow the variability of the signal (Bove et al., 1997). A time window, sized to contain at most one single spike (4ms), is shifted along the signal, sampled at the frequency of 10 kHz. Within the window, when the difference between the maximum and the minimum exceeds the threshold, a spike is found and its time stamp is saved. In this way, the resulting spike train signal is sampled at 250 Hz. The threshold is proportional to the noise standard deviation (SD) and is calculated separately for each individual channel (typically as 7 × SD) before the beginning of the actual experiment.

Blanking of stimulus artifact. Stimulus artifacts are detected when the recorded signal exceeds a second, higher threshold. The artifact is then suppressed by cancelling the first sample in the spike train occurring immediately after it, corresponding to a signal blanking of 4ms after stimulus delivery.

Post-stimulus Time Histogram. To investigate the neural activity evoked by stimulation, we computed the post-stimulus time histogram (i.e., PSTH), which represents the impulse response of each site of the neural preparation to electrical stimulation (see Fig. 3). The PSTHs were calculated by taking 400 ms time windows from the recordings that follow each stimulus. We then counted the number of spikes occurring in a 2-4 ms bin and divided this measure by the number of stimuli (Rieke et al., 1997). For our cultures, typical PSTHs show an "early" (less than 50 ms) and a "late" (50–250 milliseconds) component (Marom and Shahaf, 2002; Cozzi et al., 2006).

Functional Connectivity Index (FCI). The amplitude of the early response to stimuli can be taken as an indicator of the functional connectivity between stimulation and recording site (Cozzi et al., 2006). The PSTH area over 100ms from stimulus averages the early evoked network response and thus it represents the network Functional Connectivity Index (CFI) and helps in differentiating strong and weak functional input/output connection.

Stimulus-Triggered Speed. The stimulus-triggered speed (STS) is constructed by averaging the speed waveform due to each stimulus. The STS indicated the average wheel speed (that is proportional to the neuronal activity of the "brain" region connected to that wheel) in response to each stimulus (sensory feedback). The 4 STS curves (i.e., the variations of the speed of the left and the right wheels in response to the left stimulus are the first two curves and constitute the first STS, and the variation of the speed of the left and the right wheels in response to the right stimulus constitute the second STS with the latter two curves) are a representation of the sensory-motor pathways and also indicate the robot behavior performance and side-selectivity relationship between stimuli and speeds.

The behavior of the neurorobotic system can be evaluated by means of the *stimulus-triggered speed (STS)*, that is, the average linear speed elicited by a single electrical stimulus, and, as already showed (Novellino et al., 2007) this is the best parameters to catch the relationship between electrophysiology and robot performance and its evolution.

2.4 Experimental procedure

The experimental protocol consisted of the following steps:

1. baseline:

a. spontaneous activity recording (5 minutes);

b. pre-processing: test stimulus from 8 channels (serial stimulation). To test the response to stimulation from different sites in different areas of the neuronal network, trains of 50 electrical stimuli are delivered (±750mV amplitude, 500μs duration, and duty cycle 50%).

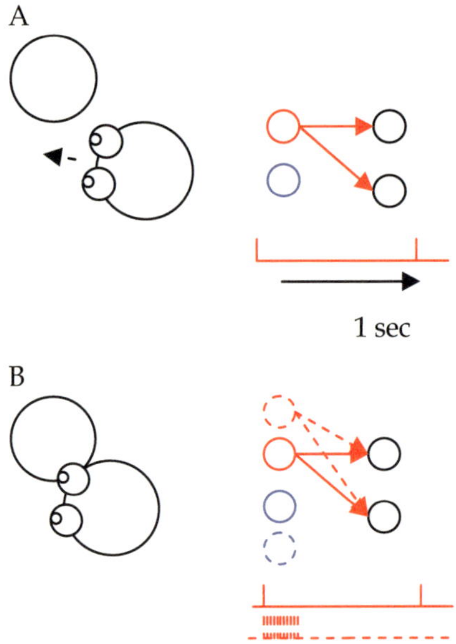

Fig. 4. Schematic representation of the sensory feedback stimulation as well as the conditioning stimulation protocol. A. Let's assume the robot is approaching the obstacle. When the robot is close enough to the obstacle, the robot's eyes detect the obstacle and the sensory feedback (1Hz) is delivered from the input neurons (red circle) to evoke activity in the output neurons (black neurons). If the robot bumps into the obstacle we can have two different stimulations: if the system is running the "non-conditioning phase" of phase 3.a and 3.c, the stimulation is only delivered from the input neuron, as shown in this panel. B. If the system is running the conditioning phase (i.e. the experimental phase is 3.b), and the robot bumps into the obstacle the co-activation neurons (i.e. red dashed circle) are enabled to deliver the "in-phase tetanic stimulation" in parallel with the normal feedback stimulation (bottom) to enhance the robot reactive behavior performance for obstacle avoidance.

2. input-output mapping:
Reactive behavior means the robot is able to immediately react to the presence of an obstacle where immediately means in a time shorter than 100ms, that is, we need input-output pathways characterized by a relatively early (up to 50 ms) and sustained response meaning a "high strength" in the functional connectivity (Novellino 2007). Practically at least 2 channels for input (sensors), 2 channels for conditioning input (sensors), and 2 channels for output (motors) were chosen according the stimuli induced activity and "selectivity maps".
3. closed-loop experiment: Robot running (5 + 5 + 5minutes):
a. free running;
b. obstacle avoidance with the application of the in-phase tetanic protocol (when the robot hits an obstacle, a conditioning stimulus is delivered from the collision side), according to our previous studies (Chiappalone et al., 2008).
c. free running.
4. Input/output mapping checking:
a. spontaneous activity recording (5minutes);
b. test stimulus from the chosen stimulating channels.

To avoid manual removal of the robot and possible damage due to wheels' motors heating in case of a collision against an obstacle, a step-back mechanism has been enabled. Figure 4 schematizes the 3rd step.

3. Results and discussion

3.1 Spontaneous activity requirements

We applied a quite restrictive selection of the experimental chips. In particular we only selected mature cells (i.e. after the third week in vitro) which presented synchronized native bursting activity (minimal bursting rate 0.2Hz) all over the network. I/O pathways (and co-activation sites) selection consisted in serial delivery of 50 stimuli from at least 7 different sites (i.e. 350 stimuli).

3.2 Identification of input-output sites

In order to obtain a reactive behavior, we need the network to promptly respond after the feedback stimulation, that is, we need input-output pathways characterized by a relatively early (up to 50 ms) and sustained response meaning a "high strength" in the functional connectivity. If the network reacts to the sensory feedback and the evoked electrophysiological response is characterized by a relatively long activation phase (up to 200–300 ms), the robot would not be able to react to the presence of an obstacle in 100 ms (i.e., the delay among successive serial communications between the system and the robot). This is one of the reasons why we need to accurately select the input-output pathways. We need the stimulus-evoked response to be fast, prolonged, reliable, and therefore effective for the entire duration of the experiments (i.e., all day long). As already said, the general aim is to have a robot that follows a specific task on the basis of the spontaneous/stimulated electrophysiological activity shown by the neuronal culture. To this end, it is a fundamental prerequisite to characterize the collective activity of the network that will be connected to the robot (i.e., analysis of both spontaneous and stimulus evoked neuronal activity).
The post-stimulus time histogram (PSTH) (i.e., the average number of spikes obtained in response to a stimulus, at different latencies) is then used for quantifying the strength of connections between a specific stimulating sites and all the other recording sites (Figure 4).

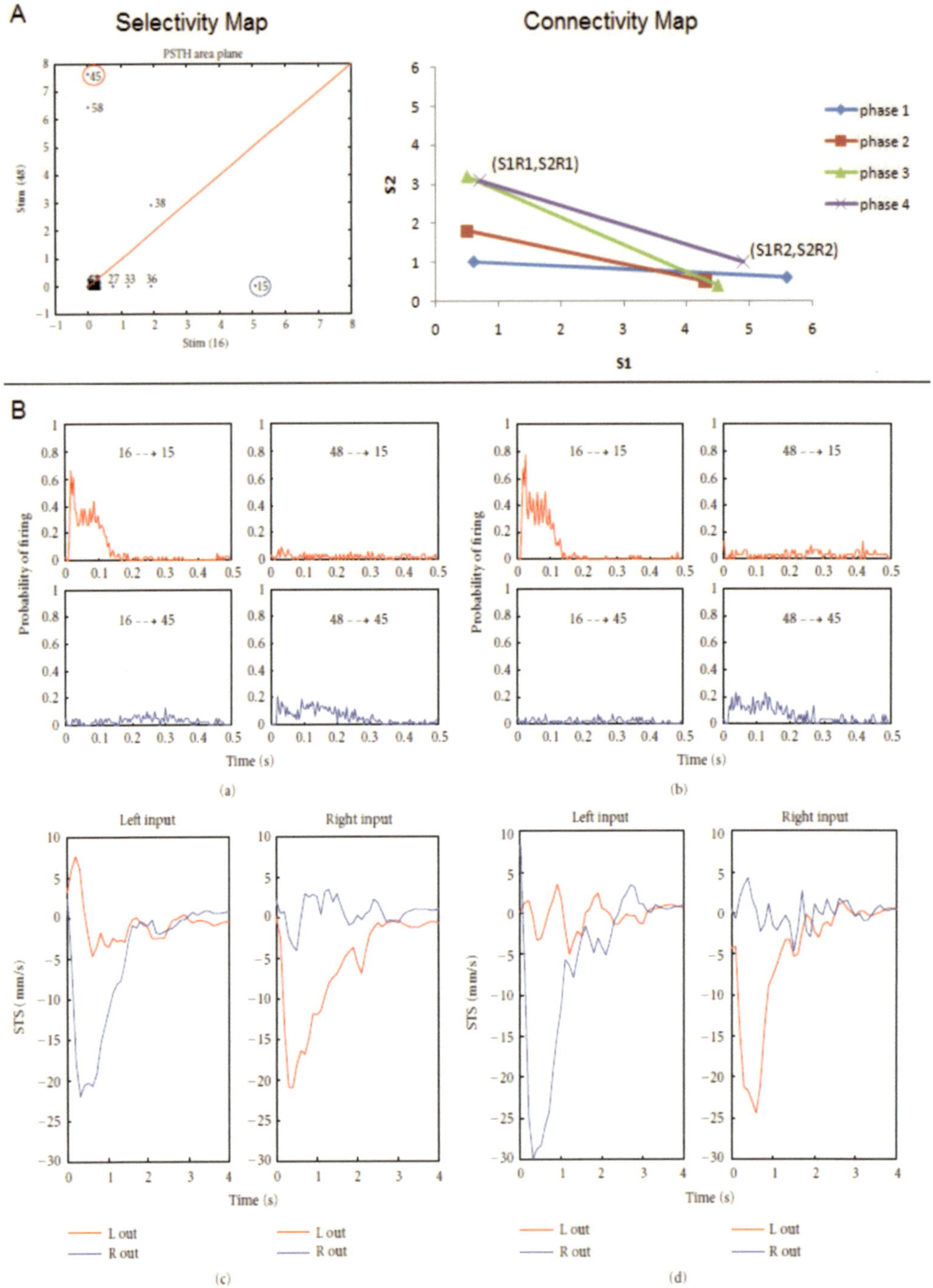

Fig. 5. **A)** (left) Selectivity Map for a representative experiment: each point represents the evoked activity in a channel from the two stimulation channels, i.e. x-coordinate is the PSTH area in response to electrical stimulation delivered from Stim 16 (S1), and y-coordinate is is the PSTH area in response to electrical stimulation delivered from Stim 48 (S2). The circled

channels are the channels that answer with the best selectivity. (right) Connectivity Map for a representative experiment: displacements of the right and left recording channels during the four acquisition phases in closed-loop on the basis of their PSTH areas with respect to the right and left stimulating channels. Phase 1 indicates the first 'unconditioned' phase. **B)** PSTH and STS in a neurorobotic experiment. (a) PSTHs for two electrodes (chosen as recording–motor electrodes) with respect to two stimulating sites. Electrode 15 responds well to stimulation from electrode 16 and poorly to stimulation from electrode 48; on the contrary electrode 45 responds well to 48 and poorly to 16. This tendency is maintained and even improved after the robotic experiment (b). The STS graphs before (c) and after (d) the robotic experiment proves again the selectivity of the chosen electrodes and the improvement in the performances (increased STS area).

It is the impulse response (in terms of instantaneous firing rate) to a single test stimulus. The algorithm for the selection of the output (motor) and input (sensory) sites supplies the I/O pairs corresponding to maximum selectivity and it is based on network effective functional interconnectivity, i.e. Selectivity Map, then it is used to check the evolution of the selected I/O pairs, i.e. Connectivity Map (Figure 5A).

The ideal case is described in the following: given two (or more) stimulating channels (e.g., S1 and S2) and two groups of recording sites (e.g., R1 and R2), the strength of the connectivity S1-R1 and S2-R2 is "high" and simultaneously, the strength of the connectivity S1-R2, and S2-R1 is "low" (i.e., good selectivity in input-output pairs). The described scheme guarantees, somehow, that the responses in the two (groups of) recording sites are different on the basis of the stimulating electrodes. Of course the above is an ideal situation and, since the mean connectivity of the network is high, also due to the high density of plated cells, it is hard to get highly specific responses in the input-output pathways.

The methodology that we developed to make a selection of the pathways is the "selectivity map". Each dot represents the PSTH area at a specific recording site given that there was a stimulation from a couple of stimulating sites (Fig. 5A). All the possible input-output combinations are explored and only the pathways producing responses lying more distant from the diagonal (i.e., closer to the axis) are selected. Those specific pathways (of sensory-motor activations) can be then conveniently utilized for driving the robot and for implementing simple reactive behaviors (e.g., obstacle avoidance), as presented in previous sections.

3.3 The robot behavior

Figure 6 shows an example of the robot behavior during three phases of the experiment. The mobile robot starting position was indicated by a green circle, robot position was sampled at 0.5Hz. The robot was moving inside its playground and occasionally it hit either obstacles or playground walls (red dots). During the learning phase when the mobile robot bumped into any obstacle the punishing-conditioning feedback stimulation (11pulses at 20Hz) from the co-activation site was delivered. The robot performed its explorative-obstacle avoidance reactive behavior into the playground for 10 minutes for any phase (see Methods).

Robot behavior performance parameters, such as number of hits, covered space etc, do not allow quantifying any relationship between the motor response and the sensory information, but, considering different phases, if the robot covered the same area and the trajectories are in the same order of length, then the two phases are comparable. A reduction of the number of hits should indicate an improvement of the robot's behavior; anyhow any improvement in the robot's behavior must correspond to an improvement in the relationship

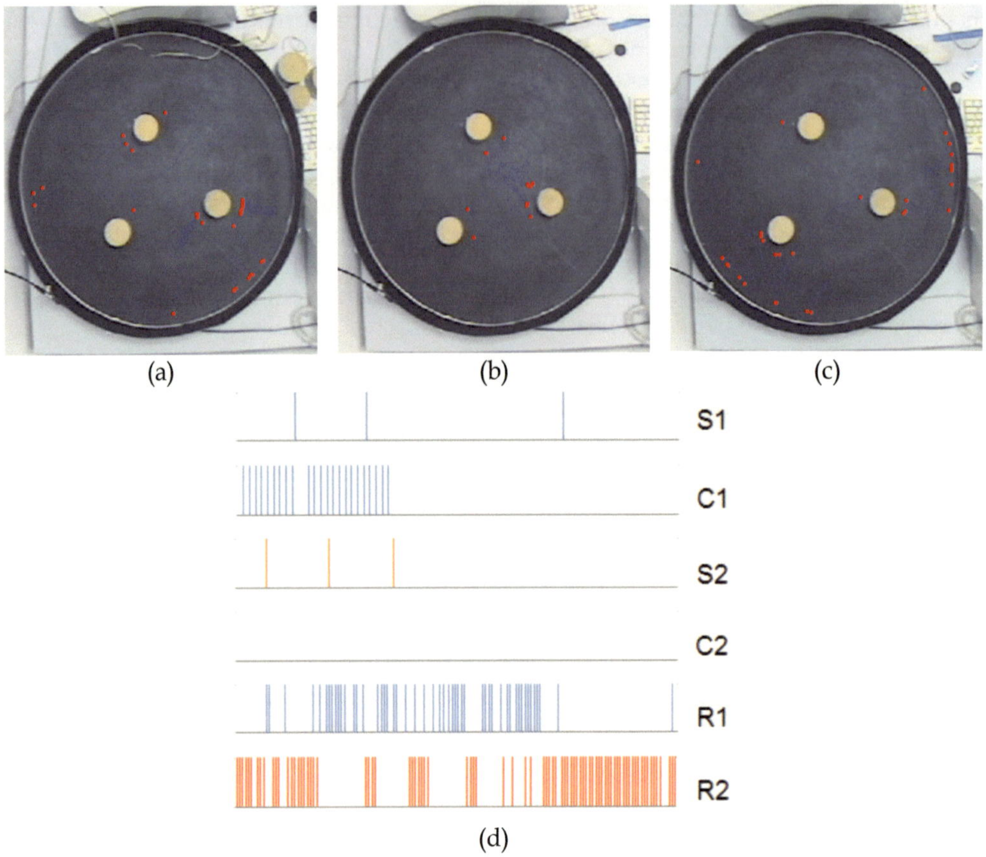

Fig. 6. Robot trajectories and performances in a neurorobotic experiment. (a) Robot trajectory during the first free running phase. (b) Robot trajectory during the learning phase. (c) Robot trajectory during the last free running phase. (d) From top: right feedback stimulus, right co-activation stimulus, left stimulus, left co-activation stimulus, two right output channels and two left channels. The window is 5 sec. It is possible to note that according to the WTA method (see Methods) only the right co-activation stimulus was delivered.

between the motor response and the feedback sensory information (i.e., the STSs). Furthermore the shapes of the PSTHs must be similar to those of the PSTHs obtained during the characterization phase in order to ensure the stability of the response of the neuronal culture. If the area of the PSTHs drastically decreases at the end of the closed loop phases it means that the neuronal network has been fatigued by excessive repeated stimulations (Shahaf and Marom, 2001). The wellness and stability of the culture are "sine qua non" conditions to be verified before describing the neurorobotic behavior by means of the robot's performance indicators. Under these conditions, the performance of the neural preparation in controlling the robot can be represented by the STSs, depicted in Figures 13(c) and 13(d). The STS is the only parameter that permits to understand and demonstrate whether a different behavior of the robot actually corresponds to a different dynamics of the neuronal activity, and for this reason it can be considered the best indicator of the performance of the overall neurorobotic system. The comparison of the STSs and connectivity maps obtained during each phase illustrates that a modification in the robot's behavior has occurred.

4. Conclusions and perspectives

The possibility to control actions from thoughts is becoming reality thanks to recent advances in Brain Machine Interfaces (Donoghue, 2008). However modern BMIs are mostly unidirectional (i.e. without providing sensory information to the subject) and aimed at restoring lost motor functions at the level of the Peripheral Nervous System. In a wider perspective, neural interfaces must be bi-directional devices that substitute motor, sensory or cognitive circuits within the brain, which might have been damaged as a result of an injury or a disease. In this context it becomes crucial to gain the necessary knowledge to be able to efficiently interact with the nervous system to realize new class of neuroprostheses and BMIs aimed at treating those diseases where a portion of CNS or PNS is damaged.

An in-vitro bi-directional closed-loop neuro-robotic interface, as the one we presented here, goes in the direction of elucidating and optimizing means to interact with the nervous system. Using simplified experiment models, the proposed approach aims at shedding some light into the mechanisms of information coding and of efficient bi-directional interaction with external artificial devices. At the same time, the proposed system has indeed some important biological limitations, such as: lacking of a specific structure-architecture (as in brain slices); lacking of any 3D structure (as the layers in a cortical brain region). One way to overcome this limit relies in the possibility to 'confine' the network in interacting cell sub-populations (Berdondini et al., 2005) that can behave as dynamically connected cell assemblies, as done by the cortical areas within the brain or to move to 3D neuronal culture (Pautot et al., 2008). Indeed, all that can be usefully tested and studied in such a simplified experimental framework then need to be translated to an in-vivo situation in which an external devices is connected to specific areas of the CNS. With specific reference to the results reported here, still there are things that need to be further addressed. The understanding of input/output mapping is fundamental and site selections is still representing a major factor of the overall system performance. Successful modification and improvement of robot behavior was recorded for 30% of performed recorded experiments. 90% of the unsuccessful experiments (i.e., 70%) did not even satisfy the preliminary requirements for input to output connection/correlation, meaning that a deeper understanding of how input-output mapping can be wired and re-wired has still to be fully understood.

In spite of the discussed limitations, the finalized experiments represent a fundamental step towards a better understanding of the biological principles of information processing that can be obtained by multi-site parallel recordings and multi-modal stimulation. Such information will be extremely valuable for developing our understanding of how real neuronal networks wire up, process information and change their connectivity (i.e. plasticity). More specifically, the use of a bi-directional neuro-robotic system allows to focus along three main research lines: (i-IN/OUT transformation, transfer function) investigate network dynamics and derive the input-output function of neuronal assemblies; (ii-CODING and DECODING) design new strategies for effectively sending appropriate information to cell assemblies, by developing stimulation protocols, and for efficiently extracting information from those assemblies; (iii-PROCESSING) derive the coding and information transmission properties of neuronal assemblies by designing computational

models which might substitute the dynamic behavior of neuronal population under observation (test).
In the field of BMI and neuroprosthesis, focused investigations by using a neuro-robotic paradigm are expected to give a relevant boost to the development of next generation of devices. Particularly, a significant improvement over current BMIs is expected from the newly gained capability to efficiently extract information from stimulus evoked activity, which will provide new algorithms and tools (i.e. optimized decoding scheme).
A second aspect relies directly in the bi-directional interface that plays a central role in the entire embodied system and from which the capability to optimally code information will constitute the basic of the new interfaces (i.e. optimized coding scheme) when missing sensory information has to be conveyed to the brain.
Finally, the paradigm of an artificially embodied network can provide a new investigation tool in the field of neuroscience where plasticity, dynamics and functional architecture can be studied in a context in which the brain tissue is not isolated but benefits of a direct interaction with the external environment. As a whole, this framework of study mainly targeting advanced BMI and neuroprosthetic applications is aimed at offering a future hope for potential therapy for the rehabilitation of severely impaired patients but also an useful platform to test various ideas on how population of neurons encode information in the brain.

5. References

Andersen RA, Musallam S, Pesaran B (2004) Selecting the signals for a brain-machine interface. Current Opinion in Neurobiology 14:720-726.

Bakkum DJ, Gamblen PM, Ben-Ary G, Chao ZC, Potter SM (2007) MEART: The Semi-Living Artist. Frontiers in Neurorobotics 1:5.

Bakkum DJ, Shkolnik AC, Ben-Ary G, Gamblen P, DeMarse TB, Potter SM (2004) Removing some 'A' from AI: Embodied cultured networks. Lect Notes Artif Int 3139:130-145.

Berdondini L, Chiappalone M, van der Wal PD, Imfeld K, de Rooij NF, Koudelka-Hep M, Tedesco M, Martinoia S, van Pelt J, Le Masson G, Garenne A (2005) A microelectrode array (MEA) integrated with clustering structures for investigating in vitro neurodynamics in confined interconnected sub-populations of neurons. Sensors and Actuators B: Chemical 114:530-541.

Bove M, Grattarola M, Verreschi G (1997) In vitro 2D networks of neurons characterized by processing the signals recorded with a planar microtransducer array. IEEE Transactions on Biomedical Engineering 44:964-977.

Brooks RA (2002) Robot: the future of flesh and machines. London, UK: The Penguin Press.

Chapin JK, Moxon KA, Markowitz RS, Nicolelis MAL (1999) Real-time control of a robot arm using simultaneously recorded neurons in the motor cortex. Nature Neuroscience 2:664-670.

Chiappalone M, Massobrio P, Martinoia S (2008) Network plasticity in cortical assemblies. European Journal of Neuroscience 28:221-237.

Chiappalone M, Bove M, Vato A, Tedesco M, Martinoia S (2006) Dissociated cortical networks show spontaneously correlated activity patterns during in vitro development. Brain Research 1093:41-53.

Chiappalone M, Vato A, Berdondini L, Koudelka-Hep M, Martinoia S (2007) Network dynamics and synchronous activity in cultured cortical neurons. International Journal of Neural Systems 17:87-103.

Chiel HJ, Beer RD (1997) The brain has a body: adaptive behavior emerges from interactions of nervous system, body and environment. Trends in Neurosciences 20:553-557.

Cozzi L, D'Angelo P, Sanguineti V (2006) Encoding of time-varying stimuli in population of cultured neurons. Biological Cybernetics 94:335-349.

Daly JJ, Wolpaw JR (2008) Brain-computer interfaces in neurological rehabilitation. Lancet Neurology 7:1032-1043.

DeMarse TB, Wagenaar DA, Blau AW, Potter SM (2001) The neurally controlled animat: biological brains acting with simulated bodies. Auton Robot 11:305-310.

Donoghue JP (2008) Bridging the brain to the world: a perspective on neural interface system. Neuron 60:511-521.

Fukayama O, Suzuki T, Mabuchi K (2010) RatCar: A vehicular neuro-robotic platform for a rat with a sustaining structure of the rat body under the vehicle. In: 32nd Annual International Conference of the IEEE EMBS, pp 4168-4171. Buenos Aires, Argentina: IEEE.

Hatsopoulos NG, Donoghue JP (2009) The science of neural interface systems. Annual Review of Neuroscience 32:249-266.

Hochberg LR, Serruya MD, Friehs GM, Mukand JA, Saleh M, Caplan AH, Branner A, Chen D, Penn RD, Donoghue JP (2006) Neuronal ensemble control of prosthetic devices by a human with tetraplegia. Nature 442:164-171.

Hogberg HT, Sobanski T, Novellino A, Whelan M, Weiss DG, Bal-Price AK (2011) Application of micro-electrode arrays (MEAs) as an emerging technology for developmental neurotoxicity: evaluation of domoic acid-induced effects in primary cultures of rat cortical neurons. Neurotoxicology 32:158-168.

Jimbo Y, Tateno Y, Robinson HPC (1999) Simultaneous induction of pathway-specific potentiation and depression in networks of cortical neurons. Biophysical Journal 76:670-678.

Jimbo Y, Kawana A, Parodi P, Torre V (2000) The dynamics of a neuronal culture of dissociated cortical neurons of neonatal rats. Biological Cybernetics 83:1-20.

Kositsky M, Chiappalone M, Alford ST, Mussa-Ivaldi FA (2009) Brain-Machine Interactions for Assessing the Dynamics of Neural Systems. Front Neurorobotics 3:1.

le Feber J, Stegenga J, Rutten WL (2010) The effect of slow electrical stimuli to achieve learning in cultured networks of rat cortical neurons. PLoS One 5:e8871.

Maes P (1994) Modeling adaptive autonomous agents. Artificial Life 1:135-162.

Marom S, Shahaf G (2002) Development, learning and memory in large random networks of cortical neurons: lessons beyond anatomy. Quarterly Reviews of Biophysics 35:63-87.

Martinoia S, Sanguineti V, Cozzi L, Berdondini L, Van Pelt J, Tomas J, Le Masson G, Davide FA (2004) Towards an embodied in-vitro electrophysiology: the NeuroBIT project. Neurocomputing 58-60:1065-1072.

Mussa-Ivaldi FA, Miller LE (2003) Brain-machine interfaces: computational demands and clinical needs meet basic neuroscience. Trends Neurosci 26:329-334.

Nicolelis MA (2003) Brain-machine interfaces to restore motor function and probe neural circuits. Nat Rev Neurosci 4:417-422.

Nicolelis MAL, Chapin JK (2002) Controlling robots with the mind. Scientific American 287:46-53.

Nicolelis MAL, Lebedev MA (2009) Principles of neural ensemble physiology underlying the operation of brain-machine interfaces. Nature Reviews Neuroscience 10:530-540.

Novellino A, Zaldivar J-M (2010) Recurrence quantification analysis of spontaneous electrophysiological activity during development: characterization of In vitro neuronal networks cultured on multi electrode array chips. Advances in Artificial Intelligence Volume 2010:10 pages.

Novellino A, Chiappalone M, Vato A, Bove M, Tedesco M, Martinoia S (2003) Behaviors from an electrically stimulated spinal cord neuronal network cultured on microelectrode arrays. Neurocomputing 52-54C:661-669.

Novellino A, D'Angelo P, Cozzi L, Chiappalone M, Sanguineti V, Martinoia S (2007) Connecting neurons to a mobile Robot: an in vitro bidirectional neural interface. Computational Intelligence and Neuroscience 2007:13 pages.

Pautot S, Wyart C, Isacoff EY (2008) Colloid-guided assembly of oriented 3D neuronal networks. Nature Methods 8:735-40.

Potter SM, DeMarse TB (2001) A new approach to neural cell culture for long-term studies. Journal of Neuroscience Methods 110:17-24.

Reger BD, Fleming KM, Sanguineti V, Alford S, Mussa-Ivaldi FA (2000) Connecting brains to robots: an artificial body for studying the computational properties of neural tissues. Artificial Life 6:307-324.

Rieke F, Warland D, de Ruyter van Steveninck R, Bialek W (1997) Spikes: exploring the neural code. Cambridge, Massachusetts: The MIT Press.

Schwartz AB (2004) Cortical neural prosthetics. Annual Review of Neuroscience 27:487-507.

Shahaf G, Marom S (2001) Learning in networks of cortical neurons. J Neurosci 21:8782-8788.

Tateno T, Jimbo Y (1999) Activity-dependent enhancement in the reliability of correlated spike timings in cultured cortical neurons. Biological Cybernetics 80:45-55.

Taylor DM, Tillery SI, Schwartz AB (2002) Direct cortical control of 3D neuroprosthetic devices. Science 296:1829-1832.

Vajda I, van Pelt J, Wolters P, Chiappalone M, Martinoia S, van Someren E, van OOyen A (2008) Low-frequency stimulation induces stable transitions in stereotypical activity in cortical networks. Biophysical Journal 94:5028-5039.

Velliste M, Perel S, Spalding MC, Whitford AS, Schwartz AB (2008) Cortical control of a prosthetic arm for self-feeding. Nature 453:1098-1101.

von Twickel A, Büschges A, Pasemann F (2011) Deriving neural network controllers from neuro-biological data:implementation of a single-leg stick insect controller. Biological Cybernetics 104:95-119.

Wagenaar DA, Madhavan R, Pine J, Potter SM (2005) Controlling bursting in cortical cultures with closed-loop multi-electrode stimulation. J Neurosci 25:680-688.

Part 3

Navigation and Path Planning

8

Parallel Computing in Mobile Robotics for RISE

Janusz Będkowski
Institute of Automation and Robotics, Warsaw University of Technology
Industrial Research Institute for Automation and Measurements
Poland

1. Introduction

RISE – Risky Intervention and Surveillance Environment is very demanding task for mobile robot where time is crucial. It can be assumed that on-line task execution is very important, therefore the research in parallel computing applied in mobile robotics is needed. Nowadays many researchers are focused on such approaches that uses GPGPU (General Purpose Graphics Processing Unit) for improvement State of The Art (SoA) algorithms. In this chapter three areas of research are shown such as 3D data registration, robot navigation and 3D cloud of points processing for normal vector computation, all are improved by GPGPU computation. The on – line data registration problem is discussed. The approach based on robust KNN k-nearest neighborhood search applied for improvement of ICP algorithm is shown. The path planning parallel approach based on modified diffusion method is shown. On – line 3D cloud of points' segmentation based on normal vector computation is presented. The set of proposed algorithms where tested on GPGPU NVIDIA CUDA GF 580, the results are satisfying. The improvement of SoA algorithms based on CUDA implementation shows on-line advantages during real task execution.

Robust ICP algorithm is needed in mobile robotics applications where data registration has to be performed on–line. New generation of Time of Flight and RGB-D cameras will offer better accuracy and resolution, therefore GPU accelerated data registration algorithms will improve robot navigation, obstacle avoidance and map building. It can be stated that commercial 3D scanners based on rotated lasers offer data acquisition time < 3 seconds, therefore ICP that works in this time will be enough for on-line map building in stop-scan fashion. For this reason robust data registration algorithm based on 3D space decomposition is proposed and the ICP (Iterative Closest Point) approach is chosen as registration method. Algorithm in current version performs matching of two cloud of points up to $512 \times 512 = 262144$ in 300ms for 30 ICP iterations (NVIDIA GF 580). The proposed solution is efficient since it performs nearest neighbor search using a bucket data structure (sorted using CUDA primitive) and computes the correlation matrix using parallel CUDA all-prefix-sum instruction. The amount of processed points can be increased by implementation on NVIDIA GPU with Compute Capability 2.1.

Robot navigation is very important task that can be improved using parallel computation since robot environment is represented by grid of cells. This report focuses on graphics processing units (GPUs) in particular applied for modified diffusion method. The experiments performed with 512 × 512 pixels grid maps show an advantage of GPU that results on−line path planning in static environment that can be expanded by dynamic obstacles.
Fast data segmentation technique based on local neighborhood search is implemented. The result of segmentation is colored map where different colors correspond to different objects such as {wall, floor...}. The research is related to the problem of collecting 3D data with DGB-D camera mounted on rotated head and 3D laser scanner, to be used in mobile robot applications. Assumed performance of data registration algorithm is achieved, therefore it can used as on−line. Robust nearest neighbor search procedure is obtaining normal vectors for each range point. Normal vectors are represented as r,g,b values, therefore similar colors correspond to one plane.
The experiments where performed using different type of 3D sensors including PMD Photonic Mixer Devices camera, X-BOX 360 Kinect sensor and rotated LMS SICK 200 (3DLSN Unit). Results are satisfying and it is planned to continue research and expand into another areas from mobile robotics such as 6D SLAM and semantic mapping.
The paper is organized as follows: in *Related work* the State of The Art concerning parallel computing applied in mobile robotics is described. In section *3D data registration* the ICP Iterative Closest Point algorithm improved by parallel KNN (k-nearest neighborhood search) approach is shown. Section *Path planning* is demonstrating parallel approach for robot navigation purpose. In *Cloud of points processing* section the implementation of parallel normal vector computation for 3D cloud of points segmentation is described. In *Experiments* the results are shown. *Conclusion and future work* finalize the paper.

2. Related work

Several researches of 3D mapping are based on the simulation of 3D laser range finder to obtain 3D cloud of points Magnusson et al. (2007). In most cases 3D laser simulator is built on the basis of a rotated 2D range finder. The rotation axis can be horizontal Nüchter et al. (2003), vertical Montemerlo & Thrun (2004) or the rotational axis in the middle of the scanner's field of view Kohlhepp et al. (2004). Another approach of obtaining 3D cloud of points using 2 orthogonal lasers is shown in the work of Thrun Thrun et al. (2000). The applications are related with urban mapping Ortega et al. (2009).
Alignment and merging of two 3D scans, which are obtained from different sensor coordinates, with respect to a reference coordinate system is called 3D registration Huber & Hebert (2003) Fitzgibbon (2001) Magnusson & Duckett (2005). Park Park et al. (2010) proposed a real-time approach for 3D registration using GPU, where the registration technique is based on the Iterative Projection Point (IPP) algorithm. IPP technique is a combination of point-to-plane and point-to-projection registration schemes Park & Subbarao (2003). Processing time for this approach is about $60ms$ for aligning 2 3D data sets of 76800 points during 30 iterations of the IPP algorithm. Fast searching algorithms such as the k-d tree algorithm are usually used to improve the performance of the closest point search Nuchter, Lingemann & Hertzberg (2007) Rusinkiewicz & Levoy (2001). GPU accelerated nearest neighbor search for 3D registration is proposed in work of Qiu Qiu et al. (2009), where the advantage of Arya's priority search algorithm described in Arya & Mount (1993) to fit

NNS in the SIMD (Single Instruction Multiple Data) model was used for GPU acceleration purpose. Purcell suggested that k-d tree and priority queue methods are efficient but difficult to be implemented on GPU Purcell et al. (2003). Garcia proves, that a brute force NNS approach using NVidia Compute Unified Device Architecture (CUDA) is 400 times faster over the CPU k-d tree implementation Garcia et al. (2008). GPU-based NNS with advanced search structures is also used in the context of ray tracing Foley & Sugerman (2005), where NNS procedure builds trees with a different manner from a triangle soup, and takes these triangles as the objects of interest. To convert k-d tree into serialized flat array that can be easily loaded into CUDA device, left-balanced k-d tree was proposed Jensen (2001) Qiu et al. (2009). Another technique for 3D registration using Fast Point Feature Histograms (FPFH) is shown in the work of Rusu Rusu et al. (2009). Rusu also proposed a way of characterizing the local geometry of 3D points, using persistent feature histograms, where the relationships between the neighbors of a point are analyzed and the resulted values are stored in a 16-bin histogram Rusu et al. (2008). The histograms are pose and point cloud density invariant and cope well with noisy datasets. An alternative concept to ICP algorithm which relies on instantaneous kinematics and on the geometry of the squared distance function of a surface is shown in the work of Pottmann Pottmann et al. (2002). The proposed algorithm exhibits faster convergence than ICP, which is supported both by results of a local convergence analysis and by experiments. The ICP algorithm is used in SLAM 6D (Simultaneous Localization and Mapping), where 6 DOF (Degree of freedom) of robot position is computed based on aliment of 3D clouds of points and loop-closing technique Sprickerhof et al. (2009).

3. 3D data registration

Classic Iterative Closest Points algorithm ICP was improved using GPGPU computation to obtain on-line execution. The implementation performs nearest neighbor search using a bucket data structure (sorted using CUDA primitive) and computes the correlation matrix using parallel CUDA all-prefix-sum instruction.

3.1 GPU architecture

NVIDIA GPGPUs are programmable multi core chips built around an array of processors working in parallel. The GPU is composed of an array of streaming multiprocessors (SM), where each of them can launch up to 1024 co-resident concurrent threads. Currently available graphics units are in the range from 1 SM up to 30 SMs for the high end products. Each single SM contains 8 scalar processors (SP) each with 1024 32-bit registers. The total of 64KB of register space is available for each SM. Each SM is also equipped with a 16KB on-chip memory that is characterized by low access latency and high bandwidth. Thread management (creation, scheduling, synchronization) is performed in hardware and the overhead is extremely low. SM works in SIMT scheme (Single Instruction, Multiple Thread), where threads are executed in groups of 32 called *warps*. CUDA programming model defines the *host* and the *device*. *Host* executes CPU sequential procedures while the *device* executes parallel GPU programs - *kernels*. A *kernel* works in a SPMD scheme (Single Program, Multiple Data). CUDA gives an advantage of using massively parallel computation for several applications. Detailed GPU architecture can be found in the original documentation *NVIDIA CUDA C Programming Guide 3.2* (2010). Useful additional programming issues are published in best practices guide CUDA (2010b).

3.2 Improved classic Iterative Closest Point

In this subsection classic Iterative Closest Point algorithm proposed by Besl & McKay (1992) and implementation proposed by Nüchter, Lingemann, Hertzberg & Surmann (2007) improved by GPGPU computing is described. Aligning two-view range images with respect to the reference coordinate system is performed by the ICP (Iterative Closest Points) algorithm. Range images are defined as a model set M and data set D, N_m and N_d denotes the number of the elements in the respective set. The alignment of these two data sets is solved by minimization with respect to **R**,**t** of the following cost function:

$$E(\mathbf{R},\mathbf{t}) = \sum_{i=1}^{N_m}\sum_{j=1}^{N_d} w_{ij} \left\| \mathbf{m}_i - \left(\mathbf{R}\mathbf{d}_j + \mathbf{t}\right) \right\|^2 \tag{1}$$

w_{ij} is assigned 1 if the i-th point of M correspond to the j-th point in D in the same bucket (or neighbor bucket). Otherwise w_{ij}=0. **R** is a rotation matrix, **t** is a translation matrix. $\mathbf{m}_i$ and $\mathbf{d}_i$ corresponds to the i-th point from model set M and D respectively. The ICP algorithm using CUDA parallel programming is given:

Algorithm 1 ICP - parallel computing approach

allocate the memory
copy data from the host(M_{host}, D_{host}) to the device(M_{device}, D_{device})
for $iter = 0$ to $max\_iterations$ **do**
  select closest points between M_{device} and D_{device}
  calculate (R, t) that minimizes equation 1
  transform D_{device} by (R, t) and put the results into $D_{deviceRt}$
  copy $D_{deviceRt}$ to D_{device}
end for
copy D_{device} to D_{host}
free memory

3.2.1 Calculation of (R,t)

Calculation of the rotation and translation (**R**,**t**) is performed using reduced equation 1:

$$E(\mathbf{R},\mathbf{t}) \propto \frac{1}{N}\sum_{i=1}^{N} \left\| \mathbf{m}_i - (\mathbf{R}\mathbf{d}_i + \mathbf{t}) \right\|^2 \tag{2}$$

where

$$N = \sum_{i=1}^{N_m}\sum_{j=1}^{N_d} w_{ij} \tag{3}$$

Rotation **R** is decoupled from computation of translation **t** using the centroids $\mathbf{c}_m$ and $\mathbf{c}_d$ of points:

$$\mathbf{c}_m = \frac{1}{N}\sum_{i=1}^{N} \mathbf{m}_i \tag{4}$$

$$\mathbf{c}_d = \frac{1}{N}\sum_{i=1}^{N}\mathbf{d}_i \tag{5}$$

and modified data sets:

$$\mathbf{M}' = \left\{\mathbf{m}_i{}' = \mathbf{m}_i - \mathbf{c}_m\right\}_{1,\ldots,N} \tag{6}$$

$$\mathbf{D}' = \left\{\mathbf{d}_i{}' = \mathbf{d}_i - \mathbf{c}_d\right\}_{1,\ldots,N} \tag{7}$$

After applying equations 4, 5, 6 and 7 to the mean square error function E(**R**, **t**), the equation 2 takes the following form:

$$E\left(\mathbf{R},\mathbf{t}\right) \propto \frac{1}{N}\sum_{i=1}^{N}\left\|\mathbf{m}_i{}' - \mathbf{R}\mathbf{d}_i{}' - \left(\mathbf{t} - \mathbf{c}_m + \mathbf{R}\mathbf{c}_d\right)\right\|^2 \tag{8}$$

Assuming that:

$$\mathbf{t} - \mathbf{c}_m + \mathbf{R}\mathbf{c}_d = \widetilde{t} \tag{9}$$

equation 8 takes following form:

$$\begin{aligned} E\left(\mathbf{R},\mathbf{t}\right) \propto \tfrac{1}{N}\textstyle\sum_{i=1}^{N}\left\|\mathbf{m}_i{}' - \mathbf{R}\mathbf{d}_i{}'\right\|^2 - \\ \tfrac{2}{N}\widetilde{t}\cdot\textstyle\sum_{i=1}^{N}\left(\mathbf{m}_i{}' - \mathbf{R}\mathbf{d}_i{}'\right) + \tfrac{1}{N}\textstyle\sum_{i=1}^{N}\left\|\widetilde{t}\right\|^2 \end{aligned} \tag{10}$$

To minimize 10 the algorithm has to minimize the following term:

$$\sum_{i=1}^{N}\left\|\mathbf{m}_i{}' - \mathbf{R}\mathbf{d}_i{}'\right\|^2 \tag{11}$$

with the assumption:

$$\widetilde{t} = 0 \tag{12}$$

The optimal rotation is calculated by $\mathbf{R} = \mathbf{V}\mathbf{U}^T$, where matrices **V** and **U** are derived from the Singular Value Decomposition (SVD) described in section 3.2.8 of a correlation matrix $\mathbf{C} = \mathbf{U}\mathbf{S}\mathbf{V}^T$ given by:

$$\mathbf{C} = \sum_{i=1}^{N}\mathbf{m}_i{}'^T\mathbf{d}_i{}' = \begin{bmatrix} c_{xx} & c_{xy} & c_{xz} \\ c_{yx} & c_{yy} & c_{yz} \\ c_{zx} & c_{zy} & c_{zz} \end{bmatrix} \tag{13}$$

where:

$$c_{xx} = \sum_{i=1}^{N} m_{ix}{}' d_{ix}{}', c_{xy} = \sum_{i=1}^{N} m_{ix}{}' d_{iy}{}', \ldots, c_{zz} = \sum_{i=1}^{N} m_{iz}{}' d_{iz}{}' \tag{14}$$

Correlation matrix elements are computed using optimized parallel reduction described in section 3.2.7. The optimal translation **t** is derived from equation 12 and 9, therefore

$$\mathbf{t} = \mathbf{c}_m - \mathbf{R}\mathbf{c}_d \tag{15}$$

3.2.2 Memory allocation on device and copy from host

Figure 1 and algorithm 2 shows main data used in presented approach. The *table_of_found_buckets*, *table_of_sorted_buckets*, *table_of_sorted_points* consist of 512×512 integer elements, *table_of_amount_of_points_in_bucket* and *table_of_bucket_indexes* consist of $256 \times 256 \times 256$ integer elements. M (reference data) and D (data to be align) data sets are stored in six 512×512 tables consisting float values stored in one dimensional array. For sorting the table of buckets routine described in section 3.2.4 and used in algorithm 2 the CUDA Radix Sort class available in CUDA (2010a) briefly described in Satish et al. (2008) and Satish et al. (2009) is used. The method initialize() called by the constructor of Radix Sort Class allocates temporary storage for the sort and the prefix sum that it uses. Temporary storage is (2*NUMBER_OF_POINTS + 3*8*M/CTA_SIZE) unsigned ints, with a default CTA_SIZE of 256 threads and $NUMBER\_OF\_POINTS = 512 \times 512$. Amount of data is large, therefore the procedure of memory allocation is done initially.

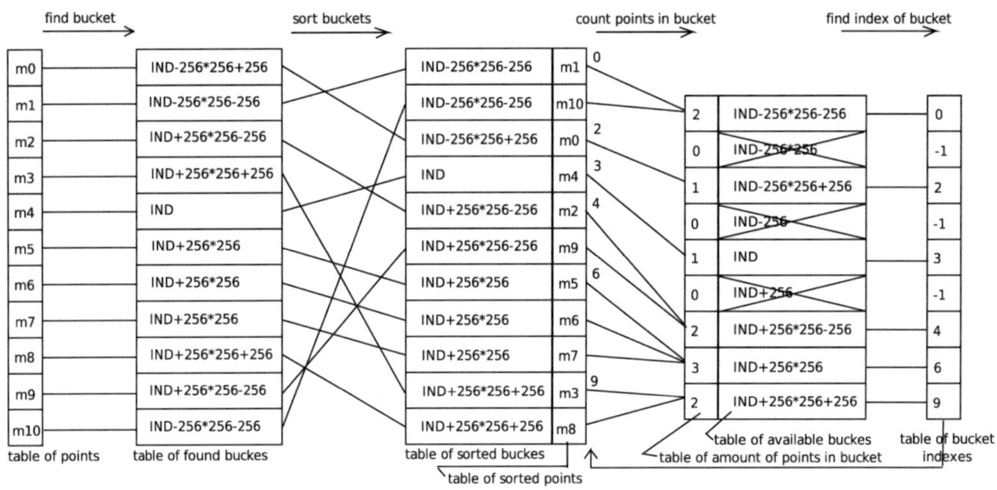

Fig. 1. Initial steps: selection of closest points procedure - example related to figure 4.

3.2.3 Selection of closest points

The distance between two points in Euclidean distance metric for point $p_1 = \{x_1, y_1, z_1\}$ and $p_2 = \{x_2, y_2, z_2\}$ is:

$$distance(p_1, p_2) = \left[(x_1 - x_2)^2 + (y_1 - y_2)^2 + (z_1 - z_2)^2 \right]^{\frac{1}{2}} \tag{16}$$

To find pairs of closest points between model set M and data set D, the decomposition of XYZ space, where x,y,z $\in < -1, 1 >$, into 2^8x2^8x2^8 buckets is proposed. The idea of the decomposition will be discussed for 2^2x2^2x2^2 case. Figure 2 shows decision tree that decomposes XYZ space into 64 buckets. Each node of the decision tree includes boundary decision, therefore points are categorized into left or right branch. Nodes that do not have branches assign buckets. Each bucket has unique index and is related to cubic subspace with length,width,height = $2/2^2$,$2/2^2$,$2/2^2$. Each bucket that does not belong to border has 26 neighbors. The 27 neighboring cubic subspaces are shown in figure 3 where also the way of indexing is given. Figure 4 demonstrates the idea of nearest neighbor (NN) search technique on 2 dimensional example. Assuming that we are looking for nearest neighbor that satisfies condition $R < 2/2^8$ and $circle_{R=2/2^8} \subset bucket_{3R}$ NN will be found in the same bucket or in neighboring bucket (in this example NN of point d is m5). Algorithm 2 describes the procedure of selection of closest points. For better explanation figure 1 shows initial steps of this algorithm where set M of 10 points from figure 4 is used for NN search. The details of the algorithm will be discussed in next subsections.

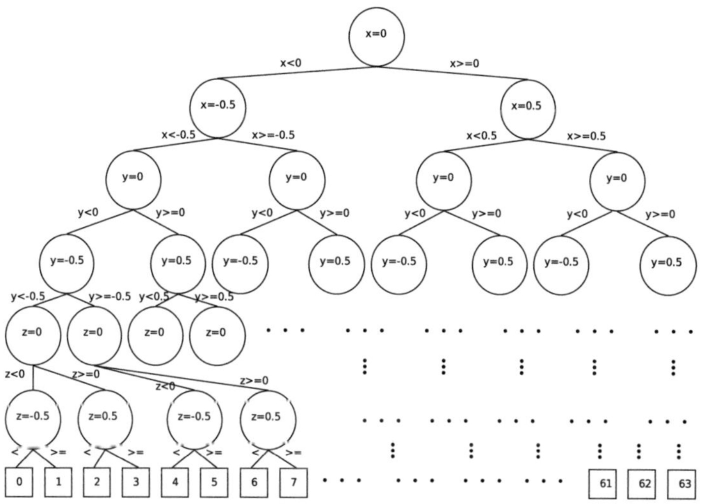

Fig. 2. Tree structure used for XYZ space decomposition into 64 buckets. From root to the leaf/bucket chosen left or right brunch depends on the current separation line.

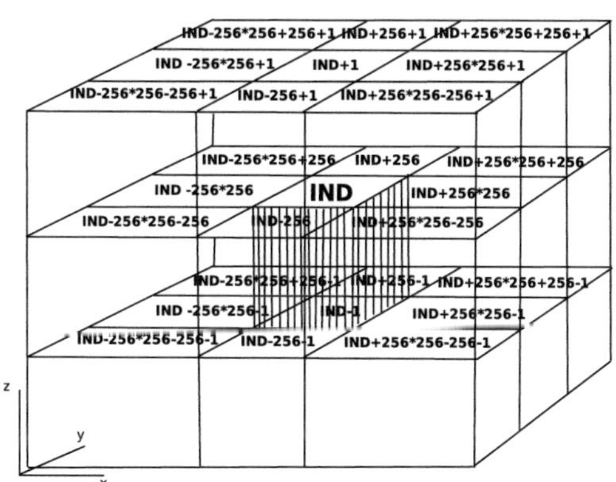

Fig. 3. Cubic subspaces - neighboring buckets, the way of indexing is explained in section 3.2.6.

3.2.4 Sort buckets

Radix sort class CUDA (2010a) is used to sort unsigned integer key-value pairs. Keys correspond to the elements of table of buckets and value corresponds to elements from table of points. Procedure outputs sorted table of buckets. Figure 1 shows an example of sorting

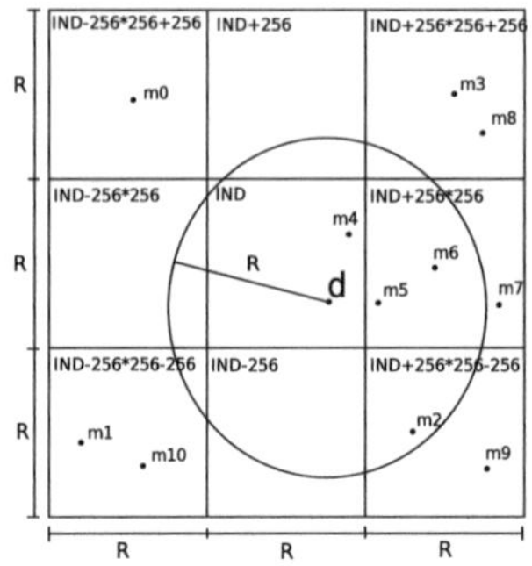

Fig. 4. 2 dimensional example of NN search in neighboring buckets.

Algorithm 2 Selection of closest points

> **for** all points m_{xyz} in parallel **do**
>   find $bucket_m$
>   update $table\_of\_found\_buckets$
> **end for**
> in parallel sort $table\_of\_found\_buckets$ {radix sort}
> in parallel count points in each bucket
> **for** all points d_{xyz} in parallel **do**
>   find $bucket_d$
>   **for** all neighbors of $bucket_d$ **do**
>   find NN for d_{xyz} {nearest neighbor is one from m_{xyz}}
>   **end for**
> **end for**

result. Radix sort is a well known sorting algorithm, very efficient on sequential machines for sorting small keys. It assumes that the keys are d-digit numbers and sorts on one digit of the keys at a time, starting from least and finishing on most significant. The complexity of sorting n keys will be O(n). The details of GPU based radix sort implementation can be found in Satish et al. (2008) Satish et al. (2009). The implementation of GPU based radix sort is robust, therefore it can be used for on-line computation.

3.2.5 Count points in bucket and find index of bucket

In the procedure of counting points that belong to the same bucket the counting is based on $table\_of\_sorted\_buckets$ (see figure 1). It is important to notice, that also the index of the found bucket is computed. This index, along with the information concerning an amount of points in the bucket, will be used for searching nearest neighbor in algorithm 2.

3.2.6 Find bucket

Figure 2 shows tree structure used for indexing of 2^2x2^2x2^2 buckets. The concept of finding bucket index for point m_{xyz} is shown on the scheme 5, where x corresponds to border for current level in the tree and 0, 1, 2, 3, ... 14, ... correspond to actual bucket index during its computation. The bucket indexing procedure is executed in parallel, where each CUDA kernel computes bucket index for different $point_{xyz}$.

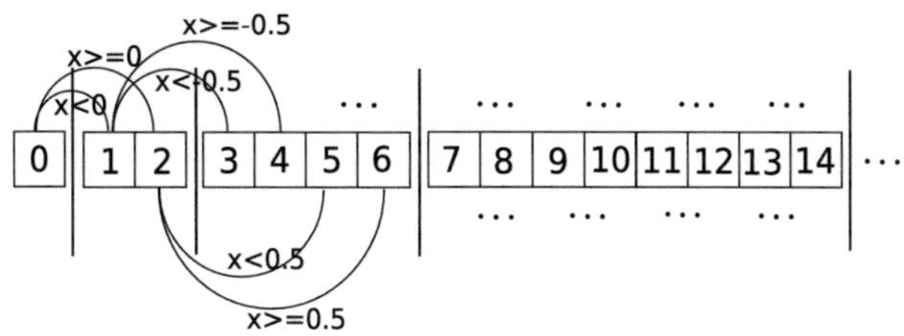

Fig. 5. The scheme of bucket indexing procedure.

3.2.7 Correlation matrix elements computation using optimized parallel reduction

For correlation matrix (equation 13) parallel prefix sum Harris et al. (2007) available in CUDA (2010a) is used. The all-prefix-sums operations take a binary associate operator $\oplus$ with identity I, and an array of n elements

$$[a_0, a_1, ..., a_{n-1}] \tag{17}$$

and returns the array

$$[I, a_0, (a_0 \oplus a_1), ..., (a_0 \oplus a_1 \oplus ... \oplus a_{n-2})] \tag{18}$$

All-prefix-sums operations on array of data is commonly known as scan. The parallel implementation uses multiple thread blocks for processing an array of 512×512 data points stored in one dimensional array. The strategy is to keep all multiprocessors on the GPU busy to increase the performance. An assumption is that each thread block reduces a portion of the array. To avoid problem of global synchronization the computation is decomposed into multi kernel invocations. Optimized kernel available in CUDA (2010a) is used in parallel computation.

3.2.8 Singular Value Decomposition (SVD)

The equation for singular value decomposition of A 3×3 matrix is the following:

$$A = U\Sigma V^T \tag{19}$$

where U is an 3×3 matrix, Σ is an 3×3 diagonal matrix, and V^T is also an 3×3 matrix. The columns of U are called the left singular vectors, $\{ u_k \}$, and form an orthonormal basis. The rows of V^T contain the elements of the right singular vectors, $\{ v_k \}$. The elements of Σ are only nonzero on the diagonal, and are called the singular values. Thus, $\Sigma = diag(\sigma_1, ..., \sigma_n)$. Furthermore, $\sigma_k > 0$ for $1 \leq k \leq r$, and $\sigma_i = 0$ for $(r+1) \leq k \leq n$. The ordering of the singular vectors is determined by high-to-low sorting of singular values, with the highest singular value in the upper left index of the Σ matrix. In this particular application we need to compute the SVD of a 3x3 matrix. For such a small matrix, generalized SVD algorithms from libraries like LAPACK (Linear Algebra PACKage) LAPACK (2011) are not beneficial especially when we have to implement it on GPGPU. Our implementation computes the singular values by solving for the roots of a cubic polynomial and then eigenvectors of $A^T A$ for V, then it uses A and V to compute U. The algorithm is executed in 5 steps.

1. Compute A^T and $A^T A$.
2. Determine the eigenvalues of $A^T A$ (by solving for the roots of a cubic polynomial) and sort these in descending order. Compute square roots to obtain singular values of A.

3. Construct diagonal matrix Σ by placing singular values in descending order along its diagonal. Compute Σ^{-1}.
4. Use the ordered eigenvalues from step 2 and compute the eigenvectors of $A^T A$. Place these eigenvectors along the columns of V and compute V^T.
5. Compute $U = AV\Sigma^{-1}$

4. Path planning

Motion planning is very important task for mobile robot working in unknown environment. Assuming that we have grid map describing robot environment, a trajectory of the robot can be computed using the modification of diffusion method described in Siemiatkowska (2008) improved by GPU computation. GPGPU implementation is using two 2 dimensional grids of 512x512 cells. One grid is used for initiation, where each cell can be free, occupied by the obstacle, occupied by a robot or occupied by a goal. Second grid is used for diffusion computation. The idea of usage GPU is to perform computation for each cell in parallel till diffusion reach robot position. To obtain robot trajectory we start from robot position and iteratively by finding local maximum in neighboring cells we are approaching the goal.

5. Cloud of points processing

Main idea is to decompose 3 dimensional space into grid of buckets. For each bucket local approximation plane is computed (if more than 3 points belong to bucket) and for each point in bucket normal vector is assigned. The advantage of proposed method is satisfactory result obtained with noisy data set. The procedure of normal vectors computation for registered range images is given in algorithm 3, it uses CUDA for robust nearest neighbors search briefly described in previous sections. The parameter $maxnumberofplanes$ in algorithm 3 is assigned experimentally as 10. This value guarantee robust procedure execution with satisfying heuristic of random planes generation.

Algorithm 3 Compute normal vectors (r,g,b)

```
for all range points (x,y,z) in parallel do
  bucket = findbucket(x,y,z)
  for allneighboringbuckets do
    add points from bucket to listofpoints
  end for
  for i = 0 to maxnumberofplanes do
    compute plane based on 3 random points
    sum_i = 0
    for all points in listofpoints do
      sum_i += distance(point,plane)
    end for
  end for
  normalvector = plane for min(sum_i)
end for
```

6. Experiments

All experiments were performed using NVIDIA GF 580 GPGPU. Following subsections demonstrate results in area of accelerated ICP, path planning and normal vector computation.

6.1 GPU accelerated ICP

Experiments where performed using different type of sensors: Time of Flight camera (see figure 6), Kinect sensor (see figure 8) and 3DLSN unit (see figure 10). Data registration result with low cost Time of Flight camera IFM O3D201 is not satisfactory because of the low resolution of sensor (figure 7). New RGB-D Kinect camera (figure 8) offers high resolution with acceptable level of noise, therefore the registration result is acceptable (figure 9). Unfortunately Kinect sensor is not designed for robotics application especially working in outdoor environments, therefore usage is limited. The most optimistic result is obtained

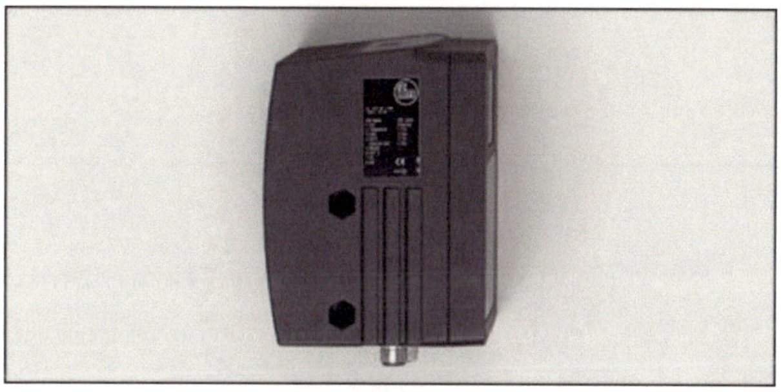

Fig. 6. Time of Flight camera type IFM O3D201.

Fig. 7. Data registration using Time of Flight camera type IFM O3D201. The result is not satisfactory because of low resolution of sensor.

Fig. 8. Robot PIONEER 3AT equipped with Kinect sensor. Photo taken during Land Robot Trial CELROB 2011.

Fig. 9. Data registration using Kinect sensor. Result is satisfactory because of high resolution of Kinect sensor with acceptable level of noise.

for commercially available 3DLSN unit (figure 10) composed of laser measurement system LMS SICK 200 offering accurate data acquisition working in INDOOR and OUTDOOR environments. The disadvantage is the limitation of data acquisition in stop-scan fashion. Data where acquired in INDOOR and OUTOOR environments shown in figures 11 and 14. Results of ICP algorithm for 30 and 100 iterations are shown in figures 12 and 13. Average time of 30 iterations of GPGPU based ICP is less than 300ms and 100 iterations is less than 1 second. Empirical evaluation shows that 30 iteration enough for accurate data registration, therefore it is assumed that ICP takes 300ms in average for alignment of two data sets containing 512 x 512 data points.

Fig. 10. Robot PIONEER 3AT equipped with 3DLSN sensor.

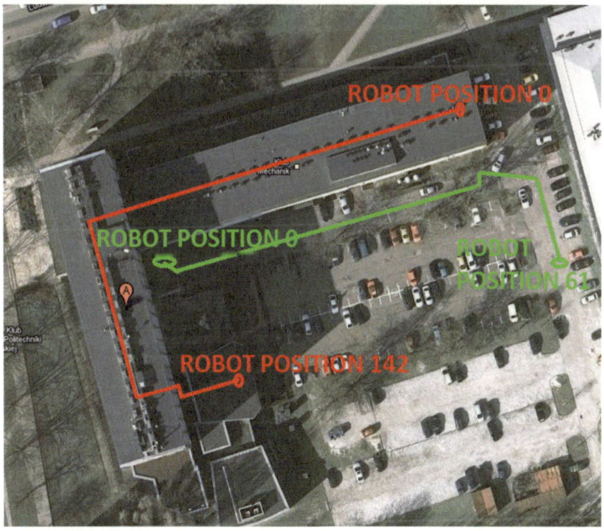

Fig. 11. Data acquisition using robot equipped with 3DLSN unit in INDOOR and OUTDOOR environments (Faculty of Mechatronics, Warsaw University of Technology).

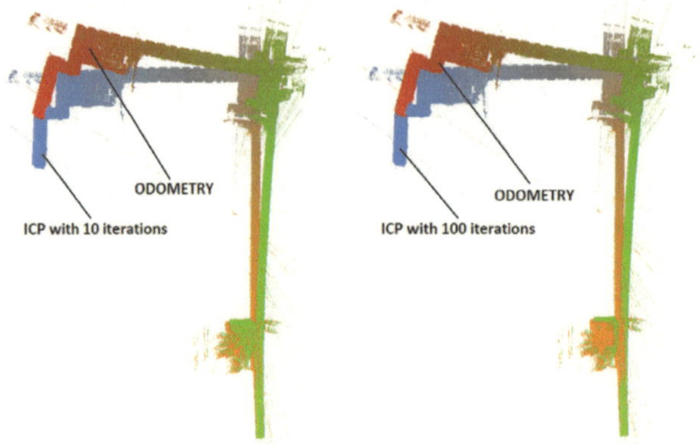

Fig. 12. Result of data registration in INDOOR environment from figure 11. It is shown a small difference of the result of ICP with 30 and 100 iterations.

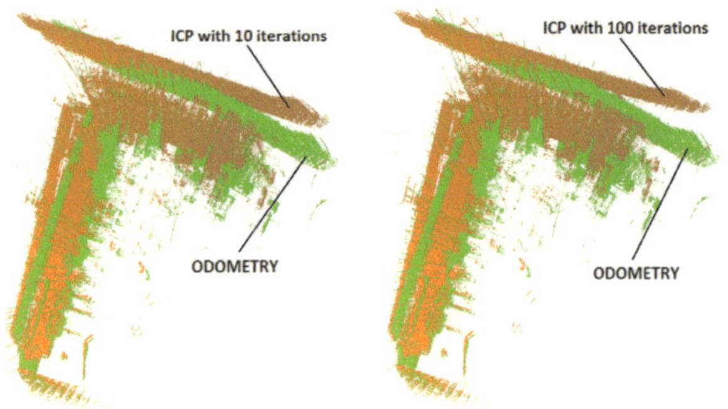

Fig. 13. Result of data registration in INDOOR environment from figure 11. It is shown a small difference of the result of ICP with 30 and 100 iterations.

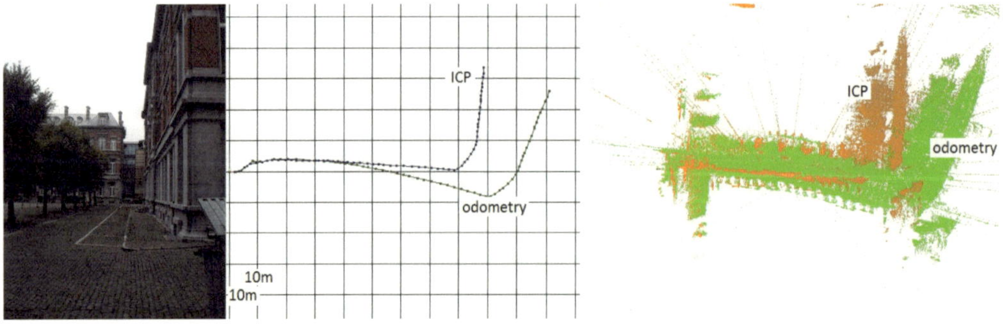

Fig. 14. Data registration result (Royal Military Academy Campus, Brussels, Belgium).

6.2 Path planing simulation results

Path planning was tested using simulated environments with different amount of random obstacles (figure 15, 16, 17, 18). Robot environment is represented by two dimensional grid of cells (512 x 512). Goal is located on the left and robot position is located on the right, path is visualized as black and white line. Diffusion process is visualized in gray scale, where white pixels correspond to max values and black pixels correspond to min values. The average time of CUDA kernel execution of elementary diffusion process takes 0.06 ms, and total diffusion

Fig. 15. Path planning result (goal- circle on the left, robot position - circle on the right). Environment with no obstacles.

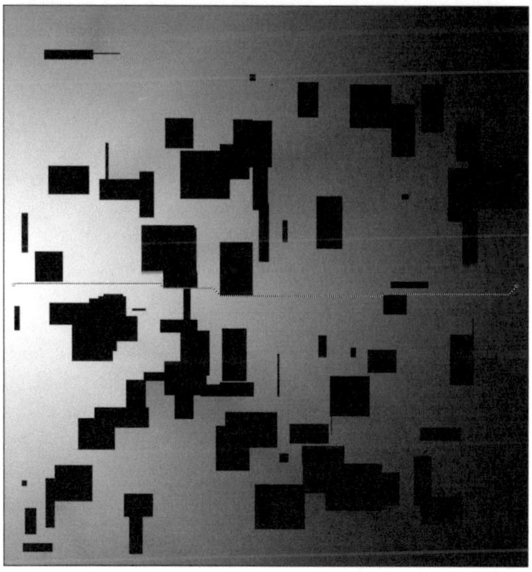

Fig. 16. Path planning result (goal- circle on the left, robot position - circle on the right). Environment with 100 random obstacles.

process for 512 iterations takes in average 30ms. The result is satisfactory because it is showing on-line capability of proposed implementation.

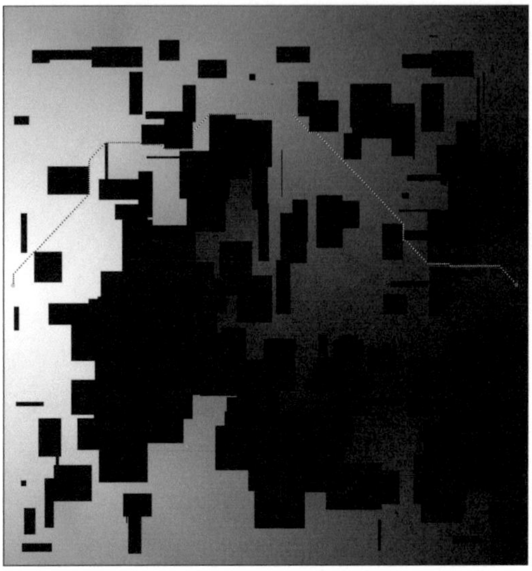

Fig. 17. Path planning result (goal- circle on the left, robot position - circle on the right). Environment with 200 random obstacles.

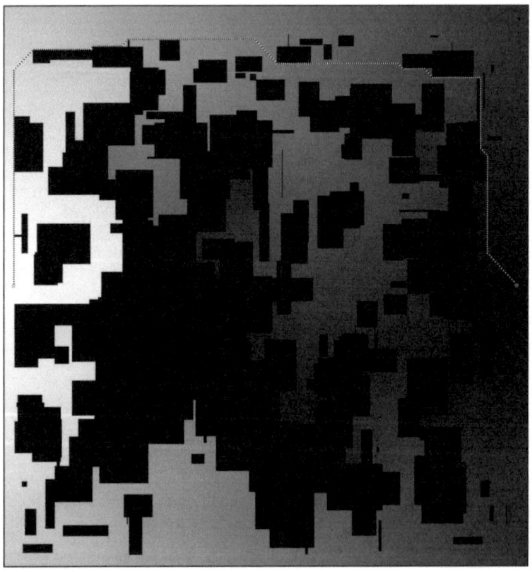

Fig. 18. Path planning result (goal- circle on the left, robot position - circle on the right). Environment with 300 random obstacles.

6.3 Normal vector computation

GPGPU accelerated normal vector computation was tested for data acquired using Kinect sensor (figure 19) and 3DLSN unit (figure 20). The average time of normal vector computation is less than 100ms for data set of 512 x 512 data points. The implementation is robust for noisy data delivered by Kinect sensor.

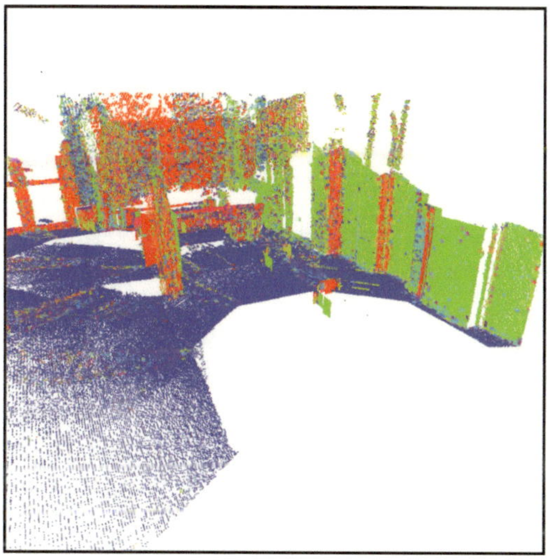

Fig. 19. Normal vector computation for data acquired by Kinect sensor (see also figure 9). Vectors (x,y,z) are represented by colors (r,g,b).

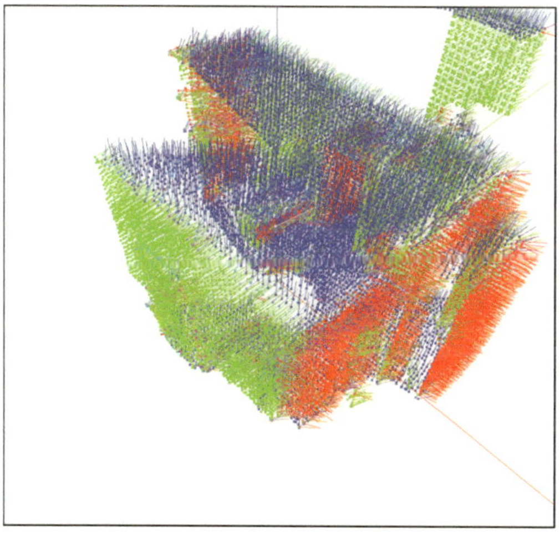

Fig. 20. Normal vector computation for data acquired by 3DLSN unit. Vectors (x,y,z) are represented by colors (r,g,b).

7. Conclusion and future work

In this chapter three areas of research where shown such as 3D data registration, robot navigation and 3D cloud of points processing. All approaches are improved using GPGPU parallel computation. The on – line data registration problem was discussed. New approach based on robust KNN nearest neighborhood search applied for improvement of ICP algorithm where shown. The implementation is using Radix Sort for bucket sorting. Data registration implementation was tested for data sets delivered by different sensors in different environments. The average ICP time computation for 30 iteration is less than 300ms.
The path planning parallel approach based on modified diffusion method was shown. It was tested using simulated environment compound from different amount of random obstacles. The result is satisfactory because 30 ms of computation guarantee on-line execution in practical application.
On – line 3D cloud of points' segmentation based on normal vector computation was presented. The result is satisfactory even for noisy data obtained by Kinect sensors. 100ms average time is optimistic to use this implementation in practical application.
The set of proposed algorithms was tested on GPGPU NVIDIA CUDA GF 580. The improvement of SoA algorithms based on CUDA implementation shows the possibility to use in real RISE applications because of decreased time of execution. Future work will be related to development of 6DSLAM using GPU-ICP as odometry correction and normal vector computation for obtaining semantic images for Loop Closing procedure.

8. References

Arya, S. & Mount, D. M. (1993). Algorithms for fast vector quantization, *Proc. of DCC '93: Data Compression Conference*, IEEE Press, pp. 381–390.

Besl, P. J. & McKay, N. D. (1992). A method for registration of 3-d shapes, *IEEE Trans. Pattern Anal. Mach. Intell.* 14: 239–256.
URL: *http://portal.acm.org/citation.cfm?id=132013.132022*

CUDA (2010a). http://www.nvidia.com/cuda.

CUDA (2010b). CUDA C Best Practices Guide 3.2, http://www.nvidia.com/cuda.

Fitzgibbon, A. W. (2001). Robust registration of 2d and 3d point sets, *In British Machine Vision Conference*, pp. 411–420.

Foley, T. & Sugerman, J. (2005). Kd-tree acceleration structures for a gpu raytracer, *Proceedings of the ACM SIGGRAPH/EUROGRAPHICS conference on Graphics hardware*, HWWS '05, ACM, New York, NY, USA, pp. 15–22.

Garcia, V., Debreuve, E. & Barlaud, M. (2008). Fast k nearest neighbor search using gpu, *2008 IEEE Computer Society Conference on Computer Vision and Pattern Recognition Workshops*, pp. 1–6.

Harris, M., Sengupta, S. & Owens, J. D. (2007). *GPU Gems 3, Parallel Prefix Sum (Scan) with CUDA*, Addison-Wesley, chapter 39, pp. 851–876.

Huber, D. & Hebert, M. (2003). Fully automatic registration of multiple 3d data sets, *Image and Vision Computing* 21(1): 637–650.

Jensen, H. W. (2001). *Realistic image synthesis using photon mapping*, A. K. Peters, Ltd., Natick, MA, USA.

Kohlhepp, P., Pozzo, P., Walther, M. & Dillmann, R. (2004). Sequential 3d-slam for mobile action planning, *Proceedings of 2004 IEEE/RSJ International Conference on Intelligent Robots and Systems, Sendai, Japan*, pp. 722–729.

LAPACK (2011). http://www.netlib.org/lapack.

Magnusson, M. & Duckett, T. (2005). A comparison of 3d registration algorithms for autonomous underground mining vehicles, *In Proc. ECMR*, pp. 86–91.

Magnusson, M., Duckett, T. & Lilienthal, A. J. (2007). 3d scan registration for autonomous mining vehicles, *Journal of Field Robotics* 24(10): 803–827.

Montemerlo, M. & Thrun, S. (2004). A multi-resolution pyramid for outdoor robot terrain perception, *AAAI'04: Proceedings of the 19th national conference on Artifical intelligence*, AAAI Press, pp. 464–469.

Nuchter, A., Lingemann, K. & Hertzberg, J. (2007). Cached k-d tree search for icp algorithms, *Proceedings of the Sixth International Conference on 3-D Digital Imaging and Modeling*, IEEE Computer Society, Washington, DC, USA, pp. 419–426.

Nüchter, A., Lingemann, K., Hertzberg, J. & Surmann, H. (2007). 6d slam for 3d mapping outdoor environments, *Journal of Field Robotics (JFR), Special Issue on Quantitative Performance Evaluation of Robotic and Intelligent Systems* 24: 699–722.

Nüchter, A., Surmann, H. & Hertzberg, J. (2003). Automatic model refinement for 3D reconstruction with mobile robots, *Fourth International Conference on 3-D Digital Imaging and Modeling 3DIM 03*, p. 394.

NVIDIA CUDA C Programming Guide 3.2 (2010). http://www.nvidia.com/cuda.

Ortega, A., Haddad, I. & Andrade-Cetto, J. (2009). Graph-based segmentation of range data with applications to 3d urban mapping, *4th European Conference on Mobile Robots ECMR 09, September 23-25, 2009, Mlini Dubrovnik, Croatia*, pp. 193–198.

Park, S.-Y., Choi, S.-I., Kim, J. & Chae, J. (2010). Real-time 3d registration using gpu, *Machine Vision and Applications* pp. 1–14. 10.1007/s00138-010-0282-z.

Park, S.-Y. & Subbarao, M. (2003). An accurate and fast point-to-plane registration technique, *Pattern Recogn. Lett.* 24: 2967–2976.

Pottmann, H., Leopoldseder, S. & Hofer, M. (2002). Registration without icp, *Computer Vision and Image Understanding* 95: 54–71.

Purcell, T. J., Donner, C., Cammarano, M., Jensen, H. W. & Hanrahan, P. (2003). Photon mapping on programmable graphics hardware, *Proceedings of the ACM SIGGRAPH/EUROGRAPHICS Conference on Graphics Hardware*, Eurographics Association, pp. 41–50.

Qiu, D., May, S. & Nuchter, A. (2009). Gpu-accelerated nearest neighbor search for 3d registration, *Proceedings of the 7th International Conference on Computer Vision Systems: Computer Vision Systems*, ICVS '09, Springer-Verlag, Berlin, Heidelberg, pp. 194–203.

Rusinkiewicz, S. & Levoy, M. (2001). Efficient variants of the ICP algorithm, *Third International Conference on 3D Digital Imaging and Modeling (3DIM)*.

Rusu, R. B., Blodow, N. & Beetz, M. (2009). Fast point feature histograms (fpfh) for 3d registration, *Proceedings of the 2009 IEEE international conference on Robotics and Automation*, ICRA'09, IEEE Press, Piscataway, NJ, USA, pp. 1848–1853.

Rusu, R. B., Marton, Z. C., Blodow, N. & Beetz, M. (2008). Persistent Point Feature Histograms for 3D Point Clouds, *Proceedings of the 10th International Conference on Intelligent Autonomous Systems (IAS-10), Baden-Baden, Germany*.

Satish, N., Harris, M. & Garland, M. (2008). Designing efficient sorting algorithms for manycore gpus, *NVIDIA Technical Report NVR-2008-001*, NVIDIA Corporation.

Satish, N., Harris, M. & Garland, M. (2009). Designing efficient sorting algorithms for manycore gpus, *Proceedings of the 2009 IEEE International Symposium on Parallel&Distributed Processing*, IEEE Computer Society, Washington, DC, USA, pp. 1–10.

Siemiatkowska, B. (2008). Coordinating the motion of mobile robots using cellular neural network, *Journal of Automation, Mobile Robotics and Intelligent Systems* 2(2): 65–69.

Sprickerhof, J., Nüchter, A., Lingemann, K. & Hertzberg, J. (2009). An explicit loop closing technique for 6d slam, *4th European Conference on Mobile Robots ECMR 09, September 23-25, 2009, Mlini/Dubrovnik, Croatia*, pp. 229–234.

Thrun, S., Burgard, W. & Fo, D. (2000). A real-time algorithm for mobile robot mapping with applications to multi-robot and 3d mapping, *ICRA*, pp. 321–328.

9

Multi-Flock Flocking for Multi-Agent Dynamic Systems

Andrew McKenzie, Qingquan Sun and Fei Hu
Department of Electrical and Computer Engineering
University of Alabama
USA

1. Introduction

There has been much research done on single-flock flocking Shaw (1975), Partridge (1984), Partridge (1982), Okubo (1986), Reynolds (1987), Vicsek et al. (1995), Toner & Tu (1998), Shimoyama et al. (1996), Mogilner & Edelstein-Keshet (1999), Helbing et al. (2000), Vicsek (2001), Parrish et al. (2002), Olfati-Saber (2006), but none done on multi-flock flocking Gazi & Fidan (2007). One might ask, "Why would we need multiple flock flocking?" Consider the following scenario: there are two groups (squads/flocks) of Unmanned Vehicles (UV), both being in between the other group and the other groups' objective/goal. If both groups had the same capabilities then all we would need to do is to swap the groups goals. Unfortunately the groups have different sensing capabilities. One group of UV's is equipped with infrared cameras and the other with high-resolution cameras. Since each group is in the way of the other, it would be great if they could move out of each other's way. This in turn would decrease the amount of time for both groups to meet their goals.

We propose a new flocking algorithm that allows flocks to maneuver around other flocks (if needed) decreasing the amount of time each flock takes to reach their respective goals. We will do this by adding an additional agent, τ, to Olfati-Saber's Olfati-Saber (2006) existing α, β and γ-agents. The resulting algorithm will be compared to Olfati-Saber's flocking algorithm. Both algorithms will be simulated in multiple scenarios using Matlab. The scenarios will consist of both flock's being in-between the other flock and the other flocks goal, using different size flocks and only 1 group for a baseline. Section 2 presents related works. Section 3 includes our approach and multi-flock flocking algorithm. Section 4 contains our simulation setup. The simulation results are in 5; followed by the analysis in Section 6. Conclusions and future directions are in Section 7.

2. Related works

2.1 Flocking

Flocking is a kind of group behavior that includes a common objective and local interactions over a large number of group members. We find the emergence of flocking from many animals such as birds, fish, penguins, bees, and crowds, as well as swarming, and schooling Reynolds (1987), Partridge (1982), Toner & Tu (1998), Shimoyama et al. (1996).

Currently the flocking technique is mainly used in massive sensing using mobile sensor networks, self-assembly of connected mobile networks, and performing military missions such as reconnaissance, and surveillance. The self-organized feature of flocks can provide a heuristic conception in the design of mobile sensor networks and robotics systems.
The development of flocking techniques has had three phases. The first phase was primarily from a theoretical perspective. The typical researchers include: Viscek et al. Vicsek et al. (1995), whose work was mainly focused on alignment in self-driven particle systems; Toner and Tu Toner & Tu (1998) proposed a new scheme called continuum mechanics; and Levine et al. Levine et al. (2000), who developed a novel algorithm called rotating swarms to simulate ant mills with the all-to-all interactions. Additionally, several other continuum models of swarms were proposed which include works by Mogilner and Eldstein-Keshet Mogilner & Edelstein-Keshet (1996), Mogilner & Edelstein-Keshet (1999), and Topaz and Bertozzi Topaz & Bertozzi (2004). Helbing et al. Helbing et al. (2000) designed an empirical particle-based flocking model to study the escape panic phenomenon.
The second phase focused on the consensus problem and network topology. The contributions were mainly made by Olfati-Saber and Murray Saber & Murray (2003) Olfati-Saber & Murray (2004), Jadbabaie et al. Jadbabaie et al. (2003), Moreau Moreau (2005), and Ren and Beard Ren & Beard (2005). Although in the alignment problem, there is no constraint on the consensus value, when used for networked dynamic systems, the objective is distributed computation of a function via agreement Saber & Murray (2003), Olfati-Saber et al. (2007). Olfati-Saber and Murray Saber & Murray (2003) Olfati-Saber & Murray (2004) created a graph-induced potential function based structural formation control. Another work that belongs to this phase is on formation control and graph Laplacian by Fax and Murray Fax & Murray (2004).
Nowadays, the stability analysis of particles or agents with all-to-all interactions draws more attention. With respect to this issue, Tanner et al. Tanner et al. (2007) proposed a centralized algorithm for a particle system which leads to irregular collapse. They also proposed a distributed algorithm that leads to irregular fragmentation. Since collapse and fragmentation are two usual pitfalls of flocking, stability analysis on collapse and fragmentation is the evaluation method used for modern flocking algorithms.

2.2 Distributed intelligence in multi-robot systems

In Parker (2008), Parker gives an overview of the distributed intelligence field and its use in multi-robot systems. She first defines distributed intelligence and then defines the domain space of distributed intelligence. She defined four types of interactions in distributed intelligence: collective; cooperative; collaborative; and coordinative.
Collective interaction is defined as when entities are not aware of other entities on their team but share the same goals. The entities also help each other even though they are not planned to do so. An example of collective interaction is swarm robotics. Cooperative interaction is defined by entities being aware of the others and share goals. They also benefit each other with their actions. An example of cooperative interaction would be a team of robots who share map information and are trying to explore an unknown area. Collaborative interaction is defined by the entities having individual goals, and being aware of other entities on the team. The entities actions help the others achieve their own goals. Coordinative interaction occurs when entities have individual goals and can communicate with other entities. They do

not have a common goal and their actions are not helpful to the other entities. Essentially all of the entities are selfish and only care about their own goal.
She also defined three paradigms for distributed intelligence, which are: bio inspired, emergent swarms paradigm; organizational and social paradigms; and knowledge-based, ontological, and semantic paradigms. The main message of her paper can only be come from her words when she wrote "The main message of these discussions is that the choice of paradigm is not always obvious, and is dependent upon the requirements of the application to be addressed. We also note that complex systems of multiple robots can make use of several different paradigms simultaneously"Parker (2008).

3. Approach

In this section we will present two distributed algorithms for multi-flock flocking algorithms. The first algorithm is Olfati-Saber's from Olfati-Saber (2006) and the second algorithm is our algorithm, which extends Olfati-Saber's but adds an additional agent that adds coordinative interaction.

3.1 Olfait-Saber's flocking algorithm

Olfati-Saber's algorithm includes 3 agents: *alpha*, *beta* and *gamma*. All entities, such as a single bird in a flock of birds, are physical agent's with dynamics $\ddot{q}_i = u_i$ called an *alpha*-agent. The other two agents, *beta* and *gamma*, are virtual agents which model the effects of obstacles and the groups collective objective. The flocking algorithm has the capability to perform multiple obstacle avoidance. The equation consists of three terms:

$$u_i = u_i^\alpha + u_i^\beta + u_i^\gamma \tag{1}$$

where u_i^α denotes the (α, α) interaction terms, u_i^β denotes the (α, β) interaction terms, and u_i^γ is the groups distributed navigational feedback. The (α, α) interaction is used to keep the agents in a lattice /flock form. The (α, β) interaction term is used for obstacle avoidance, where a virtual β-agent is on the closest point of the obstacle from the α-agent. The γ-agent is a virtual leader, used to lead the flock to the desired location. Each term in equation 1 is explained as:

$$\begin{aligned}
u_i^\alpha &= c_1^\alpha \sum_{j \in N_i^\alpha} \phi_\alpha(\left\|q_j - q_i\right\|_\sigma)\mathbf{n}_{i,j} \\
&+ c_2^\alpha \sum_{j \in N_i^\alpha} a_{ij}(q)(p_j - p_i) \\
u_i^\beta &= c_1^\beta \sum_{k \in N_i^\beta} \phi_\beta(\|\hat{q}_{i,k} - q_i\|_\sigma)\hat{\mathbf{n}}_{i,k} \\
&+ c_2^\beta \sum_{j \in N_i^\beta} b_{i,k}(q)(\hat{p}_{i,k} - p_i) \\
u_i^\gamma &= -c_1^\gamma \sigma_1(q_i - q_r) - c_2^\gamma (p_i - p_r)
\end{aligned} \tag{2}$$

where $\sigma_1(z) = \frac{z}{\sqrt{1+\|z\|^2}}$ and c_η^ν are positive constants for all $\eta = 1,2$ and $\nu = \alpha, \beta, \gamma$. Each α-agent's state is denoted by (q_i, p_i), where q_i is the position and p_i is the velocity of the agent. The pair (q_r, p_r) is the state of a static or dynamic γ-agent. The vectors $\mathbf{n}_{i,j}$ and $\hat{\mathbf{n}}_{i,k}$ are given by $n_{i,j} = \frac{q_j - q_i}{\sqrt{1+\varepsilon\|q_j-q_i\|^2}})$, $\hat{n}_{i,k} = \frac{\hat{q}_{i,k}-q_i}{\sqrt{1+\varepsilon\|\hat{q}_{i,k}-q_i\|^2}}$.

3.2 Our flocking algorithm with coordinative interaction

Our equation has four agents: α, β, γ and τ. Where the α, β and γ agents are the same as in Olfati-Saber's Algorithm. The τ-agent is a virtual agent which is used to add Coordinative Interaction between the α agents. The equation consists of four terms:

$$u_i = u_i^\alpha + u_i^\beta + u_i^\gamma + u_\tau \tag{3}$$

where u_i^α, u_i^β and u_i^γ are the same as in equation 1. The coordinative interaction is added using the τ-agent. Each term in equation 3 is explained as:

$$\begin{aligned}
u_i^\alpha &= c_1^\alpha \sum_{j\in N_i^\alpha} \phi_\alpha(\left\|q_j - q_i\right\|_\sigma)\mathbf{n}_{i,j} \\
&+ c_2^\alpha \sum_{j\in N_i^\alpha} a_{ij}(q)(p_j - p_i) \\
u_i^\beta &= c_1^\beta \sum_{k\in N_i^\beta} \phi_\beta(\|\hat{q}_{i,k} - q_i\|_\sigma)\hat{\mathbf{n}}_{i,k} \\
&+ c_2^\beta \sum_{j\in N_i^\beta} b_{i,k}(q)(\hat{p}_{i,k} - p_i) \\
u_i^\gamma &= -c_1^\gamma \sigma_1(q_i - q_r) - c_2^\gamma(p_i - p_r) \\
u_i^\tau &= -c_1^\tau \sigma_2(q_i - q_r)\begin{bmatrix}0 & 1\\ 1 & 0\end{bmatrix} \\
&- c_2^\tau \sigma_2(p_i - p_r)\begin{bmatrix}0 & 1\\ 1 & 0\end{bmatrix}
\end{aligned} \tag{4}$$

As in equation 2, each α-agent's state is denoted by (q_i, p_i), where q_i is the position (with an x and y component) and p_i is the velocity of the agent. The pair (q_r, p_r) is the state of a static/dynamic γ-agent. The other components are defined as $\sigma_1(z) = \frac{z}{\sqrt{1+\|z\|^2}}$; $\sigma_2(z)$ is 1 if $\frac{\pi}{2} \geq \theta_{i-j} \geq \frac{3\pi}{2}$ else it is 0; $z \leq d_c$; $\theta_i = \tan(p_{i_y}/p_{i_x})$, where p_{i_y} and p_{i_x} are x and y components of the agents velocity respectively; $\theta_{i-j} = \theta_i - \theta_j$; and d_c is the cooperation distance for all j α-agents in both flocks. Essentially the τ agent is applied if two agents' (i and j) trajectories are going to intersect and their distance is less then d_c. c_η^ν are positive constants for all $\eta = 1,2$ and $\nu = \alpha, \beta, \gamma, \tau$. The vectors $\mathbf{n}_{i,j}$ and $\hat{\mathbf{n}}_{i,k}$ are given by $n_{i,j} = \frac{q_j-q_i}{\sqrt{1+\varepsilon\|q_j-q_i\|^2}})$, $\hat{n}_{i,k} = \frac{\hat{q}_{i,k}-q_i}{\sqrt{1+\varepsilon\|\hat{q}_{i,k}-q_i\|^2}}$.

4. Simulation setup

Both flocking algorithms, Olfati-Saber's algorithm (equation 1) and our algorithm (equation 3), were implemented using Matlab. Matlab version 7.6.0 (R2008a) was used to run the simulations. The following parameters were used for both algorithms: $d = 7$ meters, $r = 1.2d$, the time step size is 0.01 seconds. Both algorithms used the same values for all constants: $(c_1^\alpha, c_2^\alpha, c_1^\beta, c_2^\beta, c_1^\gamma and c_2^\gamma)$.

Simulations were run using 1 and 2 groups of agents, consisting of 5, 10, 15, 20, 25, 30, 35, 40, 45 and 50 agents in each group. The groups were initially positioned into lattices with lines of 10 agents. Each group started a distance of 150 meters from their goal. If 2 groups were used, the groups were put directly in-between the other group and its respective goal. The groups were started 100 meters apart from each other. The agents were limited to a maximum speed of 7 m/s. Agents were considered to have reached their groups goal if they got within 14 meters ($2d$) of the goal. Each simulation was run until all agents reached their goal or until 480 seconds, in simulation time, had passed. Each simulation's time to finish was recorded. The time to finish recorded the time from the start of the simulation till all agents had reached their goal. One image was saved every second of all simulations.

5. Simulation results

The simulations were broken into two groups for analysis. Group A contains the results from the simulations where the agents start in a symmetric lattice (i.e. groups that had a multiple of 10 agents in them as in Figure 1 a). The results for group A can be seen in Table 1 as well as in Figure 2.

Agents per group	1 Group	2 Groups	
	with and without Tao agent	with Tao agent	without Tao agent
10	139.11	134.37	did not finish
20	143.6	167.15	190.19
30	148.01	170.93	212.19
40	152.69	183.39	218.18
50	159.23	186.28	227.9

Table 1. Simulation results showing the total time to finish for symmetric lattices, group A.

Group B contains the results from the simulations when the agents started in a non-symmetric lattice (i.e. groups that did not have a multiple of 10 agents in them, as in Figure 3). The results for group B can be seen in Table 2 and in Figure 4. The combined results of both groups can be seen in Figure 5.

Selected images from group 1 at key points in the experiments can be seen in Figures: 6, 7, 1, 8 and 9. Selected images from group 2 can be seen in Figure 3. Figure 6 shows images for 2 groups with 10 agents per group using the τ agent flocking algorithm. Images from the experiments with 2 groups with 50 agents per group using the τ agent are shown in figure 1, and without using the τ agent in figures 8 and 9.

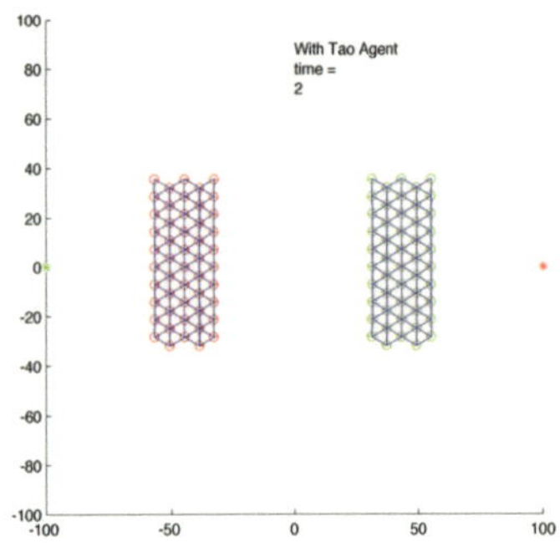

(a) With τ-agent, n=50, t=2s

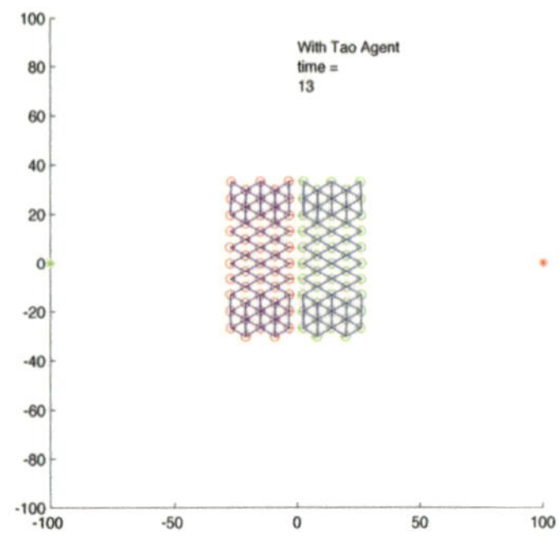

(b) With τ-agent, n=50, t=13s

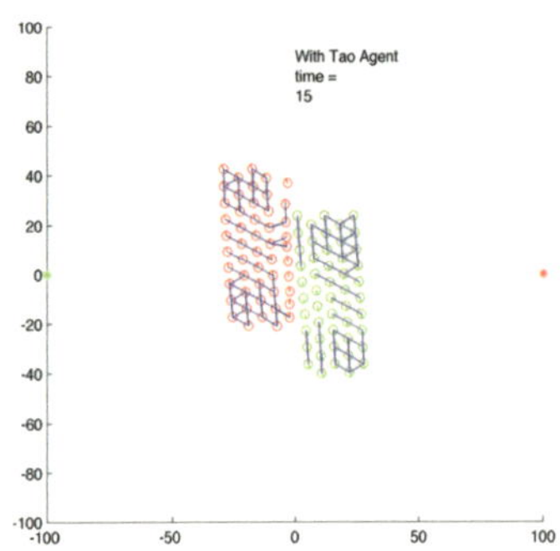

(c) With τ-agent, n=50, t=15s

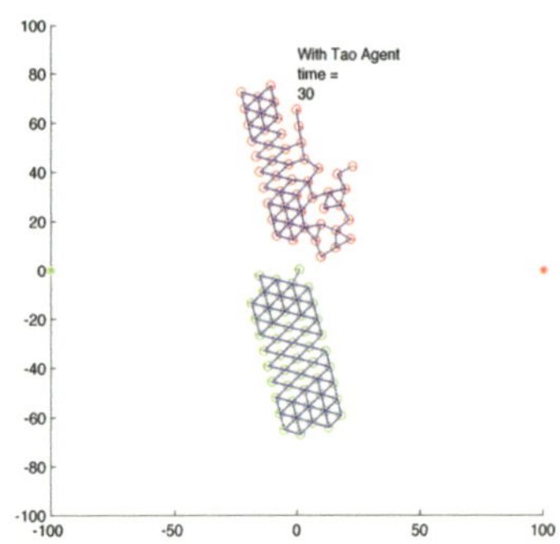

(d) With τ-agent, n=50, t=30s

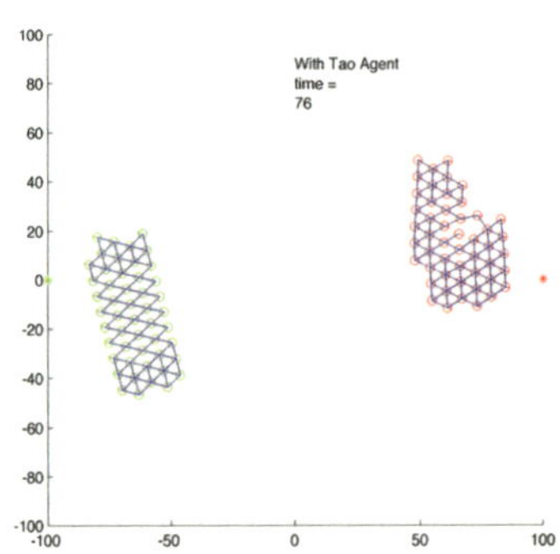

(e) With τ-agent, n=50, t=76s

Fig. 1. With τ-agent, 2 Groups, 50 agents per group

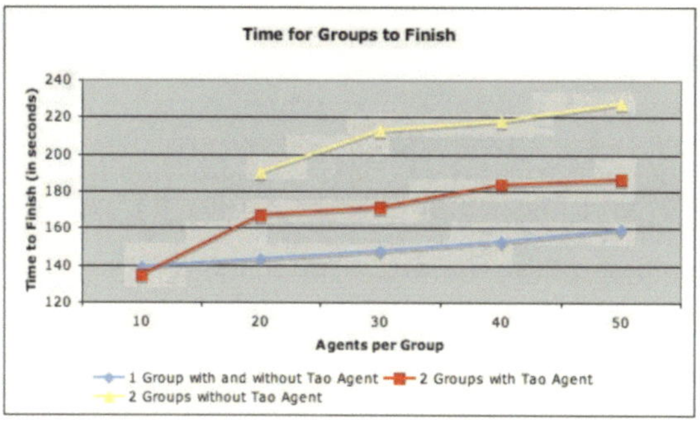

Fig. 2. Symmetric lattice finish times, Group 1

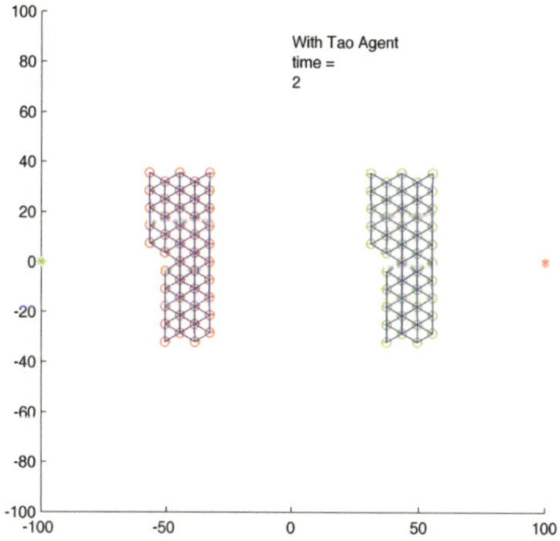

Fig. 3. Non-symmetric lattice, with τ, 2 groups and 45 agents per group

Agents per group	1 Group with and without Tao agent	2 Groups with Tao agent	2 Groups without Tao agent
5	113.23	112.61	did not finish
15	127.92	148.04	162.44
25	139.77	167	187.46
35	145.17	177.98	201.2
45	153.66	186.41	222.04

Table 2. Simulation results for Non-symmetric Lattice, group B.

Images comparing experiments with and without the τ agent, showing side by side images, can be seen in figures 10 and 11 for 10 agents per group and figures 12 and 13 for 50 agents per group.

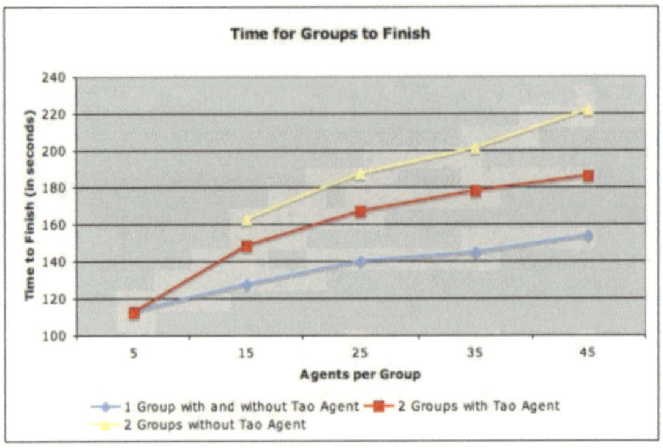

Fig. 4. Non-symmetric lattice finish times, group B

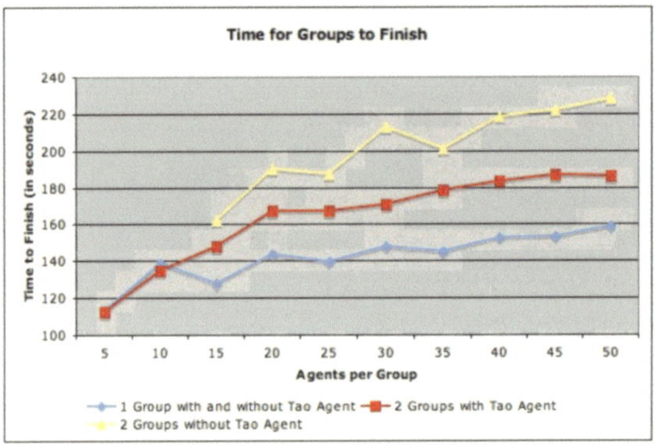

Fig. 5. Non-symmetric and symmetric lattice finish times

6. Analysis

One can deduce from figure 5, that our additional τ agent, eq 3, improves the Olfati-Saber algorithm, eq 1. This is because without the τ agent, when the two flocks come close to each other a huge traffic jam occurs because the two flocks want to go the same direction. This can be seen in figures 7, 8 and 9. What is interesting is that a deadlock occurred in the experiments with 5 and 10 agents per group. Although, the deadlock was unexpected, it is easily explainable. The reason the deadlock occurs is because the flocks do not have the additional velocity, essentially a push, from the additional rows of agents behind them as in all of the other experiments.

Another interesting result is that all of the 1 group experiments, group B finished faster then their counterpart in group A, n=15 vs. n=10, n=25 vs. n=20, n=35 vs. n=30, and n=45 vs. n=40). This also occurred in the without τ, equation 1, experiments except for n=15 vs. n=10, because the agents did not finish in n=5 and n=10 This result is due to the fact that the groups are not symmetric but rather rotated versions of the other which can be seen in figure 3. Because of the non symmetric lattices the groups actions are different from each other when the traffic jam occurs, which in turn allows the groups to get around each other faster.

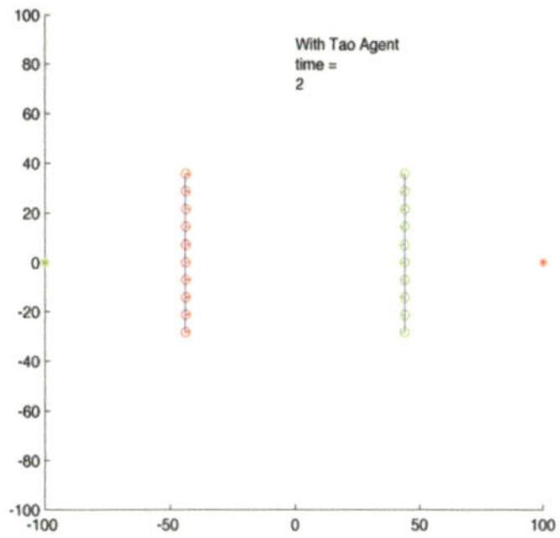

(a) With τ-agent, n=10, t=2s

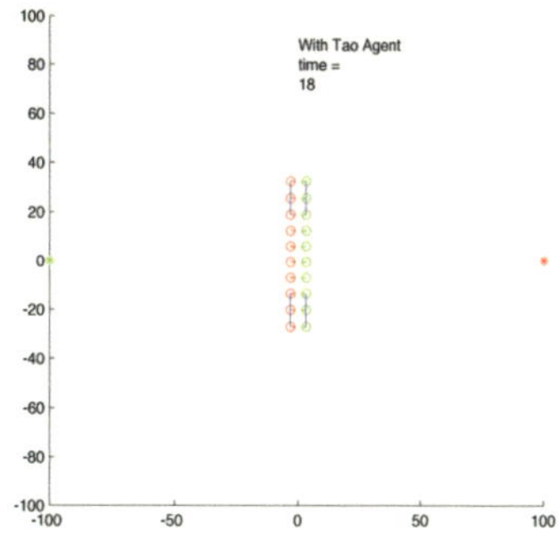

(b) With τ-agent, n=10, t=18s

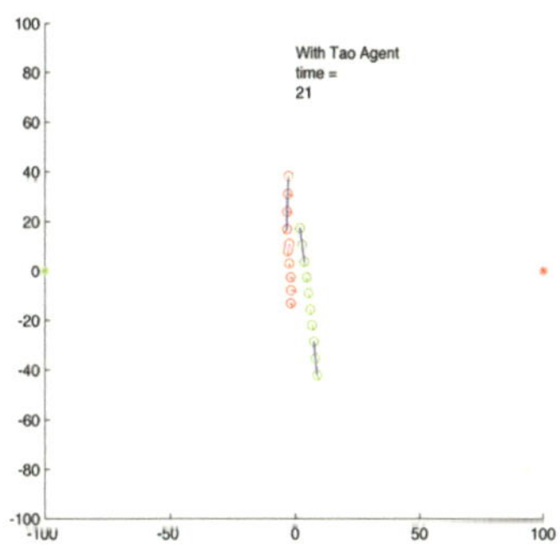

(c) With τ-agent, n=10, t=21s

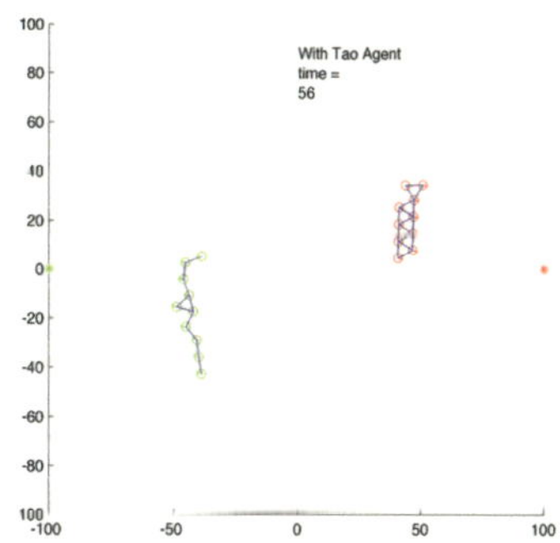

(d) With τ-agent, n=10, t=56s

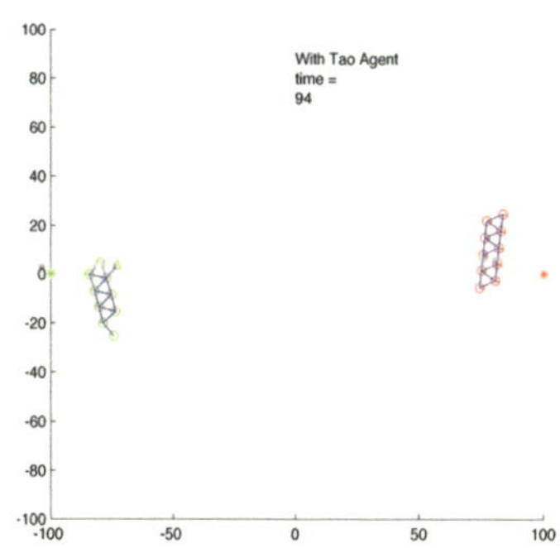

(e) With τ-agent, n=10, t=94s

Fig. 6. With Tao Agent, 2 Groups, 10 agents per group

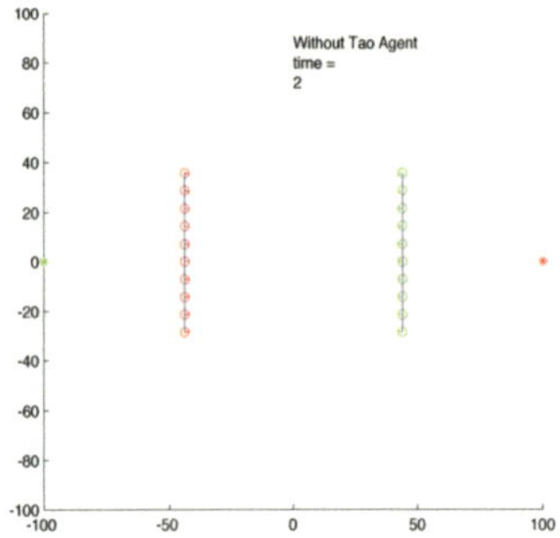

(a) Without τ-agent, n=10, t=2s

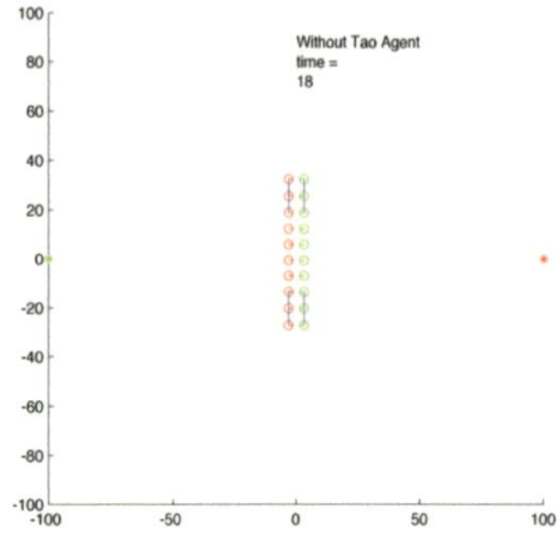

(b) Without τ-agent, n=10, t=18s

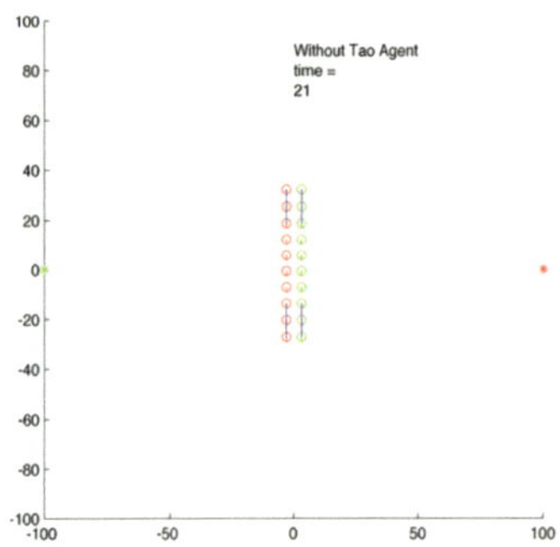

(c) Without τ-agent, n=10, t=21s

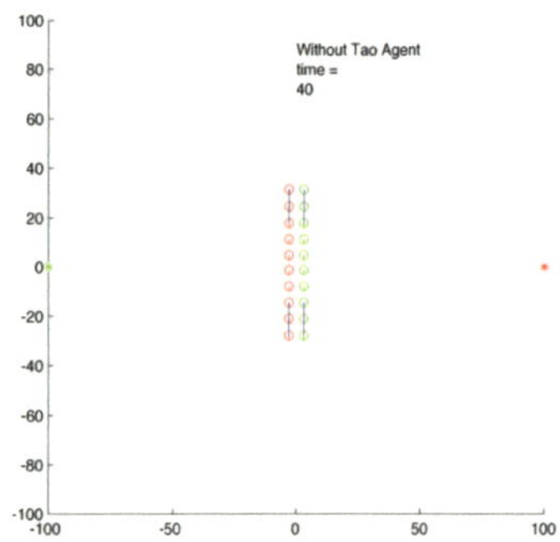

(d) Without τ-agent, n=10, t=40s

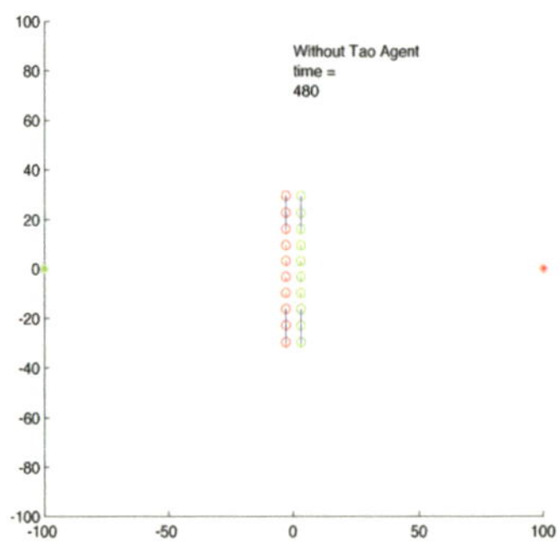

(e) Without τ-agent, n=10, t=480s

Fig. 7. Without Tao Agent, 2 Groups, 10 agents per group

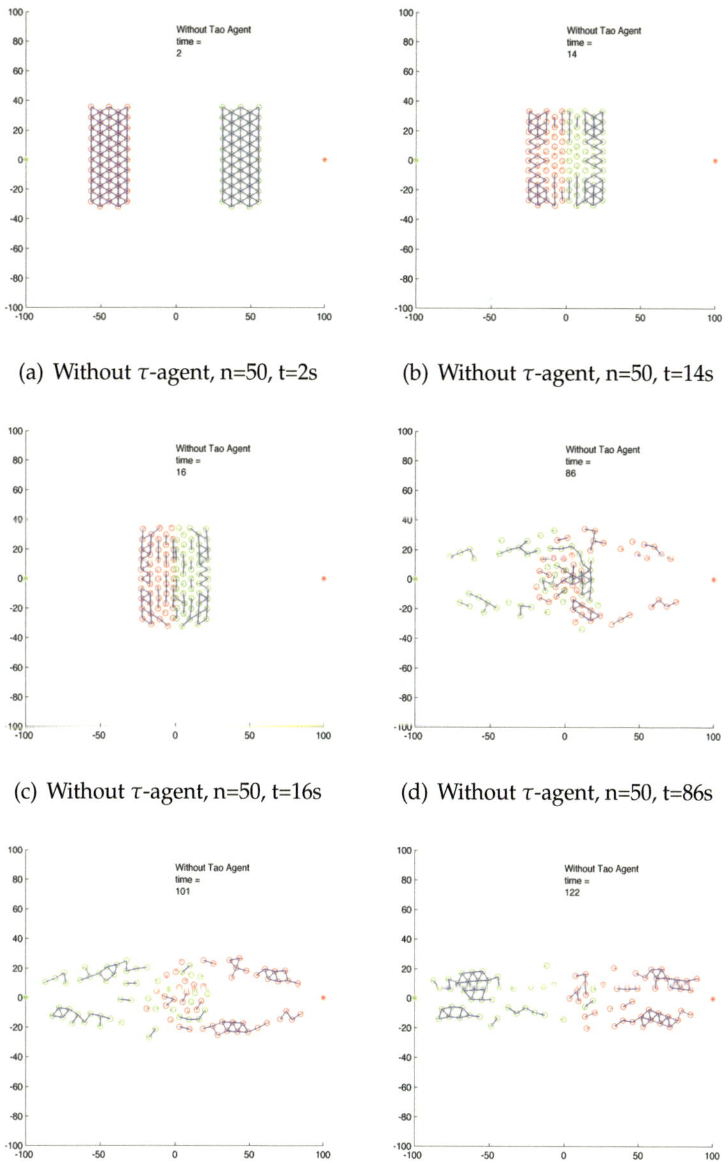

(a) Without τ-agent, n=50, t=2s

(b) Without τ-agent, n=50, t=14s

(c) Without τ-agent, n=50, t=16s

(d) Without τ-agent, n=50, t=86s

(e) Without τ-agent, n=50, t=101s

(f) Without τ-agent, n=50, t=122s

Fig. 8. Without τ-agent, 2 Groups, 50 agents per group

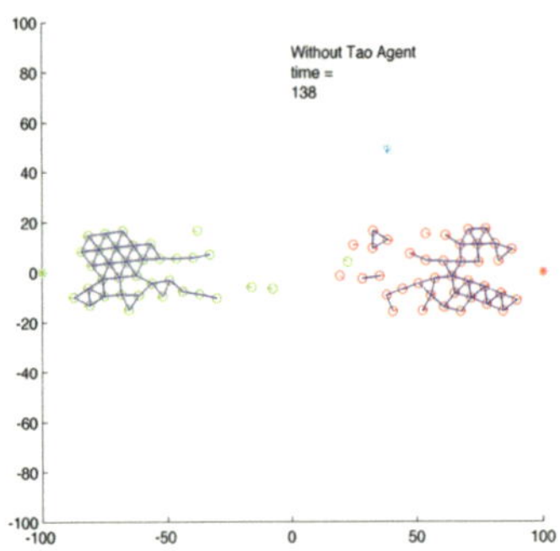

(a) Without τ-agent, n=50, t=138s

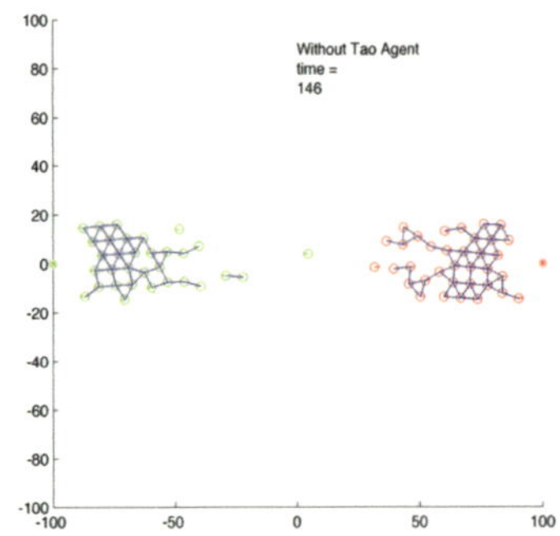

(b) Without τ-agent, n=50, t=146s

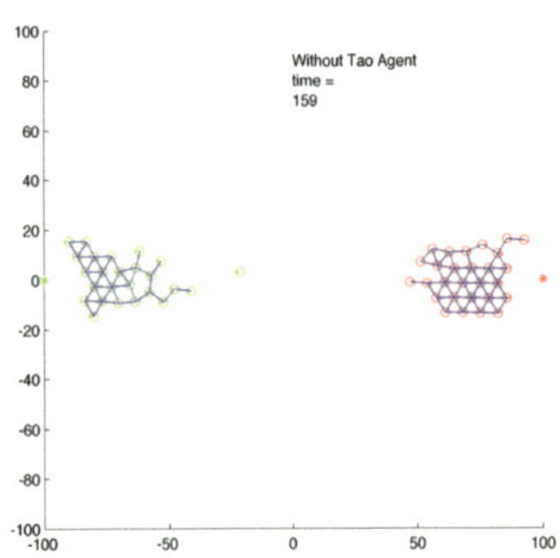

(c) Without τ-agent, n=50, t=159s

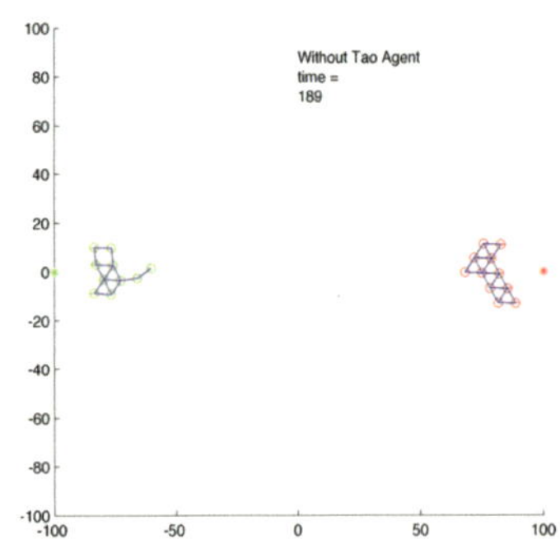

(d) Without τ-agent, n=50, t=189s

Fig. 9. Without τ-agent, 2 Groups, 50 agents per group

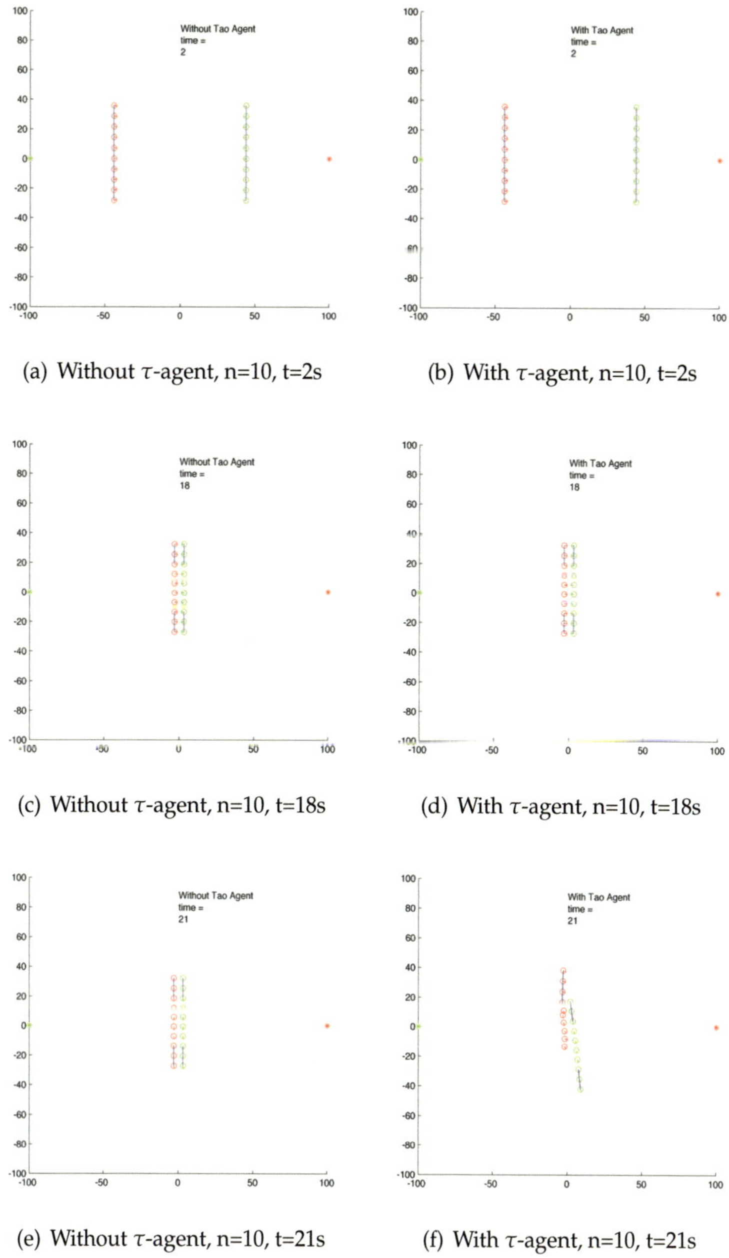

(a) Without τ-agent, n=10, t=2s

(b) With τ-agent, n=10, t=2s

(c) Without τ-agent, n=10, t=18s

(d) With τ-agent, n=10, t=18s

(e) Without τ-agent, n=10, t=21s

(f) With τ-agent, n=10, t=21s

Fig. 10. With and Without τ-agent, 2 Groups, 10 agents per group

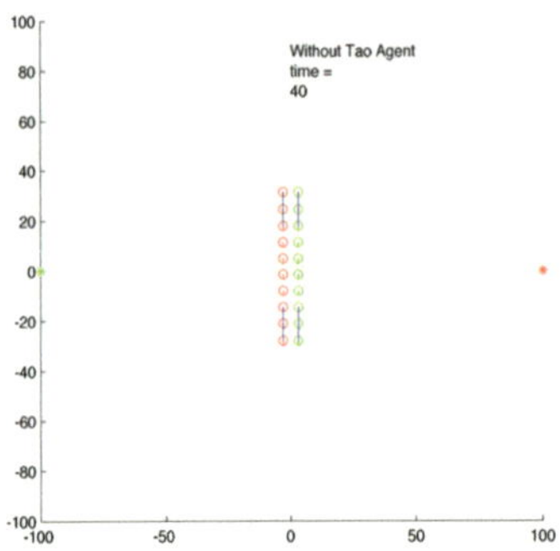

(a) Without τ-agent, n=10, t=40s

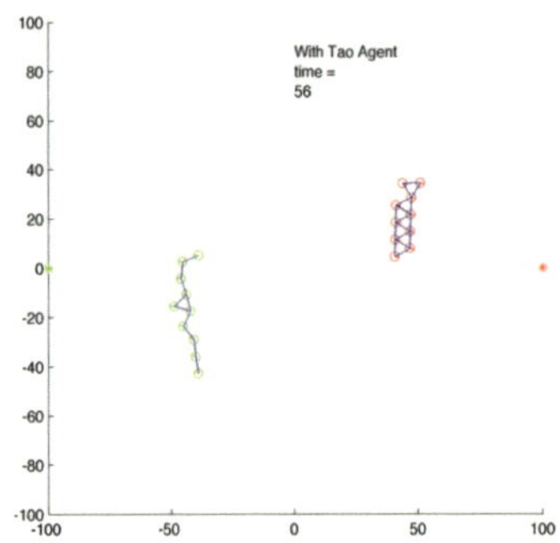

(b) With τ-agent, n=10, t=56s

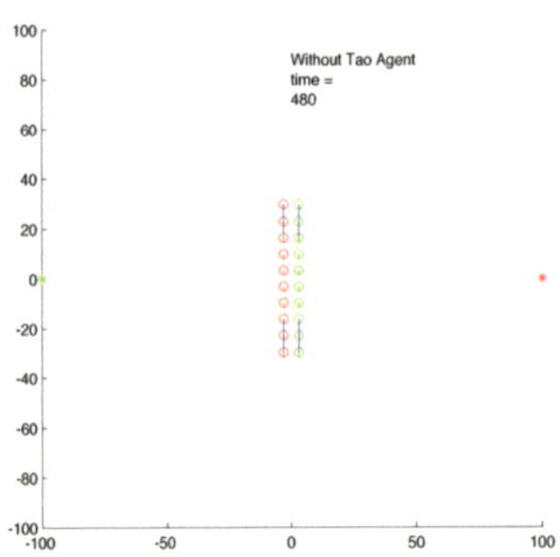

(c) Without τ-agent, n=10, t=480s

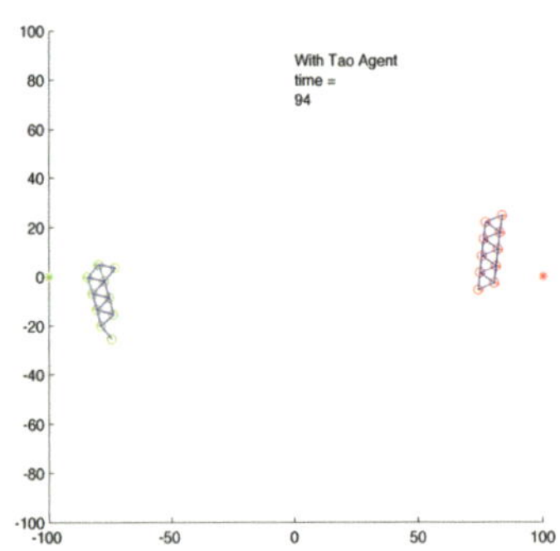

(d) With τ-agent, n=10, t=94s

Fig. 11. With and Without τ-agent, 2 Groups, 10 agents per group

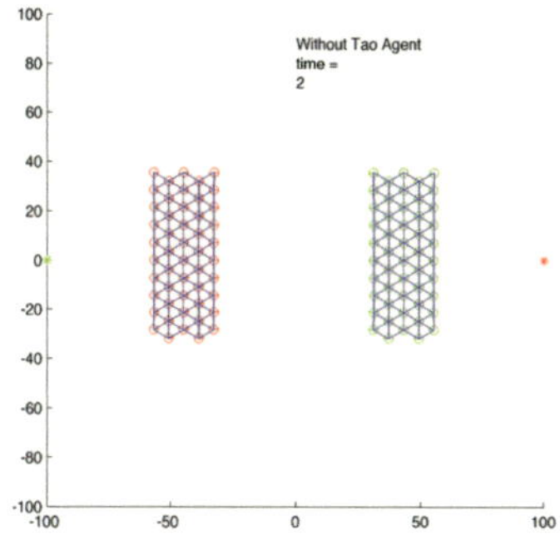

(a) Without τ-agent, n=50, t=2s

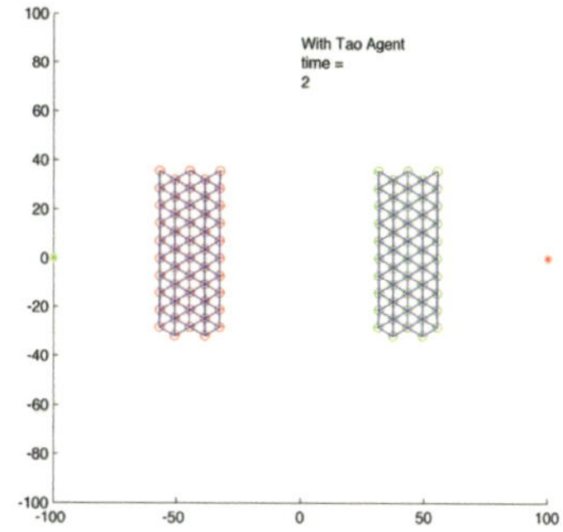

(b) With τ-agent, n=50, t=2s

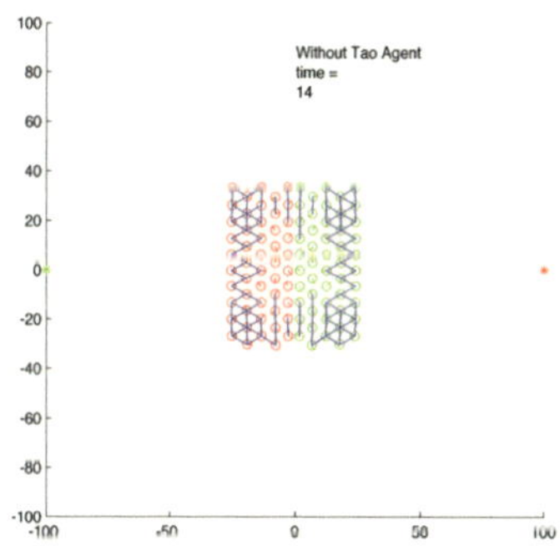

(c) Without τ-agent, n=50, t=14s

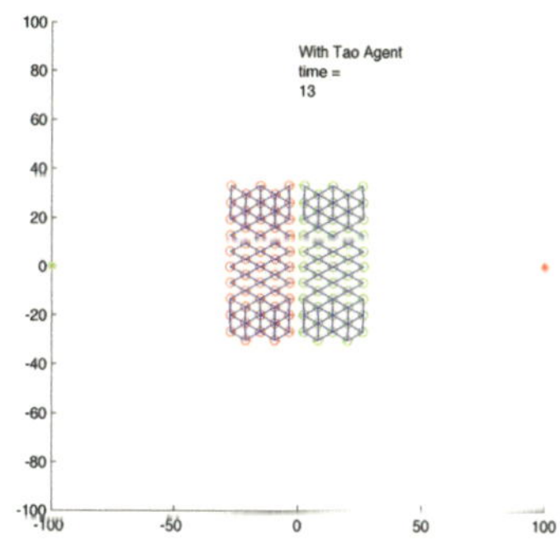

(d) With τ-agent, n=50, t=13s

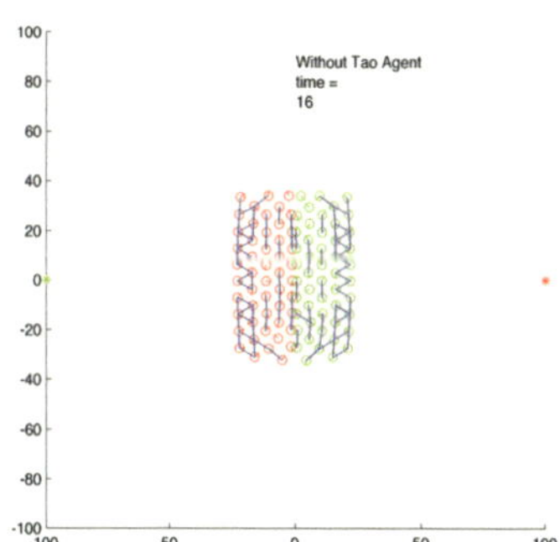

(e) Without τ-agent, n=50, t=16s

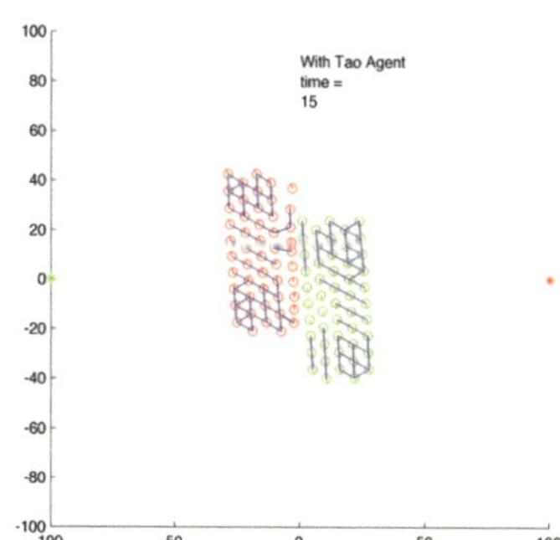

(f) With τ-agent, n=50, t=15s

Fig. 12. With and Without τ-agent, 2 Groups, 50 agents per group

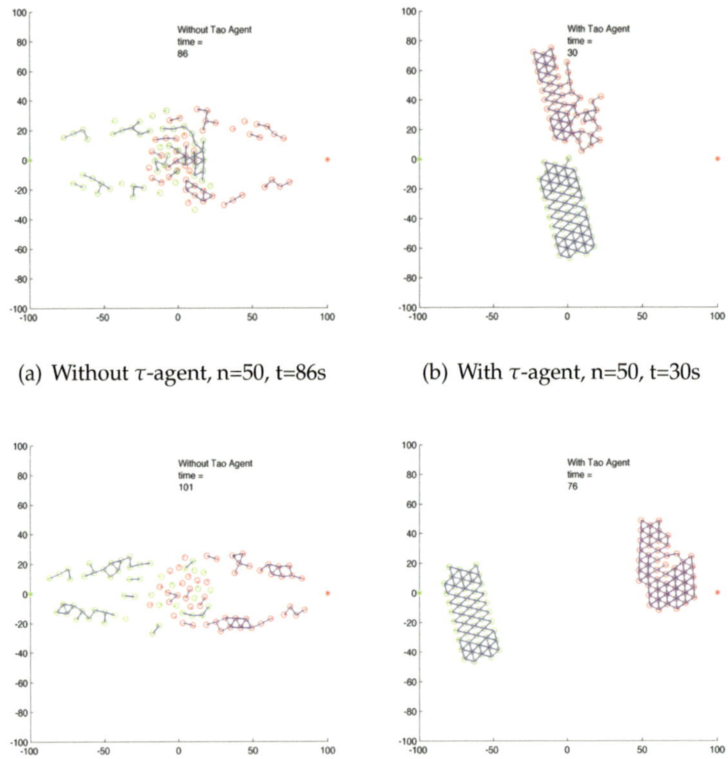

(a) Without τ-agent, n=50, t=86s

(b) With τ-agent, n=50, t=30s

(c) Without τ-agent, n=50, t=101s

(d) With τ-agent, n=50, t=76s

Fig. 13. With and Without τ-agent, 2 Groups, 50 agents per group

7. Conclusion and future work

In this paper we presented an improved multi flock flocking algorithm (equation 3), that added coordinative interaction to Olfati-Saber's flocking algorithm Olfati-Saber (2006) (equation 1). Our algorithm performed better than Olfati-Saber's algorithm in simulations where two groups were in between the other group and the other groups' goal. It performed exactly as Olfati-Saber's algorithm when there was only one flock. This was because our algorithm was based on Olfati-Saber's, and the only difference was when two flocks were close enough to the other flock and the flocks trajectories were set to intersect the other.
The next stage of this work would be to improve the algorithm by modifying our τ agent to change the agent's trajectory based on the trajectory of both agents instead of just rotating τ agent 90 degrees. Another improvement would be to weight the τ agent based on the number of agents in the other flocks. These improvements should allow for a better interaction between multiple flocks instead.

8. References

Fax, J. & Murray, R. (2004). Information flow and cooperative control of vehicle formations, *Automatic Control, IEEE Transactions on* 49(9): 1465–1476.

Gazi, V. & Fidan, B. (2007). Coordination and control of multi-agent dynamic systems: models and approaches, *Proceedings of the 2nd international conference on Swarm robotics*, SAB'06, Springer-Verlag, Berlin, Heidelberg, pp. 71–102.
URL: *http://portal.acm.org/citation.cfm?id=1763837.1763843*

Helbing, D., Farkas, I. & Vicsek, T. (2000). Simulating dynamical features of escape panic, *Nature* 407(6803): 487–490.
URL: *http://dx.doi.org/10.1038/35035023*

Jadbabaie, A., Lin, J. & Morse, A. (2003). Coordination of groups of mobile autonomous agents using nearest neighbor rules, *Automatic Control, IEEE Transactions on* 48(6): 988–1001.

Levine, H., Rappel, W.-J. & Cohen, I. (2000). Self-organization in systems of self-propelled particles, *Phys. Rev. E* 63(1): 017101.

Mogilner, A. & Edelstein-Keshet, L. (1996). Spatio-temporal order in populations of self-aligning objects: formation of oriented patches, *Physica D* 89: 346–367.

Mogilner, A. & Edelstein-Keshet, L. (1999). A non-local model for a swarm, *Journal of Mathematical Biology* 38: 534–570. 10.1007/s002850050158.
URL: *http://dx.doi.org/10.1007/s002850050158*

Moreau, L. (2005). Stability of multiagent systems with time-dependent communication links, *Automatic Control, IEEE Transactions on* 50(2): 169–182.

Okubo, A. (1986). Dynamical aspects of animal grouping: swarms, schools, flocks. and herds, *Advances in Biophysics* 22: 1–94.

Olfati-Saber, R. (2006). Flocking for multi-agent dynamic systems: algorithms and theory, *Automatic Control, IEEE Transactions on* 51(3): 401 – 420.

Olfati-Saber, R., Fax, J. & Murray, R. (2007). Consensus and cooperation in networked multi-agent systems, *Proceedings of the IEEE* 95(1): 215–233.

Olfati-Saber, R. & Murray, R. (2004). Consensus problems in networks of agents with switching topology and time-delays, *Automatic Control, IEEE Transactions on* 49(9): 1520–1533.

Parker, L. (2008). Distributed intelligence: Overview of the field and its application in multi-robot systems, *Journal of Physical Agents* 2(1).
URL: *http://www.jopha.net/index.php/jopha/article/view/18/14*

Parrish, J. K., Viscido, S. V. & GrÃijnbaum, D. (2002). Self-organized fish schools: An examination of emergent properties, *Biol. Bull* 202: 296–305.

Partridge, B. L. (1982). The structure and function of fish schools, *Scientific American* 246(6): 114–123.

Partridge, B. L. (1984). The chorus-line hypothesis of maneuver in avian flocks, *Nature* 309: 344 – 345.

Ren, W. & Beard, R. (2005). Consensus seeking in multiagent systems under dynamically changing interaction topologies, *Automatic Control, IEEE Transactions on* 50(5): 655 – 661.

Reynolds, C. W. (1987). Flocks, herds and schools: A distributed behavioral model, *Proceedings of the 14th annual conference on Computer graphics and interactive techniques*, SIGGRAPH '87, ACM, New York, NY, USA, pp. 25–34.
URL: *http://doi.acm.org/10.1145/37401.37406*

Saber, R. & Murray, R. (2003). Consensus protocols for networks of dynamic agents, *American Control Conference, 2003. Proceedings of the 2003*, Vol. 2, pp. 951–956.

Shaw, E. (1975). Fish in schools, *Natural History* 84(8): 40–45.

Shimoyama, N., Sugawara, K., Mizuguchi, T., Hayakawa, Y. & Sano, M. (1996). Collective motion in a system of motile elements, *Phys. Rev. Lett.* 76(20): 3870–3873.

Tanner, H. G., Jadbabaie, A. & Pappas, G. J. (2007). Flocking in fixed and switching networks, *Automatic Control, IEEE Transactions on* 52(5): 863–868.

Toner, J. & Tu, Y. (1998). Flocks, herds, and schools: A quantitative theory of flocking, *Phys. Rev. E* 58(4): 4828–4858.

Topaz, C. & Bertozzi, A. L. (2004). Swarming patterns in a two-dimensional kinematic model for biological groups, *SIAM Journal on Applied Mathematics* 65(1): 152–174.

Vicsek, T. (2001). A question of scale, *NATURE* 411: 421.
URL: *doi:10.1038/35078161*

Vicsek, T., Czirók, A., Ben-Jacob, E., Cohen, I. & Shochet, O. (1995). Novel type of phase transition in a system of self-driven particles, *Phys. Rev. Lett.* 75(6): 1226–1229.

10

Cooperative Formation Planning and Control of Multiple Mobile Robots

R. M. Kuppan Chetty[1], M. Singaperumal[2] and T. Nagarajan[3]
[1]*Monash University*
[2]*Indian Institute of Technology Madras*
[3]*Universiti Teknologi PETRONAS*
[1,3]*Malaysia*
[2]*India*

1. Introduction

Application of intelligent Wheeled Mobile Robots (WMR) for material handling in the manufacturing environment has been the topic of research in the past decade. Even though researchers have succeeded in applying mobile robots for material handling purpose in the shop floor environment, transporting heavy objects in the assembly line is still a challenge. The dynamical characteristics of a manufacturing environment impose particular abilities a mobile robot should have if it is to operate on the shop floor efficiently, accurately and successfully. Consequently, a WMR needs to adapt itself to everlasting changes. Under such conditions, the use of multiple WMR in closed defined geometric spatial pattern/ formation can be a solution for such applications.

One of the essential problems in guiding multiple mobile robots in such dynamically changing environments is to plan, navigate and coordinate the motion of robots, avoiding obstacles as well as each other while transporting the materials/objects towards the goal. Further it requires the robots to control their relative position and orientation between them on the fly. Hence the control of group of mobile robots performing such tasks requires coordination at different levels starting from navigation to formation (Sugar et al., 2001).

A variety of strategies and control approaches for formation control of group of coordinated robots, have been adopted in the literature such as Graph Theory (Desai, 2002), Vector Potential Field (Yamaguchi et al., 2001), Virtual Structure (Belta & Kumar, 2004), Leader Follower (Fierro et al., 2002) and Behaviour Based Approaches (Arkin, 1998; Goldberg & Matarić, 2002). Further, a comprehensive review of robotic formation, control approaches and algorithms, applications and their advantages and disadvantages have also been addressed (Chen & Wang, 2005). Among all the approaches mentioned in the literature, behaviour - based and leader - follower approach has been widely adopted and well recognized by the researchers because of their simplicity and scalability.

Even though robots are able to move in a closed defined formation when controlled using the various methods reported in the literature, the major limitation has been the difficulty to achieve a stable formation between the robots in the group in dynamically changing, unknown environments filled with obstacles. Under such circumstances, as the number of robots increases, the control methods such as the virtual structure, leader follower and graph theory fail due to their centralization and requirement of higher communication bandwidth. Further it is difficult to design and model the system in a traditional manner and necessitates the implementation of a distributed control strategy with wide communication capabilities between the group members to have knowledge about their states and actions of their teammates. Hence, the control of group of mobile robots performing such tasks requires coordination at different levels starting from navigation to formation (Sugar et al., 2001).

Further, in most of the studies found in the literature, the researchers have dealt the Formation planning and Navigation problems separately, in spite of considering them as combined entity. However, as in the case of guiding the robot group in an unknown environment, in addition to formation planning, robots also need other navigational capabilities to plan their paths to reach their particular goal by avoiding collision between themselves and obstacles in the environment of interest, which have not been addressed in the literature. Another important challenge for formation control is active obstacle avoidance on follower robots, which are not studied in detail in the literature (Chaimowicz et al., 2004; Shi-cai et al., 2007). Therefore the more challenging and important problem is to combine the formation planning and active obstacle avoidance on the follower path, because the follower robot not only needs to perform obstacle avoidance but also has to control itself to remain in the desired formation in relation to the other robots in the group.

In view of the limitations summarized above, three important issues related to distributed formation planning and control of multiple mobile robots namely i) distributed layered formation control framework ii) dynamic role switching algorithm and iii) real-time implementations are addressed in this chapter. Towards acheiving these goals, a new distributed methodology for the multi robot formation platoons of unicycle robots is presented in this chapter. The presented methodology combines the formation planning and navigation based on the hierarchical architecture composed of layers, whose components are the fundamental behaviours/motion states of the robots. Apart from combining formation planning and navigation, one of the most important problems and the major challenge is the avoidance of obstacles in the path of the robots designated other than the leader while guiding the robot group in an unknown environment. To address this problem, a dynamic switching of roles, based on the exchange of leadership between the robots, is incorporated in the control methodology. This is an peculiar feature that distinguishes the presented methodology from the others that exist in the literature.

In the first part of the chapter, the detailed description and methodology regarding the layered approach developed in this work, for solving the multi robot formation control problem is addressed. The control problem combines together formation planning, navigation and active obstacle avoidance, when operating in an unknown environment. In the subsequent parts, the detail about the method of selection of the individual layers and behaviours of the presented approach is presented. Detailed description about how the individual task achieving layers and behaviours are formulated is also provided. The theoretical formulation of formation behaviour based on the leader referenced model and the development of the tracking controller, which is used to minimize the tracking error asymptotically zero and makes the robot in the desired tight formation, is also addressed. In addition to that, the dynamic role switching methodology through the exchange of leadership between the robots and its behaviours, adopted by the robots other than leader in the group, to actively avoid obstacles in their path is presented. Finally the contributions on the development of the presented formation control methodology and its advantages over the other methods found in the literature are summarized and concluded.

2. Multi robot formation control

The detailed description and method of the formation control approach is presented in this section. The main purpose of this approach is to achieve a formation control which guides a group of mobile robots in a closed defined formation relative to each other, while navigating in an unknown unstructured environment. To perform this collective task, layered distributed control architecture whose components are the fundamental behaviours/motion states of the robots, similar to (modified form of) the pack and homogeneous controller (Harry Chia et al., 2005), is developed and presented as shown in Fig .1 (Kuppan Chetty et al., 2010, 2011). In this layered architecture, to achieve the desired objective of the formation planning and navigation in a distributed manner, the total functionality of the multi-robot system is decomposed into functional behaviours. The behaviours such as Navigation and Formation are obtained, based on the motion states of the robots, utilizing the methodology of the behavior based reactive approach, as given in (Arkin, 1998; Xiaoming Hu et al., 2003; Goldberg & Matarić, 2001).
The fundamental behaviours/motion states of the robots are derived based on the advantages of the behaviour based approach and the objective of the entire system. The states are arranged into two levels, a lower level navigation and a higher supervisor level formation, which works on individual goals concurrently and asynchronously. These two levels yield the collective task upon integration. Both these layers and behaviours are related using the priority based arbitration technique, where the entire sets of behaviours are swapped in and out of execution for achieving the goals such as navigation and formation. Therefore, the robots select a particular behaviour/motion state, based on the sensory information perceived by the robot sensors from environment during the fly.

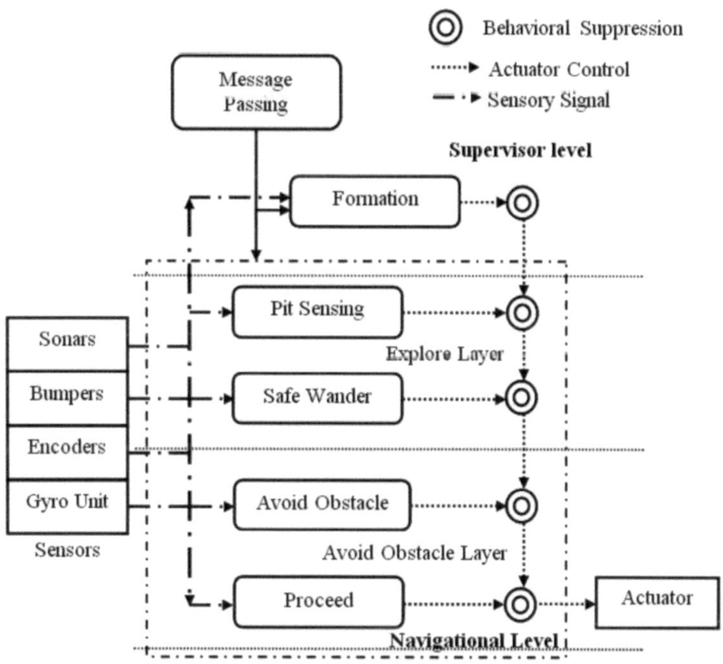

Fig. 1. Layered formation control architecture with task achieving behaviours classified into supervisor level formation and lower level navigation

A reactive controller made up of simple prepositional representation comprising of if –then –else model with task specific sensing, reasoning and planning is used to formulate the behaviours at the navigational level. This controller provides the necessary navigational capabilities and deals with the dynamic control of robots while guiding them in the environment of interest. Further, to have the theoretical formalization and the convergence of the robots into the desired formation, a closed-loop tracking controller at the supervisor level formation is realized using the kinematic model of robots employed in the group in the leader follower model. This controller handles the higher-level objective of multi robot formation. Hence, the proposed approach conquers the deficiencies of the leader follower approach and the behavior based approach, by wrapping up the former with the later, and has the advantages of both the approaches. Thus, the proposed controllers have the capability to address the combined problem of formation planning and navigation through obstacle avoidance.

Referring to Fig. 1 the rectangles in the architecture represent the robot sensors, with the sensor values being transmitted to the behaviors along the long dashed lines. The behaviors themselves are drawn as rectangles with rounded corners arranged in three levels of hierarchical layers namely avoid obstacle layer, explore layer and supervisor layer. The dotted lines represent the command signals sent by the behaviors to the actuators and the inter behavior control signals. The arrowheads in the command lines indicate the priority of the behaviors, which constitutes the pathway to the actuator. The subsumption style priority based arbitration scheme is represented by '◎' with the actuator command coming from the upper level layer taking the precedence. The next section gives the details of the layers, their corresponding behaviors, and the functionality of each behavior in the architecture.

3. Layers and behaviors

This section briefs about the selection of layers and their corresponding task achieving behaviors and how these behaviors are formulated and coordinated to achieve the objective of formation maintenance and navigation tasks.

3.1 Selection of layers and behaviours

In order to have the distributed control nature, the collective task of formation planning and navigation, is divided into three primitive tasks such as Formation, Navigation and Obstacle avoidance, based on the motion states of the robot. These primitive motion states are considered as the basic fundamental behaviors of the robot and are placed in two separate hierarchical levels called the control levels, termed as the supervisor level and the navigational level in the control architecture. The navigational level is the controller's low level. This is responsible for the robot designated as the leader to safely guide the team members towards the goal, without colliding with obstacles or with other team members in the environment of interest, using the behaviors/ motion states present in this level. The supervisor level is the controller's top level, which helps the robots designated as followers to remain in the closed defined formation with their leader using the formation behavior present in this level. The selection of control levels by the robots are based on the priority number assigned to them.

While considering the specific problem of navigation, the leader robot has to perform numerous motion states such as estimating the destination from current location, avoiding collisions with obstacles and other robots, and guiding the follower robots to move in the absence of obstacles to cover large areas. To do this the navigational behaviour is further decomposed into lower level behaviours such as Proceed, Avoid obstacles, Safe-wander and Pit-sensing. All these behaviours are placed in the avoid obstacle and explore layers in the navigational level as shown in Fig. 1. Further, it is necessary to retain the robots in a closed defined formation and to minimize the separation error between the robots. Hence, the formation behaviour is chosen to provide the necessary tracking control for the robots and placed in the supervisor layer in the controller top level. In order to retain the robot formation, it is necessary for the robots to have the postures, orientation and behaviour/state information of the other robots in the group. This helps the robots designated as follower to position themselves relative to the leader with the desired relative separation and orientation, while the leader maneuvers independently. Therefore, the behaviour of message passing is used to provide the explicit communication between the robots using the wireless TCP/IP protocol and placed in the supervisor level.

Initially by default, safe wander behaviour is activated in the controller's leader level and others are suppressed. If one of the other behaviours becomes active, transition from the safe wander to the avoid obstacle occurs accordingly. If the robot is in the follower mode, it executes the formation behaviour which is responsible to keep the robot in a desired formation by minimizing the separation error to zero. Therefore, there are totally five behaviours which are derived based on the objective of the collective task and motion states of the robots, arranged in three layers and in two levels as shown in the Fig 1.

Even though the task of maintenance of formation has the highest priority in the approach, the obstacle avoidance priority also finds the higher order of priority based on the role of the robots in the group. The formation behavior has the highest priority in the follower robot and the obstacle avoidance is considered as the critical behavior, while in the leader robot the obstacle avoidance has the higher priority and the Pit sensing behavior is considered as the critical behavior. The next section presents the details on how these layers and the behaviours on the control architecture are formulated.

3.2 Description of layers and behaviors

The lower level navigational controller of layered formation control architecture consists of two layers such as explore and avoid obstacle layer, which are necessary to provide the navigation capability to the robots. The higher-level formation control consists of supervisor layer, necessary for high-level missions such as formation and communication. The details of the behaviours in the above mentioned layers are given below.

a. Proceed: This behaviour processes the positioning data and provides approximate proceed values for the safe wandering and the obstacle avoidance behaviours. This provides the robot current position and orientation information at every instant in the two-dimensional workspace using dead reckoning principle and the kinematic configurations of the robots and makes the robot to head towards the goal. Hence this behaviour is placed under avoid obstacle layer in the architecture.

b. Avoid Obstacle: This behaviour makes the robot to avoid obstacles / objects without colliding and to manoeuvre within the workspace based on the information received from its sensors as shown in Fig 2 (a). When the robot sensors finds an obstacle/object within the workspace enclosed by the semicircle 'd', the avoid obstacle behaviour is activated and it manipulates the wheel velocities of the robot, as given by Eqn. 1 (a) and (b) necessary to avoid the obstacle.

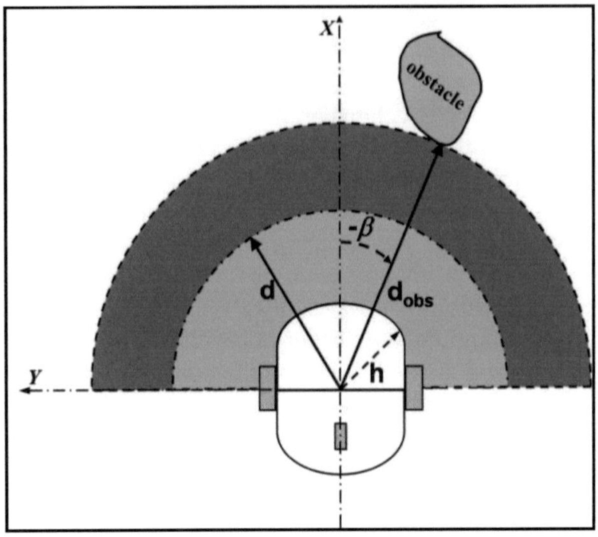

(a)

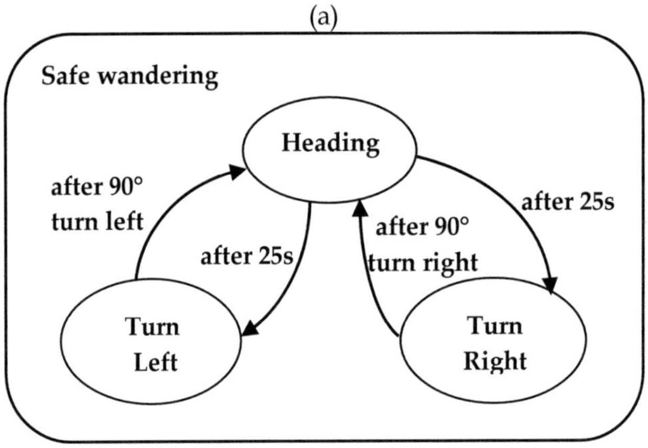

(b)

Fig. 2. (a) Emergence of Obstacle Avoidance behaviour – A top view look of robot and obstacle, (b) Safewander Behavior

$$v = v_{avoid} * (\frac{d_{obs} - h}{d}) \tag{1a}$$

$$\omega = -(\alpha_t * sign(\beta)) \tag{1b}$$

where, v and ω correspond to the translational and rotational velocities in mm/s and rad/s respectively. v_{avoid} is the avoid velocity in mm/s, d_{obs} is the distance of the obstacle with the robot, where the maximum distance is less than 5 m , d is the desired threshold distance necessary to avoid the obstacle, h is the distance between the sensor assembly from the axis of rotation of the robot. α_t is the angle of turn in radians and β is the angle of the obstacle with respect to the robot frame. This behaviour is placed in the avoid obstacle layer, whose functionality is to prevent the collision of robot with obstacle or with other robots and to ensure the safety of the robot.

c. Pit sensing: Existence of pit in the environment is a very critical issue, which has to be avoided by robots manoeuvring the environment of interest. This behaviour makes the robot to avoid pits by controlling the motion of the robot in the backward direction to a predetermined distance, based on the sensor information. Hence this critical behaviour is placed with the highest priority in the explore layer.
d. Safe-wandering: This behaviour guides the robot through the workspace/environment with piece wise constant velocity by turning left or right at regular intervals with predetermined angles as shown in Fig. 2(b). This makes the robot to wander through the environment thoroughly and to look for goals if specified. Hence this behaviour is placed under the explore layer in the architecture.
e. Message passing: Message passing behaviour provides the necessary interaction between the robots allowing them to exchange their motion states, position, orientation and velocity information, using the explicit socket communication capability through wireless links (Hu et al., 1998; Crowley & Reignier, 1993). A 5 - 10 Hz updation rate is provided making the inter-robot communication feasible. This helps other robots in the group to know the current task or behaviour of their teammates. This in turn helps the individual robot in the group to make suitable decisions. Hence this behaviour is assigned with highest priority and is placed on top of all the layers.
f. Formation: When the followers know the trajectory or plan of the leader, this behaviour manipulates the necessary wheel velocities of the follower to maintain its position relative to the leader with desired separation and orientation through the tracking controller. This behaviour is placed in the supervisor level, since our main objective is to maintain the desired formation between the leader and the follower with the robots manoeuvring the workspace/environment. Mathematical modelling of this behavioural function is given in detail in the next section.

4. Mathematical modeling of formation behavior

Formation behavior is made up of mathematical formulation of tracking controller based on the kinematics of the wheeled mobile robot. This section details the formulation of such model to keep the multiple mobile robots in a defined geometric spatial pattern. In formation control problem of multiple wheeled mobile robots, the objective of the robots is to remain in the closed defined geometrical pattern with their teammates. There are several approaches discussed in the literature for formation control. One of the prominent approaches is the leader follower approach where one or more robots acts as the leader and other robots designated as followers follow them. Therefore, the major critical task is to derive a control methodology for the followers to maintain their desired linear and angular separation with their leader to remain in the defined formation topology. This section details about the methodology adopted in deriving a control strategy for the formation behavior, which plays a major role for the robots. To derive a suitable control methodology the following assumptions are made.

4.1 Research assumptions

- Robots employed in the group are identical in their kinematic model, and have same set of sensors, actuators and control strategies.
- Tire differences, model uncertainties, odometry errors by slide and skid are ignored.
- Cartesian coordinate representation as mentioned in (Xiaohai Li et al., 2004) is utilized to develop the control algorithm.

- Robots employed are assumed as perfect velocity controlled robot without considering the dynamics. Hence, the kinematic model is used for development of multi robot leader follower formation framework.
- Robots are aware of each other through explicit inter robot communication with onboard sensing, computation and communication capability.
- The relative information between the robots is utilized rather than the global information systems.
- Leader Referenced model is used, in which other robots maintain the desired position to the leader.

4.2 Leader referenced multi robot system

Let us consider a simple system consisting of two robots in a leader-follower framework and assume that the pose vectors of both the robots are given in the global reference frame of the cartesian coordinate representation as given by Eqn. 3. Fig. 3 shows the kinematic model of non-holonomic differential drive wheeled mobile robots arranged in such a configuration. Here R_L represents the leader robot, R_F represents the follower robot and R_r represent the desired position to be reached by the follower robot to remain in closed formation with the leader by keeping the leader as its reference. In the formation control of pair or robots as shown in figure 3, there are two critical parameters l and φ that determine the geometric shape of the two-vehicle sub system as given by the following representation

$$SF_i = \begin{pmatrix} L_i & \varphi_i^d & l_i^d \end{pmatrix} \tag{2}$$

Where, i = 1, 2, 3…n denotes the robots identification number and SF_i denotes the shape of the formation.

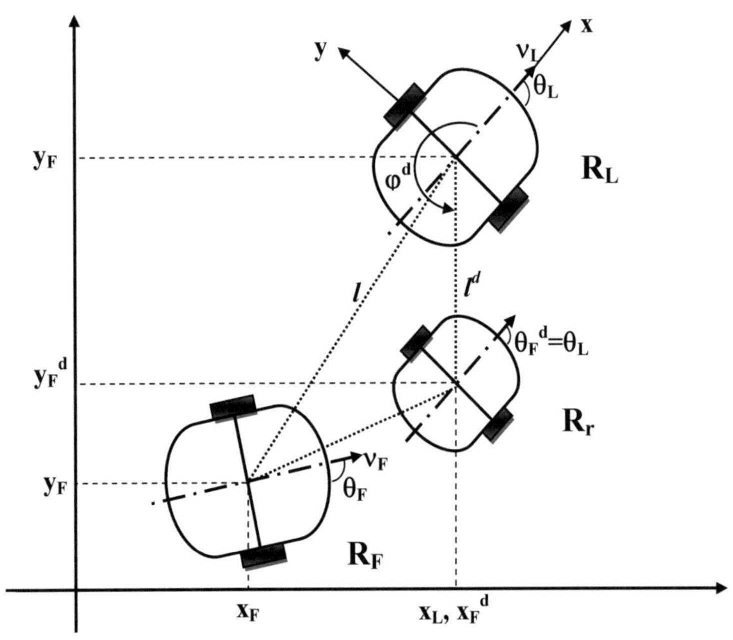

Fig. 3. Kinematic model of the robots in Leader – Follower Formation

Let l and l^d be the relative and desired linear separation and φ and φ^d be the relative and the desired angular separation between the robots respectively. To achieve the desired formation the control has to make $l \rightarrow l^d$ and $\varphi \rightarrow \varphi^d$ and to bring the separation and orientation errors asymptotically to zero. In this case, the control problem reduces to a trajectory tracking control problem rather than the regulation problem of the follower, where it plans its path to efficiently position itself relative to its leader by observing the leader's information. Hence, a tracking controller is to be designed for the follower robots to remain in the closed formation. Therefore, the objective of the tracking controller is to find the velocities of the follower robots.

4.3 Tracking controller

In formation control, the objective of the tracking controller is to find the values of the translational and rotational wheel velocities v_F and ω_F of the follower robots in such a way that the formation/separation errors (linear and angular) decay asymptotically to zero, and position the follower in the desired geometric pattern with its leader.

In order to formulate the tracking control algorithm to find out the wheel velocities of the follower, let the position of the leader and the follower robot in a unit time as in Fig. 3; be given by X_L, Y_L and X_F, Y_F respectively in the fixed ground coordinate system. Let X_r and Y_r be the position of the reference robot, which is the desired position to be reached by the follower to remain in a formation with the desired linear and angular separation with the leader. The orientation of leader and follower robots is given by θ_L and θ_F respectively, and the orientation of the reference robot θ_r is same as the orientation of the leader, which is the basic requirement for the formation platoons.

4.3.1 Generalized tracking controller in global frame

The motion for a non-holonomic differential drive wheeled mobile robot is governed by Eqn. 3,

$$\begin{bmatrix} \dot{x} \\ \dot{y} \\ \dot{\theta} \end{bmatrix} = \begin{bmatrix} \cos\theta & 0 \\ \sin\theta & 0 \\ 0 & 1 \end{bmatrix} \begin{bmatrix} v \\ \omega \end{bmatrix} = J \begin{bmatrix} v \\ \omega \end{bmatrix} \tag{3}$$

Where, J is the Jacobian matrix which defines the kinematics of the WMR. Here, the robot pose vectors are assumed in the global (inertial) reference frame of the cartesian coordinate representation.

Based on the above kinematic model, the velocity equations of the leader robot in the ground frame is given by

$$\begin{aligned} \dot{x}_L &= v_L \cos\theta_L \\ \dot{y}_L &= v_L \sin\theta_L \\ \dot{\theta}_L &= \omega_L \end{aligned} \tag{4}$$

Similarly, for the robot designated as follower, the velocity equations are

$$\begin{aligned} \dot{x}_F &= v_F \cos\theta_F \\ \dot{y}_F &= v_F \sin\theta_F \\ \dot{\theta}_F &= \omega_F \end{aligned} \tag{5}$$

In order to minimize the separation errors, the follower robot adjusts its position relative to its leader by estimating the leader's current posture and velocity information. However, as in the real case, the estimation of information in the global coordinates in the real time requires complex estimation methods. There exists a reference position, which is obtained based on the desired linear and angular separation relative with the leader, since the reference position cannot be estimated directly in the real world. This reference position is used as the desired position for the follower robot to be reached to remain in the formation and to minimize the separation error. Therefore the pose vector of the reference/desired position to be reached by the follower from its position relative with the leader is given by

$$\begin{aligned} \dot{x}_r &= v_L \cos\theta_L + l^d \sin\varphi^d \\ \dot{y}_r &= v_L \sin\theta_L - l^d \cos\varphi^d \\ \dot{\theta}_r &= \dot{\theta}_L = \omega_L \end{aligned} \tag{6}$$

Fig. 4 shows the block diagram of the tracking controller assuming that the robots are in the global coordinate frames.

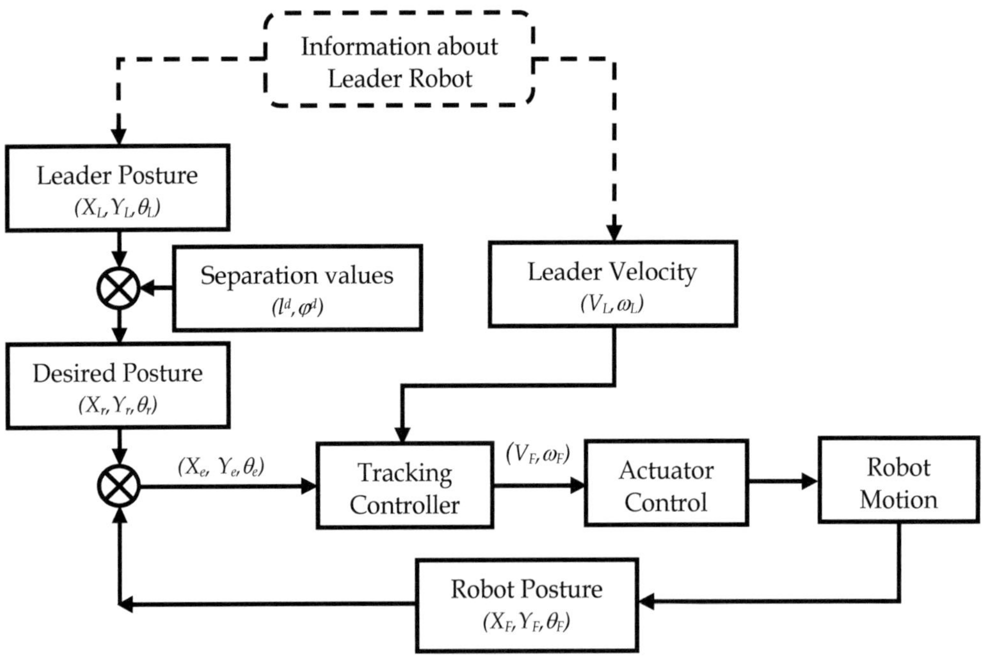

Fig. 4. Block Diagram of the Tracking Controller

The next step is to find the tracking error vector between the reference and actual position for the follower robot. This is given by the Eqn. 7.

$$\begin{bmatrix} \dot{x}_e \\ \dot{y}_e \\ \dot{\theta}_e \end{bmatrix} = \begin{bmatrix} (\dot{x}_r - \dot{x}_F) \\ (\dot{y}_r - \dot{y}_F) \\ (\dot{\theta}_r - \dot{\theta}_F) \end{bmatrix} \tag{7}$$

By substituting Eqn. 5 and Eqn. 6 in the above Eqn. 7, the tracking error becomes

$$\dot{x}_e = \dot{x}_r - \dot{x}_F = (v_L \cos\theta_L - v_F \cos\theta_F + l^d \sin(\varphi^d)) \quad (8)$$

$$\dot{y}_e = \dot{y}_d - \dot{y}_F = (v_L \sin\theta_L - v_F \sin\theta_F + l^d \cos(\varphi^d)) \quad (9)$$

$$\dot{\theta}_e = \dot{\theta}_d - \dot{\theta}_F = \omega_L - \omega_F \quad (10)$$

The above equations are nonlinear. In order to derive a controller, the non linear nature of the above equations are linearized by Input-Output linearization technique and by choosing $\dot{x}_e = -K_1 x_e, \dot{y}_e = -K_2 y_e$ and $\dot{\theta}_e = -K_3 \theta_e$

The control equation becomes

$$V_F = \frac{V_L(\cos\theta_L - \sin\theta_L) + K_1 Xe - K_2 Ye - l^d \cos\varphi^d - l^d \sin\varphi^d}{(\cos\theta_F - \sin\theta_F)} \quad (11)$$

$$\omega_F = K_3(\theta_L - \theta_F) + \omega_L \quad (12)$$

4.3.2 Tracking controller with mapping of coordinate frames

As in the first case, the state of the robot is described in relation to the global coordinate system. But in the real case of the robots, their coordinate system is different from the global coordinate system. In this case, X-axis represents the forward motion of the robot. Hence, to describe the robot motion in terms of component motion, it is necessary to map the motion along the axes of the global reference frame to motion along the axes of the robot local reference frame as shown in Fig. 3. The following orthogonal rotational transformation matrix is used to accomplish the mapping.

$$R(\theta) = \begin{bmatrix} \cos\theta & -\sin\theta & 0 \\ \sin\theta & \cos\theta & 0 \\ 0 & 0 & 1 \end{bmatrix} \quad (13)$$

Further in the case of the Multi Robot Systems (MRS), different robots in the group as members must be able to compare measurements and coordinate their actions with in a common frame of reference. Hence, it is necessary to match the coordinate frame of robots designated as the followers relative to the leader coordinate frames, using the relationship between the inertial and relative coordinate representation of the robots.

When the leader is subjected to only rotation by an angle θ, it is reflected in the reference robot (desired position to be reached by the follower) as a combination of translation and rotation, following a circular path with linear separation as the radius. Hence, the velocity equations of the reference robot in the ground frame is given by

$$\begin{aligned} \dot{X}_r &= v_L \cos\theta_L + l^d \sin(\varphi^d + \theta_L)\dot{\theta}_L \\ \dot{Y}_r &= v_L \sin\theta_L - l^d \cos(\varphi^d + \theta_L)\dot{\theta}_L \\ \dot{\theta}_r &= \dot{\theta}_L = \omega_L \end{aligned} \quad (14)$$

Similarly, the velocity equations of the follower robot is given by

$$\begin{aligned}\dot{X}_F &= v_F \cos\theta_F + \omega_F h \sin\theta_F \\ \dot{Y}_F &= v_F \sin\theta_F + \omega_F h \cos\theta_F \\ \dot{\theta}_F &= \omega_F\end{aligned} \tag{15}$$

The next step in deriving the tracking controller is to obtain the error dynamics of the system in the robot frame by choosing the error coordinates x_e in the direction of v and y_e perpendicular to this direction as depicted in Fig. 3.4. Therefore, to obtain the error in the common local reference/ robot frame, where the coordinate system of the follower is related with leaders system, the orthogonal rotation matrix given above takes the form as

$$\begin{bmatrix} x_e \\ y_e \\ \theta_e \end{bmatrix} = \begin{bmatrix} \cos\theta_F & -\sin\theta_F & 0 \\ \sin\theta_F & \cos\theta_F & 0 \\ 0 & 0 & 1 \end{bmatrix} . \begin{bmatrix} (X_r - X_F) \\ (Y_r - Y_F) \\ (\theta_r - \theta_F) \end{bmatrix} \tag{16}$$

where, $R(\theta) = \begin{bmatrix} \cos\theta_F & -\sin\theta_F & 0 \\ \sin\theta_F & \cos\theta_F & 0 \\ 0 & 0 & 1 \end{bmatrix}$ is the orthogonal rotation matrix.

Fig. 5 shows the block diagram of tracking controller in which the tracking error is represented in the new coordinate system of robot frame.

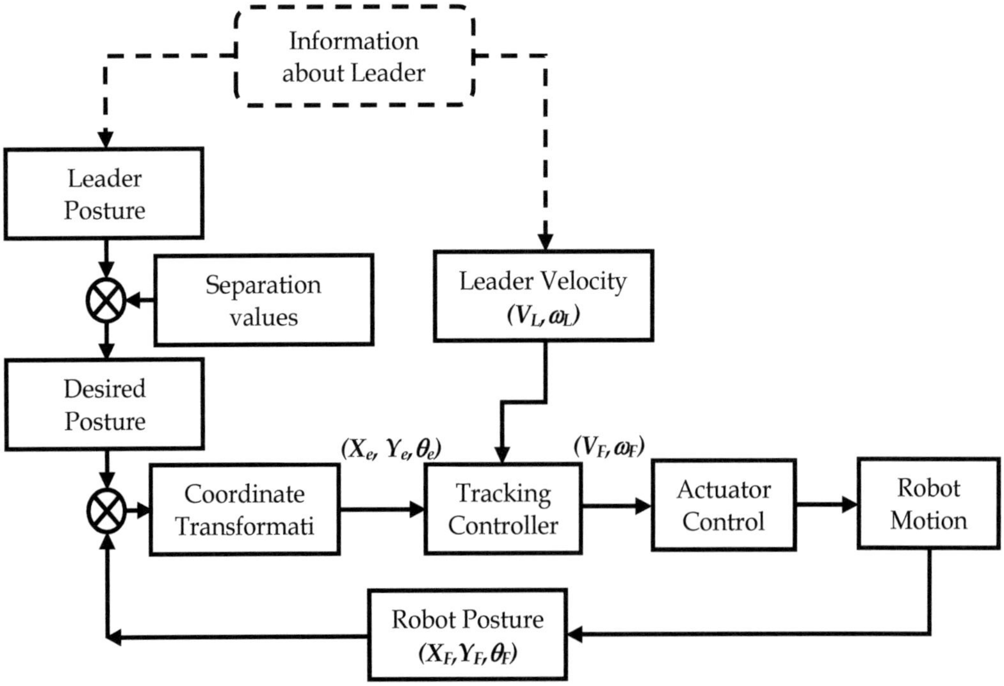

Fig. 5. Block diagram of tracking controller in the robot frame

By substituting Eqn. 14 and Eqn. 15 in the above Eqn. 16, the tracking error in the new coordinate system is obtained as,

$$\begin{bmatrix} \dot{x}_e \\ \dot{y}_e \end{bmatrix} = \dot{R}(\theta).\begin{bmatrix} (\dot{X}_r - \dot{X}_F) \\ (\dot{Y}_r - \dot{Y}_F) \end{bmatrix} = \begin{bmatrix} \cos\theta_F & \sin\theta_F \\ -\sin\theta_F & \cos\theta_F \end{bmatrix} \begin{bmatrix} (\dot{X}_r - \dot{X}_F) \\ (\dot{Y}_r - \dot{Y}_F) \end{bmatrix} + \begin{bmatrix} 0 & \omega_F \\ -\omega_F & 0 \end{bmatrix} \begin{bmatrix} x_e \\ y_e \end{bmatrix} \tag{17}$$

$$\dot{x}_e = -v_F + v_L \cos(\theta_L - \theta_F) + l^d \omega_L \sin(\varphi^d + \theta_L - \theta_F) + \omega_F y_e \tag{18}$$

$$\dot{y}_e = -v_L \sin(\theta_L - \theta_F) - l^d \omega_L \cos(\varphi^d + \theta_L - \theta_F) - \omega_F x_e - b\omega_F \tag{19}$$

The next step is to find a control law for the velocity input and to position the follower in the desired position w.r.t the leader and to minimize the tracking error to zero. An obvious complexity of the above equations of the error dynamics is that the presence of the terms $\omega_F y_e$ and $\omega_F x_e$ indirectly relating the output x and y with the inputs v_F and ω_F through the state variable. Thus, in turn makes the system model to be nonlinear in nature.

In order to find the control law for the velocity inputs of the tracking controller, the Eqn. 18 and 19 are linearized using suitable linearization technique. Here the idea is to algebraically transform the nonlinear system dynamics in to a fully or partly one, so that the linear control theory can be applied. Hence, Feedback linearization technique finds the best solution to linearize the above nonlinear system model especially for tracking controllers (Khalil, 1996; Shankar Shastry, 1999) where the approach involves coming up with a transformation of the nonlinear system into an equivalent linear system through a change of variables and a suitable control input. The difficulty of the tracking controller design is decreased by finding a simple and direct relation between the system output 'y' and the control input '$u = [v \ \omega]^T$' by applying

$$\begin{aligned} \dot{x}_e &= \omega_F y_e - k_1 x_e \\ \dot{y}_e &= -k_2 y_e - \omega_F x_e \end{aligned} \tag{20}$$

and hence, after the feedback linearization the control law for v_F and ω_F is obtained as

$$\begin{bmatrix} v_F \\ \omega_F \end{bmatrix} = \begin{bmatrix} \cos\theta_e & l^d \sin(\varphi^d + \theta_e) \\ \dfrac{\sin\theta_e}{h} & \dfrac{l^d \cos(\varphi^d + \theta_e)}{h} \end{bmatrix} \begin{bmatrix} v_L \\ \omega_L \end{bmatrix} + \begin{bmatrix} k_1 & 0 \\ 0 & \dfrac{-k_2}{h} \end{bmatrix} \begin{bmatrix} x_e \\ y_e \end{bmatrix} \tag{21}$$

where,

$$x_e = \left[\left(X_L - l^d \cos(\varphi^d + \theta_L) - X_F \right) \cos\theta_F + \left(Y_L - l^d \sin(\varphi^d + \theta_L) - Y_F \right) \sin\theta_F \right] \tag{22}$$

$$y_e = \left[-\left(X_L - l^d \cos(\varphi^d + \theta_L) - X_F \right) \sin\theta_F + \left(Y_L - l^d \sin(\varphi^d + \theta_L) - Y_F \right) \sin\theta_F \right] \tag{23}$$

stands for the position error between the desired position i.e. position of the reference robot, and the position of the follower robot in the new coordinate system and k_1 and k_2 are controller gains that are constant positive integers greater than zero, which guarantee the system stability.

4.4 Obstacle avoidance on follower

One of the most important problems and the major challenge is the avoidance of obstacles in the path of the robots designated other than the leader while guiding the robots group in an unknown environment. To avoid this problem, a dynamic role switching methodology based on the exchange of leadership between the robots is incorporated in the developed formation control methodology. The principle behind the switching strategy is that the robot designated as the leader in the real time takes the responsibility of guiding the group through the environment by executing the navigational part of the controller and the robot designated as the follower follows the leader by executing the formation controller. The Follower only leads the group during short time periods in the fly when it has to avoid the obstacle present on its path. Therefore, at any moment during the coordination motion, the robot performing the leading role can become a follower, and any follower can take over the leadership of the team and makes the robot controller to exchange their control modes from navigation to formation and formation to navigation, based on their previous roles and sensory information through explicit inter-robot communication.

As the follower perceives the obstacle on its path based on the sensory information received from its sensors, it sends a request packet to the leader to release the leadership. When the current leader receives the request of releasing the leadership, it immediately releases the leadership to the follower, through an explicit inter-robot socket communication mechanism. Once the follower attains the leadership, it switches its role from follower to leader. Hence, the follower switches from the controllers formation mode to the navigational mode and starts navigating the environment as a temporary leader. In the other side leader robot switches its control to the formation mode and plans its path to track the temporary leader in the defined spatial pattern until the obstacle has been avoided. Under such conditions, the desired linear separation of the follower remains the same and the angular separation changes based on the geometric relationship between the robots as given by Eqn. 24 and shown in Fig. 6.

$$\begin{aligned} \varphi^d &= \pi + \varphi^d\,; \qquad \varphi^d < \pi \\ &= \varphi^d - \pi\,; \qquad \varphi^d > \pi \end{aligned} \tag{24}$$

Fig. 7 shows three robots R_1, R_2 and R_3 in a wedge shaped formation topology, with R_1 designated as leader and R_2 and R_3 designated as followers. When any one of the follower robots R_2 and R_3 perceives the obstacle on its path, the analogous robot attains the leadership temporarily.

Let us consider that the follower robot R_2 perceives the obstacle in it paths. Under such conditions, the Robot R_2 attains the leadership, the desired formation parameter between R_1 and R_2 is obtained as given by the Eqn. 25, and the other robot R_3 tracks the robot R_1 as its reference, without changing the formation parameters between them, which in turn tracks the temporary leader R2 using the simple geometrical relationship between the robots. Similarly, when robot R_3 attains the leadership temporarily, desired formation parameter between R_3 and R_1 is obtained as given by the Eqn. 26, and the other robot R_2 tracks the robot R_1 as its reference. This methodology reduces the computational and inter-robot communication complexity, and helps to minimize the transitory errors between the robots, due to the abrupt change in the formation parameters and the switching of leadership between the robots.

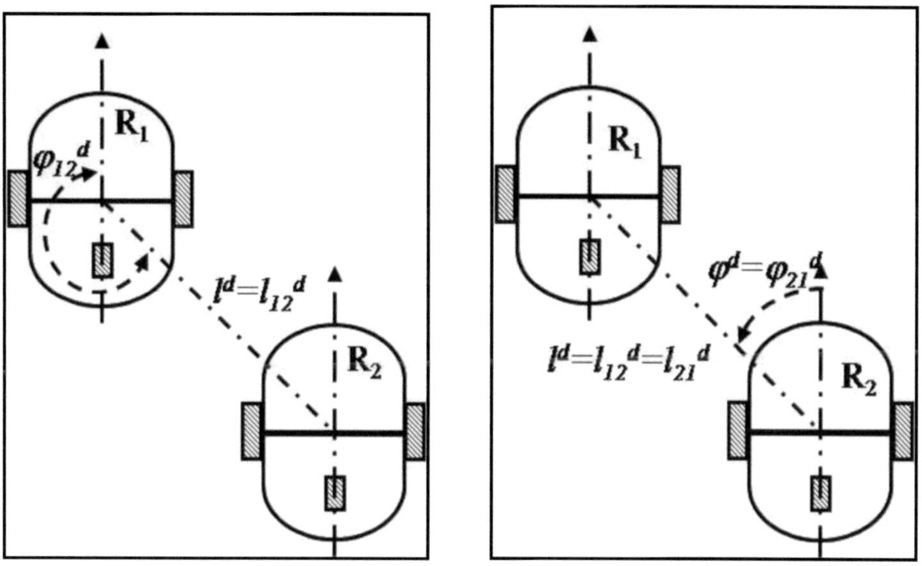

(a) (b)

Fig. 6. Role of angular separation while role switching when (a) R1 as leader and R2 as follower, (b) R2 as leader and R1 as follower

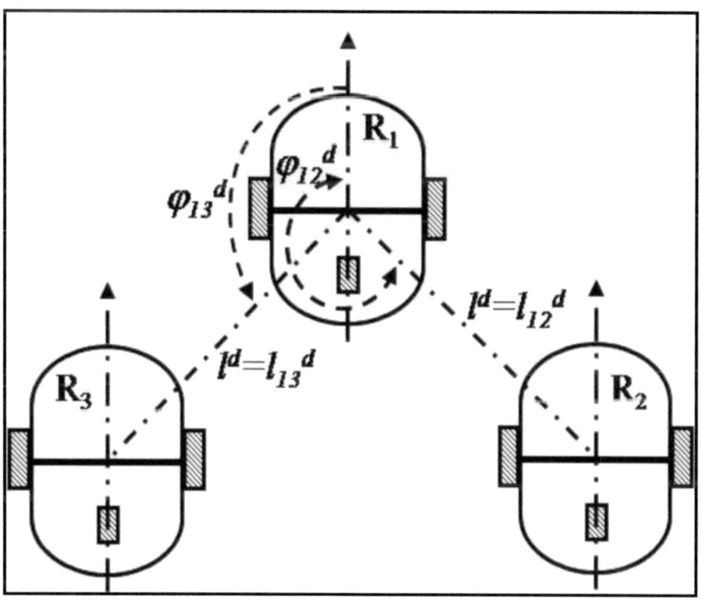

Fig. 7. Role of angular and linear separation while role switching when more than two robots in the group

$$\begin{aligned} \varphi_{12}{}^{d} &= \pi + \varphi_{12}{}^{d}\,; \quad \varphi_{12}{}^{d} < \pi \\ &= \varphi_{12}{}^{d} - \pi\,; \quad \varphi_{12}{}^{d} > \pi \end{aligned} \quad ;\ \text{When obstacle is on } R_2 \tag{25}$$

$$\begin{aligned} \varphi_{13}{}^{d} &= \pi + \varphi_{13}{}^{d}\,; \quad \varphi_{13}{}^{d} < \pi \\ &= \varphi_{13}{}^{d} - \pi\,; \quad \varphi_{13}{}^{d} > \pi \end{aligned} \quad ;\text{ When obstacle is on } R_3 \qquad (26)$$

Another important problem in the three-robot formation is, when the follower robots R_2 and R_3 perceive the obstacle in their path at the same time. In these circumstances, Robot R_1 retains the leadership by itself and commands the follower robots to change the type of formation to inline formation from their initial formation. Therefore, the required change in the desired formation parameters is obtained based on the simple geometric relationship between the robots, and are given by Eqn. 27 and 28.
For Robot R_2

$$\begin{aligned} \varphi_{12}{}^{d} &= \Delta\varphi + \varphi_{12}{}^{d}\,; \ \varphi_{12}{}^{d} < \pi \\ &= \pi\,; \quad \varphi_{12}{}^{d} >= \pi \end{aligned} \quad \text{and} \quad l^{d} = l_{12}^{d} \qquad (27)$$

and for Robot R_3

$$\begin{aligned} \varphi_{13}{}^{d} &= \pi\,; \qquad \varphi_{13}{}^{d} <= \pi \\ &= \varphi_{13}{}^{d} - \Delta\varphi\,; \varphi_{13}{}^{d} > \pi \end{aligned} \quad \text{and} \quad \begin{aligned} l_{13}^{d} &= l_{13}^{d} + \Delta l\,; \varphi_{13}{}^{d} \neq \pi \\ &= l_{13}^{d}; \qquad \varphi_{13}{}^{d} = \pi \end{aligned} \qquad (28)$$

This kind of change in formation topology is preferred to avoid the deadlock situation in the release of leadership, when more than one robot requests the leadership simultaneously. After avoiding the obstacle, temporary leader releases the leadership back to the previous leader, starts executing the formation behaviour, and plans its path according to the leader. Hence, the dynamic switching roles/behaviours in the control architecture allow the robots to trade their roles between them and to actively avoid obstacles on the robots designated as follower's path while maintaining the desired formation.

5. Simulation studies

5.1 Simulation description

The main objective of the simulation studies is to evaluate the different aspects of the control architecture, i.e. to measure the emergence of all possible reactive behaviours/AFSM of the layered approach. The response of the active behaviour yield the motor control output to the robot actuator, which determines the motion of the robot.
Simulation studies are carried out in a phased manner. Before going into the study of multi Robot formation systems, simulation studies to measure the response of the reactive task achieving behaviours of the navigational controller are carried out. The relationship between the formation parameters such as the Instantaneous Centre of Radius/Curvature (ICR/ICC), linear separation (l^d) and the angular separation (φ^d), and their effects on the tracking controller are investigated. Then the error dynamics of the tracking controller which are responsible for positioning the follower robots in the desired separation and orientation with the leader is tested for various formation topologies. In this simulation study, the leader robot is made to move in a predefined trajectory such as 'straight line', 'arc', 'S-shaped' and 'eight shaped' trajectories, with the generalized parameters obtained from the simulation studies. Formation strategies such as in-line, parallel and wedge shaped

have been simulated for both two and three robots in the group. Finally, the dynamic switching of roles for active obstacle avoidance in the follower is incorporated along with the formation planning and navigation.
Matlab - SIMULINK/ Stateflow environment as the simulation tool to evaluate the different aspects of the proposed formation control architecture, since it is an interactive graphical design and development tool that works with Simulink to model and simulate complex systems modeled as finite state machines, also called reactive event driven systems (Stormont & Chen, 2005; Dougherty et al., 2004). In this simulation model, simulation is carried out for 160 units in the time frame with two robots R1 and R2 performing the Leader - Follower formation. Leader is made to navigate the environment using the lower level navigational behaviour, with a piecewise constant wheel velocity of 100 mm/s and the Instantaneous Centre of Rotation (ICR) of 5732mm. Follower is made to track the leader with the desired separation and orientation using the supervisor level formation behaviour. Wheel velocities, wheel direction of rotation and the position and orientation information of the robots are taken as the behavioural outputs and are logged into the data loggers. Simulation parameters and the threshold values for the behavioural activation are provided in the simulation environment taking the kinematics of the robots into account. The wheel velocities obtained as behavioural outputs are constrained and bounded by the conditions $v < v_{max} < 300$ mm/s and $\omega < \omega_{max} < 50°/s$

5.2 Simulation results

5.2.1 Relationship between the formation parameters and ICR / ICC

Fig. 8 shows the simulation results that are obtained to address the generalized relationship between the formation parameters such as the linear separation (l^d), angular separation (φ^d) and the Instantaneous center of curvature ICC/ICR. The critical value of the ICC/ICR needed for the robots to remain in the closed defined formation is obtained for various values of linear separation (l^d) starting from 500mm to 3000 mm, in several formation topologies starting from parallel line to inline formation, using the eqn. 21. The various values of angular separation (φ^d) represent the type of formation topology employed between the leader and follower robots as shown in Fig. 9. The desired angular separation values of 90° and 270° represents the parallel line, 180° represents the inline and the values between 91° and 179° & 181° and 269° represents the collateral line spatial pattern between the robots.
It is observed from Fig. 8, that for a particular type of formation topology either for a parallel, inline or collateral spatial pattern, as the value of the linear separation increases the value of the ICC/ICR also increases making it to be directly proportional in nature. It is also observed that the maximum ICR/ICC is at the collateral formation topology with the desired angular separation of $((^d)$ = 157° and 202°, in which the follower robots are placed in the II and IV quadrant of the cartesian coordinates w.r.t the robot frame as shown in Fig. 9. For the further investigations, the line representing the collateral formation of φ^d = 157° and 202° is taken as the reference based on the consideration of the length and the diameter of the real robots, which are 420mm and 345mm respectively, used in the experimental setup. With this information, in order to have the closed defined stable formation between the robots in all formation topologies, the value of the linear separation and ICC/ICR of 1000mm and 2.3m is chosen as generalized values, by considering the wheel base of the robot.

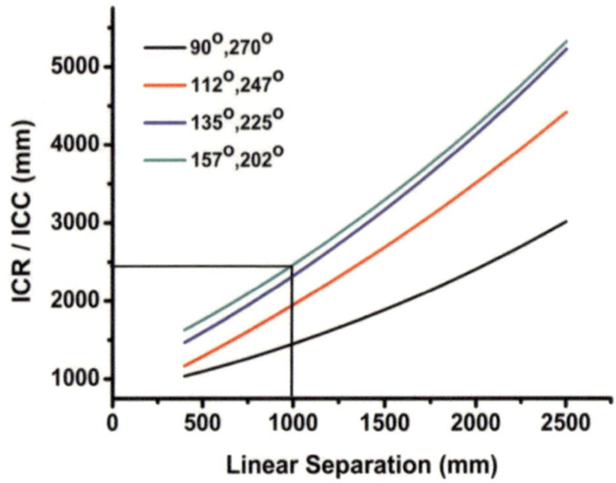

Fig. 8. Relationship between Formation Parameters

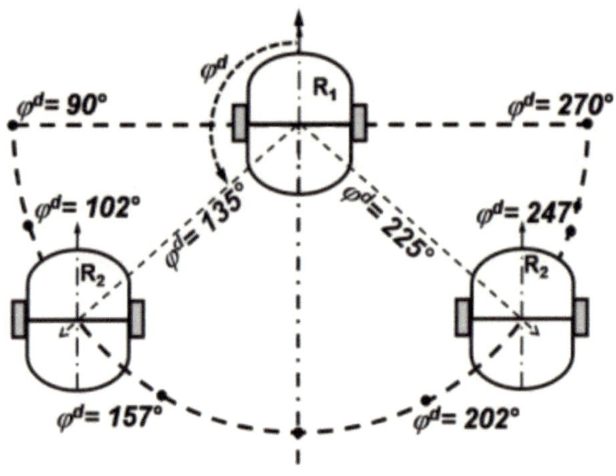

Fig. 9. Robot formation topologies

5.2.2 Formation control combined with dynamic switching of roles

From the several simulation studies carried out, the results obtained from the state based simulation studies, in which the dynamic switching control strategy is incorporated along with formation planning and navigation between the robots is presented. The input parameters for simulation studies are given in Table 1

Fig. 10 shows the trajectory of robots R_1 and R_2, where the leader robot navigates the environment by switching between safe wandering, avoid obstacle, and pit sensing behaviours marked as 'S', 'O' and 'P' respectively. The dynamic switching of roles and exchange of leadership between the robots are indicated by the dotted circles in the figures. At these places R_2 encounters obstacles/pit. Hence robot R_1 acts as the follower and adjusts its wheel velocities to track its temporary leader 'R_2' maintaining the linear and angular separations as required. Fig. 11 shows the orientation profile of both the robots, where it is observed that both robots remain in the same orientation throughout the workspace. This

Parameter	Value
Formation topology	Parallel
Desired Linear separation (l^d)	1000 mm
Desired Angular separation (φ^d) R_1 acts as leader R_2 acts as leader	 270° 90°
ICR/ICC	2.3 m
Translational and Rotational velocity	100 mm/s; 2.5° /s
Initial Positions of the Robots Robot R_1 Robot R_2	 (0,0)mm (-1000, -2000)mm
Simulation time	1600 s

Table 1. Simulation Parameters

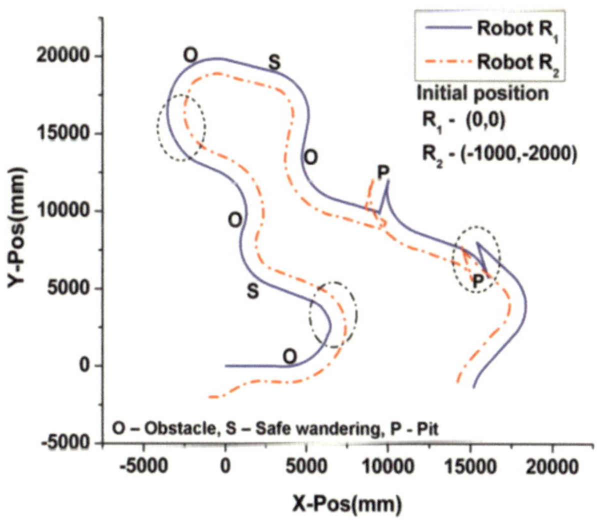

Fig. 10. Trajectory of two robots in parallel line formation.

type of formation is best suited for the application of platoon of vehicles in search and surveillance and industrial transport systems. Fig. 12 (a) and (b) provides the individual behavioural/state output of the robots R_1 and R_2 for the given set of sensory information, in which 'logical 1' represents the behaviours that are active at that time instant to have control over the robot actuator. Fig. 13 shows the relative linear and angular separation between the robots being traded to meet the desired one, where φ^d = 270° and 90 shows the desired angular separation of the robots when R_1 and R_2 acts as the leader respectively. The dynamic motions of the follower robots to remain in the tight spatial pattern with its leader were observed from the translational and rotational velocity profiles plots obtained during simulation as shown in Fig. 14. A positive spike in the follower's wheel is due to the transitory error that persisted in the controller when a sudden reversal of the wheel velocity are involved in the non-holonomic robot systems, when pit sensing behaviour is executed to overcome a pit in both the robots.

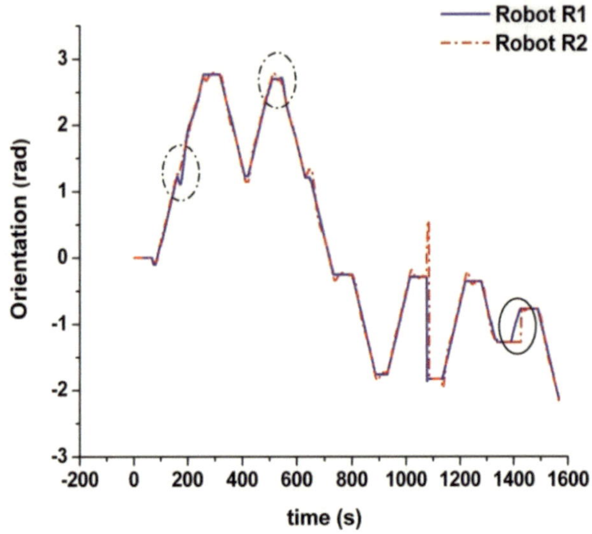

Fig. 11. Orientation profiles of robots in parallel line formation

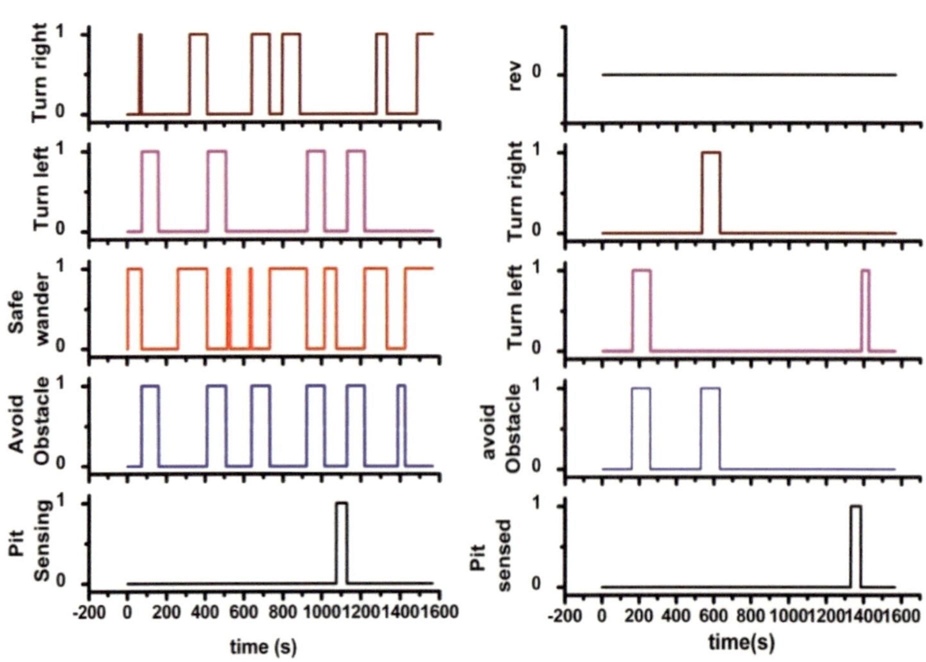

(a) (b)

Fig. 12. Individual Behavior/state output of (a) Robot R_1 and (b) Robot R_2

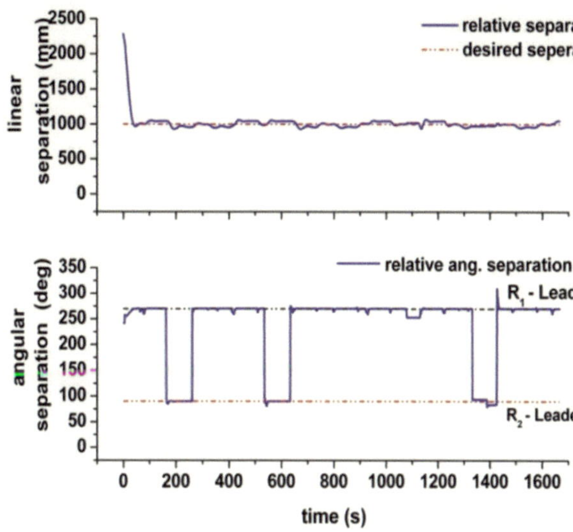

Fig. 13. Formation errors with role switching.

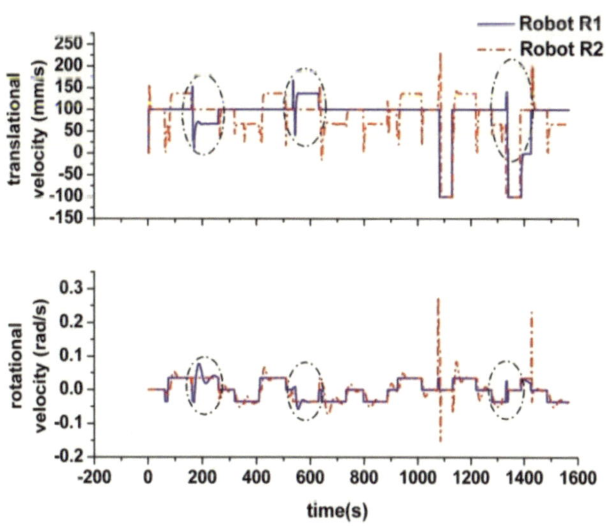

Fig. 14. Velocity profiles of robots with role switching

The dynamic switching of roles and exchange of leadership between the robots are marked by a dotted circle in the above figures, which can also be observed from the encircled portions of the Fig. 14, where robot R_1 acts as the follower and adjusts its wheel velocities to track its temporary leader 'R_2' by keeping the linear separation of 1000mm and angular separation as given by Eqn. 25; to avoid the obstacle found in the path of 'R_2'. Similarly, the formation errors in both linear and angular separation for several formation topologies are tabulated in Table 2. From the simulation studies, it is seen that the tracking error between the robots is around 1.8% and 0.44% in the linear and angular separation respectively, even though the roles and behaviours of the robots are interchanged dynamically

Input Parameters for Leader Robot R_1: K_1=0.55, K2=-0.55; ICC/ICR of 2.3m						**% Error**		
Formation Topology	**l^d (mm)**	**φ^d (deg)**	**θ_e (deg)**	**$\mid l^d - l \mid$ (mm)**	**$\mid \varphi^d - \varphi \mid$ (deg)**	**l**	**φ**	**θ**
Parallel	1000	270	1	4	0.5	0.4	0.2	0.45
Parallel	1000	90	2	6	0	0.6	0	0.55
Wedge	1000	247	1	10	0.5	1	0.2	0.70
Wedge	1000	112	2	8	0.5	0.8	0.45	0.30
Wedge	1000	225	2	25	1	2.5	0.44	0.55
Wedge	1000	135	2	30	1	3	0.74	0.55
Wedge	1000	202	1	36	0.7	3.6	0.35	0.30
Wedge	1000	157	1	32	0.5	3.2	0.32	0.36
Mean Value of error						**1.88**	**0.34**	**0.43**

Table 2. Formation errors in simulation

5.2.3 Switching of formation topology

Fig 15 (a) and (b) shows the results for the simulation studies carried out to measure the error dynamics and formation convergence of the controller when the formation between the robots are switched from one topology to another. This study has been carried out to investigate the application of switching of formation topologies between the robots to avoid the deadlock situation that occurs when more than two follower robots requests for leadership. In this case the follower robots are made to change their initial formation into an inline formation with the leader, to avoid the deadlock situation encountered while taking up the leadership. Two different configurations are simulated which can be used to solve the deadlock when two or more than two robots are used in the formation framework.

In the first configuration, the formation topology between the robots is varied by changing the angular separation at a defined rate w.r.t time and the linear separation between the robots is kept constant. This is used when only two robots are in the group. When more than two robots are involved in the formation framework, the linear separation between the robots are also varied at a defined rate along with the angular separation, inorder to avoid the positioning of the robots to overlap each other.

Fig. 15 (a) shows the results of the simulation study, where the leader is made to move in a straight line trajectory and the follower is made to switch from parallel line to inline formation by varying its angular separation at a continuous rate. In this simulation study, initial positions of the leader and follower robots are taken as (0,0) and (-200, -1000) and the desired linear separation between the robots is set as 800 mm. The change in the formation topology is initiated as a continuous change in the angular separation at various rates to find the optimum value that can be used in the experimental studies. It is also clearly observed that the design of the tracking controller makes the follower robot to switches its formation topology from parallel line to inline spatial pattern with the leader. Fig. 15 (b) shows the linear separation errors between the robots for various rate of change of the angular separation values that determines the change in formation topology. It can be clearly observed from the above results that the continuous rate of change of angular separation at a value of 3°/s, makes the robot to gradually change its formation topology to inline topology with minimum error even though it takes more time when compared with other values. Further, it takes around 39s to settle in the desired linear separation with the leader which includes 30 s taken by the controller to change its angular separation at a rate of 3°/s until it changes from 270° to 180°.

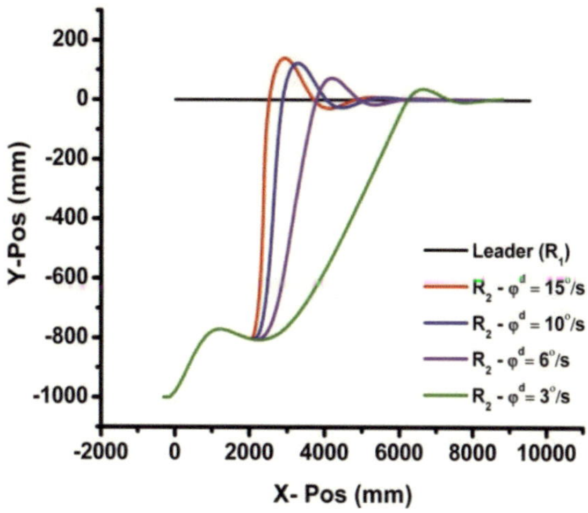

(a)

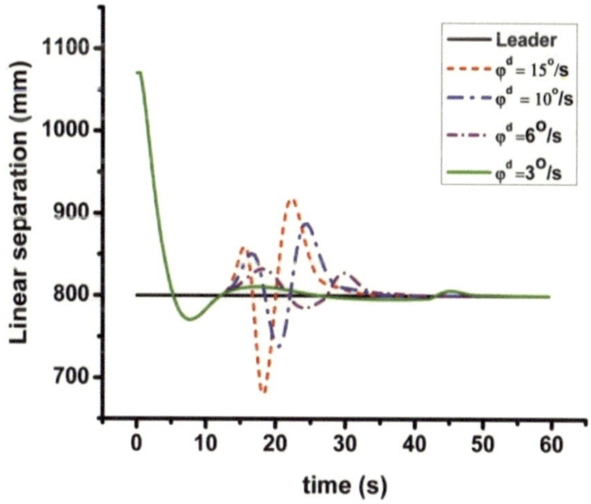

(b)

Fig. 15. Switching of Formation topology with variations in angular separation (a) trajectory of robots switching from Parallel line (b) Separation plot of robots

6. Real time implementations and experimental investigations

6.1 Experimental environment

Experiments are carried out with two commercially available Pioneer P3DX Robot research platforms, named as PEIL R_1 and R_2 as shown in Fig. 16. These have identical sensor, actuator and kinematic configurations. A Bi-directional multiple Client-Server architecture incorporating the Pioneer platforms is developed as an experimental architecture. Robot R1 is designated as server/leader and Robot R2 as client/follower performing the combined task of navigation and formation.

Fig. 16. Experimental Robot Research Platforms

The individual task achieving behaviours are programmed as parallel motion functions, and they are integrated as the functional classes/library files in the application interfaces of the mobile robots. An explicit wireless socket communication mechanism is developed by using the 'Net packet' functional library classes of the mobile robots, to provide the necessary inter-robot communication and to initiate the exchange of leadership between the robots. A trapezoidal acceleration function with the acceleration and deceleration value of 50% is used as the velocity function. During startup, the robots are considered to be in the origin position of 0, 0, 0 in X, Y and θ. The initial velocities of the robots are considered to be zero. Further, the Leader robot is made to initiate its motion, only if the connection between the follower robots is established by the wireless socket communication protocol. Hence, when two robots are employed in the formation framework, leader robot is powered first and then the followers are started up.

Several experiments have been carried out to measure the performance of the proposed formation control approach for formation convergence and as a supplement to the investigation of the same in the simulation studies reported in the previous section.

First experiment is carried out to investigate the performance of the tracking controller and the formation errors when applied to multiple WMRs moving as a whole in a tightly coupled coordination in all formation topologies. In this experiment, the robot R_1 designated as leader is made to move in the environment in predefined trajectories such as 'arc', 'S-shaped' and 'eight shaped trajectories, with the generalized parameters obtained from the simulation studies. These trajectories are preferred, since they provide both translational and rotational profiles to the robot and the translational and rotational changes can be considered as step changes in the input to the controller as considered in the simulation studies.

Second experiment is carried out to measure the performance of the proposed formation control architecture combined with lower level navigation and supervisor level formation. In this experiment, the initial and desired separation values are taken as 1414 mm, 262° and 1000 mm, 270° respectively and both the robots are employed with the layered control architecture. The leader is made to navigate an environment of 12m by 10m rectangular workspace filled with rectangular obstacles of size 200 x 180 mm and 550 x 400 mm. Leader velocities are fixed at 160 mm/s and 4°/s, obstacle distance of 800 mm and safe wander time of 6 s, and the follower is made to follow the leader with the desired separation and orientation as mentioned earlier.

Experimental study on dynamic switching of roles is carried out as the third experiment, to test the performance of the approach on combined formation and navigation capability of multi-robot systems, to have active obstacle avoidance in the follower path. Experiment is conducted in a similar environment as used in the second experiment with the robots moving in a parallel line formation. The initial and desired separation values are taken as 1019 mm, 258° and 800mm, 270° between the robots. In this experiment, both the robots are employed with the same control configuration, having the capability to perform both navigation and formation and they are allowed to trade their roles between themselves using the leadership exchange method. These experiments have been carried out in a workspace of 12 m by 10 m with three obstacles of size 200 x 180 mm, 200 x 180 mm and 550 x 400 mm as shown in Fig. 17.

Fig. 17. Arrangement of robots and obstacles in the 2D workspace

Finally, experiments on switching of formation topology between the robots are carried out to measure the capability of the proposed approach to avoid from the deadlock situation when three robots are used in the formation study. The idea is to measure the adaptability of the proposed approach and convergence of the tracking controller, when the robots designated as followers encounter obstacle on their path simultaneously. In this case the follower robots are made to change their initial formation into an inline formation with the leader, to avoid the deadlock situation encountered while taking up the leadership. This experiment is conducted with two robots instead of three robots due to the limited availability of research platforms, and the robots designated as leader is made to navigate a straight line in the workspace and the follower is made to switch from the parallel line to the in-line formation to measure the formation convergence. In this experiment, the initial and desired separation between the robots are taken as 0, 0 and -1044 mm, 264° and 800 mm, 180° respectively and are given by the following relationship. The change in formation is initiated with constant linear separation and varying the angular separation at the rate of

3°/s. The results of all the above experiments are reported and discussed in detail in the following sub section.

6.2 Experimental results

Fig. 18 to Fig. 21 shows the results of the experiments illustrating the good performance of the control algorithm (Kuppan Chetty et al., 2010). For the first experiment, Fig. 18 (a) shows the trajectory of the robots in the parallel line formation, where the leader robot navigates the environment by switching between safe wandering, avoid obstacle behaviours marked as 'S', 'O' respectively whenever it perceives the obstacle information from the environment. Moreover the performance of the formation controller is also observed from figure 18, where follower tracks the behaviour of the leader form its initial linear and angular separation and remains in the desired formation relative to the leader throughout the workspace. It also shows that, the formation controller minimizes the separation error to zero, keeping the robots in the tight coupled formation in the environment filled with obstacles.

Fig. 18 (b) shows the orientation plot of the above experiments, where it can be observed that the tracking controller makes the follower to remain in the same orientation with the leader throughout the fly, making the system suitable for transporting objects from one location to the goal in the dynamic industrial environments.

Fig. 19 and 20 shows the experimental results on dynamic switching of roles, carried out to test the obstacle avoidance in the follower path. Fig. 19(a) shows the trajectory of leader and follower robots performing the combined task of navigation, formation and obstacle avoidance where 'O' represents the obstacles in the path of the robots and the dotted rectangles represents the switching of roles and leadership between the robots to reach the desired goal. Fig. 19 (b) shows the behavioural / state output of the robots R_1 and R_2 involved in the experiments. Initially, robot R_1 is designated as the leader and leads R_2 which follows the motion of the leader in the parallel line formation with the desired separation of 800 mm and 270°. At t = 8.4 and 26s, follower finds the obstacle on its path, request for exchange of leadership with the leader, acquires the leadership and performs obstacle avoidance on its path. After avoiding the obstacle, robot R_2 relinquishes the leadership back to the leader at t=14 and 30s. This can be clearly observed from Fig. 19 (b), where the role switching behaviour takes precedence which intern activates obstacle avoidance behaviour present in the control architecture of robot R_2. This clearly indicates that at any moment during the coordinated motion, the follower robot, which experiences obstacle on its path, can take over the leadership from the leader and the robot performing the lead role can become a follower by switching their leadership between themselves. Fig. 20 shows the variation in the linear and angular separation errors, where the robot R_2 performs obstacle avoidance on its path by switching its roles with R_1 and preserves the formation by adjusting its angular separation as given by Eqn. 25. It is also observed; that the formation controller takes 9s to settle in the desired formation with the leader and the formation errors is estimated to be less than 1.4% and 0.5% in both linear and angular separation between the robots.

Figs. 21 (a) and (b) show a comparison between the simulation and experimental results indicating the behaviour of the tracking controller to position the follower robots even after switching of formation topology from parallel line to inline spatial pattern. The angular separation (φ^d) is changed at a rate of 3°/s, which is the critical parameter to initiate the formation change.

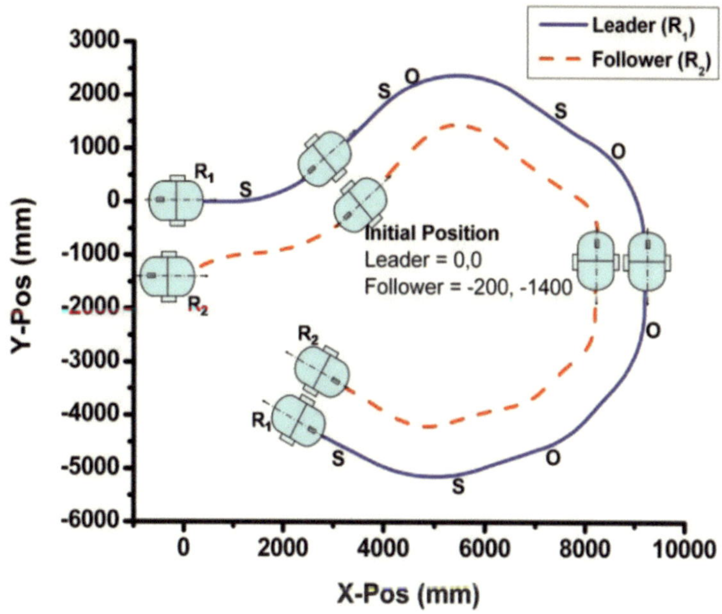

(a)

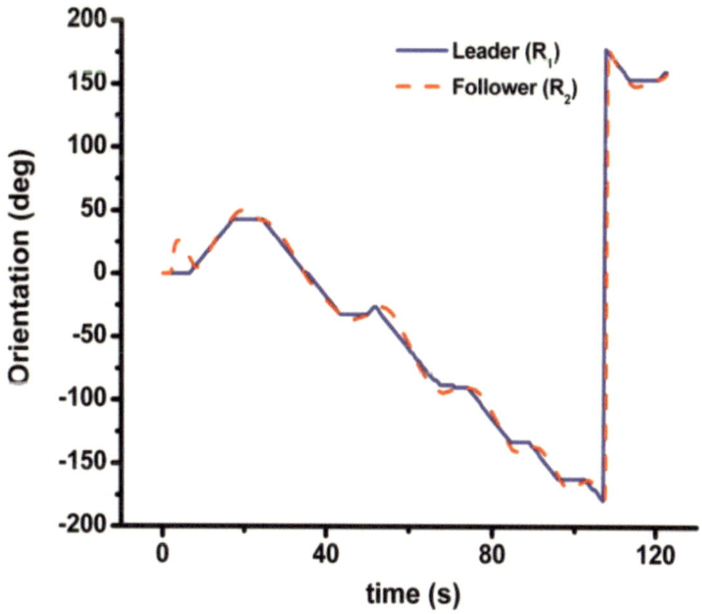

(b)

Fig. 18. Performance of Formation Controller combined with Navigation and Formation behaviors (a) Trajectory of robots in parallel line formation (b) Orientation profile between the robots

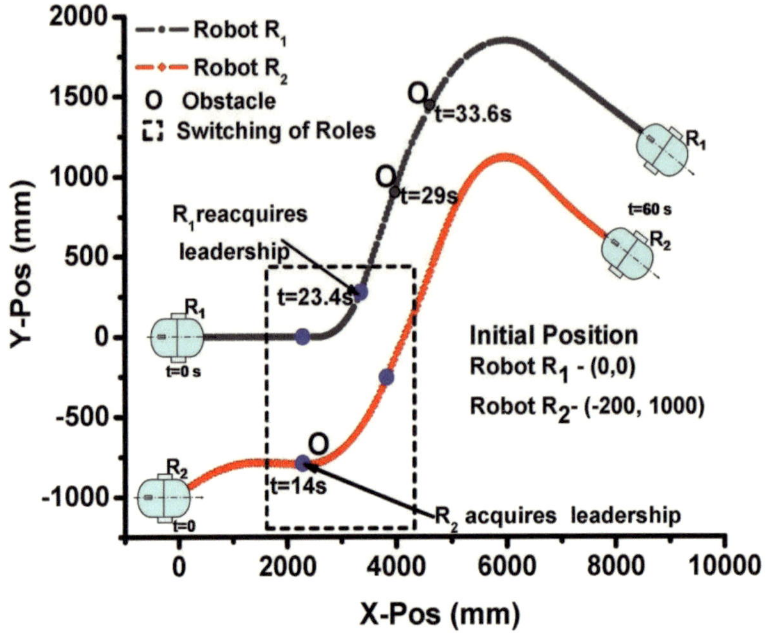

(a)

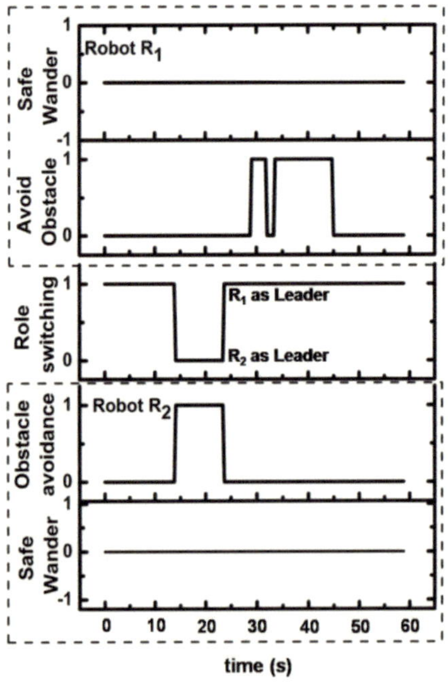

(b)

Fig. 19. (a) Trajectory of robots performing obstacle avoidance on both leader and follower robots in a parallel line formation (b) Behavioral/State outputs of the Robots

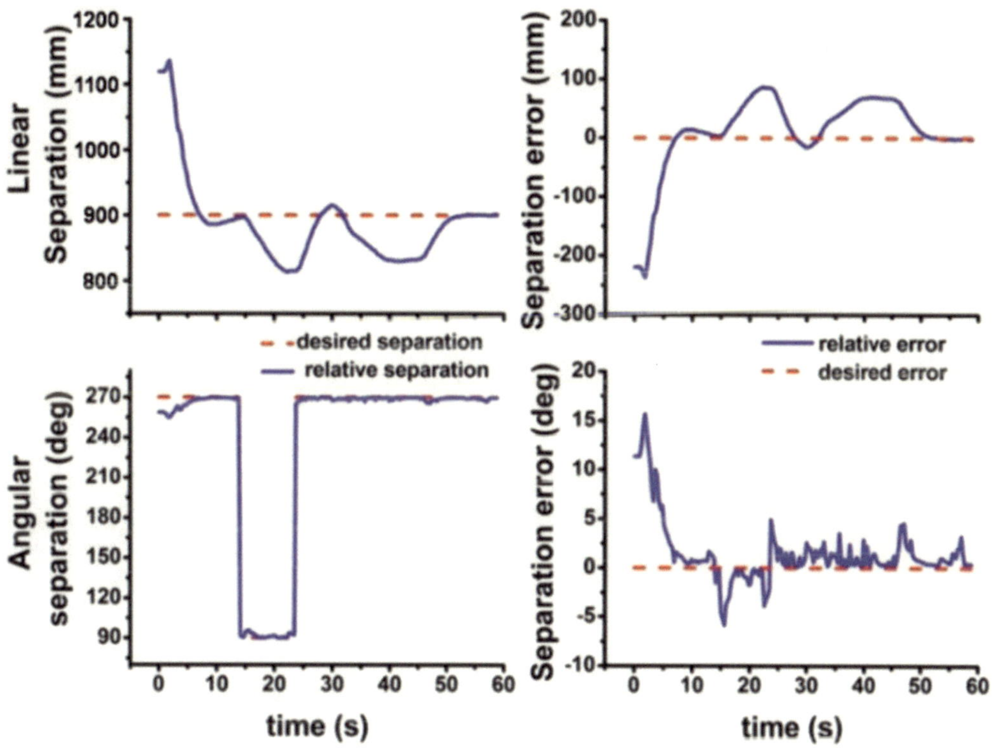

Fig. 20. Formation plot of in Parallel line formation along with switching of roles

Input Parameters for Leader Robot R_1: K_1=0.55, K_2=-0.55; ICC/ICR of 2.3m						**% Error**		
Formation Topology	l^d (mm)	φ^d (deg)	θ_e (deg)	$\lvert l^d - l \rvert$ (mm)	$\lvert \varphi^d - \varphi \rvert$ (deg)	l	φ	θ
Parallel	900	270	1.5	22	1.6	2.4	0.6	0.45
Parallel	900	270	2	18	1	2	0.40	0.55
Parallel	1000	90	2	12	0.6	1.2	0.7	0.70
Parallel	1000	90	1	16	1	1.6	1.11	0.30
Collateral	1000	202	2	48	1	4.8	0.50	0.55
Collateral	1000	135	2	30	2	3.0	1.50	0.55
Collateral	1000	157	1	46	2	4.6	1.26	0.30
Mean value of Error						**2.81**	**0.82**	**0.71**

Table 3. Formation errors (experimental analysis)

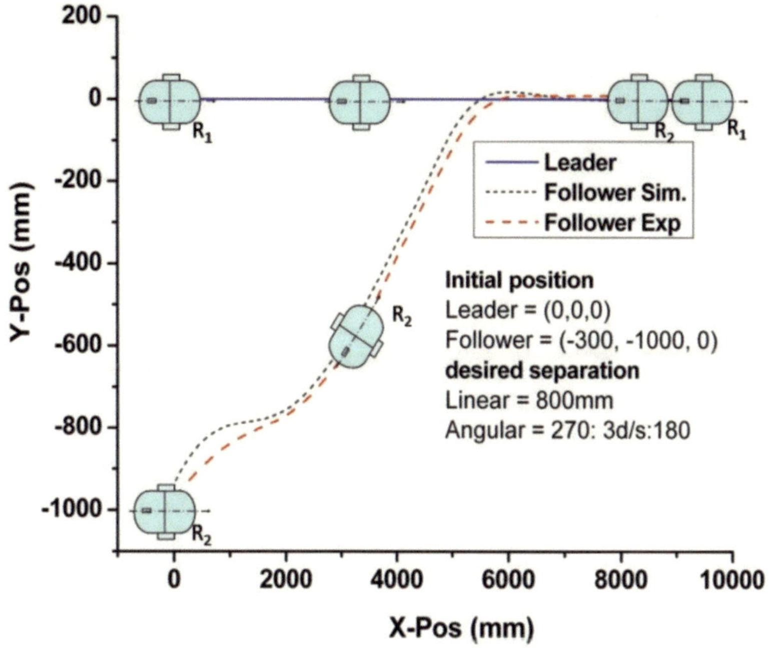

(a)

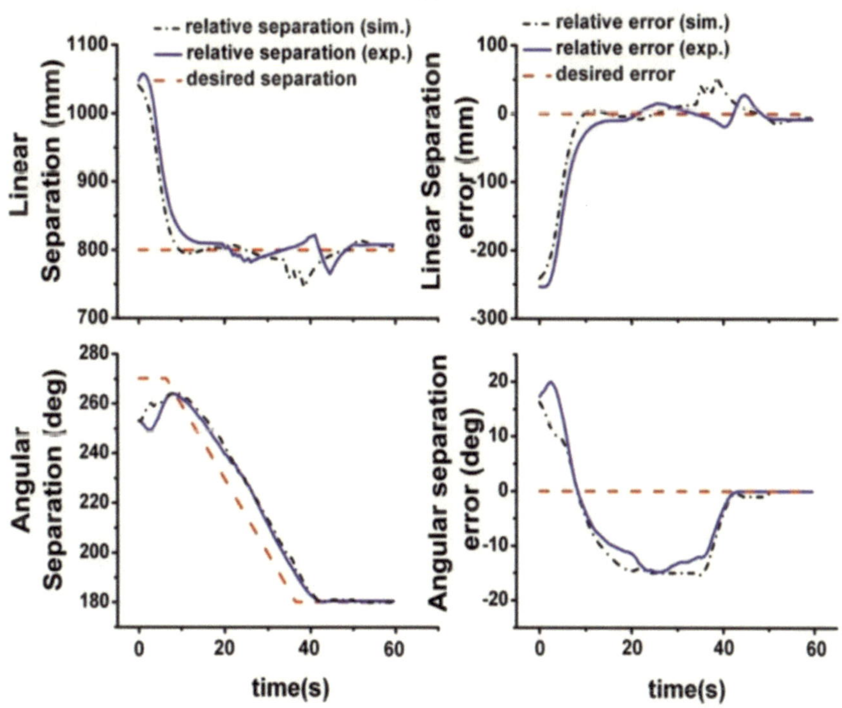

(b)

Fig. 21. (a) Trajectory of robots switching from parallel to straight line formation (b) Formation plot of robots switching from parallel to straight line.

It is observed from the results, that the tracking controller takes less than 12 s to settle in the desired linear separation of 800mm between the robots from their initial position and it takes another 40s to switch the formation topology between the robots which includes the 30s time taken for the rate of change of desired angular separation. This method of changing the formation topology is one of the remedy to avoid the deadlock caused when one or more robots in the group request for leadership simultaneously, to avoid the obstacle on their path and the group leader is free to move in the environment.

7. Conclusion

A combined layered formation control approach to conquer the deficiencies of the leader follower approach and the behavior based approach, by wrapping up the former with the later is developed. The developed approach has the advantages of both the approaches, which have the capability to address the combined problem of formation planning and navigation through obstacle avoidance. In order to have the distributed control nature collective task of formation planning and navigation is divided into three major (primitive) tasks such as Formation, Navigation and Obstacle avoidance based on the motion states of the robot. These primitive motion states are considered as the basic fundamental behaviors of the robot according to the reactive behavior based approach and are placed in three separate layers, which yields the collective task upon integration. Further, to have the theoretical formalization and the convergence of the robots into the desired formation, and the closed loop tracking controller based on the leader follower model and the kinematics of the unicycle mobile robots is formulated to keep the robots in the team in a closed defined spatial pattern, where the separation errors are ,minimized asymptotically to zero.
A unique feature of the developed formation control approach is the incorporation of dynamic role switching methodology through the exchange of leadership between the robots and their behaviours, to actively avoid the obstacle in the path of the robots designated other than the leader in the group. The results show the capability to avoid the obstacle in the follower's path. This clearly indicates the capability of the approach to make the robots to remain in a closed defined stable formation between them even though the roles are switched dynamically during the fly. Under these circumstances, the tracking controller takes less than 12s to make the follower to remain in the formation with the leader and the formation errors are less than 2.81% and 0.82% in both linear and angular separation between the robots respectively. As a whole the formation error for the projected formation control approach combined with formation planning, navigation and obstacle avoidance capability between the robots, the formation errors are estimated at less than ±3% and ±0.81% in linear and angular separation between the robots; when compared to ±5.5% and ±3.6% reported in the literature (Fierro et. al., 2002 and LIU Shi-Cai et. al., 2007).
In future, we plan to test the capability of the developed control methodology by conducting the experiments carrying a real object in the real dynamic industrial environment filled with obstacles. The switching control strategy between the behaviours and robots arises many interesting questions such as deadlocks between the robots when they exchange the leadership. There are many cases to be solved to answer this deadlock phenomenon. We also plan to extend into the issues to overcome the deadlock situation in the multi robot systems through simulation and experimentation studies.

8. Acknowledgement

We would like to acknowledge Dr. Eng. Tetsunari Inamura, of Interactive Intelligent Systems Laboratory, National Institute of Informatics, Tokyo, Japan, for offering visiting researcher position and providing the necessary laboratory research facilities to perform preliminary experimental investigations at Chiba Annexe. We would also like to acknowledge National Institute of Informatics for the financial assistance provided during the course of stay in Japan.

This work has been supported by and conducted at Precision Engineering and Instrumentation Laboratory, Department of Mechanical Engineering, Indian Institute of Technology Madras (IIT Madras), Chennai, INDIA.

9. References

Arkin, R. (1998). *Behaviour-Based Robotics*. Chap 3 & 4, The MIT Press, Cambridge, Massachusetts, USA.

Belta, C. & Kumar, V. (2002). Trajectory design of formations of robots by kinetic energy shaping, *Proc. of. The IEEE International Conference on Robotics and Automation*, Washington, pp. 2593 - 2598.

Chaimowicz,L., Vijay kumar, R. & Mario F.M. Campos (2004). A mechanism for Dynamic coordination of Multiple Robots. *Int. J. Autonomous Robots*, 17, pp. 7-21.

Chen, Y.Q. & Wang, Z. (2005). Formation control: A Review and Consideration. *In Proceedings of the IEEE/RSJ international Conference on Intelligent Robots and Systems*, Alberta, Canada, August, pp. 3181- 3186.

Crowley, J. & Reignier, P. (1993). Asynchronous Control of Rotation and Translation for a Robot Vehicle. *International Journal of Robotics and Autonomous Systems*, 10, pp. 243–251.

Desai, J.P. (2002). A graph theoretic approach for modeling mobile robot team formations. *Journal of Robotic Systems*, 19(11), pp. 511-525.

Dougherty, R., Ochoa, V., Randies, Z. & Kitts, C. (2004). A Behavioral Control approach to formation keeping through an obstacle field. *In Proceedings of the IEEE Aerospace conference*, Big Sky MT, March, 1, pp. 168-175.

Fierro, R., Das, A., Spletzer, J., Exposito, J., Kumar, V., Ostrowski, J.P., Taylor, C.J., Hur, Y., Alur, R., Lee, I., Grudic, G. & Southall, B. (2002). A framework and architecture for multiple-robot coordination. *International Journal of Robotics Research*, 21(10 -11), pp 977-995.

Goldberg, D. & Matarić, M.J., (2002). Design and Evaluation of Robust Behavior-Based controller for distributed multi robot collection tasks, 315 - 344, In T. Balch and L. E. Parker, *Robot Teams: From Diversity to Polymorphism*, A.K. Peters, Natick, M.A, USA.

Harry Chia, Hung Hsu, & Alan Liu (2005). Multiagent-Based Multi team Formation Control for Mobile Robots. *Int. Journal of Intelligent and Robotic Systems*, 42 (1), pp. 337-360.

Hu, H., Kelly, I., Keating, D. & Vinagre, D. (1998). Coordination of Multiple Mobile robots via Communication. *Proceedings of SPIE'98 Mobile Robots XIII Conference*, Boston, USA, pp. 94-103.

Khalil, H. (1996). *Nonlinear Systems*, Prentice - Hall International, UK.

Kuppan Chetty, RM., Singaperumal, M. and Nagarajan, T. (2010). Behavior Based Planning and Control of Leader Follower Formations in Wheeled Mobile Robots. *International Journal of Advanced Mechatronics Systems*, 2(4), pp. 281-296.

Kuppan Chetty, RM., Singaperumal, M. & Nagarajan, T. (2011). Distributed Formation planning and Navigation Framework for Wheeled Mobile Robots. *Journal of Applied Sciences*, 11 (9), pp. 1501 -1509.

LIU Shi-Cai, TAN Da-Long & LIU Guang-Jun (2007). Formation Control of Mobile Robots with Active Obstacle Avoidance. *International Journal of Acta Automatica Sinica*, 33(5), pp. 529 - 535.

Shankar Shastry., (1999). *Nonlinear systems: analysis, stability, and control*, Springer Verilag.

Stormont, D.P. & Chen, Y.Q. (2005). Using mobile robots for controls and Mechatronic education. *International Journal of Engineering Education*, 21(3), pp. 1039-1042.

Sugar, T., Desai, J.P., Kumar, V. & Ostrowski, J.P. (2001). Coordination of multiple mobile manipulators. *Proceedings of the IEEE International conference on Robotics and Automation*, 3(1), pp. 3022-3027.

Xiaohai Li, Jizhong Xiao & Jindong Tan (2004). Modeling and Controller design for multiple mobile robots formation control. *Proceedings of the IEEE International conference on Robotics and Biometrics*, Shenyang, China, Aug, pp. 839-843.

Xiaoming Hu, Alarcon, D. F. & Gustavi, T. (2003). Sensor Based Navigation Coordination for Mobile Robots. *Proceedings of the IEEE International Conference on Decision & Control*, 6375-6380.

Yamaguchi, H., Arai, T. & Beni, G. (2001). A distributed control scheme for multiple mobile robotic vehicles to make group formations, *Int. Journal of Robotics and Autonomous systems*, 36 (4), pp. 125 - 147.

11

Cooperative Path Planning for Multi-Robot Systems in Dynamic Domains

Stephan Opfer, Hendrik Skubch and Kurt Geihs
University of Kassel
Germany

1. Introduction

A stationary robot in an assembly line expects to execute all its hard-coded movements without any collision. Its recognition is limited to its tasks and the corresponding area of operation. In case of a failure an alarm is given and it stops working until a human being corrects the fault. The assembly robot does not react independently, which is the exact opposite to autonomous mobile robots. An autonomous mobile robot has to solve many problems an assembly robot is never confronted with. While the latter moves according to its hard-coded patterns, a mobile robot on the contrary moves in a potentially unknown and dynamic environment. Operating in such environments requires the mobile robot to localise its position relative to the environment, sense its surroundings continuously, and move accordingly. In order to achieve its goals in a changing environment, a mobile robot also has to adapt its actions according to the changes. Hard-coded action sequences are too limited and error-prone to reliably achieve a specific goal. Finally, a secure execution demands the robot to plan its movements and their consequences. Collision-avoidance is an important topic in this context.

The requirements for an autonomous mobile robot which is part of a team of robots are even more challenging. A robot has to recognise its team members separately, include their positions and abilities in its own planning and consider their paths for its own path planning. Nevertheless, many reasons for accepting this overhead exist. One robot handling complicated tasks is possibly more expensive and complex than several single robots. Furthermore, several robots are able to solve a given task faster than a single robot, by dividing the task into subtasks and assigning them to different robots. Moreover, a team of robots exhibits reliability, as malfunctioning robots can be replaced by team members at runtime. Finally, a large spectrum of tasks can be covered by a group of robots at once, while a single robot has to be adapted for each specific task.

A central requirement to benefit from these advantages is the ability to coordinate multiple autonomous mobile robots. For this chapter we will confine ourselves to the problem of path planning in a team of autonomous mobile robots. Refining the requirements, we assume the following conditions:

Highly Dynamic Environment Assigned tasks are very critical with regard to time, because of a highly dynamic environment. Therefore robots must hove at speeds in the order of 2-6 meters per second and smooth paths are crucial to complete tasks in time and without accidents such as rollovers.

Neutral or even antagonistic entities It is possible that robots and other moving obstacles are crossing or blocking the intended path. Handling these situations demands reactive behaviour.

Limited Sensor Range The sensor range of the robots is limited, so that the environment is only partially observable. Hence, destinations are often beyond the local sensor range.

Unreliable Communication Team members are able to communicate their sensor information between each other by means of unreliable network communication, such as W-LAN. Therefore, sensor information of team members are only available with a certain delay or not at all. Moreover, due to the dynamic and time critical nature of the domain, robots cannot wait for communication to take place before acting.

Noisy Sensors All sensor information is subject to a certain amount of noise.

The RoboCup Middle Size League[1] is a domain that features these conditions. Figure 1 shows robots competing in that league. Among other scenarios, we use this domain as a test bed.

Fig. 1. RoboCup Middle Size League 2009

These conditions require careful crafting of the whole process, ranging from sensor fusion, validation of sensed information, and tracking of objects over time, to information abstraction, planning and conflict resolution. In this chapter we will address all of these issues.
In Section 2, we will discuss the foundational techniques as well as the prior work on which our approach relies. Section 3 explains our approach in detail. It is, same as the approach itself, divided into two parts, namely a data-driven part, in which sensor fusion, tracking and information abstraction occurs, and a command driven part, in which planning and reactive conflict resolution occurs. The data-driven part is explained in Section 3.1, while the command-driven computations are described in Section 3.2. Afterwards, in Section 4, we discuss the presented framework and conclude in Section 5.

2. Foundations

This chapter explains the foundations of cooperative path planning. Examples and scenarios are drawn from the robotic soccer domain. Section 2.1 gives an overview of existing path planning algorithms and techniques in the multi-robot context. In the following, in Section 2.2 to 2.3, the algorithms used in our approach are explained in more detail.

[1] `www.robocup.org`

2.1 Path planning in multi-robot systems

There are various approaches to path planning in multi-robot system, however, finding the optimal solution is NP-hard. Hopcraft et al. (1984) simplify the planning problem to the problem of moving rectangles in a rectangular container. They proved the NP-hardness of finding a plan from a given configuration to a goal configuration with the least amount of steps. Hence, all feasible approaches to path planning are a compromise between efficiency and accuracy of the result.

Parker (2009) classifies path planners in three groups: central, or coupled, approaches, decoupled approaches and coordinated approaches. In the following, each of them is described.

2.1.1 Central approaches

This form of planning considers the individual robots as part of a single, coupled system, and uses a single, central planning unit for calculating the plans. The employed algorithms are then directly drawn from single-robot versions:

Potential Fields Using derivable functions, such as gaussians, each point in space is mapped onto a real value. These values can be interpreted as a force pushing a robot away or drawing it in. In the most simple case, a robot just needs to follow the gradient to arrive at its goal. Figure 2(a) shows an example, depicting a robot with a blue dot and its goal with a cross. (Bruijnen et al., 2007; Topor, 1999)

Space Decomposition Here, the space between obstacles is divided into small areas. A path is represented as a set of neighbouring areas. Should multiple robots try to enter the same area, the area is sub-divided. The subdivisions are then assigned to the different robots. (Parsons & Canny, 1990)

Combinatorial Approaches In order to simplify the search space, it is abstracted to road maps. A road map consists of a set of globally available paths and intersections. Each robot has to follow one of the paths and exit the road map close to its goals (Peasgood et al., 2008). Voronoi graphs, such as the example in Figure 2(b) are a road map variant. We will discuss advantages of Voronoi graphs in Section 2.5.

2.1.2 Decoupled approaches

In contrast to coupled approaches, decoupled approaches consider each robot on its own. Thus, no central planning unit is necessary and every robot can calculate its own path. There are two possibilities to resolve conflicts:

- Each robot is assigned a priority. Robots with lower priority have to wait for the planning result of robots with higher priority and consider the received paths when planning.
- All paths are planned without regard for each other. Only afterwards potential conflicts are solved to avoid collision.

In contrast to central approaches, the complexity of decoupled approaches is only linear in the number of participating robots. (Parker, 2009)

2.1.3 Coordinated approaches

Coordinated approaches differ strongly from the two previously discussed groups. Favouring time efficiency over completeness, a complete path planning is omitted. Parker (2009) mentions three different kinds of coordinated approaches:

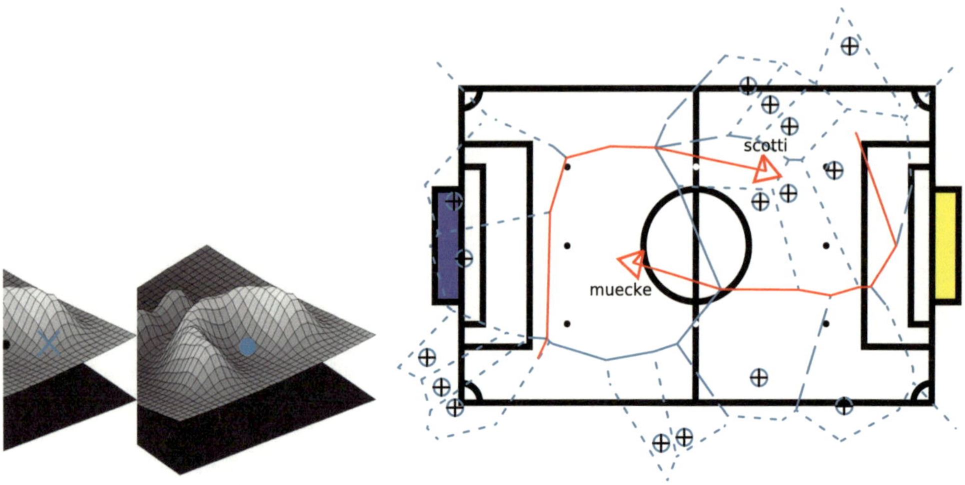

(a) A Potential Field for path planning (b) Two robots path planning using a Voronoi graph

Fig. 2. Different Path Planning Methods

Traffic Control Collisions and coordination problems are resolved by globally known traffic rules (i.e., right-before-left or vice versa). Possible applications include storage facilities, where autonomous vehicles transport goods on tracks. Cooperation is only necessary when multiple vehicles meet at an intersection. Usually simple rules such as the aforementioned right-before-left suffice. (Asama et al., 1991)

Reactive Behaviour Here, the focus is on collision avoidance. Using highly reactive behaviours, long-term planning is simplified and collisions are avoided shortly before they would occur. (Mataric, 1992)

Swarm Behaviour This group of approaches does not value that each robot reaches its goal. Instead, the swarm should exhibit a certain behaviour. For instance, one robot leads while the others follow in a certain formation. Hence, the following robots only need to consider their relative positions (Mourikis & Roumeliotis, 2006). There are various adaptations of animal swarm behaviours, such as ant swarming, buffalo migration or the following of geese.

2.2 A-star

The A-Star search algorithm is probably the most well-known informed search. It was firstly described by Peter Hart, Nils Nilsson, and Bertram Raphael (Hart et al., 1968) in 1968. It is employed to find the optimal path between a start and a goal node in a road map, given a consistent heuristic function. Since it is essential for the following chapter, we will briefly describe it.

Assuming positive costs for travelling along each edge in a graph, A-Star finds the cheapest path from start node a to goal node b. An arbitrary node x is evaluated using both the costs of arriving at x from a and a heuristic estimate of the costs from x to b. If the heuristic function is consistent, then A-Star is complete and optimal (Dechter & Pearl, 1985). A heuristic h is consistent if and only if

$$h(x, x') \leq c(x, x')$$

and

$$h(x,b) \leq c(x,x') + h(x',b)$$

where c is the cost functions, and x and x' arbitrary nodes.

2.3 Clustering

Cooperative path planning requires consistency at some level of abstraction. That is, at some step of the calculations, resulting data has to be fused throughout the participating robots. This can be at the level of paths or sooner, at the map level. Here we employ a clustering algorithm to fuse obstacle information of all robots. On the basis of the resulting clusters, a Voronoi graph is constructed. Each robot detects some obstacles and cannot perceive others. Moreover, sensor noise is present, so obstacles are not perceived at their true locations. Thus, data representing the same obstacle should be merged, ideally yielding a more precise location then any single robot's data alone. However, data representing different obstacles should not be merged, in order to prevent the disappearance of perceived obstacles. Network latencies is one of the issues that make this process difficult, as data received from remote robots is generally older than local data and represents past situations. The problem of merging these obstacles is also called data-association problem.
Clustering algorithms merge data in groups or clusters, such that objects in each cluster are as similar as possible and objects of different clusters are as different as possible. (Ester & Sander, 2000). We limit our discussion to the two-dimensional case, where obstacles can be abstracted to points. Every clustering algorithm needs a similarity, or distance measurement $dist(p,q)$. For obstacles on a plane, the euclidean distance suffices.
A cluster is a set of points and is represented by a single point, the centroid. A measurement, which describes the divergence of a set of points from its centroid is given by:

$$TD^2(C) = \sum_{p \in C} dist(p, \mu_C)^2$$

$TD^2(C)$ is also a measurement for the compactness of a cluster. The smaller $TD^2(C)$, the closer the individual points are to each other, and the smaller the variance within the cluster (Ester & Sander, 2000). *K-MEANS* (MacQueen, 1967) is a common algorithm for minimising the variance of a set of clusters. K-MEANS divides the given points into K clusters, such that $TD^2 = \sum_{i=1}^{K} TD^2(C_i)$ is minimised. However, it requires prior knowledge of the number of clusters, K.
This disadvantage is lifted by hierarchical clustering algorithms, which represent data in a hierarchical manner to derive clusters (Ester & Sander, 2000). This requires the distance measurement to be extended to set of points. The following three variants of distance measurements are derived from the smallest, largest and average individual distance between points in P and Q, respectively:

$$\text{Single-Link: } dist_{sl}(P,Q) = \min_{p \in P, q \in Q} d(p,q)$$

$$\text{Complete-Link: } dist_{cl}(P,Q) = \max_{p \in P, q \in Q} \{d(p,q)\}$$

$$\text{Average-Link: } dist_{al}(P,Q) = \frac{1}{|P||Q|} \sum_{p \in P, q \in Q} d(p,q)$$

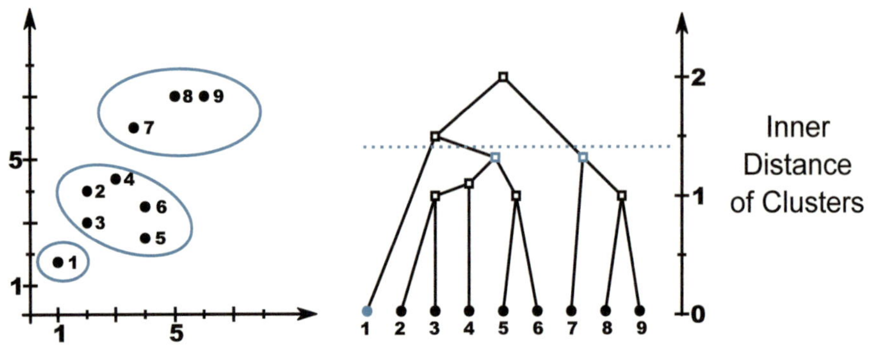

Fig. 3. Single-Link Dendrogram

Another method, based on the work of Ward (1963), uses the difference between the variance before and after fusing two clusters P and Q.

$$\text{Increase in variance: } \Delta\sigma^2(P,Q) = TD^2(P\bigcup Q) - (TD^2(P) + TD^2(Q))$$

Independent of the distance function, hierarchical clusterings can be described as either dividing (top-down) or collecting (bottom-up) algorithms. Top-down algorithms start with a single cluster containing all points and divide it into subsets until the resulting clusters fulfil a termination criterion. Bottom-up techniques start with single points and fuse them to clusters, again, until a termination criterion is met. The resulting graph is called a dendrogram (see Figure 3).

Figure 3 shows the distances within each cluster versus the number of clusters. Starting at zero, the threshold (dotted line) is increased step-by-step. As soon as the threshold becomes larger than the distance between two clusters, the clusters have to be merged. All clusters which have a larger distance remain separate. In Figure 3, the threshold is set such that three clusters remain.

Other approaches, such as the work by Schubert & Sidenbladh (2005), based on the work of Schubert (1993) and Bengtsson & Schubert (2003), use a Bayesian filter for clustering. Here, points are processed sequentially and no hierarchy is built, instead a set of hypotheses is maintained, out of which the maximum a posteriori estimate is seen as the clustering result. The advantage here is that, due to the sequential nature of the algorithm, observations can be added over time. Thus, it lends itself to algorithms tracking moving obstacles over time (see Section 3.1.2).

2.4 Delaunay triangulation

A triangulation is a triangle mesh without overlaps. Figure 5(a) shows an example for a specific triangulation, the Delaunay triangulation. Since this triangulation can be used to derive a Voronoi graph (see Section 2.6), we present some details here.

The Delaunay triangulation of a set of points S yields the Delaunay graph $DG(S)$. $DG(S)$ fulfills the *circumcircle* property: The circumcircle of every triangle does not contain any point of S which is not also one of the three points belonging to the triangle. Figure 4 shows on the left an enlarged part of the Delaunay graph. The circumcircle does not contain any other point of P. On the right side of Figure 4, the circumcircle property is violated, thus it does not show a Delaunay graph. A triangulation fulfilling this property also maximises the minimal internal angle of all triangles.

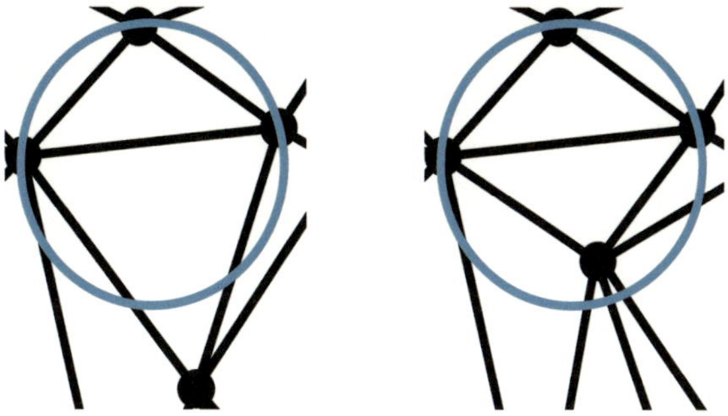

Fig. 4. Circumcircle Property

2.5 Voronoi graph

The Voronoi graph by Woronoi (1908) is a subdivision of space into so-called Voronoi regions. Figure 5(b) shows an example. We will use Voronoi graphs as road maps (see Section 2.1.1). In the following, we are limiting ourselves to the two-dimensional case, as we assume robots to be moving on a plane or sufficiently planar surface. Hence, every point is a member of $\mathbb{R}^2$. Every point p in the set S defines a Voronoi-region $VorR(p, S)$:

$$VorR(p,S) = \bigcap_{q \in S \setminus \{p\}} D(p,q)$$

where $D(p, q)$ is a open half-plane:

$$D(p,q) \stackrel{def}{=} \{x \in \mathbb{R}^2 : |p - x| < |q - x|\}$$

The Voronoi graph $Vor(S)$: is then defined by the set of all regions $VorR(S)$:

$$VorR(S) \stackrel{def}{=} \bigcap_{p \in S} VorR(p,S)$$

$$Vor(S) \stackrel{def}{=} \mathbb{R}^2 \setminus VorR(S)$$

One of the main properties of the Voronoi graph is that its edges have the maximal distance to the neighbouring points in S. Thus following the edges yields itself to obstacle avoidance, given that the obstacles are represented by the points in S.

2.6 Duality between Voronoi and Delaunay graph

The Voronoi graph of S is a connected graph $VG(S)$. We will call the set of edges of $VG(S)$ E_{VG}, the set of nodes or vertices V_{VG}, and the set of regions R_{VG}. Analogously, the Delaunay graph $DG(S)$ has the edges E_{DG}, vertices V_{DG}, and the set of regions R_{DG}. Figure 5(b) shows that some edges lead outside of the picture. Assuming the existence of another point in infinite distance, one can connect all these edges to said point, without losing any Voronoi property (de Berg et al., 2000). Figure 6 illustrates this.

Voronoi graph and Delaunay graph are dual to each other, assuming the existence of the point in infinite distance. Thus, there is an isomorphism between the vertices of one graph and the regions of the other. In other words, every point in S yields a vertex in V_{DG} and a region in R_{VG}.

$$V_{DG} \leftrightarrow F_{VG}$$

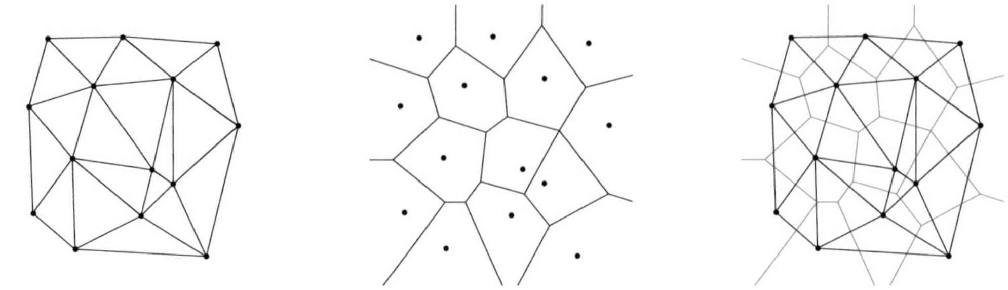

(a) Delaunay triangulation of 14 points on a plane

(b) Voronoi graph of 14 points on a plane

(c) Voronoi graph and Delaunay graph

Fig. 5. Duality between Voronoi graph and Delaunay triangulation

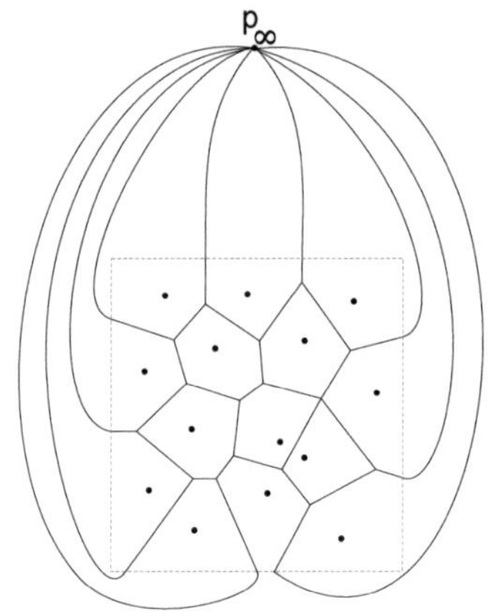

Fig. 6. Additional point in infinite distance

Moreover, every circumcentre of a region in R_{DG} is also a vertex in V_{VG}. The infinitely large area encompassing the triangulation represents the node in infinite distance.

$$F_{DG} \leftrightarrow V_{VG}$$

Finally, there is an isomorphism between the edges of both graphs. For every edge in one graph, there is one in the other at a right angle. However, they do not necessarily overlap.

$$E_{DG} \leftrightarrow E_{VG}$$

Figure 5(c) illustrates the dual relationship between the graphs. This duality can be exploited to construct a Voronoi graph by Delaunay triangulation, as there exist efficient algorithms to construct Delaunay graphs (Bhattacharya & Gavrilova, 2007; Guibas et al., 1990; Leach, 1992).

3. Approach

Our approach is divided into two main parts, data fusion and planning. Figure 7 depicts an overview of the architecture. It is meant to be used in conjunction with other frameworks or engines, which manage other parts of the robotic architecture. For instance, we assume that a behaviour issues a command to move to a certain point. What a behaviour constitutes and

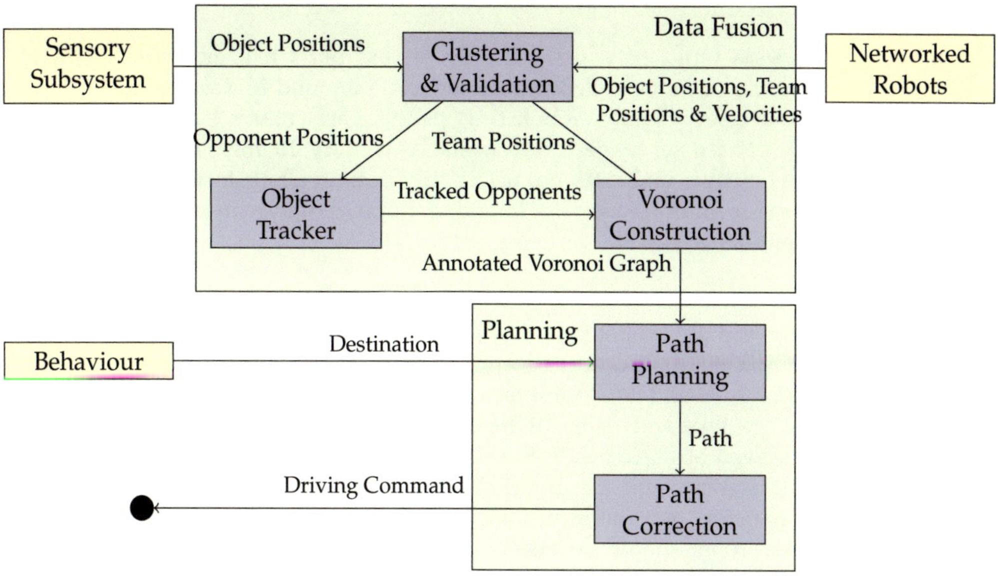

Fig. 7. Algorithm Architecture

how it comes to a decision is intentionally left open. A treatment of multi-robot behaviour control and cooperation is given in Skubch et al. (2011). Furthermore, we assume that the cooperative robots are able to communicate, for instance following an approach as described in Baer (2008).

During the data fusion step, observations from the sensory subsystem, such as image processing, is integrated with observations received from other cooperative robots. This includes the truthfully emitted positions and velocities of these cooperative robots. The data is validated and clustered, yielding a set of adversarial and a set of cooperative robots.

Positions and velocities of adversarial robots are tracked using a Bayesian filter, while positions and velocities of cooperative robots are assumed to be correct. This is due to the fact that each robot's own localisation and velocity estimate is in most cases more accurate than observations of the same information from other sources. Finally, both sets are used to generate a Voronoi Graph of the current scene.

Based on the Voronoi Graph and the current intended destination given by a behaviour, a path is planned by the robot in question. Hence, each robot calculates its preferred path individually. Thus, path planning is done decoupled. This planning step is both efficient and optimal with respect to the roadmap. Since the path planning does not account for moving obstacles, these are reacted to in a final step. Tackling this problem, we use the cooperative and reactive method for collision avoidance introduced by Fujimori et al. (2000). If collisions between cooperative robots become imminent, the current driving directions are adapted in speed and angle according to globally defined rules. In order to avoid other moving obstacles similar rules are applied, taking their potential adverseness into account. The resulting driving command is given to the actuator subsystem. The whole process is repeated every 30ms.

The next section explains the data fusion step in detail, while Section 3.2 describes both path planning and reactive behaviour.

3.1 Data fusion

Sensor fusion is a wide topic, of which we only discuss parts that are relevant for our path-planning approach. A more in-depth treatment can be found in Reichle (2010). In our setting we assume that a set of known robots $\mathcal{C}$ cooperate. Each robot $r \in \mathcal{C}$ emits its known position p, velocity v and a set of perceived obstacles o every 100ms. The obstacle set is noisy, contains false positives and does not necessarily contain all obstacles in line-of-sight. Furthermore, there exists an unknown, but bounded, number of moving adversarial robots, $\mathcal{A}$, which do not communicate at all.

3.1.1 Clustering

During the clustering step each robot fuses its own position and its perceived obstacles with the positions and obstacles of all cooperative robots. The deployed clustering is based on the method according to Ward (1963), described in Section 2.3. Furthermore the following adaptations take the age, type and source of the information into account.

The precision of perceived obstacle positions reduces with increasing age and distance. Due to potential network latencies and the lower frequency of 10Hz, the information given by cooperative robots is older than the local information. Therefore, near obstacle positions perceived by the local robot should be treated preferential to the obstacle positions of the cooperative robots. In our approach the clustering is omitted in a $1.5m$ radius around the local robot. Inside this radius only locally perceived obstacles are taken into account.

Another important adaptation is to distinguish between positions of cooperative robots and adversarial robots. If a cooperative robot is merged into a cluster, the cluster itself represents the cooperative robot. Therefore, it is important to avoid inconsistent clusters, which can be created by merging two different cooperative robots into one cluster. Another source of inconsistency is the possibility of merging an obstacle's position into the cluster of the cooperative robot which sent the obstacle position itself, because cooperative robots cannot see themselves as obstacles.

At the end of the clustering step, we are able to distinguish the clusters into the two groups of cooperative robots and adversarial robots. As mentioned before the received velocities of the cooperative robots are considered quite precise, therefore only the adversarial robot clusters are given to the object tracker, described in Section 3.1.2.

3.1.2 Tracking

Tracking is the process of relating observations over time. This process not only smooths the data, removes infrequent false positives and compensates for missing observations, but it also estimates velocities. In Section 2, we briefly mentioned the applicability of Bayesian filters to the tracking and clustering problem. Recently, van de Molengraft (2009) introduced a dynamic tracking algorithm for an unknown number of objects based on the work of Schubert & Sidenbladh (2005). Schubert's approach maintains a set of hypotheses of cluster centres. For each new observation, every hypothesis h is expanded to $n + 2$ to new ones, where n is the number of cluster centres maintained in h. That is, given h, the new observation is considered to be faulty data, belonging to a new cluster, or belonging to one of the clusters maintained in h. Schubert then estimates the a posteriori probability of all newly generated hypotheses, resamples them using Monte Carlo sampling and merges highly similar hypotheses. Due to the resampling with a constant number of particles, the amount of hypotheses does not grow beyond a constant threshold.

Schubert's method was originally only applicable to non-moving targets. Van de Molengraft (2009) extended this approach by maintaining a Kalman Filter for each cluster, using the

constant velocity assumption. Furthermore, they introduced a lower threshold, independent of the Monte Carlo resampling to the number of maintained hypotheses.
Our approach differs from the aforementioned in that we exchanged the Kalman Filters for a linear ridge regression, which proved in practice to require less precise parameter optimisation. Kalman Filters with a constant velocity assumption tend to produce estimations which are either to susceptible to noise or too inert to changing conditions if the probability parameters do not match the scenario (Lauer, 2007). Furthermore, we limited the maximum amount of tracked objects, such that the runtime complexity is now in $O(n)$, where n is the number of observations, instead of $O(nk)$, where k is the number of objects tracked. Thus, we assume a constraint number of objects instead of an unknown one. Finally, we omit the Monte Carlo sampling and directly use the likelihood as an estimation for the a posteriori probability, further reducing the computation time.
The downside of using a regression as opposed to a Kalman Filter is the lack of a probability estimate, which the Kalman Filter produces directly. As Schubert, we assume Gaussian observation noise and estimate the probability of an observation z at time t caused by object o as $P(z_t|o) \propto e^{\frac{(z-m_o(t))^2}{2\sigma^2}}$, where $m_o(t)$ is the estimated position of object o at time t.
The Monte Carlo sampling served the purpose of avoiding too fast convergence. However, since in dynamic environments the input data changes rapidly over time and the fact that similar hypotheses are collapsed into a single one, early convergence is not a problem in our approach, thus the Monte Carlo sampling can be omitted, saving computation time. This yields an algorithm that integrates 10 observations into 10 tracked objects using 30 hypotheses in under 5ms on current PCs (Intel i7 2.8GHz, single core).

3.1.3 Voronoi construction

For generating a Voronoi graph, we use the same method as Bhattacharya & Gavrilova (2007). At first, the Delaunay triangulation is created by the randomised incremental algorithm introduced by Guibas et al. (1990) and afterwards the Voronoi graph is derived from the Delaunay triangulation. Obviously, the fused data serves as input for this process, but we also insert some artificial obstacles along the side lines of the football field. The football field, being the containing rectangle of the whole scenario, is sampled similar to the sampling method of Bhattacharya & Gavrilova (2007). This yields Voronoi edges between obstacles inside the containing rectangle and the borders of that rectangle. Otherwise, the roadmap would miss edges for driving around obstacles at the sideline of the football field. In the resulting Voronoi graph each robot, including the local one, represents one Voronoi cell. In other scenarios, similar bounding boxes can be sampled this way. Alternatively, in case the bounds are unknown or unlimited, a sufficiently large bounding circle can be used, containing the current focus of interest for the robot.

3.2 Path planning

As mentioned before the path planning is divided into two clearly separated steps. The first step is described in Section 3.2.1, it searches for the best path according to the given destination and Voronoi graph. The second step comprises the adjustment of the initial driving direction. The initial driving direction is derived from the path found in the first step. To adjust the driving direction globally defined rules are used. These details are described in Section 3.2.2.

3.2.1 Path generation

The Voronoi graph and the destination, given by a behaviour, form the input of the path search algorithm. We utilize the A-Star search algorithm, described in Section 2.2. Considering the

common prominence of the A-Star search algorithm, here we only describe the used heuristics and the special cases at the start and end of the path. Besides the start and the end of a path, each path is a sequence of Voronoi edges. Therefore, the applied heuristics have to estimate the costs of a sequence of Voronoi edges.
Let p denote a path with edges $E = e_0, \ldots, e_n$, and for every edge e, let $e.s$ denote its starting node, $e.e$ its ending node, such that $e_i.e = e_{i+1}.s$, $e_0.s$ the robot's current position, and $e_n.e$ the destination D if p is complete. Further, let $l(e)$ be the length of edge e, $w(e)$ its width, $\Delta\alpha(a,b)$ the angle between edges or vectors a and b, and $\vec{g}_i$ be the calculated driving direction in the robot's i-th cycle.
Given additionally the robot's radius r, the cost function $c(p)$ and the heuristic function $h(p)$ for a path p calculated in cycle i are then computed in the following way:

$$c(p) = c_{dist}(p) + w_w c_w(p) + w_\alpha c_\alpha(p) + w_s c_s(p)$$

with

$$c_{dist}(p) = \sum_{i=0}^{n} l(e_i)$$

$$c_w(p) = \sum_{i=0}^{n} \frac{1\text{mm}}{\max(1\text{mm}, w(e_i) - r)}$$

$$c_\alpha(p) = \sum_{i=1}^{n} \Delta\alpha(e_i, e_{i-1})^2$$

$$c_s(p) = \Delta\alpha(e_0, \vec{g}_{i-1})$$

where all w_x are constant weights. Thereby, the cost function is a weighted sum of four factors. Firstly, the length of the path, the longer a path, the less valuable it is. Secondly, a function depending on the width of the occurring edges, as small gaps should be penalised as opposed to wide openings. The third factor sums up the square angle difference between adjacent edges, as sharp turns are less preferably than wide turns, which can be driven faster. Finally, the last summand depends on the previously calculated driving direction. It stabilises the decision of the robot under sensor noise, as it penalises larger changes in the driving direction over time.
The weights depend on the specific scenario and capabilities of the robot. For different manoeuvres the weights might even be set differently. For instance in RoboCup, a robot is less agile if it dribbles, as it has to maintain control of the ball, thus it would greatly prefer paths without sharp turns.
Each summand gives rise to a summand in the heuristic function for path, $h(p)$. The heuristic function we use is defined by:

$$h(p) = \begin{cases} 0 & \text{if } e_n.e = D \\ dist(e_n, D) + w_\alpha \Delta\alpha(e_n, \overline{e_n.e\, D})^2 + w_w \max_{e \in E_{VG}} \frac{1\text{mm}}{\max(1\text{mm}, w(e) - r)} & \text{otherwise} \end{cases}$$

The path length to the destination is approximated by the linear distance between the last node in the path and the destination. The costs due to angle differences are approximated by the angle between the last edge and the line connecting the last node and the destination, as this is the very least the robot will need to turn to arrive at the destination. The stabilisation summand only factors in the first node and thus it's heuristic is 0. The width costs are

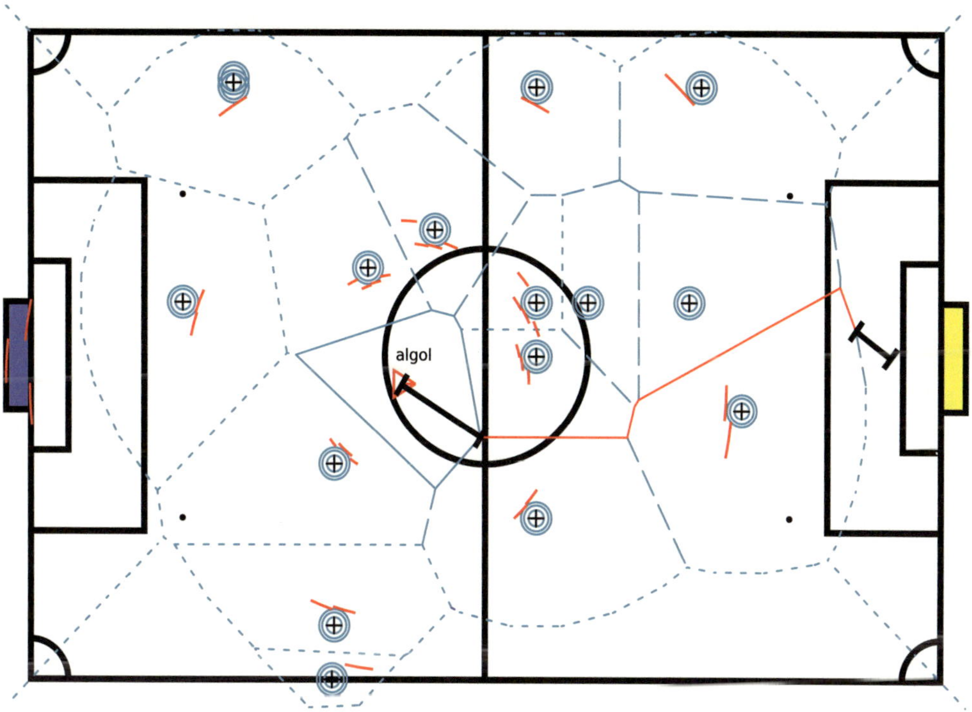

Fig. 8. Simple Path Example

approximated by the maximal width of any edge in the graph. This is a weak heuristic and could be improved by limiting the set of considered edges, however this introduces more overhead than it would save expansion steps in the search, unless the width related costs are dominating.

The width of a Voronoi edge is determined by its distance to an adjacent cell defining object. The width of a starting edge, leading from the robot's position to a Voronoi node equals the minimum distance to any object spanning a neighbouring Voronoi cell. The width of a final edge, leading to the destination, depends only on the position of the object defining the cell which contains the destination. If the destination is on a Voronoi edge, the final edge is part of this Voronoi edge, thus has equal width.

Figure 8 shows a robot planning to drive to the right goal. The resulting path is shown, with start and end edges emphasised. As each robot is inside its own convex Voronoi cell the path search can be initialized with one path for each Voronoi node of the Voronoi cell. If no initial Voronoi node is reachable, the robot is surrounded by obstacles and cannot drive anywhere. The drawback of this initializing method is that it creates unnecessary complicated paths to destinations which could be reached directly, without approaching a Voronoi node. Comparing the sequences of Voronoi edges with the direct line to the destination prevents unnecessary complicated paths.

The destination is either inside a Voronoi cell, or on the edge of a cell. If a path is expanded during the search to one of the Voronoi nodes of the cell, which includes the destination, we also expand the path to the destination if possible. This is done in a similar way as the path search was initialized. If the destination is blocked by an obstacle the path stops as near as the robot can get to the destination. It is plausible, that an obstacle overlaps with the destination,

so that it is impossible to reach the destination. In this case, the path which ends closest to the destination is returned by the search.
The driving direction according to the resulting path is not necessarily the direction of the first edge. Just as we test to drive directly to the given destination, we also test if it is possible to reach any of the Voronoi nodes on the path directly. Starting at the second to the last Voronoi node on the path, each point is considered as short cut on the path. Only if no Voronoi node serves as short cut, the driving direction corresponds to the first path edge.

3.2.2 Reactive collision avoidance behaviour

The collision avoidance behaviour is based on the work of Fujimori et al. (2000) and uses globally announced rules to make additional communication between the robots unnecessary. The needed data and its computation is described in Section 3.1.
A collision between two robots is imminent if they are near and further approaching each other. In this case, the expected minimal distance between the two robots is computed. Furthermore, if the driving directions intersect, the angle between them, δ is noted. As described by Fujimori et al., this intersection angle δ is very important. In Figure 9(a) the collision can be avoided only if both robots change their directions. The robots in Figure 9(c) would have to change their directions much more than the in Figure 9(a). Therefore, changing their velocities would be much more effective. Figure 9(e) shows a situation requiring both, changing velocity and direction. For example, Robot 1 could pass Robot 2 on its right side by changing the direction and accelerating appropriately. Meanwhile Robot 2 should reduce its velocity to finish the procedure faster. So there are different kinds of collisions and each of them needs an adjustment of directions, velocities, or both.
Given the current situation, each robot is assigned a priority to coordinate their collision avoidance behaviour. For example, the decision which robot should accelerate and which should decelerate is made according to the robot's priorities. A high prioritized robot should arrive quickly at its destination and has the right of way. Lower prioritized robots should clear the way for higher prioritized robots. Therefore, in Figure 9(c) the robot with the higher priority will accelerate and the other one will decelerate. The priority ω of robot i towards robot j is defined as follows:

$$\omega_{ij} = \begin{cases} \frac{v_i}{\overrightarrow{R_i P_{ij}}} & \text{if intersection point } P_{ij} \text{ exists,} \\ v_i & \text{if no intersection point exists and } v_i \neq v_j \end{cases} \tag{1}$$

In the first case the higher priority is assigned to the robot which first arrives at the intersection point. The latter case is applied if the driving directions are parallel to each other and a collision as shown in Figure 9(a) is imminent. Here the higher priority is assigned to the faster robot. If the priorities ω_{ij} and ω_{ji} are the same, the robot's ID decides.
According to the priorities and the angle of collision the velocities and directions can be calculated. The velocity adjustment Δv of robot i to avoid a collision with robot j is defined as follows:

$$\Delta v_{ij} = \begin{cases} sat(|\delta|, \frac{\pi}{2}, \pi, v_{acc}, 0) & \omega_{ij} > \omega ji, \\ sat(|\delta|, \frac{\pi}{2}, \pi, v_{dec}, 0) & \omega_{ij} < \omega ji \end{cases} \tag{2}$$

The corresponding direction adjustment $\Delta\gamma$ is defined by:

$$\Delta\gamma_{ij} = \begin{cases} k_\gamma \, sgn(\delta) \left|1 - \frac{2}{\pi} |\delta|\right| & \omega_{ij} > \omega ji, \\ sgn(\delta) \, sat(|\delta|, \frac{\pi}{2}, \pi, 0, k_\gamma) & \omega_{ij} < \omega ji \end{cases} \tag{3}$$

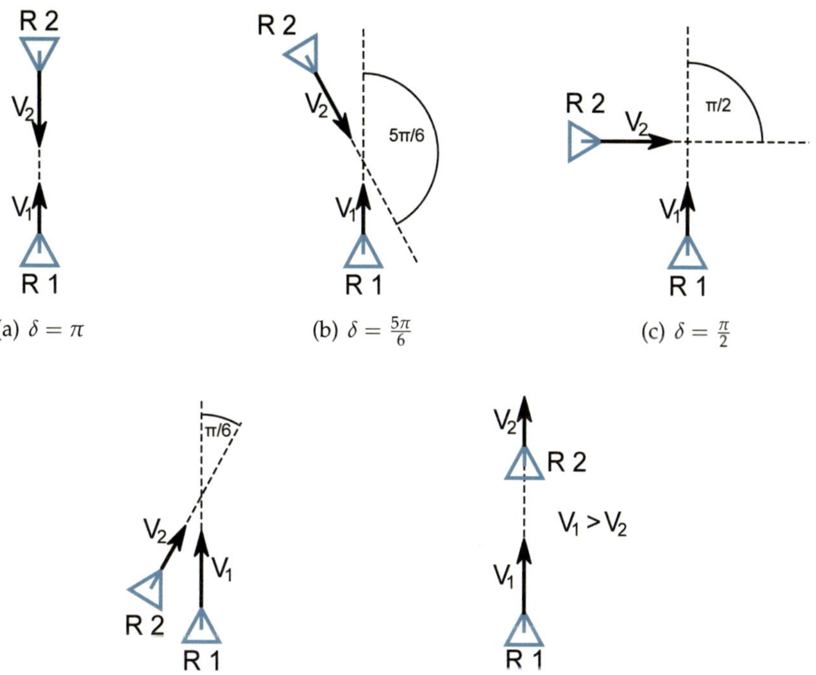

(a) $\delta = \pi$ (b) $\delta = \frac{5\pi}{6}$ (c) $\delta = \frac{\pi}{2}$

(d) $\delta = \frac{\pi}{6}$ (e) $\delta = 0$

Fig. 9. Different Angles of Collision

where $sgn(x)$ and $sat(x, a, b, c, d)$ are defined as:

$$sgn(x) = \begin{cases} 1 & x \geq 0 \\ -1 & x < 0 \end{cases} \tag{4}$$

$$sat(x, a, b, c, d) = \begin{cases} c & x > a, \\ \frac{d-c}{b-a}(x-a) + c & a \leq x < b, \\ d & x \geq b \end{cases} \tag{5}$$

Figure 10 shows the functions to calculate the adjustment of velocity and driving direction. Index H means high and L means low priority. The amount of adjustment to the driving direction can be defined through the parameter k_γ. In the opposite of Fujimori et al., we do not assign a fix value to k_γ. Instead, k_γ depends on the remaining time to collision (ttc):

$$k_\gamma = min(\eta, \eta \frac{0.3\,[sec]}{\text{ttc}\,[sec]})$$

If ttc is less than 0.3 seconds, which corresponds to 10 iterations of the robots calculation loop, the adjustment to the driving angle is set to η.
The amount of velocity change can be set by the parameters v_{acc} and v_{dec}. At most a higher prioritized robot accelerates about v_{acc} , while a lower prioritized robot decelerates about v_{dec}. Instead of the differential velocity, Fujimori et al. calculates the absolute velocity of the robots, which is less reactive concerning the situation. If the angle of collision δ is exactly π, only

the driving directions are affected and the velocities stay unchanged. If the angle of collision δ is exactly $\frac{\pi}{2}$, only the velocities are affected and the driving directions stay unchanged. At the range of $\frac{\pi}{2} \leq \delta \leq \pi$ the change to the velocities and driving directions are linearly interpolated. At the range of $0 \leq \delta \leq \frac{\pi}{2}$ the velocity changes of both robots stay at their maximum, but the change of the driving direction of the higher prioritized robot increases to its maximum again. This corresponds to the situation shown in Figure 9(e).

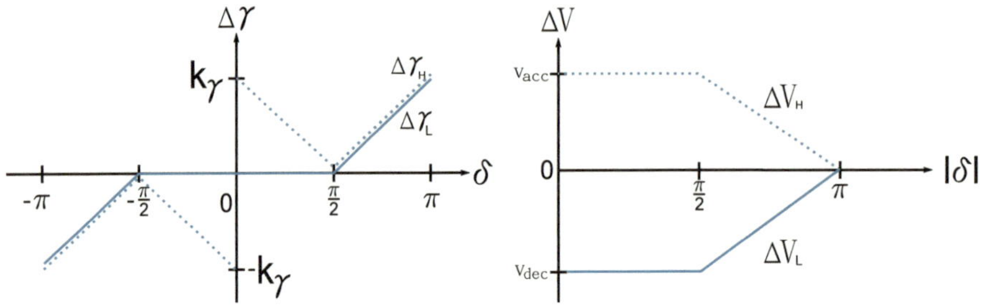

(a) change of direction (b) velocity change

Fig. 10. Change of direction and velocity

In case of collisions with more than two robots, the overall change of direction and velocity of robot i are given by the weighted sum of changes of each collision:

$$\Delta\gamma_i = \frac{\sum_{j=1}^{n} \frac{1}{ttc_{ij}} \Delta\gamma_{ij}}{\sum_{j=1}^{n} \frac{1}{ttc_{ij}}} \tag{6}$$

$$\Delta v_i = \frac{\sum_{j=1}^{n} \frac{1}{ttc_{ij}} \Delta v_{ij}}{\sum_{j=1}^{n} \frac{1}{ttc_{ij}}} \tag{7}$$

Due to the fact that each change is weighted by the corresponding inverse of ttc, earlier collisions have more influence than later collisions.

Adapting the introduced collision avoidance technique to a specific team of robots, the values of the mentioned parameters have to be chosen appropriately. These parameters include the maximum distance between two robots for collision avoidance to apply, as well as a minimum velocity, below which no collision avoidance is considered. Altogether, there are two activation parameters: the minimum velocity two robots approach each other and the distance between them. How much the collision avoidance behaviour changes the calculated driving direction and velocity of a robot depends on the parameters η, v_{acc} and v_{dec}. These three parameters should be chosen as large as necessary and as small as possible to avoid collisions with as little effort as possible. The evaluation of the approach of Fujimori et al. includes simulator tests as well as tests on real robots. According to the results a collision is avoided in most cases and problems arise if the robots have to move in confined spaces. Our evaluation follows in the next section.

4. Discussion

Although our architecture may appear to be overly complex at first glance, each part, shown in Figure 7 of Section 3, has clearly defined interfaces and makes an essential contribution to the overall performance. Furthermore, the modular design enables exchangeability of each module and increases the reusability of our architecture. Each module comes with its own set of parameters, hence adapting our architecture to other types of scenarios might raise suspicions that tuning these parameters is a lot of work. However, the modules described are very robust against small parameter changes, such that rough adjustment of the parameters already produces good results. Although a centralised approach can produce better results, due to such an approach avoiding additional coordination problems, it also has a single point of failure, which cannot be tolerated in many domains. Our combination of different algorithms with efficiency in mind yields a real-time path planning architecture, which even provides the possibility to plan paths of other cooperative robots. In the following sections, we discuss the main advantages of each module.

4.1 Clustering

We evaluated the clustering algorithm using two cooperative robots. One is located at the centre of the football field recording its own data together with the data received from the other robot. The other robot is controlled and sends its own data to the robot in the middle. The evaluation is based on the recorded data of the robot located at the centre. The data includes 5 minutes with a recording frequency of 30Hz for the local data and 10Hz for the data received from the remote controlled robot. The recorded data is subject to the conditions mentioned in Section 1.

Clustering is subject to two kinds of errors. Either data describing the same obstacle is not fused (first case), or data describing different obstacles is fused nevertheless (second case). The frequency of occurrence of both cases depends directly on the variance threshold. The variance threshold defines the maximum variance of a cluster. The bigger the variance threshold is the bigger the distance between fused data can be. In other words, the occurrences of the first case errors increase and the occurrences of the second case errors decrease. Reducing the variance threshold inversely influences the two cases of errors. Determining the occurrences of errors near the controlled robot the second case of errors is negligible, because from the recording robot's point of view there is only the controlled robot on the football field. Furthermore, obstacle positions received from another robot will not be fused with the position of this robot, because the robot cannot see itself. The frequency of occurrence against the variance threshold is shown in Figure 11.

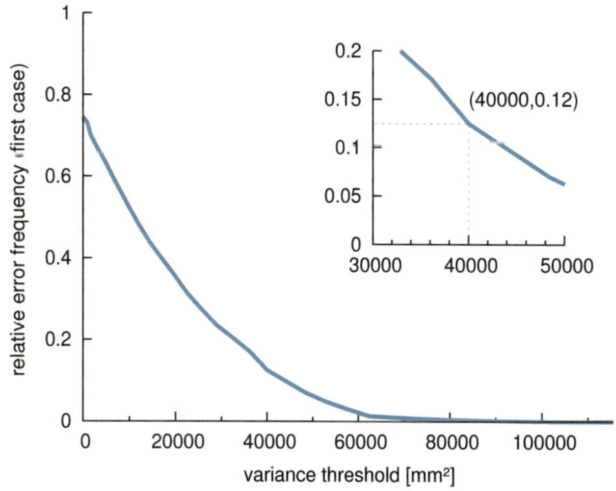

Fig. 11. Relation of error frequency against variance threshold

In Figure 11 a small part of the graph is shown magnified. The emphasised point (40000, 0.12) indicates that an obstacle position must not exceed the standard deviation of $\sqrt{40000\text{mm}^2} = 20\text{cm}$ to its cluster centre. In this case, 12 percent of the clustered situations contain a first case error. Assuming a robot radius of 25cm a variance threshold of 40000mm^2 means that two robots have to be perceived overlapping by 10cm to be erroneously fused. Moreover, a local observation of a robot, moving with a relative speed of over 9m/s with respect to the local robot, might not be fused with its broadcast position, given the average age of 50ms of any remotely received information. A relative speed of 9m/s is reached if two robots move in opposite direction with 4.5m/s, which is common in RoboCup Middle Size League games.
Both statements hold only in case of noiseless sensor information and without network latencies. Factoring in sensor noise, erroneous fusing occurs earlier. However, in most domains robots hardly move faster than 4.5m/s, thus high velocities are usually not a cause for badly clustered data.
Therefore, we can state that the clustering with a variance threshold of 40000mm^2 is fairly robust against high robot velocities. Figure 12(a) further underlines this, as the error is independent of the relative velocity for small values of said velocity.

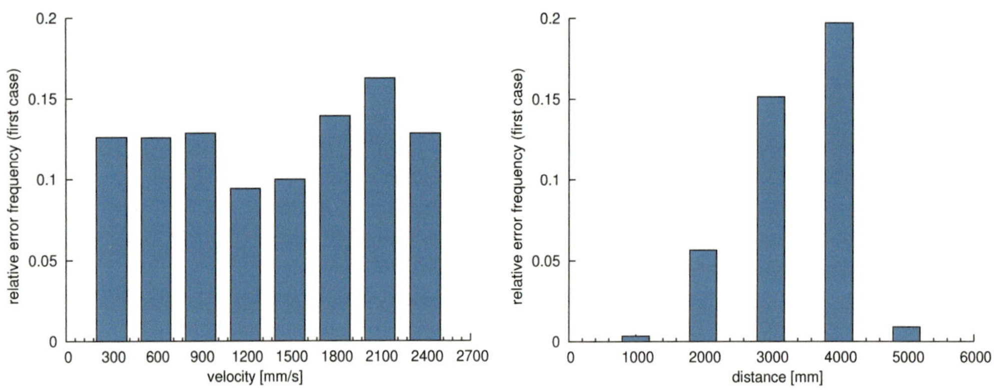

(a) velocity relation (b) distance relation

Fig. 12. Relation of error frequency against velocity and distance

Imprecision introduced by sensors noise is much more influential. Figure 12(b) shows a correlation between distance and error frequency. This correlation can be explained by the fact, that the feature extraction algorithm of the robot is much more imprecise in estimating the position of robots, which are further away than of robots which are closer by. The small number of error occurrences at a distance above 5m arose, because the feature extraction algorithm often did not recognize the remote controlled robot at these distances.
Summarized, the clustering is robust against the dynamics of a RoboCup Middle Size League game, hence it unifies the data of all cooperative robots to the same single point of view on all cooperative robots. The simple clustering algorithm used here is much more time efficient than the Bayesian filter used for tracking over time. Thus, using the clustered data as input for the tracking algorithm instead of using the raw observations reduces the latency of the whole computation.

4.2 Tracking

In this Section, we briefly demonstrate the performance of the tracking algorithm discussed in Section 3.1.2. In the following simulated experiment, 10 objects moved with a randomly

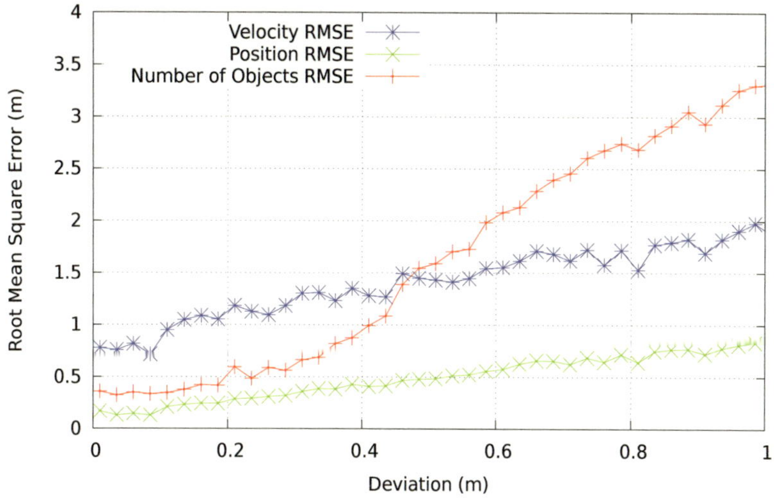

Fig. 13. Tracking performance with increasing observation noise

chosen constant velocity between 1 and 4m/s on a field of 20m × 20m size. In case they collided with the boundaries, they bounced back perfectly. The tracker was set up with a maximum hypotheses number of 30, easily allowing it to process 10 observations in less than 5ms. Each test was performed for 500 iterations.

In each iteration, 10 observations were given to the tracker, each with a probability $p = 0.1$ uniformly distributed over the field, else normally distributed around an object. Figure 13 shows the resulting tracking errors, against an increasing amount of observation noise. Note that the tracking algorithm assumes a constant deviation of 0.25m in all experiments.

One can see that position and velocity errors increase linearly with increasing deviation, while the error in the number of objects increases rapidly for deviations higher than the assumed 0.25m. At the expected deviation of 0.25m, an average of 0.5 objects are wrongly tracked. Note that this also includes the start up phase, in which the tracker gradually increases the amount of objects tracked. The average position error is 0.3m and the average velocity error is 1.2m/s. Large velocity errors mostly occur after an object has bounced off the boundary of the field, as this clearly violates the constant velocity assumption.

Figure 14 shows the same setup, where the objects additionally have a rotational speed between 10°/s and 90°/s, and thus move on circles. As expected, the velocity error is in general higher than in the case of a constant velocity. Moreover, the number of tracking errors increases earlier, as observations not fitting the model are more common. Interestingly, the position error is only marginally affected.

4.3 Path generation

Like every path planning approach using some kind of abstraction, such as a roadmap, the paths generated by our architecture are not optimal. Approaches trying to generate optimal paths are faced with a PSPACE-hard problem (see Section 2.1) and as a result these approaches are not feasible in domains similar to the one addressed by our architecture.

Avoiding stationary obstacles can easily be done using a Voronoi graph as roadmap. A robot following the edges of the Voronoi graph is as far away from the surrounding obstacles as possible. A path planning algorithm using this graph can easily adapt to the current

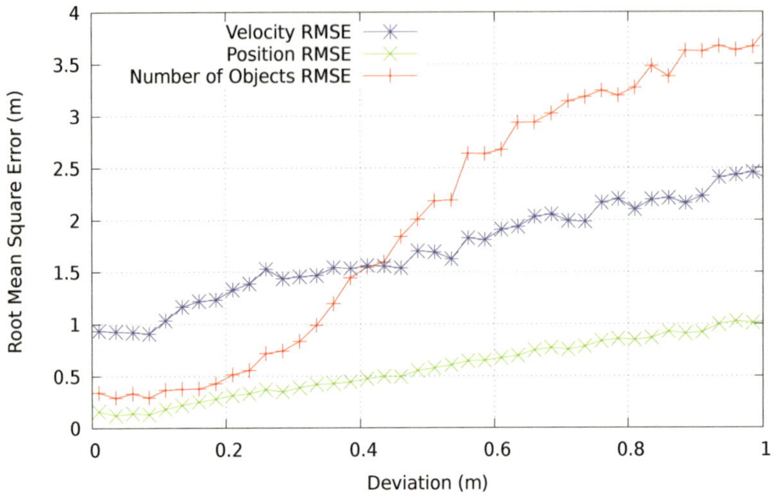

Fig. 14. Tracking performance with objects moving on circles

situation, using differently weighted cost functions, thus allowing robots to switch between an aggressive behaviour, using short, narrow paths and a safe behaviour using longer and wider paths, for instance.
Moreover, a Voronoi graph can be used to answer various spatial queries other than path planning, such as finding free areas or the robot closest to a given point.
Bhattacharya & Gavrilova (2007) demonstrated that it is possible to calculate Voronoi graphs to approximate arbitrary environments.

4.4 Path correction

The reactive nature of our collision avoidance behaviour forms a contrast to the forward planning of the path planning algorithm and it might be more intuitive to avoid collisions along the whole path. However, in order to avoid collisions along the whole path obstacle positions, including adversarial robots, need to be predicted for time periods of up to $5 - 10$s. Predicting the dynamic system that is formed by multiple cooperative and adversarial robots for more than 200ms is completely infeasible. Thus, collisions need to be handled reactively without relying on comparatively long-term predictions.
Due to the globally known avoidance rules, there is no need for an additional communication effort. Furthermore, the discussed rules scale linearly with the number of robots involved. Similar to the adaptivity of the planning algorithm, the parameters of the collision avoidance allows to switch between different behaviours during runtime, using different grades of sensitivity. Summarized, the collision avoidance technique is applicable to any number and any kind of autonomously driving robots.

5. Conclusion

The described architecture for cooperative real-time path planning and collision avoidance, consisting of various adapted algorithms, was evaluated on robots of the Carpe Noctem RoboCup team of the University of Kassel and used in tournaments since 2011. The architecture demonstrates that it is possible to use a decentralised path-planning approach combined with a conflict resolution inspired by the reactive behaviour approach. The only requirement is to fuse and unify the information among the group of cooperative robots, so that they come to similar decisions. Fused and unified information enable path planning beyond the sensor range of single robots up to the sensor range of the team, without any additional communication for conflict resolution. Furthermore, the complete calculation loop is real-time capable. More precisely, the complete recognition, fusion and path planning cycle can be done at a frequency of 30Hz. Furthermore, the architecture meets the requirements of the highly dynamic RoboCup domain of the Middle Size League described in Section 1. The stability and similarity of fused information and calculated paths among the group of cooperative robots can be analysed over time and compared with other path planning and sensor fusion approaches in future.

6. References

Asama, H., Ozaki, K., Itakura, H., Matsumoto, A., Ishida, Y. & Endo, I. (1991). Collision avoidance among multiple mobile robots based on rules and communication, *Intelligent Robots and Systems '91. Intelligence for Mechanical Systems*, Vol. 3, pp. 1215 –1220.

Baer, P. A. (2008). *Platform-Independent Development of Robot Communication Software*, Phd thesis in computer science, University of Kassel. URL: *http://www.upress.uni-kassel.de/publi/abstract.php?978-3-89958-644-2*

Bengtsson, M. & Schubert, J. (2003). Dempster-shafer clustering using potts spin mean field theory, *Soft Computing - A Fusion of Foundations, Methodologies and Applications* 5: 215 – 228. 10.1007/s005000100084.

Bhattacharya, P. & Gavrilova, M. (2007). Voronoi diagram in optimal path planning, *4th International Symposium on Voronoi Diagrams in Science and Engineering*, pp. 38 –47.

Bruijnen, D., van Helvoort, J. & van de Molengraft, R. (2007). Realtime motion path generation using subtargets in a rapidly changing environment, *Robotics and Autonomous Systems* 55(6): 470–479.

de Berg, M., van Kreveld, M., Overmars, M. & Schwarzkopf, O. (2000). *Computational Geometry: Algorithms and Applications*, second edn, Springer. URL: *http://www.cs.uu.nl/geobook/*

Dechter, R. & Pearl, J. (1985). Generalized best-first search strategies and the optimality of a, *Journal of the ACM* 32: 505–536.

Ester, M. & Sander, J. (2000). *Knowledge Discovery in Databases - Techniken und Anwendungen*, Springer.

Fujimori, A., Teramoto, M., Nikiforuk, P. N. & Gupta, M. M. (2000). Cooperative collision avoidance between multiple mobile robots, *Journal of Robotic Systems* 17(7): 347–363.

Guibas, L. J., Knuth, D. E. & Sharir, M. (1990). Randomized incremental construction of delaunay and voronoi diagrams, *Proceedings of the seventeenth international colloquium on Automata, languages and programming*, Springer-Verlag New York, Inc., New York, NY, USA, pp. 414–431.

Hart, P. E., Nilsson, N. J. & Raphael, B. (1968). A formal basis for the heuristic determination of minimum cost paths, *Systems Science and Cybernetics, IEEE Transactions on* 4(2): 100–107. URL: *http://dx.doi.org/10.1109/TSSC.1968.300136*

Hopcraft, J. E., Schwartz, J. T. & Sharir, M. (1984). On the complexity of motion planning for multiple independent objects; pspace- hardness of the "warehouseman's problem", *The International Journal of Robotics Research*, Vol. 3, pp. 76–88.

Lauer, M. (2007). Ego-motion estimation and collision detection for omnidirectional robots, *in* G. Lakemeyer, E. Sklar, D. G. Sorrenti & T. Takahashi (eds), *RoboCup 2006: Robot Soccer World Cup X*, Springer-Verlag, pp. 466–473.

Leach, G. (1992). Improving worst-case optimal delaunay triangulation algorithms, *In 4th Canadian Conference on Computational Geometry*, p. 15.

MacQueen, J. B. (1967). Some methods for classification and analysis of multivariate observations, *in* L. M. L. Cam & J. Neyman (eds), *Proc. of the fifth Berkeley Symposium on Mathematical Statistics and Probability*, Vol. 1, University of California Press, pp. 281–297.

Mataric, M. (1992). Designing emergent behaviors: From local interactions to collective intelligence, *In Proceedings of the International Conference on Simulation of Adaptive Behavior: From Animals to Animats*, Vol. 2, pp. 432–441.

Mourikis, A. I. & Roumeliotis, S. I. (2006). Optimal sensor scheduling for resource-constrained localization of mobile robot formations, *IEEE TRANSACTIONS ON ROBOTICS*, Vol. 22.

Parker, L. E. (2009). Path planning and motion coordination in multiple mobile robot teams, *in* R. A. Meyers (ed.), *Encyclopedia of Complexity and System Science*, Springer, Knoxville, Tennessee, USA, chapter 13, pp. 5783–5800.

Parsons, D. & Canny, J. (1990). A motion planner for multiple mobile robots, *IEEE International Conference of Robotics and Automation*, pp. 8–13.

Peasgood, M., Clark, C. & McPhee, J. (2008). A complete and scalable strategy for coordinating multiple robots within roadmaps, *IEEE Transactions on Robotics* 24(2): 283 –292.

Reichle, R. (2010). *Information Exchange and Fusion in Heterogeneous Distributed Environments*, Phd thesis in computer science, University of Kassel, Kassel.

Schubert, J. (1993). On nonspecific evidence, *International Journal of Intelligent Systems* 8(6): 711 – 725.

Schubert, J. & Sidenbladh, H. (2005). Sequential clustering with particle filters – estimating the number of clusters from data, *8th International Conference on Information Fusion*, Vol. 1, ISIF, pp. 1 – 8.

Skubch, H., Wagner, M., Reichle, R. & Geihs, K. (2011). A modelling language for cooperative plans in highly dynamic domains, *Mechatronics* 21(2): 423 – 433. Special Issue on Advances in intelligent robot design for the Robocup Middle Size League.
URL: *http://www.sciencedirect.com/science/article/pii/S0957415810001868*

Topor, A. (1999). *Roboterfußball: Pfadplanung in dynamischer umgebung*, Master's thesis, Universität Freiburg.

van de Molengraft, R. (2009). Techunited robocup – the object tracking problem, online. [Online; accessed 22-July-2011].

Ward, J. (1963). Hierarchical grouping to optimize an objective function, *Journal of the American Statistical Association* 58: 236–244.

Woronoi, G. F. (1908). Nouvelles applications des paramètres continus à la théorie des formes quadratiques. premier mémoire. sur quelques propriétés des formes quadratiques positives parfaites., *Journal für die reine und angewandte Mathematik* 133: 97–102.

12

Motion Planning for Multiple Mobile Robots Using Time-Scaling

István Komlósi and Bálint Kiss
Department of Control Engineering and Information Technology
Budapest University of Technology and Economics
Hungary

1. Introduction

Recent developments in robotics have raised the interest in mobile robots due to their wide range of applications requiring high level of autonomy of individual mobile agents. This increased level of autonomy needs more sophisticated path-planning and motion control methods, which have been studied extensively in the recent years [10].

This paper focuses on a typical industrial application in which several mobile robots share the same workspace. These units may carry materials or be capable of carrying out various repair jobs at some locations, etc. Due to the tight schedules to achieve high productivity in such applications, it is imperative to synchronize the motion of the units so that they arrive at the required location on time without collision with obstacles or with other units. Coordinated planning of such multi-robot systems is addressed in the literature based on some search methodology [7] or using neural network based techniques [8].

Since a high number of units can be involved, it may not be possible to design collision free trajectories for all robots, i.e. avoiding each other and all static obstacles cannot be guaranteed only by the path geometry. Instead of constructing trajectories with complex geometries so that they have only a few or no intersections at all, one can design simple paths which disregard moving obstacles (i.e. other robots) and take only the static obstacles into account, so the path geometries are supposed to be fixed for each robot. Such motion planning algorithms are readily available in the literature [1,3,4] for simple polygonal obstacle representations. This initial path geometry designed and assigned to each unit ensures the avoidance of static obstacles.

The avoidance of the moving obstacles can be achieved by choosing a suitable velocity profile along the fixed-geometry trajectories. In order to obtain the proper velocity input, we assign first a default velocity profile which the unit would use without the dynamic obstacles. Such a profile can be designed to be optimal in some sense (minimal time or cost) and may satisfy additional constraints, if necessary [2]. The default velocity profile consists of intervals with constant or zero acceleration. Acceleration parameters change at the boundaries of these intervals. This default velocity profile is then modified such that a unit decelerates or accelerates its motion to let other units pass through potential collision areas (identified from the path geometries).

We suppose that a priority level is assigned to each robot so that a unit with a given priority regards all units with higher priority level as dynamic obstacles. The ranking of the units is

arbitrary, e.g. faster units may receive higher priority in order not to slow them down. Other considerations can be also incorporated in the choice of priority order. The velocity distributions of the robots are thus determined in a serial way according to the priority ordering of the robots. Recall that one may get a globally better solution in terms of the overall time used to complete all trajectories if the priority order of the robots is relaxed but such a relaxation would imply the need of a high level search algorithm. The study of such a high level search algorithm is beyond the scope of this paper.
The velocity tuning method presented in this paper for an individual robot is based on time-scaling. The default time-scale determined by the default velocity profile is parameterized by a *virtual time parameter* which has to be then mapped to the real or global time such that no collision occurs. The mapping between the virtual and global time parameters is referred to as the time-scaling in the sequel. The time-plane is the space spanned by the virtual and global time parameters.
If a collision is possible between two units according to the path geometries (so that they intersect each other), the collision to be avoided can be converted into a so called static time-obstacle in the time-plane [5,6]. A mapping avoiding the static time-obstacles shall be used to define the time-scaling of the path of the current unit. Once the path is expressed by the global time parameter, this procedure can be repeated for all units down to the one with the lowest priority.
The remaining part of the paper is organized as follows. Section 2 presents an overview on time-scaling, Section 3 presents our path-planning method in the time-plane, Implementation results are presented in Section 4 and Conclusions are drawn in Section 5.

2. An overview on time-scaling

In this section some mathematical background is given for the time-scaling. The default velocity that is assigned to a robot and so the time distribution of its path is expressed using a virtual time parameter τ. We denote this reference path by $\Gamma(\tau)$ and define the default velocity as

$$\frac{d}{d\tau}\Gamma(\tau) = v(\tau) \tag{1}$$

The time distribution of the reference trajectory is constructed regardless to any moving obstacles. Motion of dynamic obstacles (i.e. other robots) is expressed with the global time parameter t, thus if $O_{dyn}(t)$ represents the trajectory of a dynamic obstacle and

$$\Gamma(\tau) \cap O_{dyn}(t) \neq \varnothing \tag{2}$$

then the current unit collides with the dynamic obstacle. To avoid this collision we must find a function $\tau = \theta(t)$ that alters the time distribution of the reference path such that

$$\Gamma(\theta(t)) \cap O_{dyn}(t) = \varnothing \tag{3}$$

The function $\theta(t)$ maps the virtual time values to global or real time. A mapping that ensures avoidance of all dynamic obstacles will be the time-scaling function. Such a time-scaling function may be constructed from several functions over different time intervals.

Methods for finding an adequate time-scaling function in our application will be presented in the next section. Based on the real time and virtual time parameters, we can define the $t-\tau$ time-plane. If a collision is possible between two units according to the path geometries, the collision to be avoided can be converted into a time-obstacle O_{time} which appears as a static obstacle in the time-plane. A time-obstacle is assigned to a collision area in the workspace and it is the set of those virtual and real time pairs where any parts of the unit and the moving obstacle are located at the same place within the collision area corresponding to the time-scaled units default velocity. In most of the cases they will appear as rectangular shaped time-obstacles; in special cases, however, they might be enclosed by nonlinear curves.

Avoiding dynamic obstacles in the workspace is analogous to avoiding the corresponding time-obstacles on the time-plane. Based on this analogy, the time-scaling function is the path on the time-plane that avoids all time-obstacles. After constructing this path, the scaled reference path can be generated. Since $\tau = \theta(t)$ we have

$$\frac{d}{dt}\Gamma(\tau) = \frac{d\Gamma(\tau)}{d\tau}\frac{d\tau}{dt} \tag{4}$$

which, using (1), yields to

$$\frac{d}{dt}\Gamma(\tau) = v(\tau)\dot{\theta}(t) \tag{5}$$

The velocity according to the scaled reference path can be generated by multiplying the original speed by the derivative of the time-scaling function w.r.t. the real time

$$\tilde{v}(t) = v(\theta(t))\dot{\theta}(t) \tag{6}$$

where $\tilde{v}(t)$ is the scaled velocity.

2.1 Criteria for the time-scaling function

A proper time-scaling function must not only avoid the time-obstacles, but must also satisfy criteria according to the kinematic constraints prescribed for the control (velocity) inputs. A collision-free course requires a time-scaling function not mapping any time value into a time-obstacle, which means that $(t,\theta(t)) \not\subset O_{time}$ must be satisfied for all time values and all time-obstacles.

Since time cannot rewind and we allow the robot to stop only at its final position, the scaling function $\theta(t)$ must be strictly monotonically increasing. It might also be required that the robot's velocity does not decrease under a specified minimal value except at the start and goal positions, where the velocity is assigned to be zero. We assume that the default velocity is the maximal possible speed along the trajectory and its values cannot be exceeded at any time. This implies that the scaled velocity must not be greater at any real time instance than the velocity at the corresponding virtual time instance. It is also assumed that the acceleration and the deceleration are also bounded both in negative and positive directions. The acceleration (deceleration) limit is considered to be independent of time. These aforementioned constraints define the inequalities

$$v_{\min} \leq \tilde{v}(t) \leq v(\tau) \tag{7}$$

which, together with (6), results in

$$v_{\min} \leq v(\tau)\dot{\theta}(t) \leq v(\tau) \tag{8}$$

for the scaled velocity where $v_{\min}$ is the specified minimum speed value. After dividing with $v(\tau)$ it follows that

$$\frac{v_{\min}}{v(\tau)} \leq \dot{\theta}(t) \leq 1 \tag{9}$$

which defines a global upper bound for the time-scaling function meaning that the $\tau = t$ curve in the time-plane must not be intersected. The lower bound will be the solution of the differential equation

$$\dot{\theta}(t) = \frac{v_{\min}}{v(\tau)} \tag{10}$$

The constraint on the acceleration reads

$$a_{\min} \leq \frac{d}{dt}\tilde{v}(t) \leq a_{\max} \tag{11}$$

where $a_{\max}$ and $a_{\min}$ are the admissible maximal positive and negative acceleration values. Since

$$\frac{d}{dt}\tilde{v}(t) = \frac{d^2\tau}{dt^2}v(\tau) + a(\tau)\left(\frac{d\tau}{dt}\right)^2 \tag{12}$$

one gets

$$a_{\min} \leq v(\tau)\ddot{\tau} + a(\tau)\dot{\tau}^2 \leq a_{\max} \tag{13}$$

The solutions for the maximal values will result in two curves which will be used by the path planning algorithms in the time-plane.

2.2 Construction of the time-obstacles

A time-obstacle is always assigned to a collision area around the intersection of two geometric paths. Each collision area along the path of one unit generates a time-obstacle. The time-obstacle assigned to a collision area is the set of those virtual-time and real-time pairs where any parts of two units are located at the same place. We have to take into account the real physical dimensions of the units, which means that the entering time to a collision area is when the front of the unit arrives at the boundary of the collision area, and the exit time is when its rear point leaves it. If the time-scaled unit cannot enter the area until the higher priority unit has passed through, the time-obstacle will be rectangle shaped with vertexes $(t_{enter}, \tau_{enter})$, (t_{enter}, τ_{exit}), (t_{exit}, τ_{exit}), (t_{exit}, τ_{enter}) where τ_{enter} and τ_{exit} are the time values of entering and exiting the collision area of the time-scaled unit, while t_{enter} and t_{exit}

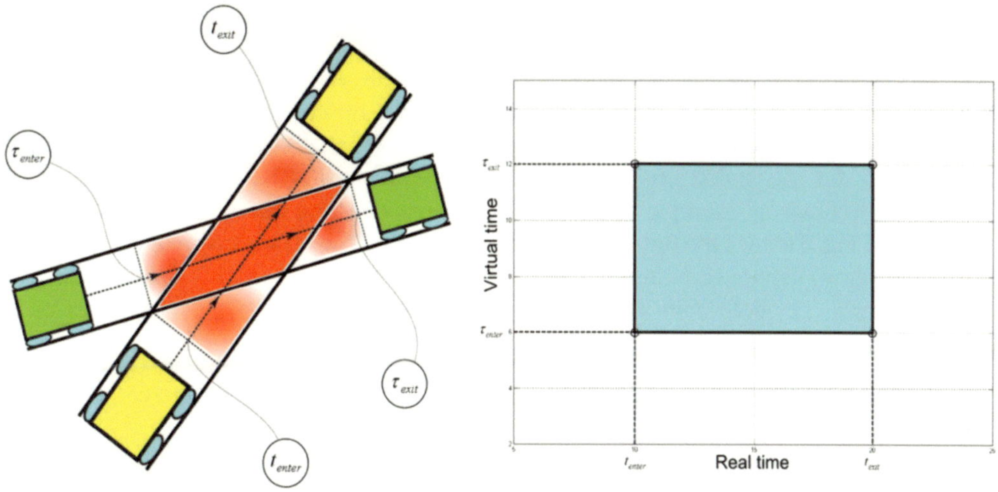

Fig. 1. Colliding units in the workspace (on the left) and the time-obstacle of the colliding units (on the right)

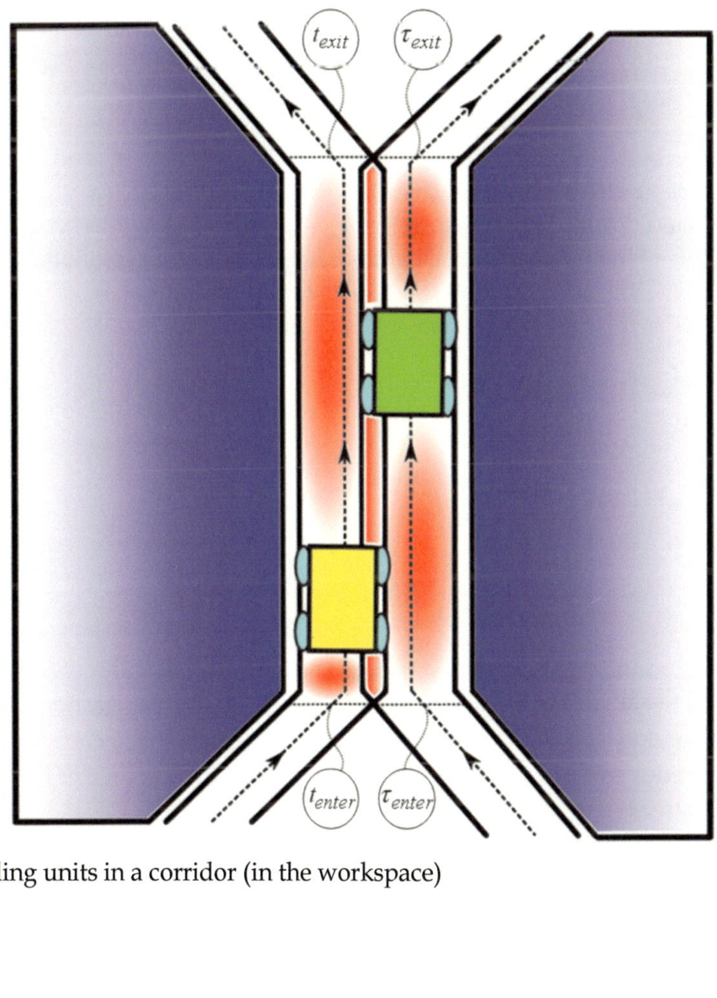

Fig. 2. Colliding units in a corridor (in the workspace)

are the similar values for the given dynamic obstacle. Fig. 1 shows a collision situation in the workspace and on the time plane.
In some special cases the path geometries may have common sections (e.g. the units follow each other in a narrow corridor). In such cases, when the motion directions of the colliding units are identical, one has to examine the time-values when certain parts of the collision area are free for the scaled unit. In this case the time-obstacle is enclosed by nonlinear curves. Fig. 2 shows an example of two robots proceeding along the same way in a narrow corridor and Fig. 3 shows the time-obstacle representation of the collision area.
Suppose that we have accelerating units in the corridor. Equation (14) defines the area which is occupied by both units at the same time instance:

$$\frac{1}{2}a_{unit}(\tau-\tau_{enter})^2+v_{unit}(\tau-\tau_{enter})+\Gamma(\tau_{enter})=\frac{1}{2}a_{obs}(t-t_{enter})^2+v_{obs}(t-t_{enter})+O_{dyn}(t_{enter}) \quad (14)$$

where $\Gamma(\tau_{enter})\in O_{dyn}(t_{enter})$. From (14) one can derive the equations of the curves that enclose the time-obstacles in case of accelerating motions (narrow corridor case).

$$\tau(t)=\tau_2-\frac{v_{unit}}{a_{unit}}+\frac{1}{a_{unit}}\sqrt{v_{unit}^2+2v_{obs}a_{unit}(t-t_1)+a_{obs}a_{unit}(t-t_1)^2} \quad (15a)$$

$$\tau(t)=\tau_1-\frac{v_{unit}}{a_{unit}}+\frac{1}{a_{unit}}\sqrt{v_{unit}^2+2v_{obs}a_{unit}(t-t_2)+a_{obs}a_{unit}(t-t_2)^2} \quad (15b)$$

In (15a) the fronts of the units arrive at the border of the corridor at τ_1 and t_1 , respectively, while their rear sides arrive to the same location at τ_2 and t_2.
In the simpler case, when the time-scaled unit moves with constant speed, Equation (16) defines the points of the time-obstacle which is enclosed by the curves defined by Equations (17a) and (17b).

$$v_{unit}(\tau-\tau_{enter})+\Gamma(\tau_{enter})=\frac{1}{2}a_{obs}(t-t_{enter})^2+v_{obs}(t-t_{enter})+O_{dyn}(t_{enter}) \quad (16)$$

$$\tau(t)=\tau_1+\frac{v_{obs}}{v_{unit}}(t-t_2)+\frac{1}{2}\frac{a_{obs}}{v_{unit}}(t-t_2)^2 \quad (17a)$$

$$\tau(t)=\tau_2+\frac{v_{obs}}{v_{unit}}(t-t_1)+\frac{1}{2}\frac{a_{obs}}{v_{unit}}(t-t_1)^2 \quad (17b)$$

In (15a), (15b), (17a) and (17b) the fronts of the units arrive at the border of the corridor at τ_1 and t_1 , respectively, while their rear sides arrive at the same location at τ_2 and t_2. Similarly at τ_3 and t_3 the front of the units arrive at the exit of the corridor, respectively, while their rear sides arrive at the same location at τ_4 and t_4. A nonlinear time-obstacle has thus six vertices (see Fig. 3).
In the sequel we focus mainly on rectangle shaped time-obstacles but the results can be extended for obstacles presented in Fig. 3.

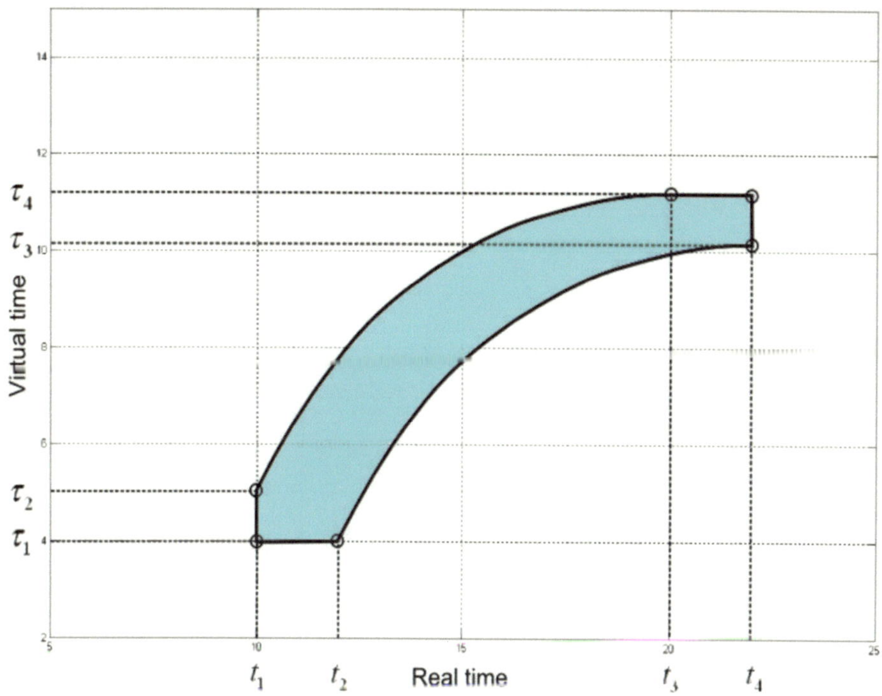

Fig. 3. Nonlinear time-obstacle

3. Path-planning in the time-plane

In this section the method of trajectory-planning in the time-plane will be presented, involving the solution of differential equations derived from the kinematic constraints prescribed for the robots motion, connecting dedicated points on the time-plane and avoiding time-obstacles.

Let us emphasize again that the default velocity profile consists of intervals with constant or zero accelerations. Acceleration parameters change at the boundaries of these intervals. We denote these points by τ_i where i refers to the index of the time instance where the acceleration change occurs. At a certain time τ where its value is assumed to be

$$\tau_i \leq \tau < \tau_{i+1} \tag{18}$$

the expression of the default velocity is

$$v(\tau) = a_i(\tau - \tau_i) + v_i \tag{19}$$

where

$$a_i = a(\tau_i)\,, \qquad v_i = v(\tau_i) \tag{20}$$

are the acceleration and velocity parameters at the border point τ_i. Fig. 4 shows an example for such a velocity function.

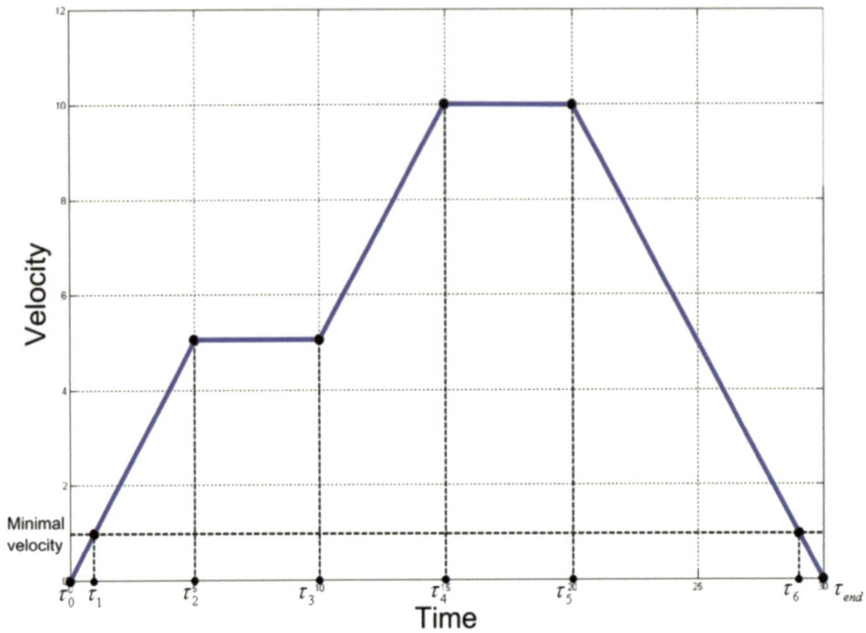

Fig. 4. Default velocity profile

3.1 Solving the differential equations

Regarding the admissible values of the accelerations specified by (13) and (19) the following equations can be obtained for the motion of the time-scaled unit:

$$a_i\tau\ddot{\tau} + a_i\dot{\tau}^2 + v_i\ddot{\tau} = a_{\max} \tag{21}$$

$$a_i\tau\ddot{\tau} + a_i\dot{\tau}^2 + v_i\ddot{\tau} = a_{\min} \tag{22}$$

$$a_i\tau\ddot{\tau} + a_i\dot{\tau}^2 + v_i\ddot{\tau} = 0 \tag{23}$$

which means that solutions of (21) and (22) result maximal and minimal accelerations in the scaled velocity, while the solution of (23) results zero acceleration. The solutions of (21) and (22) span an area on the time-plane. From a certain starting point on the time-plane only the points within this area can be reached. Let us denote such a starting point with $T_s = (t_s, \tau_s)$ and assume that the value of the default velocity at the corresponding instant is $v_s = v(\tau_s)$, while its scaled value at the corresponding t_s time instance is $\tilde{v}(t_s) = \delta_s v_s$. The $\tau = \theta(t)$ function that satisfies (21) is denoted by ${}^{a}\theta_{a\max}(t)$ in case of non-zero acceleration or ${}^{v}\theta_{a\max}(t)$ in case of zero acceleration. The subscript refers to the effect of the function on the velocity, which is $a_{\max}$ in this case.

The exact formula for ${}^{a}\theta_{a\max}(t)$ reads

$${}^{a}\theta_{a\max}(t) = \tau_s - \frac{v_s}{a_i} + \frac{1}{a_i}\sqrt{v_s^2 + 2a_i v_s \delta_s (t - t_s) + a_i a_{\max} (t - t_s)^2} \tag{24}$$

and its derivative is

$$ {}^{a}\dot{\theta}_{a\max}(t) = \frac{v_s\delta_s + a_{\max}(t-t_s)}{\sqrt{v_s^2 + 2a_i v_s \delta_s (t-t_s) + a_i a_{\max}(t-t_s)^2}} \tag{25} $$

while its inverse is

$$ {}^{a}\theta_{a\max}^{-1}(\tau) = t_s - \frac{v_s\delta_s}{a_{\max}} + \frac{1}{a_i a_{\max}}\sqrt{a_i^2 v_s^2 \delta_s^2 + a_i a_{\max}(v^2(\tau) - v_s^2)} \tag{26} $$

Applying the function ${}^{a}\theta_{a\max}(t)$ to the default velocity $v(\tau)$, the resulting scaled velocity will be $\tilde{v}(t) = {}^{a}\dot{\theta}_{a\max}(t)v({}^{a}\theta_{a\max}(t))$ that has a simple form

$$ \tilde{v}(t) = a_{\max}(t-t_s) + \delta_s v(\tau_s) \tag{27} $$

These expressions are valid for those time values where $t \in (t_s, {}^{a}\theta_{a\max}^{-1}(\tau_{i+1}))$, i.e. it is only valid over the interval for which the equations were solved. In order to span the solution for several intervals, the initial conditions τ_s and δ_s must be updated. When reaching any of the border points, the following changes have to be made regarding the parameters: $\tau_s = \tau_{i+1}$, $t_s = {}^{a}\theta_{a\max}^{-1}(\tau_{i+1})$ and $\delta_s = {}^{a}\dot{\theta}_{a\max}(\tau_{i+1})$ in the formula of ${}^{a}\theta_{a\max}(t)$. The formula of ${}^{v}\theta_{a\max}(t)$ reads

$$ {}^{v}\theta_{a\max}(t) = \frac{1}{2}\frac{a_{\max}}{v_i}(t-t_s)^2 + \delta_s(t-t_s) + \tau_s \tag{28} $$

its derivative is

$$ {}^{v}\dot{\theta}_{a\max}(t) = \frac{a_{\max}}{v_i}(t-t_s) + \delta_s \tag{29} $$

while its inverse is

$$ {}^{v}\theta_{a\max}^{-1}(\tau) = t_s - \frac{\delta_s v_i}{a_{\max}} + \frac{v_i}{a_{\max}}\sqrt{\delta_s^2 + 2\frac{a_{\max}}{v_i}(\tau - \tau_s)} \tag{30} $$

The function ${}^{v}\theta_{a\max}(t)$ results the same scaled velocity as ${}^{a}\theta_{a\max}(t)$ does, so

$$ \tilde{v}(t) = {}^{v}\dot{\theta}_{a\max}(t)v({}^{v}\theta_{a\max}(t)) = a_{\max}(t-t_s) + \delta_s v(\tau_s) \tag{31} $$

Similarly these are valid for the time values where $t \in (t_s, {}^{v}\theta_{a\max}^{-1}(\tau_{i+1}))$. The solutions of (22) which result maximal negative acceleration, are ${}^{a}\theta_{a\min}(t)$ and ${}^{v}\theta_{a\min}(t)$. Expressions of these functions, the derivatives and the inverses read

$$ {}^{a}\theta_{a\min}(t) = \tau_s - \frac{v_s}{a_i} + \frac{1}{a_i}\sqrt{v_s^2 + 2a_i v_s \delta_s (t-t_s) + a_i a_{\min}(t-t_s)^2} \tag{32} $$

$$^{a}\dot{\theta}_{a\min}(t)=\frac{v_s+a_{\min}(t-t_s)}{\sqrt{v_s^2+2a_iv_s\delta_s(t-t_s)+a_ia_{\max}(t-t_s)^2}} \tag{33}$$

$$^{a}\theta_{a\min}^{-1}(\tau)=t_s-\frac{v_s\delta_s}{a_{\min}}+\frac{1}{a_ia_{\min}}\sqrt{a_i^2v_s^2\delta_s^2+a_ia_{\min}(v^2(\tau)-v_s^2)} \tag{34}$$

where $t\in(t_s,{}^{a}\theta_{a\min}^{-1}(\tau_{i+1}))$ and

$$^{v}\theta_{a\min}(t)=\frac{1}{2}\frac{a_{\min}}{v_i}(t-t_s)^2+\delta_s(t-t_s)+\tau_s \tag{35}$$

$$^{v}\dot{\theta}_{a\min}(t)=\frac{a_{\min}}{v_i}(t-t_s)+\delta_s \tag{36}$$

$$^{v}\theta_{a\min}^{-1}(\tau)=t_s-\frac{\delta_sv_i}{a_{\min}}+\frac{v_i}{a_{\min}}\sqrt{\delta_s^2+2\frac{a_{\min}}{v_i}(\tau-\tau_s)} \tag{37}$$

where $t\in(t_s,{}^{v}\theta_{a\min}^{-1}(\tau_{i+1}))$. Both $^{a}\theta_{a\min}(t)$ and $^{v}\theta_{a\min}(t)$ result the scaled velocity function

$$\tilde{v}(t)=a_{\min}(t-t_s)+\delta_sv(\tau_s) \tag{38}$$

The solutions of (23) are $^{a}\theta_v(t)$ and $^{v}\theta_v(t)$, which can be used to keep the velocity constant once a desired value is reached. It can be any velocity within the robots' reach, but these functions are used for not letting the velocity decrease under a certain minimal value. The expressions of the functions, their derivatives and their inverses are as follows:

$$^{a}\theta_v(t)=\tau_s-\frac{v_s}{a_i}+\frac{1}{a_i}\sqrt{v_s^2+2a_iv_s\delta_s(t-t_s)} \tag{39}$$

$$^{a}\dot{\theta}_v(t)=\frac{v_s\delta_s}{\sqrt{v_s^2+2a_iv_s\delta_s(t-t_s)}} \tag{40}$$

$$^{a}\theta_v^{-1}(\tau)=t_s+\frac{v^2(\tau)-v_s^2}{2a_iv_s\delta_s} \tag{41}$$

where $t\in(t_s,{}^{a}\theta_v^{-1}(\tau_{i+1}))$. Similarly,

$$^{v}\theta_v(t)=\frac{v_s}{v_i}(t-t_s)+\tau_s \tag{42}$$

$$^{v}\dot{\theta}_v(t)=\frac{v_s}{v_i} \tag{43}$$

$$ {}^{v}\theta_{v}^{-1}(\tau) = \frac{v_i}{v_s}(\tau - \tau_s) + t_s \tag{44} $$

where $t \in (t_s, {}^{v}\theta_{v}^{-1}(\tau_{i+1}))$. These solutions enforce the scaled velocity to be

$$ \tilde{v}(t) = v_s \tag{45} $$

Since the constraint $\dot{\theta}(t) \le 1$ must be respected (see (9)), it has to be checked whether the value of the derivative of an applied $\theta(t)$ function reaches the unity at a certain point. We denote this point by $T_{ch} = (t_{ch}, \tau_{ch})$ where the control has to be changed to constant time-scaling and remain on the curve

$$ \theta_{const}(t) = m(t - t_{ch}) + \tau_{ch} \tag{46} $$

where in this case $m = 1$. Fig. 5 shows an example of applying different time scaling functions.

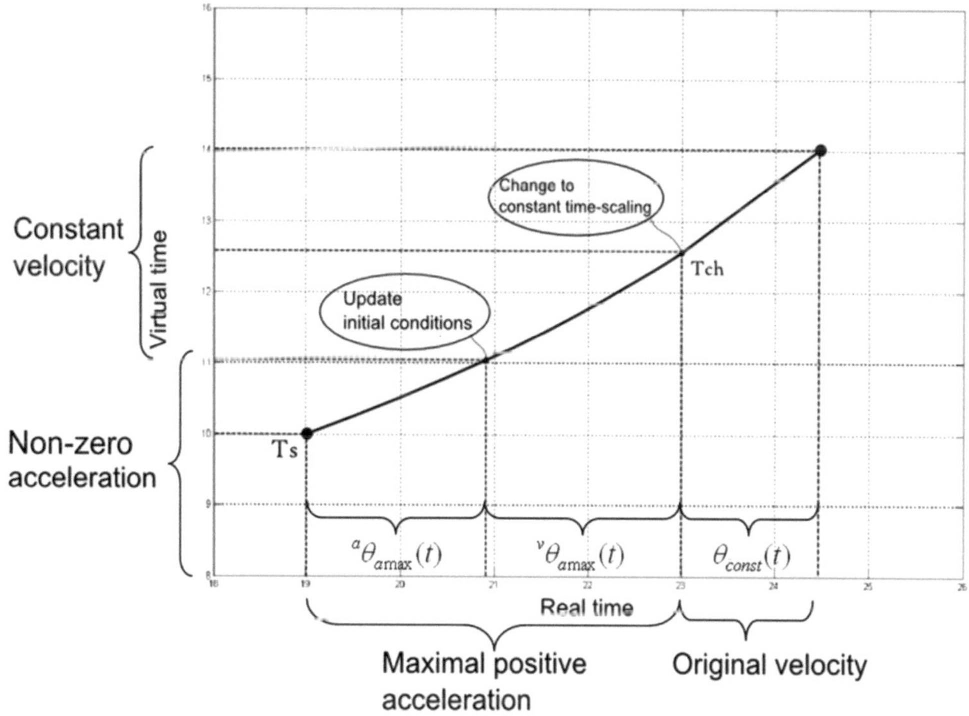

Fig. 5. Finding the changing point

To find the point T_{ch} the equation

$$ {}^{a}\dot{\theta}_{a\max}(t_{ch}) = 1 \tag{47} $$

or

$$ {}^{v}\dot{\theta}_{a\max}(t_{ch}) = 1 \tag{48} $$

should be solved, depending on the actual interval. Solving (47) results a quadratic equation

$$a_{\max}(a_{\max}-a_i)(t_{ch}-t_s)^2+2v_s\delta_s(a_{\max}-a_i)(t_{ch}-t_s)+v_s^2(\delta_s^2-1)=0 \tag{49}$$

The one satisfying $t_{ch}\in(t_s,{}^a\theta_{a\max}^{-1}(\tau_{i+1}))$ of the two roots of (49) should be selected, which results $\tau_{ch}={}^a\theta_{a\max}(t_{ch})$.
In case of (48), the solution for t_{ch} is much simpler

$$t_{ch}=(1-\delta_s)\frac{v_i}{a_{\max}}+t_s \tag{50}$$

and $\tau_{ch}={}^v\theta_{a\max}(t_{ch})$. Note that the formulae (51) and (52) only give accurate value if ${}^a\dot{\theta}_{a\max}({}^a\theta_{a\max}^{-1}(\tau_{i+1}))\geq 1$. The corresponding condition for (50) is ${}^v\dot{\theta}_{a\max}({}^v\theta_{a\max}^{-1}(\tau_{i+1}))\geq 1$, i.e. it has to be checked whether the value of the derivative at the next point determined by the function corresponding to the next border point is greater than one.
For negative accelerations, we must guarantee that the scaled velocity does not drop below $v_{\min}$. Therefore, the point $T_{ch}=(t_{ch},\tau_{ch})$ where the time scaling is changed to ${}^a\theta_v(t)$ or ${}^v\theta_v(t)$ (with $v_s=v_{\min}$ in their formulae) can be determined easily:

$$t_{ch}=\frac{\delta_s v(\tau_s)-v_{\min}}{a_{\min}}+t_s \tag{51}$$

and $\tau_{ch}={}^a\theta_{a\min}(t_{ch})$ or $\tau_{ch}={}^v\theta_{a\min}(t_{ch})$ respectively. Fig. 6 shows an example of finding this point.

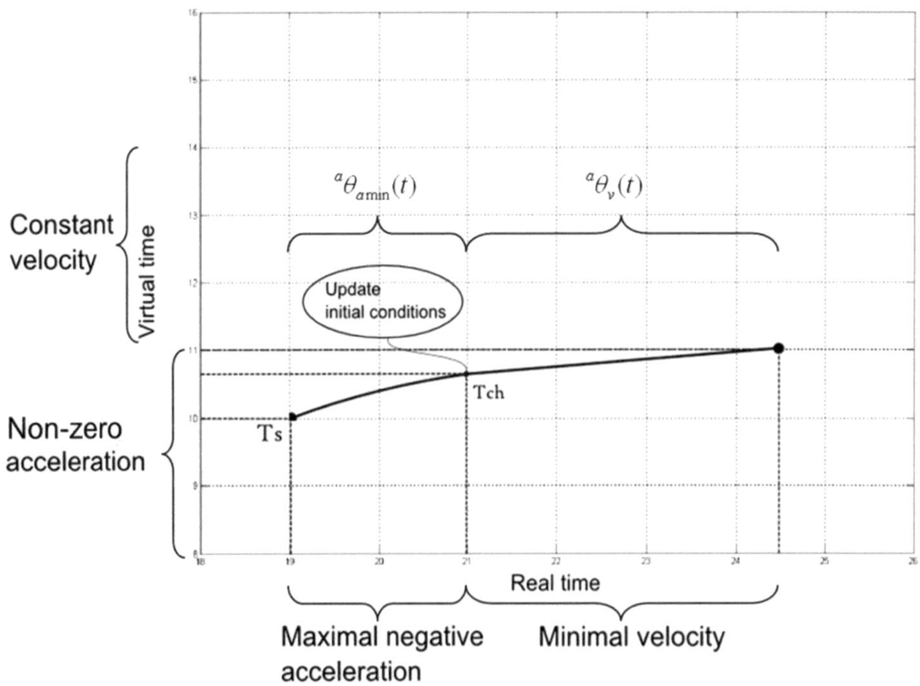

Fig. 6. Finding the changing point

When moving from a given point $T_s = (t_s, \tau_s)$ to another point $T_g = (t_g, \tau_g)$ on the time-plane, it has to be checked first whether T_g is within the area of reachable points from T_s. In order to do so, the boundaries of the reachable area have to be determined. The upper boundary will be the union of curves resulting maximal acceleration and then maximal speed, and the lower boundary will be the union of ones resulting maximal deceleration and then minimal velocity. Let us denote the upper boundary that origins in T_s and ends at t_g by $\Theta_{[Ts \to Tg],MAX}$ and the lower boundary by $\Theta_{[Ts \to Tg],MIN}$. T_g is reachable from T_s if and only if the inequalities below are both satisfied

$$\tau_g \le \Theta_{[Ts \to Tg],MAX}(t_g), \qquad \tau_g \ge \Theta_{[Ts \to Tg],MIN}(t_g) \tag{52}$$

Fig. 7 shows the area enclosed by these boundaries.

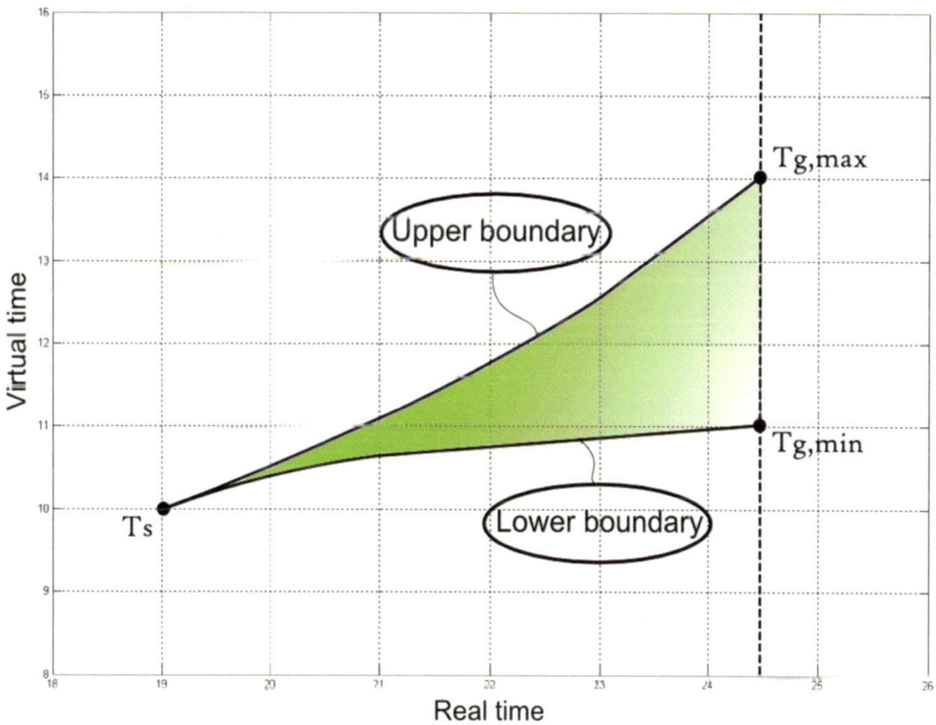

Fig. 7. The area of reachable points

If (52) is satisfied, T_g is within the reach of T_s and a trajectory connecting these two points can be planned. The main idea of trajectory planning is to travel along the upper or the lower boundary depending on the initial conditions in T_s and then switch to a straight line i.e. constant time-scaling. The switching point $T_{sw} = (t_{sw}, \tau_{sw})$ can be determined by solving the equation

$$\frac{\tau_g - \tau_{sw}}{t_g - t_{sw}} = \dot{\Theta}_{[Ts \to Tg],MAX}(t_{sw}) \tag{53}$$

or

$$\frac{\tau_g - \tau_{sw}}{t_g - t_{sw}} = \dot{\Theta}_{[Ts \to Tg],MIN}(t_{sw}) \tag{54}$$

depending on the selected boundary. In case of (53), the switching point can be on a ${}^{a}\theta_{a\max}(t)$ curve or on a ${}^{v}\theta_{a\max}(t)$ curve, so (53) transforms to

$$\frac{\tau_g - \tau_{sw}}{t_g - t_{sw}} = {}^{a}\dot{\theta}_{a\max}(t_{sw}) \tag{55}$$

or

$$\frac{\tau_g - \tau_{sw}}{t_g - t_{sw}} = {}^{v}\dot{\theta}_{a\max}(t_{sw}) \tag{56}$$

Solving (55) results in a quadratic equation, introducing the notation $\gamma = a_i v_s \delta_s$ and $v_g = a_i(\tau_g - \tau_s) + v_s$

$$\begin{aligned}&\left((\gamma + a_i a_{\max}(t_g - t_s))^2 - v_g^2 a_i a_{\max}\right)(t_{sw} - t_s)^2 + \\ &+\left(2(a_i a_{\max}(t_g - t_s)^2 + \gamma)(\gamma(t_g - t_s) + v_s^2) - 2v_g^2\gamma\right)(t_{sw} - t_s) + \\ &+\left(\gamma(t_g - t_s) + v_s^2\right)^2 - v_g^2 v_s^2 = 0\end{aligned} \tag{57}$$

Of the roots of (57) one has to select the solution that satisfies $t_{sw} \in ({}^{a}\theta_{a\max}^{-1}(\tau_i), {}^{a}\theta_{a\max}^{-1}(\tau_{i+1}))$, after that $\tau_{sw} = {}^{a}\theta_{a\max}(t_{sw})$. Solving (56) also leads to a quadratic equation

$$\frac{1}{2}\frac{a_{\max}}{v_i}(t_{sw} - t_s)^2 - \left(\frac{a_{\max}}{v_i}(t_g - t_s)\right)(t_{sw} - t_s) + \tau_g - \tau_s - \delta_s(t_g - t_s) = 0 \tag{58}$$

One has to select the solution that satisfies $t_{sw} \in ({}^{a}\theta_{a\max}^{-1}(\tau_i), {}^{a}\theta_{a\max}^{-1}(\tau_{i+1}))$, after that $\tau_{sw} = {}^{v}\theta_{a\max}(t_{sw})$.

Determining the switching point in case of travelling on the lower boundary the following equations should be solved

$$\frac{\tau_g - \tau_{sw}}{t_g - t_{sw}} = {}^{a}\dot{\theta}_{a\min}(t_{sw}) \tag{59}$$

$$\frac{\tau_g - \tau_{sw}}{t_g - t_{sw}} = {}^{v}\dot{\theta}_{a\min}(t_{sw}) \tag{60}$$

$$\frac{\tau_g - \tau_{sw}}{t_g - t_{sw}} = {}^{a}\dot{\theta}_{v}(t_{sw}) \quad \text{with} \qquad v_s = v_{\min} \tag{61}$$

depending on the possible location of the switching point. The solution of (59) and (60) are similar to (55) and (56) except that at the expressions $a_{\max}$ is replaced by $a_{\min}$. Solving (61) leads to a quadratic equation

$$\gamma^2(t_{sw}-t_s)^2+\left(2\gamma^2(t_g-t_s)+2\gamma(v_s^2-v_g^2)\right)(t_{sw}-t_s)+\left(\gamma(t_g-t_s)+v_s^2\right)^2-v_g^2v_s^2=0 \tag{62}$$

One has to select the solution that satisfies $t_{sw}\in({}^a\theta_v^{-1}(\tau_i),{}^a\theta_v^{-1}(\tau_{i+1}))$, after that $\tau_{sw}={}^a\theta_v(t_{sw})$. After determining the switching point T_{sw} the following equation can be used to travel along a straight line until reaching T_g:

$$\theta_{const}(t)=\frac{\tau_g-\tau_{sw}}{t_g-t_{sw}}(t-t_{sw})+\tau_{sw} \tag{63}$$

Here $t\in(t_{sw},t_g)$. Fig. 8 shows an example of finding such a switching point.
T_{g0} can be reached directly from T_s. It is possible to reach T_{g1} by travelling along the upper boundary and then switching to a straight line at T_{sw1}. Similarly, T_{g2} can be reached by travelling along the lower boundary and then switching to a straight line at T_{sw2}.

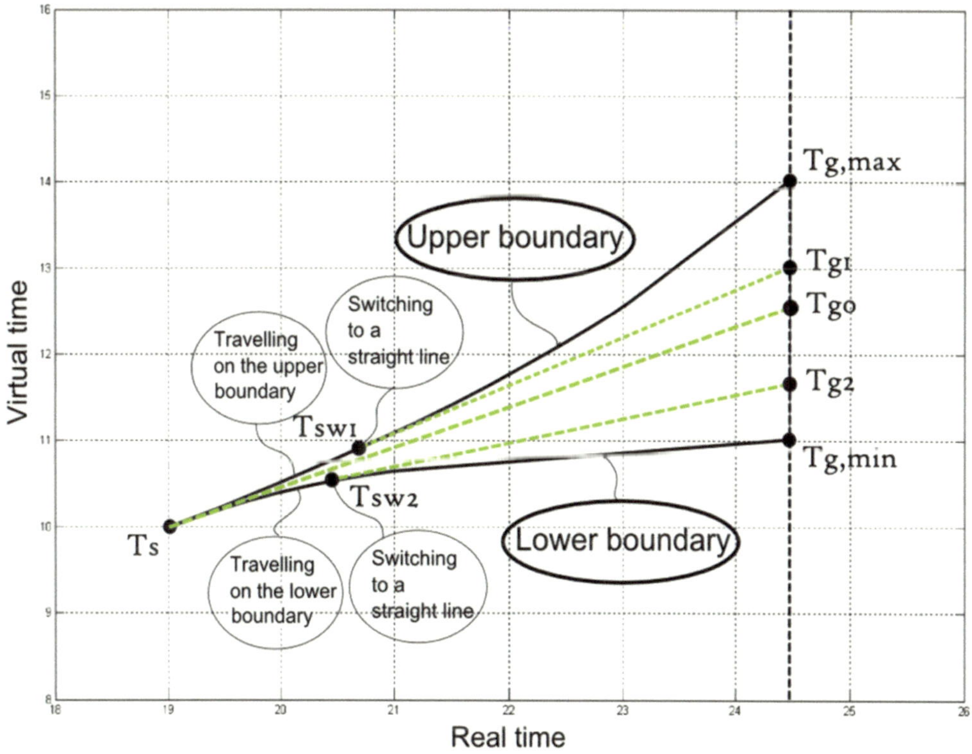

Fig. 8. Travelling on a border than switching to a straight line

Note that constant time-scaling with the general formula

$$\theta_{const}(t) = m(t - t_s) + \tau_s \tag{64}$$

results the scaled velocity function

$$\tilde{v}(t) = m^2 a_i (t - t_s) + m v_s \tag{65}$$

Denoting the curve that origins in T_s and ends in T_{sw} by $\Theta_{[Ts \to Tsw]}$ and the one that origins in T_{sw} and ends in T_g by $\Theta_{[Tsw \to Tg]}$, the path from T_s to T_g is

$$\Theta_{[Ts \to Tg]} = \Theta_{[Ts \to Tsw]} \cup \Theta_{[Tsw \to Tg]} \tag{66}$$

One should follow the following steps when constructing this path:

1. From a dedicated starting point construct the boundaries of the area of the reachable points. Update initial conditions at the border points and apply the appropriate type of curve for every interval between the starting point and the one desired to reach. When reaching a point where the scaled velocity would overrun the limits change to constant time-scaling or to the solution that results minimal speed, respectively.
2. Examine whether the desired point to reach is within the area of the reachable points.
3. Depending on the initial conditions at the starting point travel along the upper or the lower boundary until switching to a straight line is possible. Concatenate this line to the curve sequence between the starting and the switching point.

3.2 Avoiding time-obstacles

Avoiding time-obstacles includes collision detection with time-obstacles and a specific path planning method based on the algorithms mentioned above. Collision detection is simple in case of a rectangular time-obstacle. Since the solutions of the differential equations are given in analytic form, the exact time-values have to be substituted into those formulae in order to check whether a certain point of the curve belongs to a time-obstacle. Let us denote the vertices of a rectangular time-obstacle by its four corner points on the time-plane:

$$\begin{aligned} T_{enter,enter} &= (t_{enter}, \tau_{enter}) \qquad T_{enter,exit} = (t_{enter}, \tau_{exit}) \\ T_{exit,enter} &= (t_{exit}, \tau_{enter}) \qquad T_{exit,exit} = (t_{exit}, \tau_{exit}) \end{aligned} \tag{67}$$

The edges of this time-obstacle are the intervals between these points. In case of the horizontal edges, if one of

$$\tau_{enter} \le \Theta_{[Ts \to Tg]}(t_{enter}) \le \tau_{exit}, \qquad \tau_{enter} \le \Theta_{[Ts \to Tg]}(t_{exit}) \le \tau_{exit} \tag{68}$$

is satisfied, then a collision occurs. A similar formula can be applied for the vertical edges, thus a collision occurs when one of

$$t_{enter} \le \Theta^{-1}_{[Ts \to Tg]}(\tau_{enter}) \le t_{exit}, \; t_{enter} \le \Theta^{-1}_{[Ts \to Tg]}(\tau_{exit}) \le t_{exit} \tag{69}$$

is satisfied.

Note that one has to substitute the exact time-values into the appropriate element (i.e. curve sequence) of $\Theta_{[Ts \to Tg]}$ and $\Theta^{-1}_{[Ts \to Tg]}$ corresponding to the given time-value.

In case of nonlinear time-obstacles, the simplest way to determine a collision is to regard the scaled velocities. If

$$\left(\int_{tenter}^{texit} \tilde{v}(t)dt + \Gamma(t_{enter})\right) \cap O_{dyn}(t_{enter}, t_{exit}) \neq \varnothing \tag{70}$$

then the unit collides with a dynamic obstacle.
In the sequel, a path planning method based on the algorithm of connecting points on the time-plane and determining collisions with time-obstacles will be presented. Because of the presence of time-obstacles, an exact path between dedicated points might not exist, even if the desired point of reach is within the reachable area of the starting point T_s. Fig. 9 shows an example of several time-obstacles lying between the starting and the goal positions, while Fig. 10 shows that T_s cannot be connected with T_g directly because of a collision with a time-obstacle.

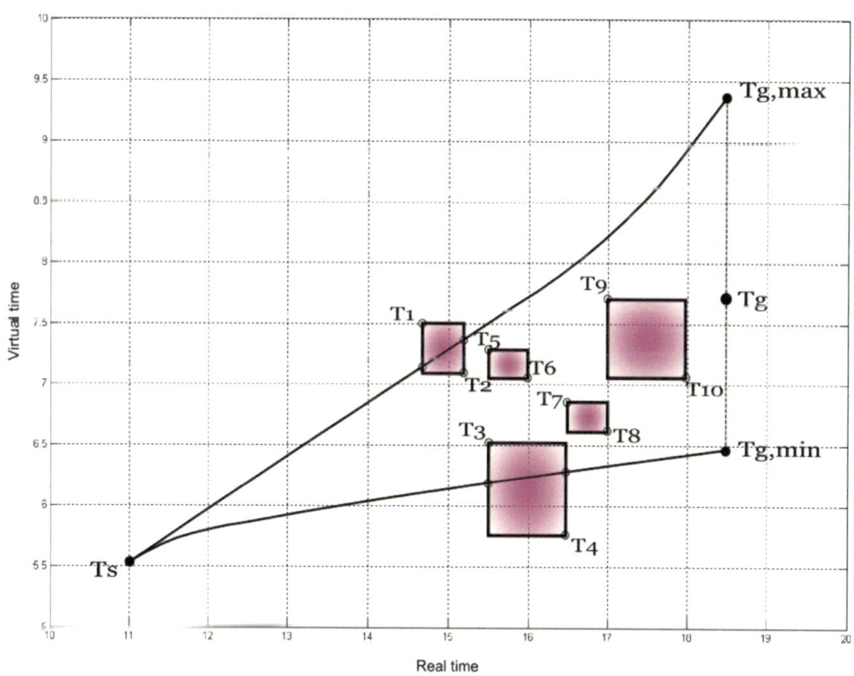

Fig. 9. Time-obstacles between two points

A path from the starting point avoiding the time-obstacles might only guarantee an endpoint with the same real-time coordinate t_g as the desired T_g , but the resulting virtual-time coordinate may differ from the desired τ_g . It is advised to examine which points with the real-time coordinate t_g can be reached from this starting point T_s. At first the corner points of the time-obstacles which can be reached without colliding with any other time-obstacles have to be determined. To do so, a graph (i.e. tree) building approach is proposed. At first the parent node is selected to be the actual point, desired to be reached. Child nodes are the upper left and lower right vertices of the time obstacle that the specific route hits first

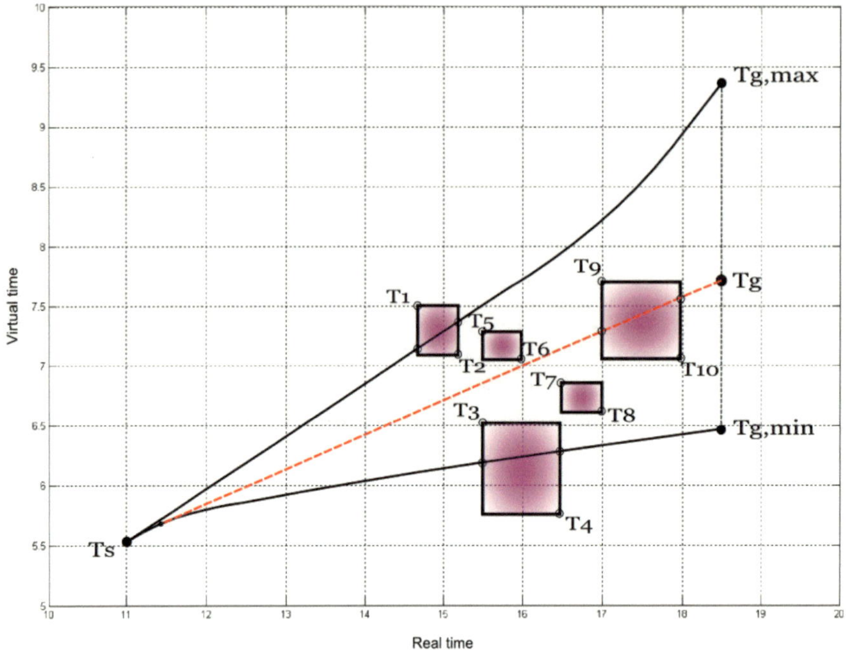

Fig. 10. The goal point cannot be reached without collision

(i.e. the one that is the nearest to the starting point and has an intersection with the actual path). The graph building can be done sequentially. Fig. 11 shows steps of a sequence of determining the child nodes and connecting them to their parent nodes.

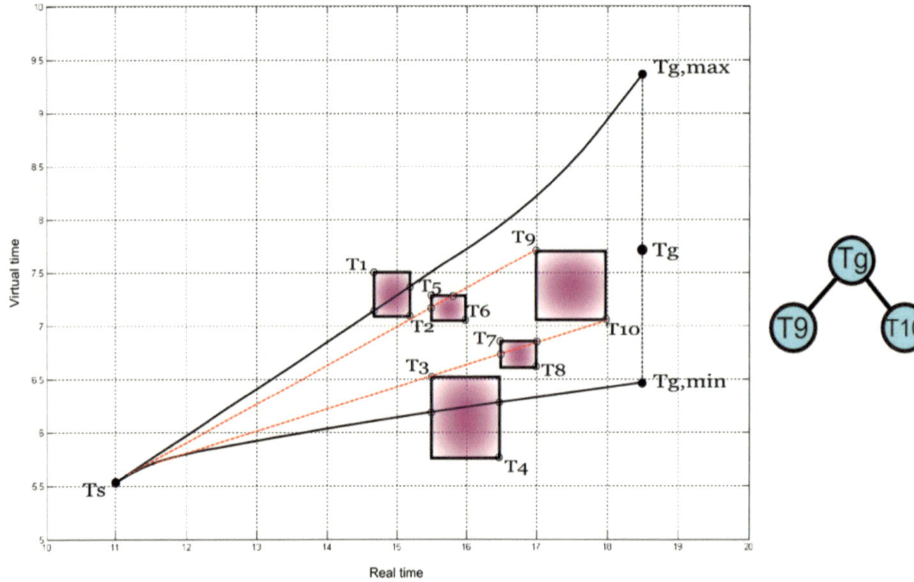

Fig. 11. Selecting child nodes by determining collisions on the time-plane (on the left) and building the graph (on the right)

The path from T_s to T_g intersects with the time-obstacle with corner points T_9 and T_{10}, thus in the graph these become the children of T_g. Now the desired point of reach will be T_9 or T_{10}, respectively. Fig. 12 shows that none of them can be reached without collision, thus they will also have child nodes in the graph.

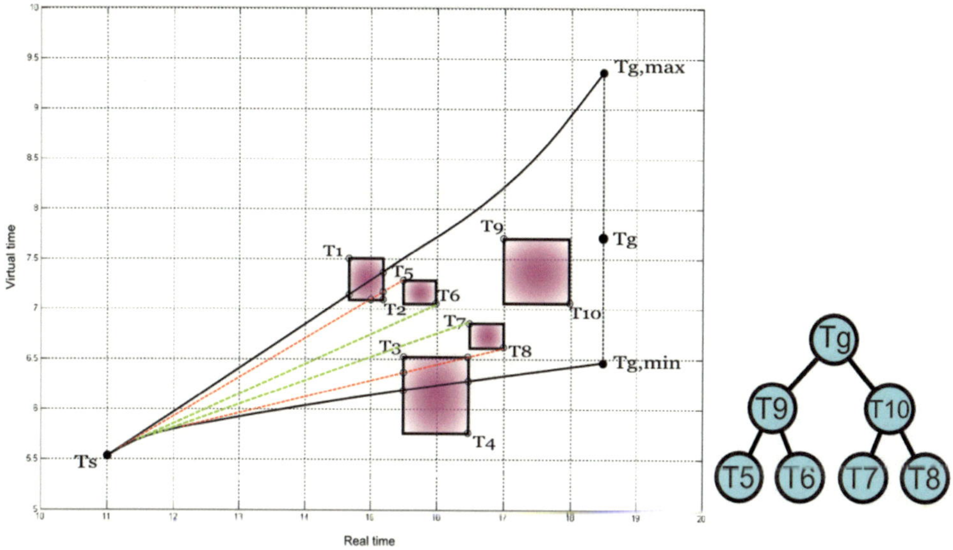

Fig. 12. Selecting child nodes by determining collisions (on the left) and building the graph (on the right)

Final steps of graph building is presented by Fig. 13:

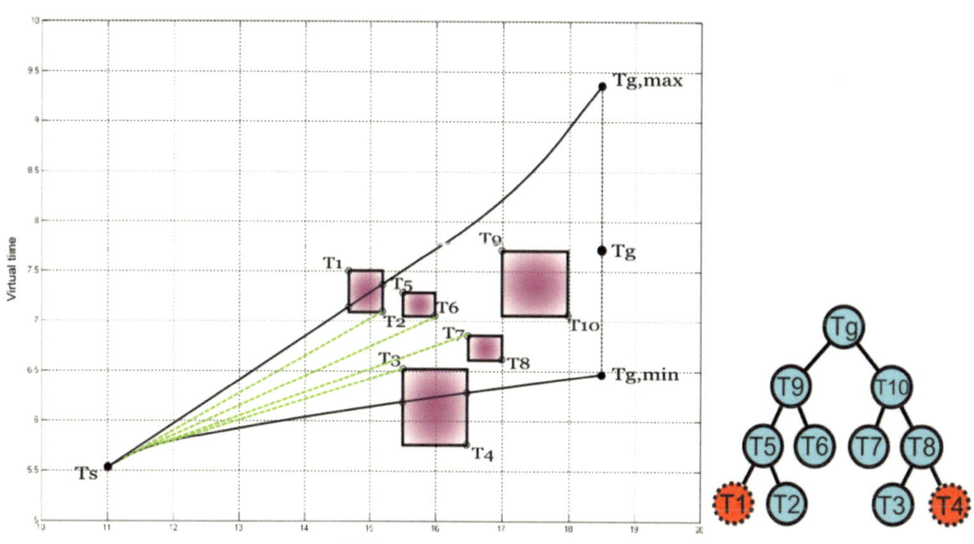

Fig. 13. The corner points that can be reached with no collision (on the left), building the graph (on the right)

On Fig. 13 the dotted contour denotes that T_1 and T_4 are located outside of the area of reachable points. Fig. 14 represents the final graph.

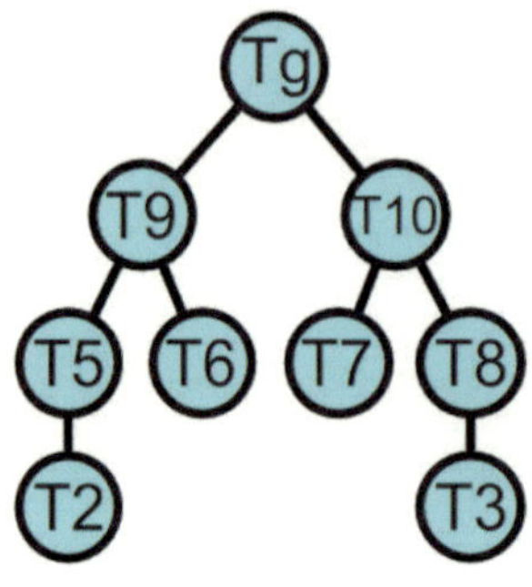

Fig. 14. The final graph

Nodes with no children provide the vertices that can be reached from the starting point without collision. After determining these points, a new starting point must be assigned from the ones determined afore, and the algorithm must be carried on until finding all possible routes from the original starting point to the ones with real-time coordinate t_g . Fig. 15 shows all possible routes from the original starting point to the points with real-time coordinate t_g while Fig. 16 shows its graph representation. Dot-contoured nodes denote that no route exists from their parent node that reaches them without collision or exceeding the kinematic constraints.

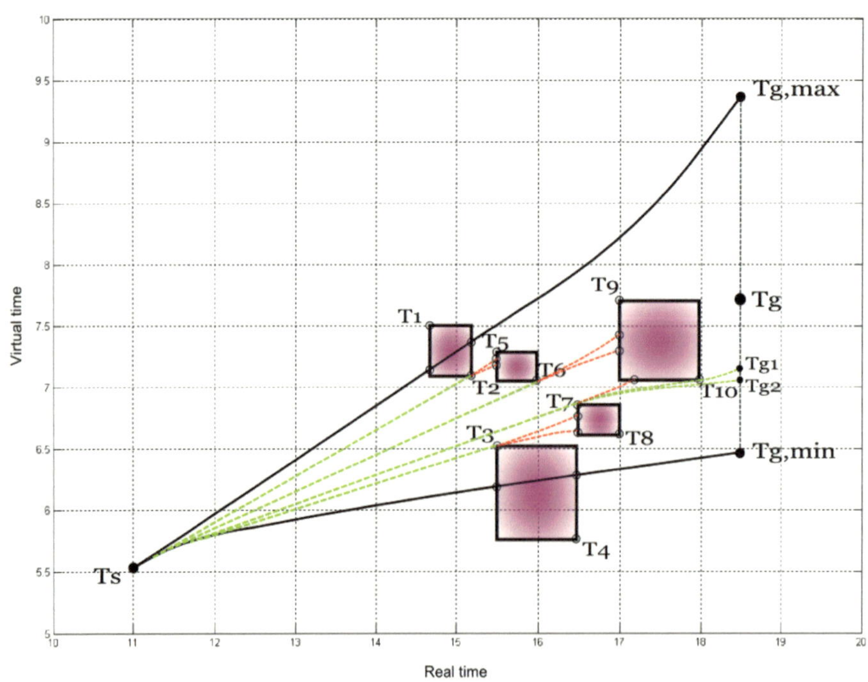

Fig. 15. Possible routes from Ts

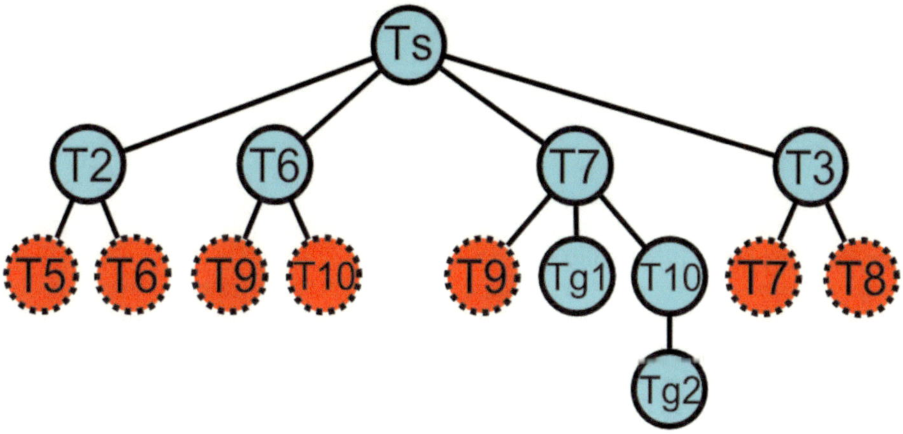

Fig. 16. Graph representation of the possible routes

3.3 The global path-planning method

Path-planning on the time-plane starts from the point at which the original velocity reaches a desired minimal value that was referred to as $v_{\min}$. The curve $\tau = t$ and the curve that spans through all intervals and results the minimal velocity enclose the area that the robot can reach during the time of its motion. Any other points outside this area are unreachable due to the kinematic constraints. Fig. 17 shows an example for such an area.

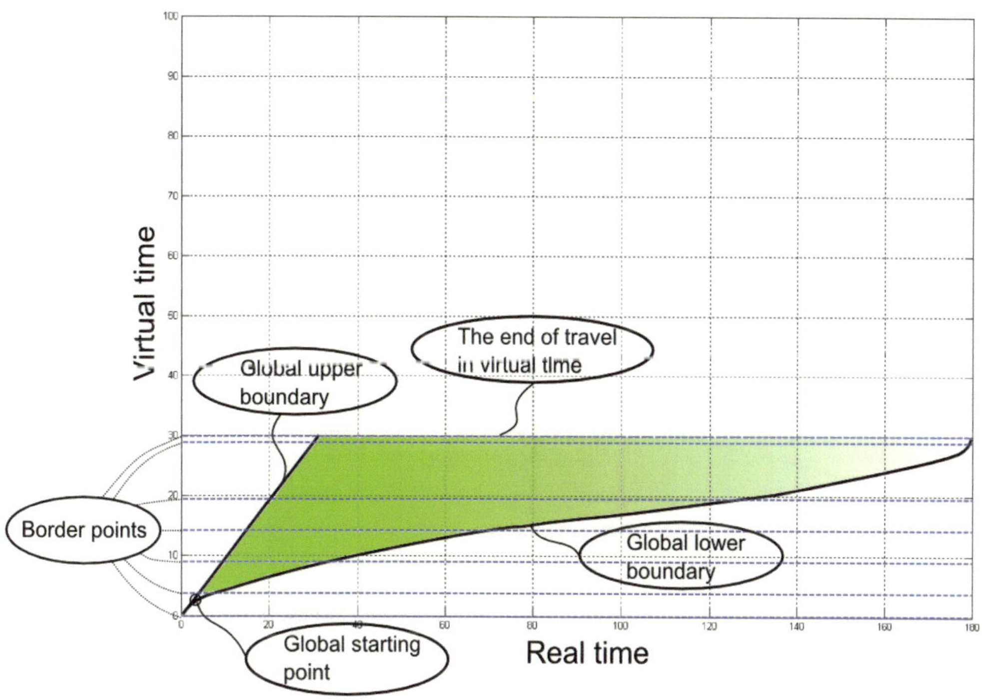

Fig. 17. Global area of reachable points

The goal of path-planning is to reach a point on the time-plane with virtual time coordinate τ_{end} (the end of travel in virtual time) and with a minimal value of real-time coordinate, i.e. to minimize the duration of travel in real time. In order to reach such a point, the following considerations should be followed:

1. When starting a path from a point, apply maximal control inputs and maintain this course until reaching a point with virtual time coordinate τ_{end}.
2. In case of collision, determine all time-obstacle corner points that can be reached from the starting point with no collision.
3. Assign a new starting point from these corner points and continue from step 1. If no further movement can be made, go back to a previous starting point and select another corner point to reach.

Navigating through all time-obstacles may not give an optimal solution. The possibility of delaying the departure of a certain robot should also be considered, which allows time-obstacles to disappear from its time-plane. Depending on the certain area of application, it can cause difficulties due to tight schedules. Fig. 18 shows an example of delaying a unit's departure so it can use its default velocity and arrive at its desired destination earlier then by using time scaling-based obstacle avoidance.

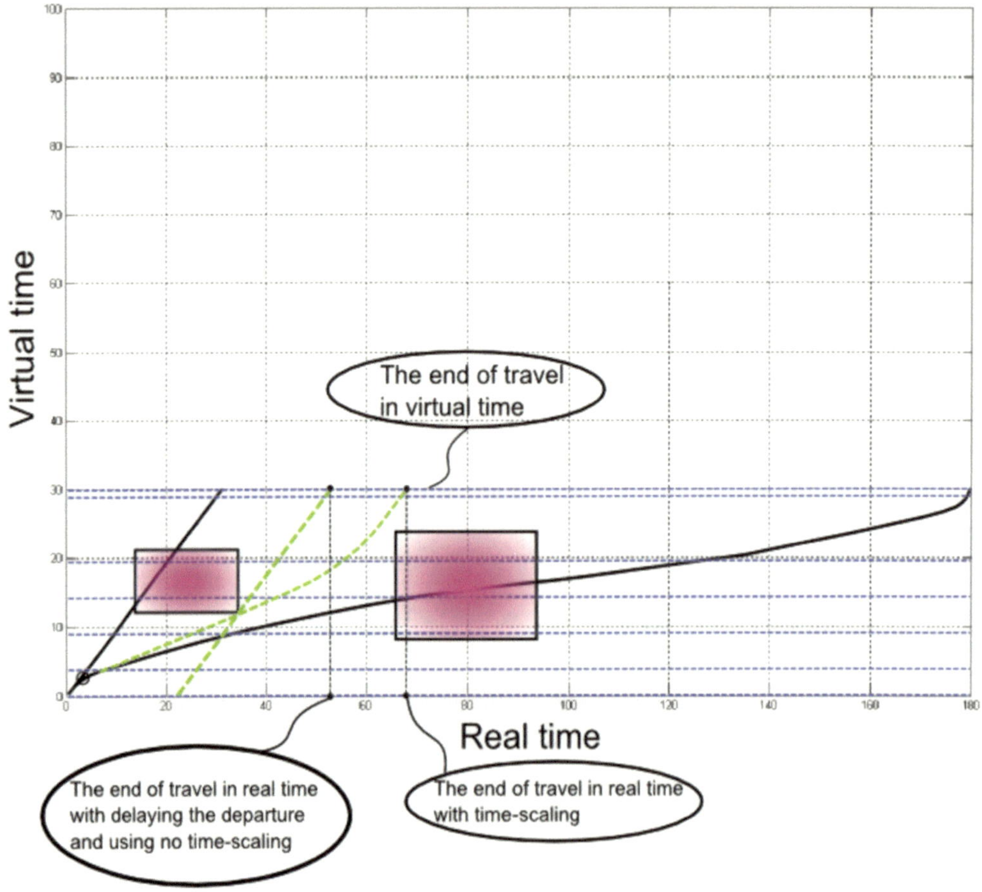

Fig. 18. Arrival times differ when delaying departure and using no time-scaling

3.4 Generation of the scaled velocity

The scaled velocity can be generated by multiplying the original velocity function and the derivative of the time-scaling function (see (6)). It is only necessary to calculate the values of the new velocity in those real-time instants where parameters in the original velocity have changed and where the time-scaling function changes its nature. The latter occurs typically at switching points where constant time-scaling is applied after maximal inputs and changing points where maximal or minimal value of velocity is reached. After these calculations, these points have to be connected by linear segments.

3.5 Synchronizing the motion of the robots

As it has been already mentioned, it is imperative to synchronize the units' motion in an area crowded by robots. A straightforward solution is to define a hierarchy between the robots so that higher priority units disregard ones with lower priority level. Such a priority order can be defined regarding several factors, and priority levels may also be redistributed regularly in order to meet the actual requirements. The general algorithm managing the multi-unit fleet is the following:

1. Consider the highest priority level.
2. Take a robot with a given priority level and design an optimal path avoiding static obstacles.
3. Assign an optimal velocity to this robot.
4. If the robot collides with any other robots with higher priority, determine the possible time-obstacles.
5. Construct the time-scaling function and generate the scaled velocity.
6. Take the next robot with lower priority and continue from step 1.

4. Implementation results

We present a robot system with four robots, the workspace and the trajectories are shown in Fig. 19.

The length of the robots are assumed to be 1 meter while their width are assumed to be 0.8 meters. (These values may be also different for each robot.) The paths consist of linear segments and they are extended with the widths of the robots. The coordinates of the points where the robots enter or leave the collision areas are calculated using these geometric parameters. On Fig. 19 the points S1, S2, S3 and S4 denote to the initial locations of the robots while G1, G2, G3 and G4 determines the goal locations respectively. The highest priority robot is Robot#1, the next one is Robot#2 then Robot#3 and Robot#4 has the lowest priority level. Table 1 contains information about the boundaries of the path elements (x and y coordinates respectively). We designed simple default velocites which are shown in Table 2 together with the scaled velocity. The admissible values of the maximal positive and negative accelerations for Robot#1 and Robot#3 are 0.2 m/s^2 and -0.2 m/s^2 while for Robot#2 and Robot#4 the same values are 0.5 m/s^2 and -0.5 m/s^2 respectively. The minimal velocity is 0.2 m/s for all robots. Since Robot#1 has the highest priority it does not need to be scaled. The time-scaling process for Robot#2, Robot#3 and Robot#4 are shown on Fig. 20, 22 and 24. Datatips on the time-scaling function indicate the relevant time instances where the scaled velocity values have to be calculated. The original and the resulted scaled velocity functions (with blue and green colors respectively) are drawn on Fig. 21, 23 and 25.

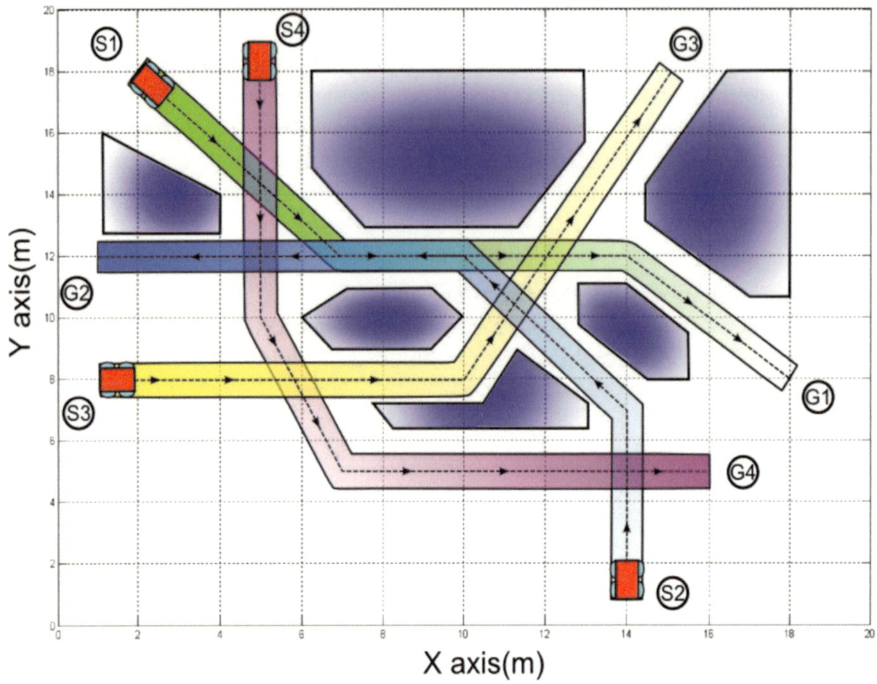

Fig. 19. Workspace

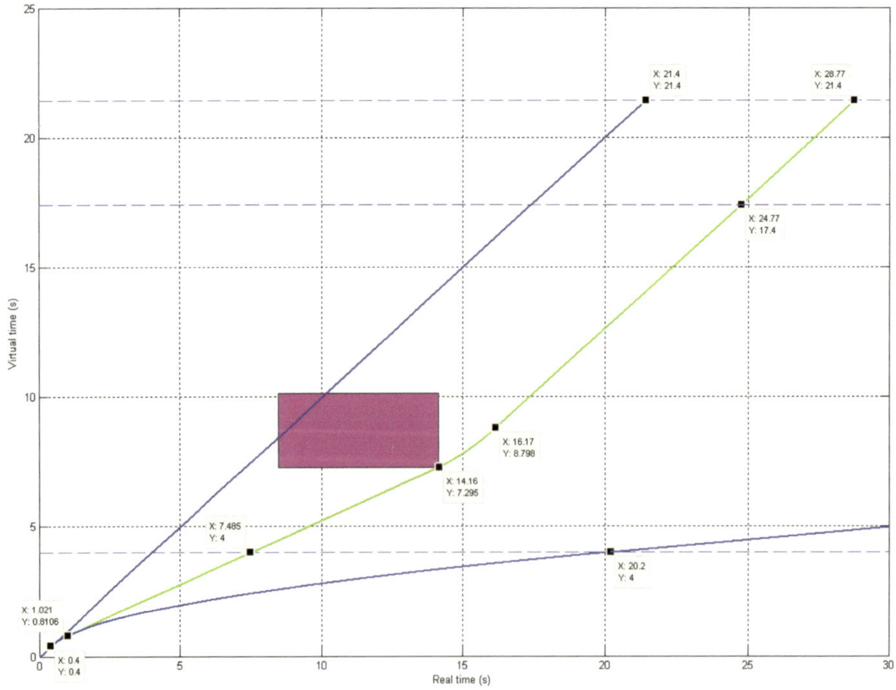

Fig. 20. Time-scaling for Robot#2

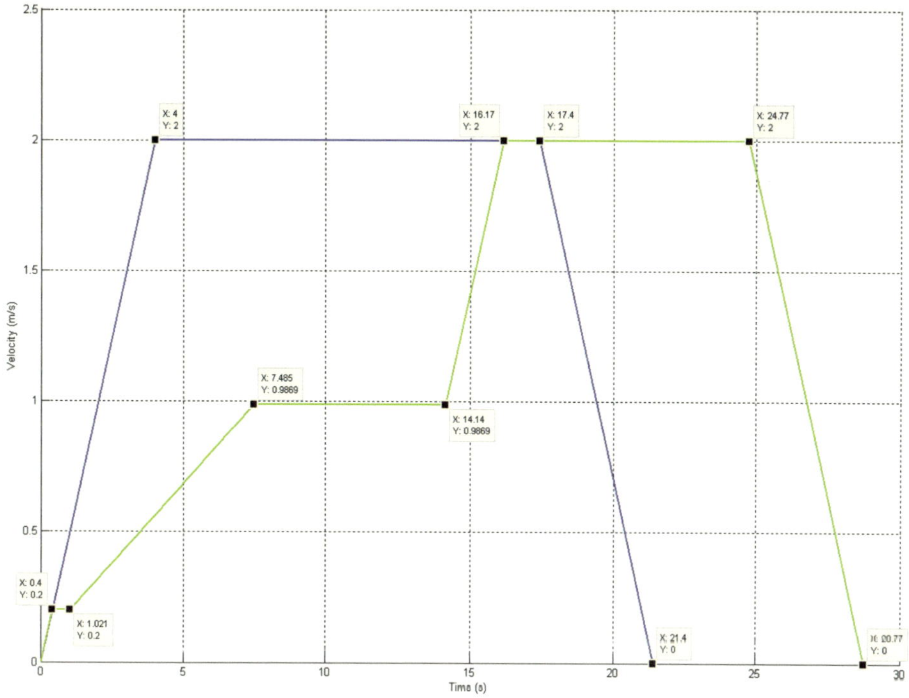

Fig. 21. Scaled velocity for Robot#2

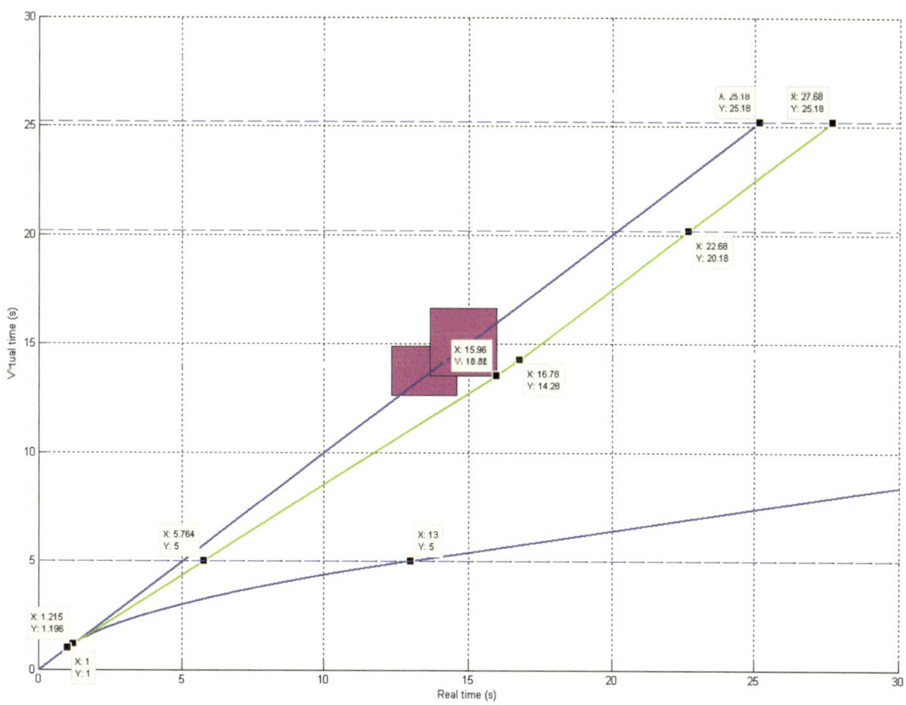

Fig. 22. Time-scaling for Robot#3

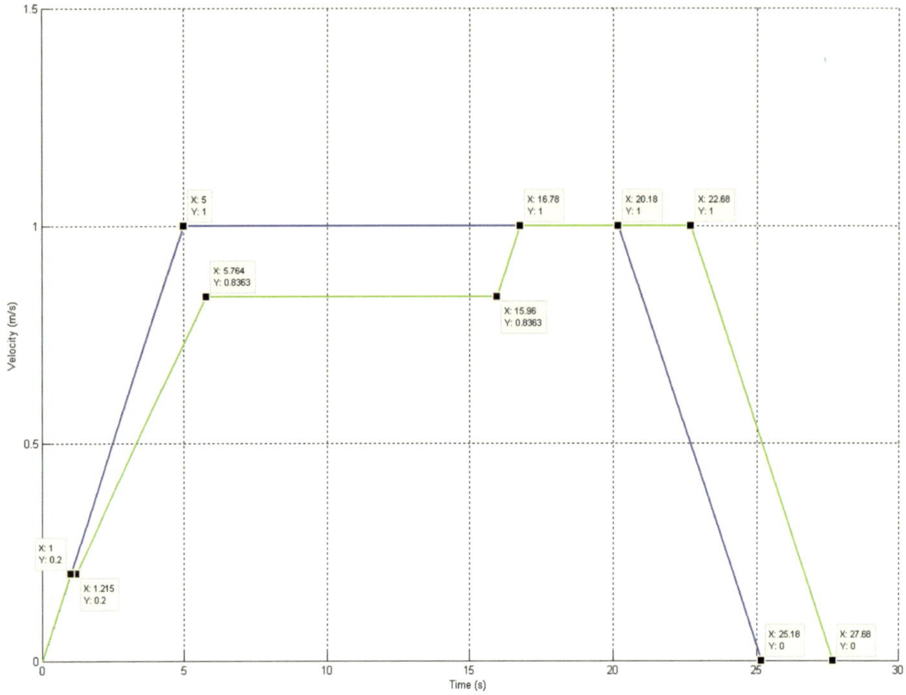

Fig. 23. Scaled velocity for Robot#3

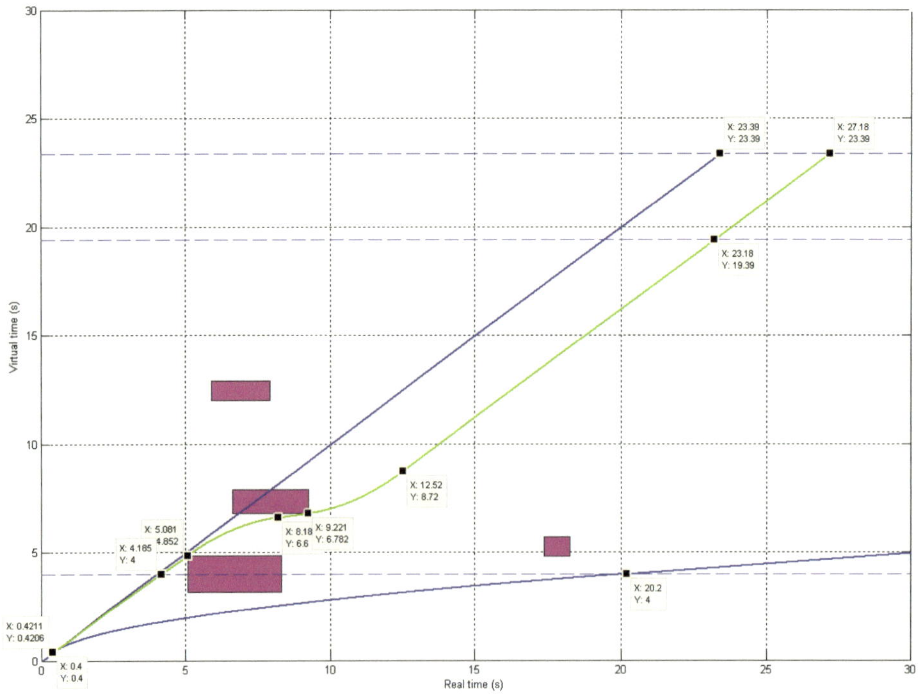

Fig. 24. Time-scaling for Robot#4

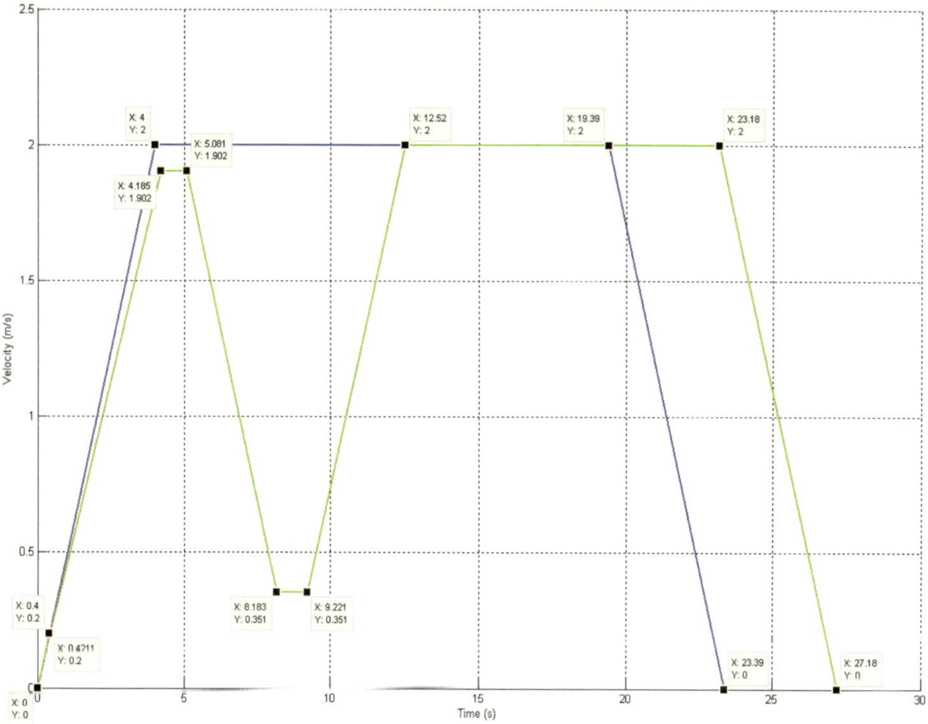

Fig. 25. Scaled velocity for Robot#4

Information on the relevant time instances, the derivate of the time-scaling function at those instances and the calculated scaled velocity values are shown in Table 2.

Robot#	x coordinate(m)	y coordinate(m)
Robot#1	2	18
	7	12
	14	12
	18	8
Robot#2	14	1
	14	7
	10	12
	1	12
Robot#3	1	8
	10	8
	15	18
Robot#4	5	19
	5	10
	7	5
	16	5

Table 1. Path of the robots

Robot#	Virtual time(s)	Real time(s)	$\dot{\theta}(t)$	Default velocity (m/s)	Scaled velocity(m/s)
Robot#2	0	0	1.0000	0	0
	0.4000	0.4000	1.0000	0.2000	0.2000
	0.8106	1.0213	0.4935	0.4053	0.2000
	4.0000	7.4846	0.4935	2.0000	0.9869
	7.2854	14.1425	0.4935	2.0000	0.9869
	8.7984	16.1686	1.0000	2.0000	2.0000
	17.4031	24.7733	1.0000	2.0000	2.0000
	21.4031	28.7733	1.0000	0	0
Robot#3	0	0	1.0000	0	0
	1.0000	1.0000	1.0000	0.2000	0.2000
	1.1957	1.2149	0.8363	0.2391	0.2000
	5.0000	5.7638	0.8363	1.0000	0.8363
	13.5249	15.9575	0.8363	1.0000	0.8363
	14.2764	16.7760	1.0000	1.0000	1.0000
	20.1803	22.6799	1.0000	1.0000	1.0000
	25.1803	27.6799	1.0000	0	0
Robot#4	0	0	1.0000	0	0
	0.4000	0.4000	1.0000	0.2000	0.2000
	0.4206	0.4211	0.9510	0.2103	0.2000
	4.0000	4.1850	0.9510	2.0000	1.9020
	4.8524	5.0813	0.9510	2.0000	1.9020
	6.5996	8.1832	0.1755	2.0000	0.3510
	6.7816	9.2205	0.1755	2.0000	0.3510
	8.7200	12.5184	1.0000	2.0000	2.0000
	19.3852	23.1836	1.0000	2.0000	2.0000
	23.3852	27.1836	1.0000	0	0

Table 2. Results of time-scaling for the robots

5. Conclusions

In this chapter a time-scaling based obstacle avoidance method was presented that is able to avoid dynamic obstacles as well as static obstacles. We designed and described an algorithm being able to find a time-scaling function that avoids all time-obstacles in the time-plane and thus ensures a collision free path in the robots' workspace.
The determined time-scaling function also satisfies the criteria according to the kinematic constraints prescribed for the control (velocity) inputs. The solutions of the differential equations describing the motion of the time-scaled units were presented with a method that makes finding a path between dedicated points in the time-plane possible.
Since the solutions of the differential equations are given in analytic form there is no need for time-consuming computations and determining the scaled velocity function is simple.
Planning on the time-plane is based on a novel and fast tree-building method. The path-planning algorithm is considered to be complete since it gives a definite answer in a finite time. All components of the algorithm are well suited for real time implementation.
Should time-scaling fail to work we still have a method to avoid collision with delaying the robots. Fig. 24 shows a scenario where a better solution is achieved by using time-scaling than simply delaying the robot.

6. Acknowledgments

This work is connected to the scientific program of the "Development of Quality-Oriented and Harmonized R+D+I Strategy and Functional Model at BME" project. This project is supported by the New Hungary Development Plan (Project ID: TÁMOP-4.2.1/B-09/1/KMR-2010-0002). The work is partially supported by the Hungarian Scientific Research Fund under grant OTKA K71762.

7. References

[1] J.C. Latombe. *Robot Motion Planning*. Kluwer Academic Publishers, Boston, MA, 1991.

[2] Yongguo Mei; Yung-Hsiang Lu; Hu, Y.C.; Lee, C.S.G. Energy-efficient motion planning for mobile robots. In: Proceedings of the IEEE Int. Conf. on Robotics and Automation, pages: 4344-4349. 2004.

[3] V. Kunchev, L. Jain, V. Ivancevic, and A Finn. Path planning and obstacle avoidance for autonomous mobile robots: A review. Volume 4252 of Lecture Notes in Artificial Intelligence, pages 537–544. SpringerVerlag, 2006.

[4] F. Cuesta and A. Ollero, Intelligent Mobile Robot Navigation, ser. Springer Tracts in Advanced Robotics. Heidelberg: Springer, 2005, vol. 16.

[5] J. Reif and M. Sharir, "Motion planning in the presence of moving obstacles", Journal of the Association for Computing Machinery, vol. 41, no. 4, pp. 764–790, 1994.

[6] S. B. Moon and S. Ahmad, "Time scaling of cooperative multirobot trajectories", IEEE Transactions on Systems, Man, and Cybernetics, vol. 21, no. 4, pp. 900–908, 1991.

[7] M. Peasgood, C.M. Clark and J. McPhee, "A Complete and Scalable Strategy for Coordinating Multiple Robots Within Roadmaps", IEEE Transactions on Robotics, vol. 24, no 2, pp. 283-292, 2008.

[8] H. Li, S. X. Yang and M. L. Seto, "Neural-Network-Based Path Planning for a Multirobot System With Moving Obstacles", IEEE Transactions on Systems, Man and Cybernetics-Part C: Applications and Reviews, vol. 39, no. 4, pp. 410-419, 2009.

[9] K. G. Jolly, R. Sreerama Kumar, and R. Vijayakumar, "A Bezier curve based path planning in a multi-agent robot soccer system without violating the acceleration limits", Robotics and Autonomous Systems, vol. 57, pp. 23-33, 2009.

[10] Steven M. LaValle, "Planning algorithms" , Cambridge University Press, 2006

13

Cooperative Reinforcement Learning Based on Zero-Sum Games

Kao-Shing Hwang[1], Wei-Cheng Jiang[1], Hung-Hsiu Yu[2] and Shin-Yi Lin[2]
National Chung-Cheng University / Industrial Technology Research Institute
Taiwan

1. Introduction

Multi-Agent systems (MAS), developed from the artificial intelligent field, include many independent agents, each of which has its own behavior and can achieve a certain goal. The system can be used to solve a complex problem by cooperation and coordination between agents, because most complex problems can be divided into many sub-problems. Therefore, cooperation is central to the topics and operations of MAS (Vassiliades et al., 2011)(Marden et al.,2009).

The use of MAS in teams to play a competitive game such as soccer is major challenge, where agents have to accomplish the goal of scoring or prevent scoring by opponents in real time in a highly varying environment. Being good players, they must be able to adapt the strategy on which they are behaving responding to actions of the opponents. In other words, agents of MAS need to operate appropriately in real time to varying states of the environment. Therefore, it is difficult to establish a well-defined state-action mapping for the agents unless the whole state space has been visited or searched exhaustedly. Instead, a learning mechanism adapting to a highly dynamic environment is more appropriate for the situation (Busoniu et al., 2008)(Xiao et al., 2007).

Reinforcement learning is probably the best-known method of learning due to the similarity of empirical learning in mammals (Gosavi, 2009)(Kaelbling et al., 1996). The learning takes place by trial and error, following by performance evaluation but always provided after a sequence of actions. Besides, the correct solution or behavior for a given trial is not provided. The goal of learning agents is to establish a strategy optimizing a scalar cost or evaluation function in long term. The strategy, also known as action policy, can choose an appropriate behavior or action based on the stated perceived by robots (Barto et al., 1983)(Bingulac et al., 1994)(Tan, 1993). Furthermore, reinforcement learning on Markov games have also been proposed for multi-agent reinforcement learning to solve cooperation problems in social dilemmas under the assumption of Nash equilibrium (Yanco et al., 2002) (Flache et al., 2002) (Littman, 1994). These general-sum stochastic games has been adopted as a framework for multi-agent reinforcement learning and proved to converge to a Nash equilibrium under specified conditions (Hu et al., 1998). That method is adoptable to find an optimal strategy only when there exists a unique Nash equilibrium in a game. However, in

some problems, such as task assignment in robot soccer games, it is not enough to consider only attributes of the agent itself because the whole state space is composed of the ones of partners and components.
Eventually, task assignment is always a major research area in MAS where a number of autonomous agents, either in heterogeneous or homogeneous fashion, work together with individual assignment. Although centralized optimization provides opportunity for efficient optimization, the coordination and the transfer of information among agents are costly and often infeasible. Hence it is important to develop decentralized optimization schemes which permit the individual agents take control of the actions that contribute towards the optimization of a global performance criteria. The motivation for using game theory for assigning tasks are driven by the fact that decentralized optimization in game theory is achieved by the agents (players of the game) acting selfishly. We utilize techniques from game theory to produce an algorithm for matching player and roles (tasks) based on situations on the field. As a result, we have created a deterministic algorithm for task assignment with built-in feedback mechanisms for reinforcement learning.
This chapter presents a cooperation strategy system with self-improving abilities for task assignment for multi-robot systems. The strategy system is eventually a formation mechanism based on reinforcement learning and the game theory through task assignment to individual robot. The mechanism coordinates robots, in they which learn to behave primitively, to work as a team by means of perceptual information with their respective roles such as, attacker, defender, or goalie, to produce appropriate behaviours. The robots with homogeneous behavioral architecture accomplish cooperatively global goals on their local sensory information cooperatively. These robots carrying out tasks of a helper or defender improve action policy at play based on Q-learning inspired by the game theory. Results of experiments demonstrate the effective performance of the proposed method in comparison with renown methods.

2. Task assignment mechanism

Eventually, the assignment of specific teaks to individual robot on a team is a major research in MAS. No matter whether the teams are homogeneous or heterogeneous, different robots will generally be required to perform different task. In our proposed system, individual robots are able to solve certain problem, such as ball shooting, obstacle avoidance, and positioning through reinforcement learning before playing a game. The system makes use of a task assignment mechanism to robots with learned capabilities. The task assignment mechanism further divided into two sub-mechanisms: a task assignment system to decide each robot's temporal task corresponding to perceived geometrical state and a reinforcement learning system to command robots to the most appropriate position for the task assigned. Each robot with the same architecture behaviors on local information accomplishes global goals; that is, each robot chooses and executes it own actions according to its task and the environmental states. Meanwhile, the choosing action strategy associated to a task is modified by means of reinforcement learning.

2.1 Dynamic task assignment for players

The assignment system assigns dynamically four tasks, an attacker, goalie, helper, and defender, respectively, to a robot one at a time.

2.1.1 Tasks of an Attacker

The task of the Attacker is aimed to kick a ball into the opponent's field to score. It is a very important task in a soccer game because the goal of a soccer game is to gain more scores. Thus, the first assigned task in the assignment mechanism is finding an attacker. The position of a robot is a major factor in considering which robot should take this assignment because a fast and precise attack is key to score a point. Therefore, a robot who can spend the shortest time to kick the ball in the correct direction should be an attacker. The procedure to identify which robots are at an advantageous position is defined as follows. With respect to a coordinate frame where origin locates at the center of the ball, and x-axis is the line from the opponent's goal center to the ball as shown in Fig. 1. The good side is defined on the side opposite to opponent's goal of y-axis. In other words, robots on the region are favorable to be an attacker. In turn, the robot on the good side and closest to the ball is assigned as an attacker. Unfortunately, if there is no robot on the good side, any robot is selected as an attacker as long as it is closest to the ball.

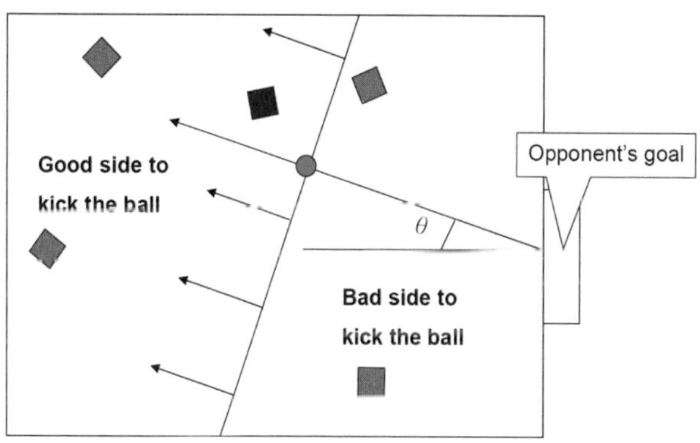

Fig. 1. The rule to select the attacker.

2.1.2 Tasks of a goalie

A goalie is required to move along the direction in parallel to the y-axis of the frame with respect to the ball in the front of its goal. This task is assigned to a robot after the attacker is selected because stopping the opponent's attack in front of the goal is a passive action in a soccer game. A robot is assigned to play this task when it is nearest to its team's goal and does not need to be assigned as an Attacker.

2.1.3 Tasks of a helper

A helper assists the attacker or the defender to accomplish the imminent efforts. This task is assigned in turn thirdly, and two free robots are assigned as helpers. The helpers are capable of learning in the game. Depending on the various situations occurring in the game field, the helpers move to suggested positions according to the formation being taken to ensure the attacker or the defender can get good support. There are three formations, attack formation, normal formation, and defense formation to deal with the highly dynamic situations on the field. Helpers learn empirically to ensure that a most appropriate formation is always taken. The details of the learning mechanism are depicted in the next section.

a. In the attack formation, the helpers are dispatched to both side of the attacker on the half way to the opponent's bottom line, as shown in Fig 2(a).

b. In the case of the normal formation, the helpers are aligned with the attacker as shown in Fig 2(b).
c. If the defense formation is taken, helpers are assigned to two sides of the attacker but at the midpoint between their own goal and the attacker as shown in Fig 2(c). Meanwhile, a robot is assigned as a helper when it is not assigned as either an attacker or a goalie and is nearest to one of suggested positions of the adopted formation.

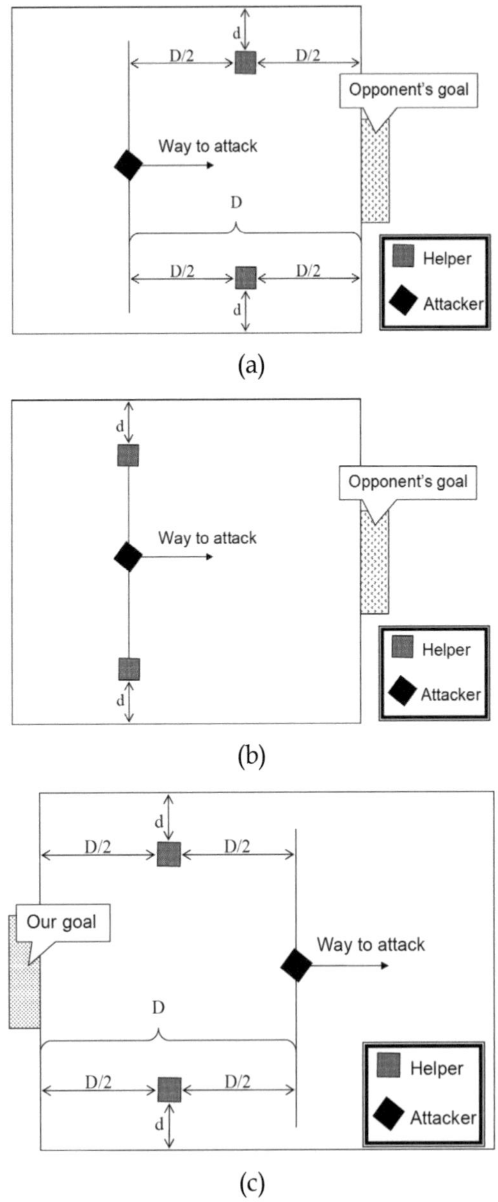

Fig. 2. Rules to select helpers: (a) attack formation, (b) normal formation, (c) defense formation.

2.1.4 Tasks of a defender

The major function of a defender is to block and prevent opponents from scoring. Hence, it needs to move to appropriate positions where it can be most helpful to the goalie. Similar to the task for helpers, there are three optional positions for the defender as shown in Fig 3. The defender has learning ability, and details of the learning system will be described later.

2.2 Learning to position

As aforementioned, each helper and defender has three prospective formations or positions to choose. A reinforcement learning system is applied to determine which formation the helper should take with respect to the position of the attacker, the defender, and the position of the soccer ball.

The reinforcement learning system associated with an individual is implemented by the Q(λ) algorithm. For computation efficiency, the state space ought to be partitioned to a reasonable amount in the learning state-action mapping function. Therefore, a linear tile-coding function approximation is used for state reception (Hu et al., 1998)(George A. Bekey, 2005) where each instance of decoded state vectors includes the information of position, speed, and direction of the soccer. Furthermore, a ball position is presented by the coordinate on the field which is divided into $6 \times 9 = 54$ grids; ball speed is mapped to a bipolar value; the direction of ball is also presented by eight different symbols. The output of the reinforcement system decides which formation should be taken.

In addition, in order to respond and adapt timely to various situations on the field, the learning system should take the opponents' current actions into consideration. Unfortunately, it is impossible, in practice, to know which action is being and going to be taken by the opponent. Thus, observing the sequence of opponents' behaviors to predict the situations is a good way to get related information. Eventually the action set of opponents cannot be obtained but temporal actions executed by the opponents are observable. Instead of obtaining the action chosen directly, positions where each opponent is standing at are followed up, because the action being executed is regarded as the output of the function of current robot formation. The quantization of the opponent's action is done by a Cerebellar Model Articulation Controller (CMAC) decoder (K. S. Hwang et al., 1998). Since the combinational state space of opponents in the opposing team will be extremely large if all the information for each opponent is taken into account. Thus, only the state of the attacker in opposing the team is considered, and that attacker is an opponent closest to the ball than the other players.

For discretizing the receptive space into a state space, a filter partitions the region around the ball into 16 cells, as shown in Fig. 4. The center of the circle is at the position of a ball, and the radii of the inner and outer circles are 3 and 6 units, respectively. Instead of Euclidian distance presentation, the distance between the opponent and ball is a scale value assigned by this filter.

The reward to each individual learning system is different. The sparse rewards of the helpers' learning system are as follows:

- Get Point: get reward +1;
- Lose Point: get reward -1;
- Become the Attacker: get reward +0.5.

The rewards for the defender are:

- Lose Point: get reward -1

- Become the Attacker: get reward +1

When the period of finding a new action is too short, the previous action will have not enough time to execute well because executing it needs enough time. This will provide wrong results when the system is just beginning to learn. To ensure sufficient time to execute each action command, a new action is triggered only after the task of an attacker is shifted to the teammates by the assignment system. According to the rule of task selection, the attacker is changed under certain situations. Therefore, the previous action can have enough time to execute.

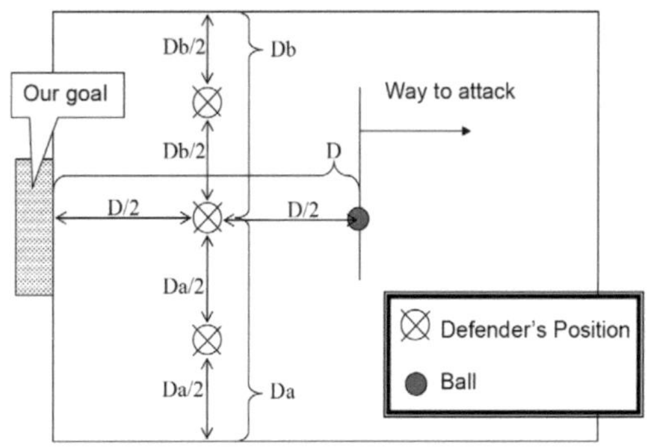

Fig. 3. Three positions of defender.

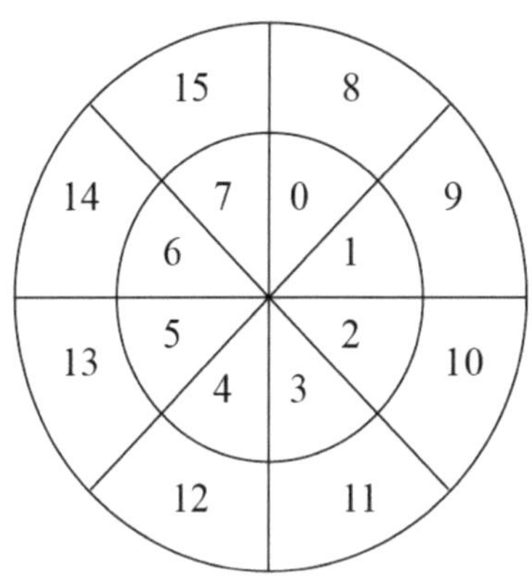

Fig. 4. Decompositions of an opponent's position.

3. Q-learning for MAS

Spontaneous behaviors of the helpers or defenders for positioning are activated by a local learning system. A Q-learning for MAS derived from the game theory is proposed to obtain a better policy in a team. In the two-player zero-sum game, the action is taken according to the opponent's behavior. If the opponent makes an action to get their best benefit, the action may result in the worst situation to our team. On the other hand, if we want to win the game in this situation, best actions to deal with the action being taken by the opponent must be always taken. Therefore, the Q-learning is modified to accommodate the principle of game theory. The original Q-learning rule is as follows:

$$Q(s,a) \leftarrow Q(s,a) + \alpha[r + \gamma \max_{a' \in A} Q(s',a') - Q(s,a)], \tag{1}$$

where **s** is the state; ***a*** is the action; ***A*** is an action set; ***a*** is the learning rate; γ is the discount factor; ***r*** is the reward. This equation tries to maximize the Q-value for the future and only consider the action that we take. So, it uses $\max Q(s',a'), a' \in A$ to update its $Q(s,a)$. In game theory, because we need to consider the opponent's action ***o***, the notation, ***o***, denotes the opponent's action set. We redefine $Q(s,a)$ to $Q(s,a,o)$ and redefine the $\max Q(s',a'), a' \in A$ to $\max\min Q(s',a',o'), a' \in A, o' \in O$. Therefore, we get the following equation:

$$Q(s,a,o) \leftarrow Q(s,a,o) + \alpha[r + \gamma \max_{a' \in A} \min_{o' \in O} Q(s',a',o') - Q(s,a,o)]. \tag{2}$$

Eligibility traces are also taken into account. The algorithm is further modified by the *replace-trace* method:

$$Q(s,a,o) \leftarrow Q(s,a,o) + \alpha[r + \gamma \max_{a' \in A} \min_{o' \in O} Q(s',a',o') - Q(s,a,o)]. \tag{3}$$

where $\delta = r + \gamma \max_{a' \in A} \min_{o' \in O} Q(s',a',o') - Q(s,a,o)$

Actually, this algorithm is derived from the classic Q-learning algorithm but inspired the game theory and eligibility traces. Fig.5 shows the complete algorithm in pseudo code.

```
Initialize Q(s, a, o) arbitrarily
Repeat (for each episode):
    Initialize s
    Repeat (for each step of episode):
        Choose a from s, o using policy derived from Q
    (e.g., like ε-greedy, but not only consider the max a, but also consider the min o)
        Take action a, observe r, o, s'
        δ = r + γ max_{a'∈A} min_{o'∈O} Q(s',a',o') − Q(s,a,o)
        e(s,a,o) ← 1
        For all s, a, o:
            Q(s,a,o) = Q(s,a,o) + αδe(s,a,o)
            e(s,a,o) ← γλe(s,a,o)
            s ← s'
        Until s is terminal
```

Fig. 5. The proposed Q-Learning.

4. Soccer robot behaviors

A robot has been required to learn three basic types of behaviors, ball shooting, obstacle avoidance, and target reaching. These primitive behaviors can be combined with more complex behaviors, such as defending, or attacking. Ball shooting is a simple but essential behavior. Through this behavior, before a robot comes to the ball position, it follows an imaginary shooting line, and switches the kicking motor on. The imaginary shooting line is a line passing through the ball and the target door. Obstacle avoidance is necessary because during the game, a robot is not allowed to collide with other robots. The temporal information about the positions and moving direction of other robots is provided by a vision system. The robot moves in a direction to avoid collision according to the distance between other robots and the current position by means of a reinforcement learning mechanism (K. S. Hwang et al., 1998). The outputs of all available actions of the learning mechanism are regarded as reference signals for driving motorizing wheels. The behavior of target reaching drives a robot to the desired position and orientation.

5. Simulations

There are two simulations, one is for the basic behaviors, and the other is for the strategy algorithm.

5.1 Behavior demonstration

There are three different basic behaviors that to be learnt: collision avoidance, target moving, and ball shooting. The simulation environment to emulate the soccer field is defined by the Federation of International Robot-soccer Association (FIRA) (FIRA, 2009). A robot is initially placed at the center but slightly to the right. The robot motion track is shown by a series of white dots. Fig. 6 shows the behavior of obstacle avoidance learned by a robot. The robot walks randomly on the field without going over the boundary, which is regarded as an obstacle. Fig. 7 shows a robot's capability of moving to a target. In this figure, the robot moves as closely as possible to the assignment posture (position with a certain orientation) with a slight tolerance represented by a hidden region centered by the assigned position. Fig. 8 shows the learnt ball shooting behavior. The robot moves to an aiming position where the robot aligns the line between the center of the opponent's gate and ball, but with a certain distance from the ball in preparation for kicking.

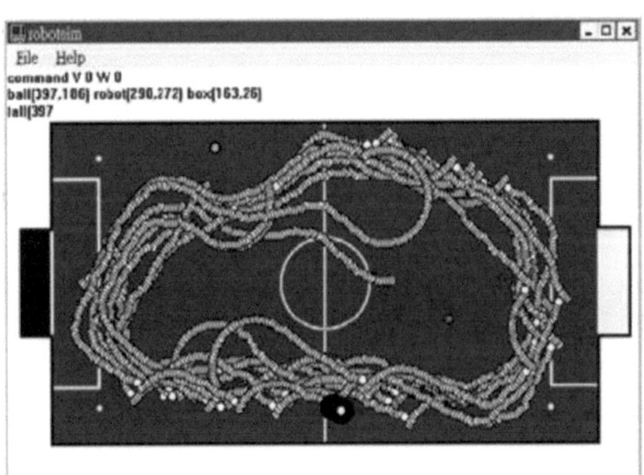

Fig. 6. Obstacle avoidance behavior.

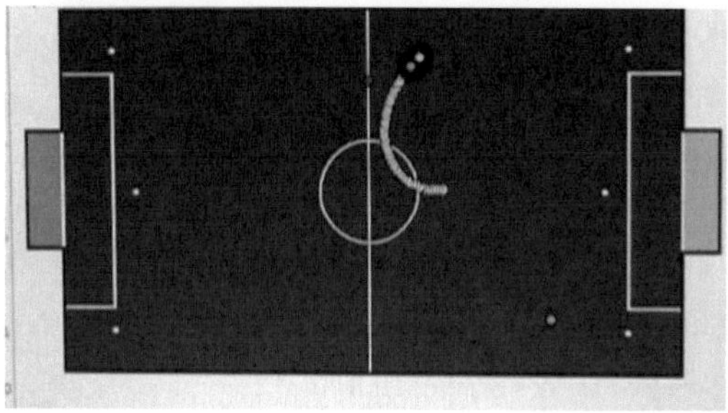

Fig. 7. Target moving behavior.

Fig. 8. Ball shooting behavior.

5.2 Strategy demonstration

The Proposed Q-learning algorithm is compared with the renowned Q(λ)-learning and minimax-Q under a 5-vs-5 simulation environment provided by FIRA (FIRA, 2009). The Q(λ)-learning is a modified Q-learning methods with the eligibility skill, and the minimax-Q further combines Q-learning with a simple game theory. These three algorithms are arranged, respectively, to play against a test-bed algorithm. In the first game, the team with the Q(λ)-learning algorithm plays against the team using the default algorithm, Lingo. Lingo is a well-established landmark algorithm along with the FIRA robot soccer simulator. Meanwhile, the minimax-Q algorithm is used to play against the team using Lingo in the other game. In another game, the proposed algorithm, Proposed Q, competes with Lingo. The scores gained by the competitors are recorded in Fig. 9. The horizontal axis of the chart expresses the number of rounds each of which lasts 10 minutes, and the vertical axis expresses the total scores the comparing team gets. The solid line is a regression of the saw-toothed line to represent the trend of scores gained along the number of rounds.

As shown in Fig. 10, where the regression lines in Fig. 9 are redrawn, the proposed algorithm beats the opponent by about 16 to 18 points. This result is better than the results of Q(λ)-learning algorithm and the minimax-Q algorithm.

In the second experiment, which is similar to the first experiment, Lingo is replaced by a rule-based strategy introduced in (K. S. Hwang et al., 2007). This rule-based strategy opponent has beaten Lingo before. The comparison between Q(λ)-learning algorithm, minimax-Q algorithm, and the proposed algorithm is indicated on Fig. 11. Intriguing enough, the opponent's score decreases more obviously when the challenger uses the proposed algorithm.

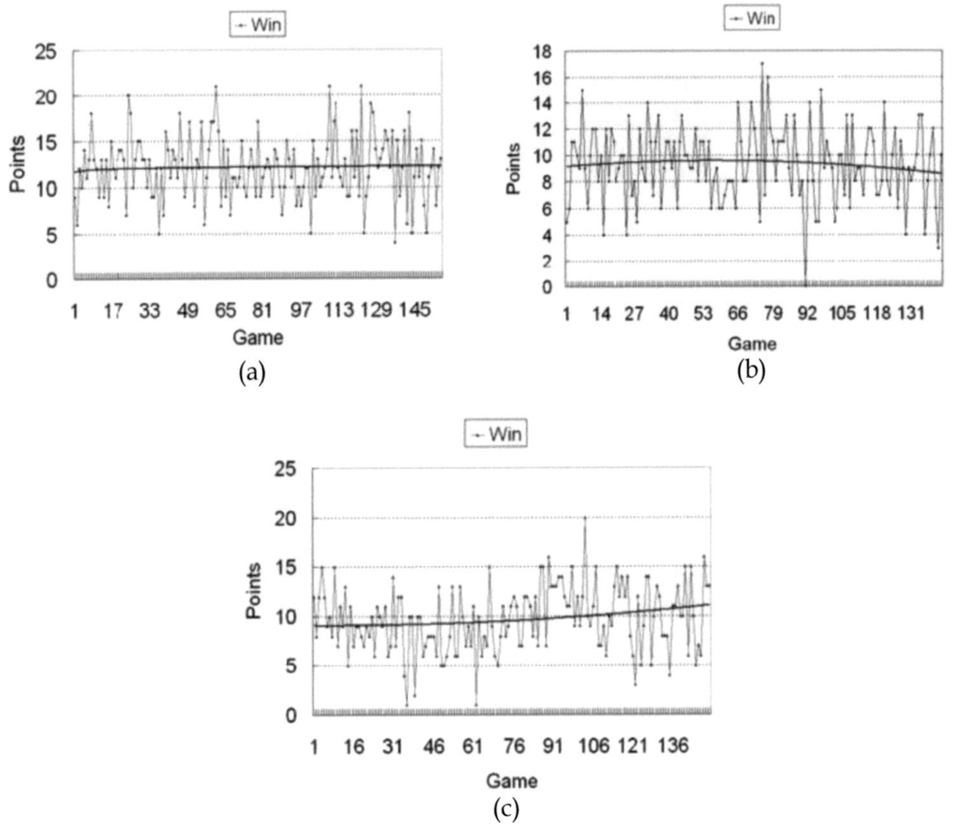

Fig. 9. (a) Q(λ)-learning vs. Lingo: Q(λ)-learning win scores log and trend-line, (b) minimax-Q vs. Lingo: Minimus-Q win scores log and trend-line, (C)Proposed Q vs. Lingo:Zero-Sum-Q(λ) win scores log and trend-line.

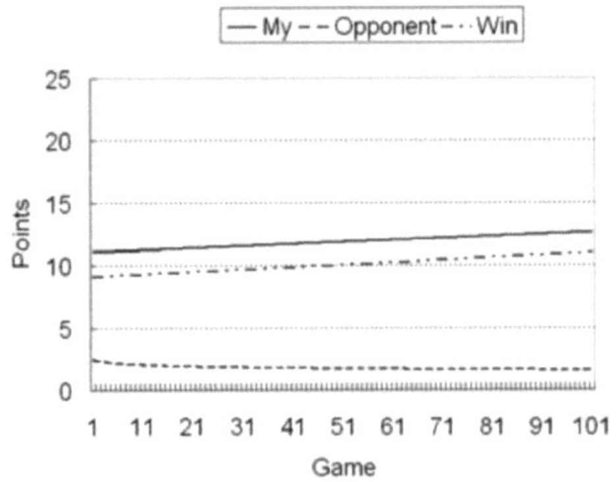

Fig. 10. Proposed Q vs. Lingo: Three trend-lines in the last 100 games.

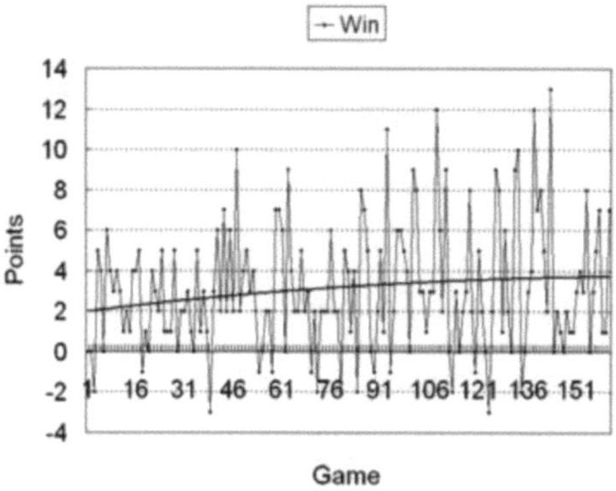

(a)

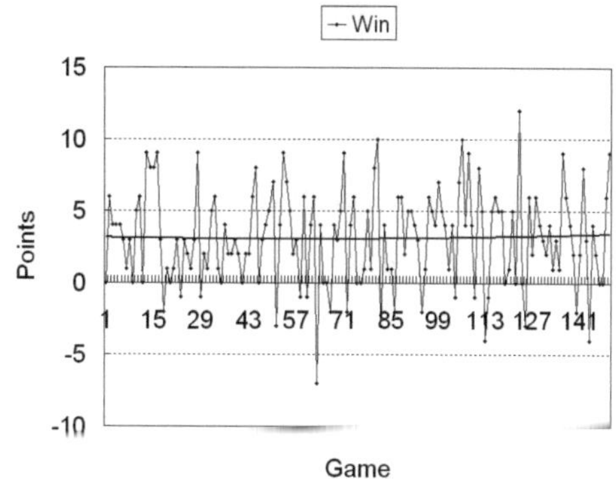

(b)

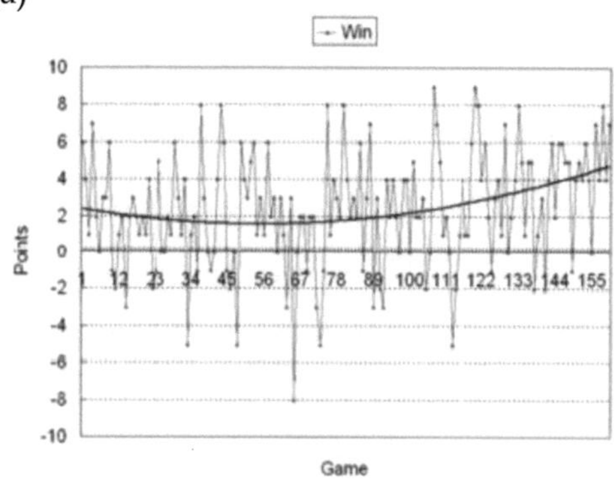

(c)

Fig. 11. (a) Q(λ)-learning vs. hand code C: My scores, (b) minimax-Q vs. hand code C: My scores, (c) Proposed Q vs. hand code C: My scores.

5.3 Communication protocols

The proposed method has been further implemented into an actual soccer robot system to demonstrate the performance in a real environment. The robot soccer system is composed of a control center, vision system, and three soccer robots. The vision system recognized the ball, the location of our team's three robots and the opponent robots and their movement. The control system consists of Bluetooth dongles and a software system. A personal computer (PC) system is used here because of its low cost and high performance. The software system includes a control subsystem, a strategy subsystem, and a communication subsystem, called the subsystem of robot protocol (SRP). A soccer robot is based on an embedded system, where the system controls the rotating speed of the wheels according to the commands from the control center. Thus, the embedded system is connected to a motor driving system that drives wheel motors and a communication system that communicates with the control center. An SRP is a self-defined protocol for communication between the PC and soccer robots through Bluetooth. The protocol regulates how the control system sends commands to robots. The protocol is very simple because the communication is only one-way from the center to robots. Each robot command is sent in turn and repeatedly as shown in Fig. 12.

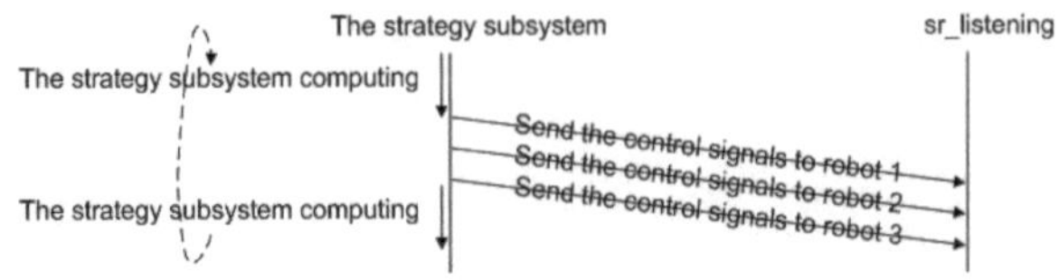

Fig. 12. The SRP protocol.

The strategy subsystem, learnt empirically in the simulator, decides how the robot moves by receiving environmental situations from the vision system and makes decisions through the proposed mechanism.

The control subsystem is initially in charge of the robots (before starting the soccer robot game), and also in charge of beginning and maintaining the connection between control system and robots through Bluetooth. In addition, it also has the responsibility of controlling how robots act when the strategy subsystem does not control the robots in order to handle exceptions. The connection beginning procedure is described in the flowchart shown in Fig 13. In the figure, the service program, sr_listening, is a program running in each robot's embedded system to receive commands from the control center.

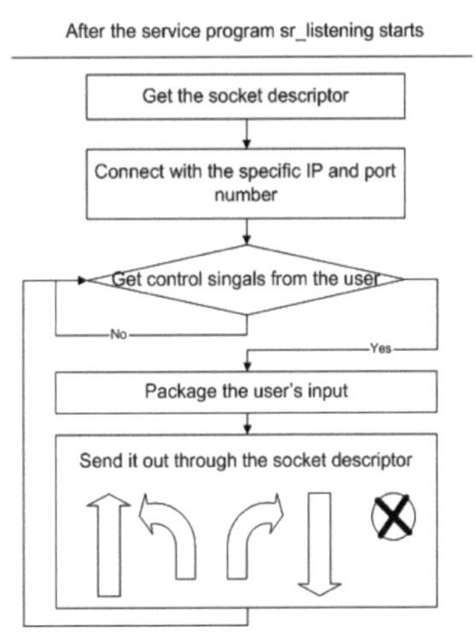

Fig. 13. The flowchart of making and maintaining the connection between the control center and robots.

The embedded system is the core of soccer robots. The duty of this system is to both receive the commands from the control center, and to control locomotion. The overall system architecture of soccer robots is shown in Fig. 14. A program called sr_listening implements the receiving command function. It is a non-stop procedure and begins to run when the robot power is switched on. The flowchart of the procedure is shown in Fig. 15. A sr_listening process invokes a threat, as the flowchart shows. The threat of each robot has its own port number and service number. According to these numbers the procedure can determine whether or not the message sent needs to be processed. The Motorola 68K core series MC68VZ328 processor has been used. Since it has low power consumption and high performance, it is appropriate for use in a soccer robot. The driving subsystem receives reference signals from the embedded system. Thus, the control system of the wheel is in the embedded system and is a software implementation. The advantage is that the control rule can be modified without changing the hardware, and so is more flexible. The control rule here is a simple open loop control, which means that there are no sensors on the wheels to feedback speed or other information.

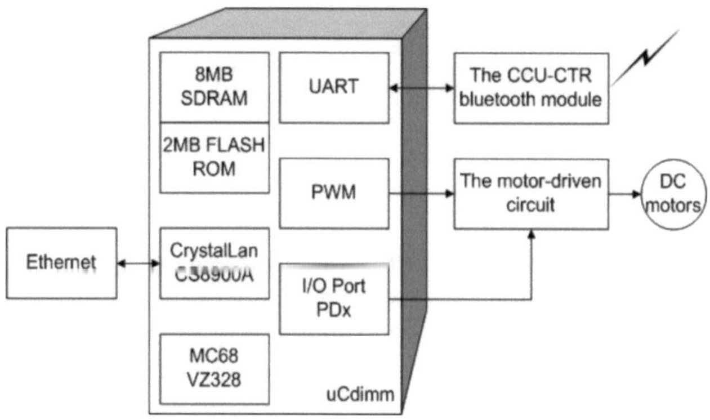

Fig. 14. The architecture of a soccer robot system

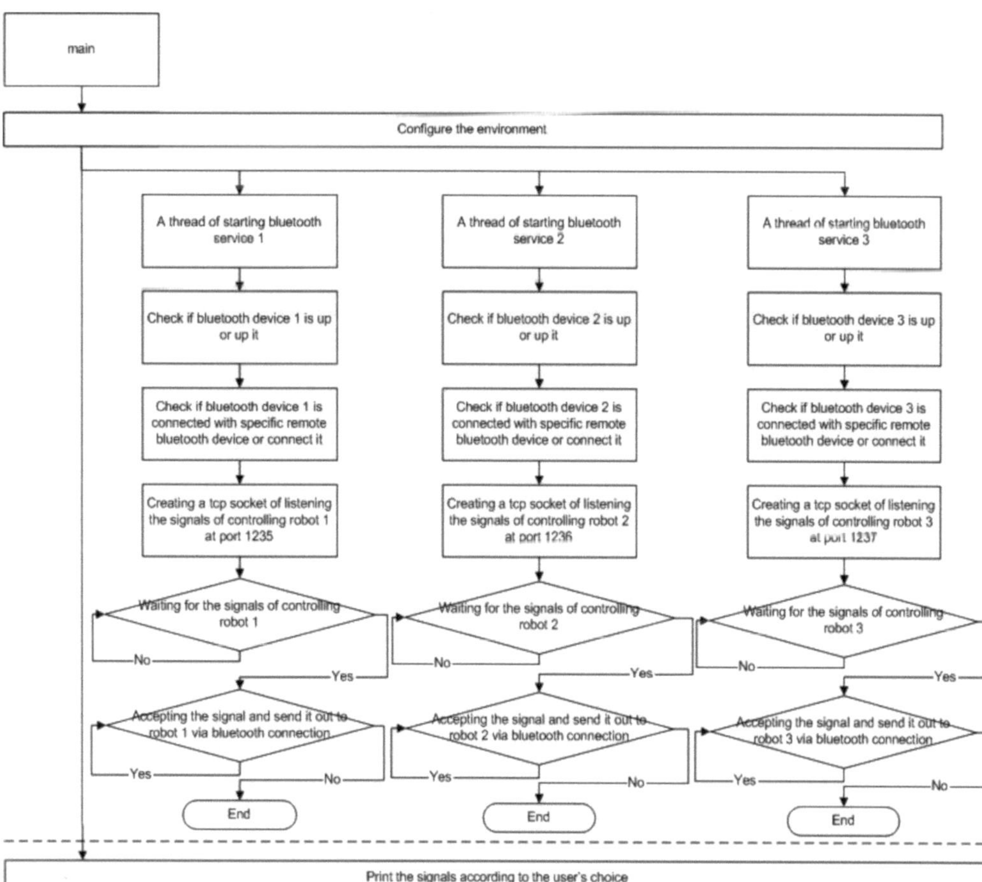

Fig. 15. The flowchart of the procedure sr_listening

5.4 Experiments

The experiments are executed on a standard robot soccer game field. As Fig.16, there are two teams in these experiments. One team in right side has been trained by a simulator, but the other has had no training at all. The names, Green, Pink and Purple will be referred to the robot which is with the green, pink and purple mask, respectively. The opponents just stand still on the field to emulate a fair competition situation for comparison. The experiments are designed to compare the results of two teams, each of which is playing against the same opponent for ten rounds with around five minutes time elapsed such that the performance of the proposed algorithm is evaluated by comparing the performance of these two teams. Table 1 shows results of algorithms versus Lingo. The results of each strategy do not improve very much. The Q(λ)-learning algorithm gets the highest scores and a little improvement after learning, because it is designed to find the maximum value and it uses eligibility traces. The minimax-Q algorithm is not work very well in this situation, because the opponent is too simple. The zero-sum game theory does not work very well when the opponent is not smart. The scores of Zero-Sum-Q(λ) algorithm can get a little improvement after learning, because it uses eligibility traces.

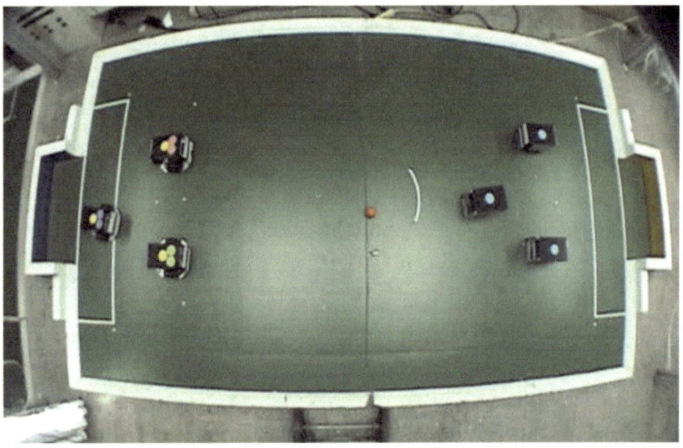

Fig. 16. The experiment platform

Scores	Q(λ)-learning Vs. Lingo	Minimax-Q Vs. Lingo	Zero-Sum-Q(λ) Vs. Lingo
My	13~14	11	11~13
Opponent	2	1.5~2	2~1.5
Win	11-12	9	9~11

Table 1. Results of algorithms vs. Lingo

The Fig. 17 and Fig. 18 illustrate the results of experiment that compete Zero-Sum-Q(λ) team(ZSQ(λ)) with Lingo team. In Fig. 17, the Zero-Sum-Q(λ) team does not have any experiences, the team trains policy in the game process. The difference of score is not obvious. In Fig. 18, the Zero-Sum-Q(λ) team that has trained can show the ability of offenseand defence. The complete score is recorded in Table 2, which indicates that the non-trained team scored fewer points (22) than the trained team (25). In addition, the non-trained team lost more scores (21) than the trained team (5). This shows that the learning mechanism has influenced the behaviours of soccer robots and the hardware has worked effectively.

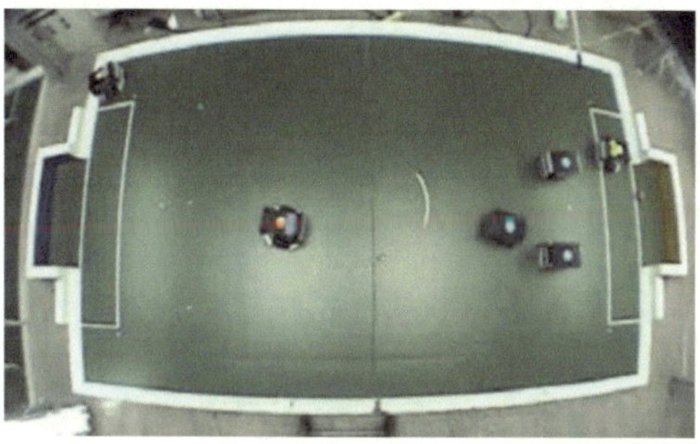

Fig. 17. (a) Game 1: ZSQ(λ) scored at the time 01:10 when Green dribbled the ball into the goal straight.

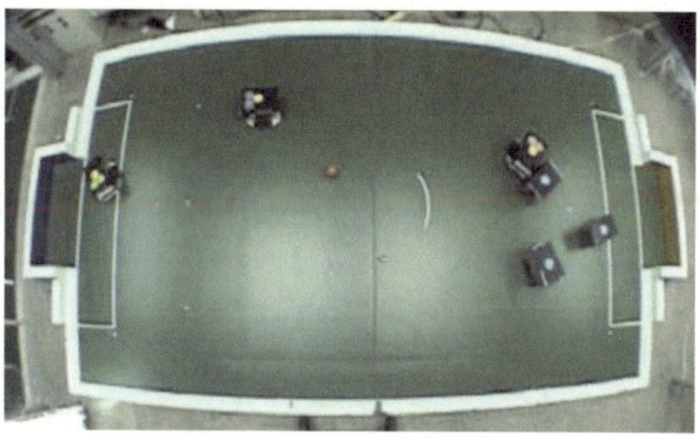

Fig. 17. (b) Game 2: In this game, ZSQ(λ) didn't score any point.

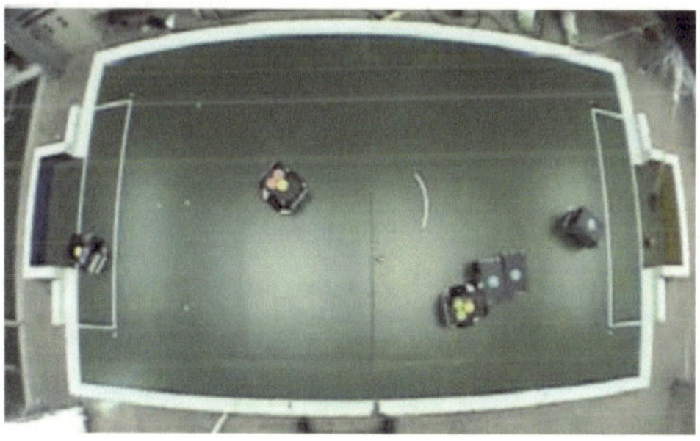

Fig. 17. (c) Game 3: ZSQ(λ) scored at the time 00:15 and 03:07. The ball rebounded toward the goal by Green.

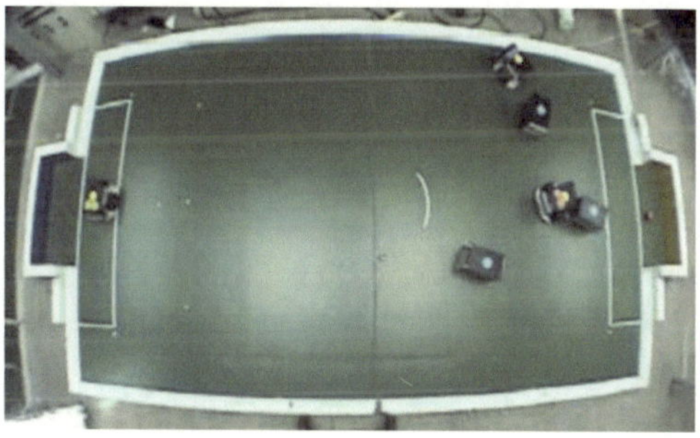

Fig. 17. (d) Game 4: ZSQ(λ) scored at the time 02:48, 02:57 and 03:48. The Pink dribbled the ball into the goal.

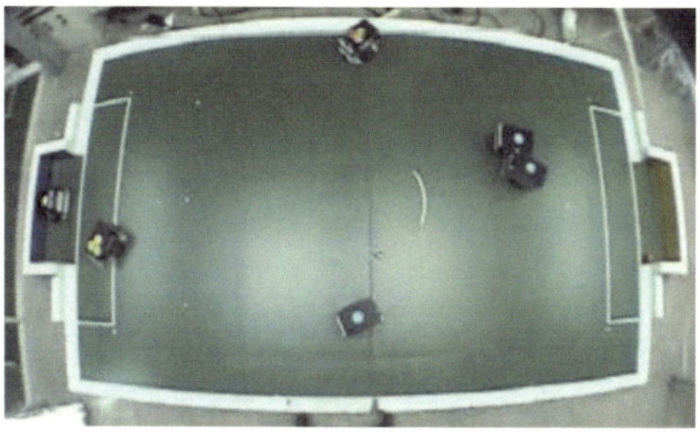

Fig. 17. (e) Game 5: ZSQ(λ) scored at the time 00:10, 01:52, 02:26 and 03:00. The ball rebounded toward the goal by Pink.

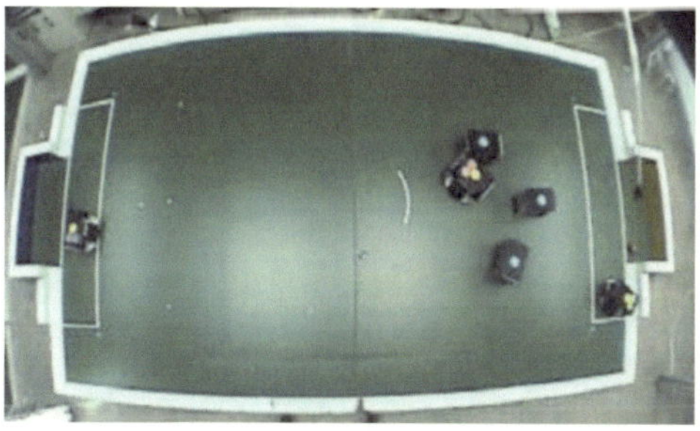

Fig. 17. (f) Game 6: ZSQ(λ) scored at the time 01:05, 02:48 and 04:00. The Green kicked the ball into the goal.

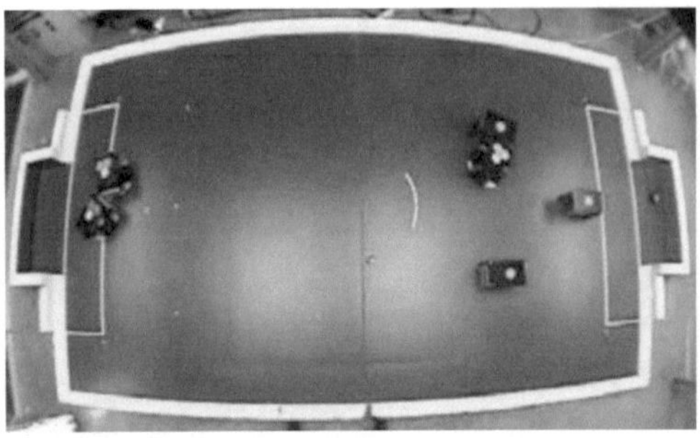

Fig. 17. (g) Game 7: ZSQ(λ) scored at the time 00:09 and 01:38. The Green kicked the ball straight toward the goal.

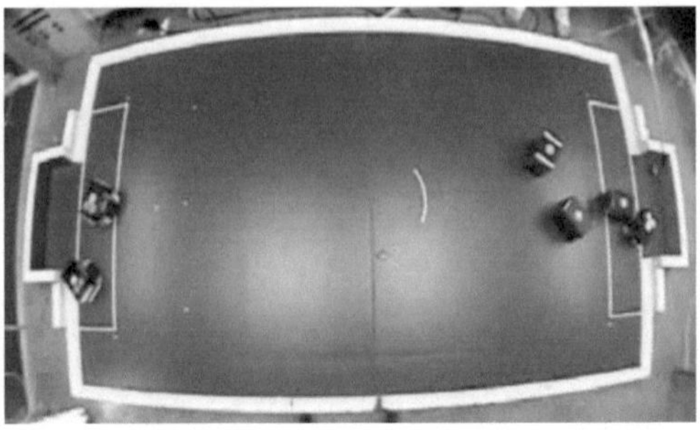

Fig. 17. (h) Game 8: ZSQ(λ) scored at the time 00:31, 01:28, 02:38 and 04:18. The Green got the fourth point.

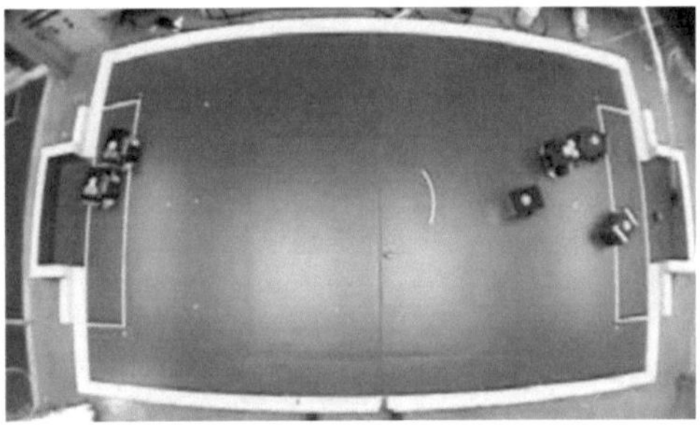

Fig. 17. (i) Game 9: ZSQ(λ) scored at the time 00:26 and 00:28. The ball was intercepted into the goal by Green.

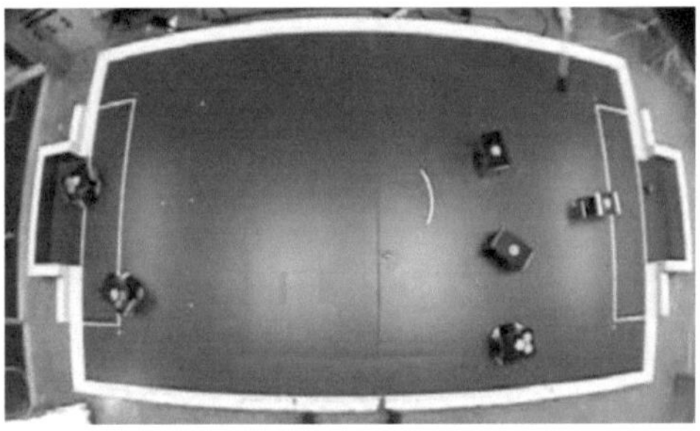

Fig. 17. (j) Game 10: ZSQ(λ) scored at the time 02:02. The ball rebounded toward the goal by Green.

Fig. 17. The snapshots of the competitions between the ZSQ(λ) team that had not trained and the Lingo team.

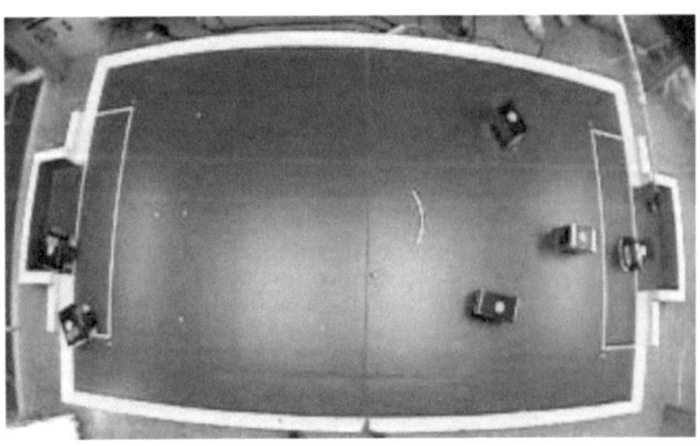

Fig. 18. (a) Game 1: ZSQ(λ) scored at the time 01:42. The Green dribbled the ball toward the goal.

Fig. 18. (b) Game 2: ZSQ(λ) scored at the time 00:08, 00:15, 00:42, 00:53 and 01:49. The Purple kicked the ball toward the goal.

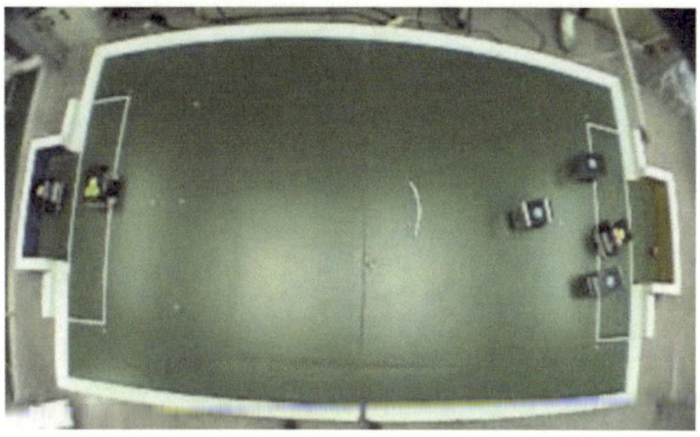

Fig. 18. (c) Game 3: ZSQ(λ) scored at the time 00:36 and 02:42. The Pink dribbled the ball toward the goal.

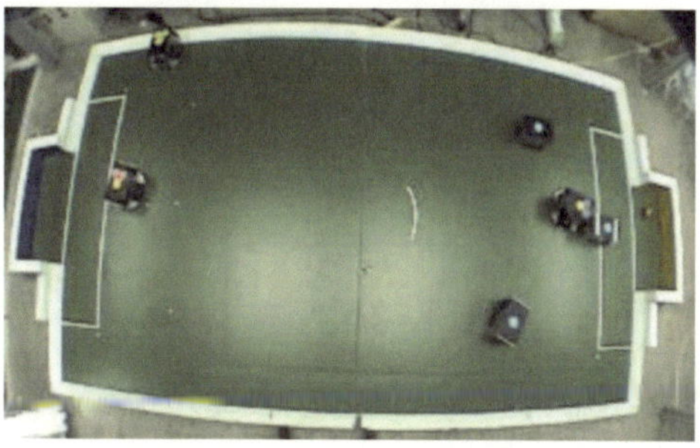

Fig. 18. (d) Game 4: ZSQ(λ) scored at the time 02:03, 02:15 and 03:42. The Purple dribbled the ball toward the goal.

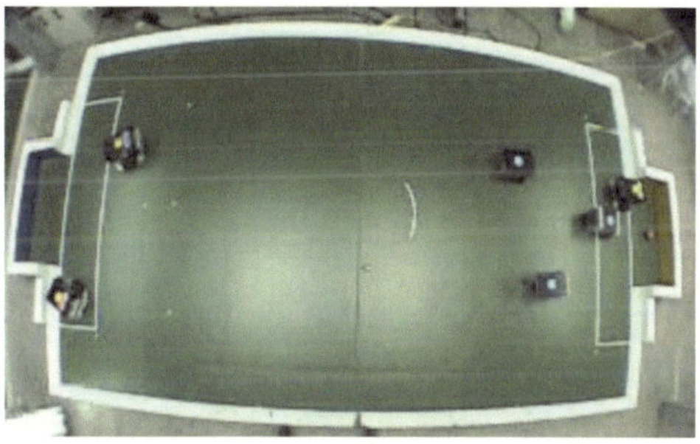

Fig. 18. (e) Game 5: ZSQ(λ) scored at the time 01:26, 03:20 and 04:38. The Green got the third point.

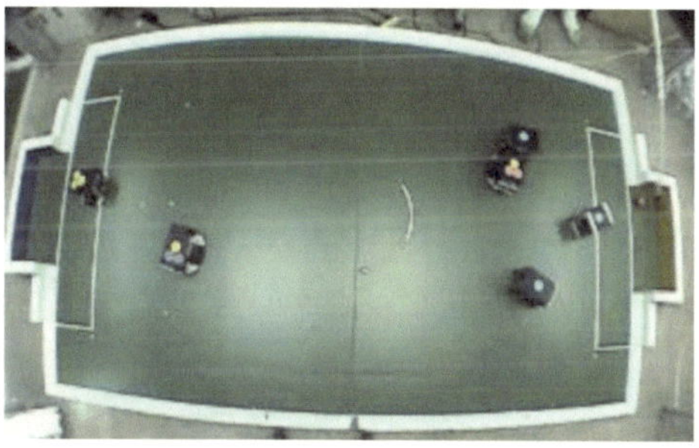

Fig. 18. (f) Game 6: ZSQ(λ) scored at the time 02:00 and 02:54. The Pink kicked the ball toward the goal.

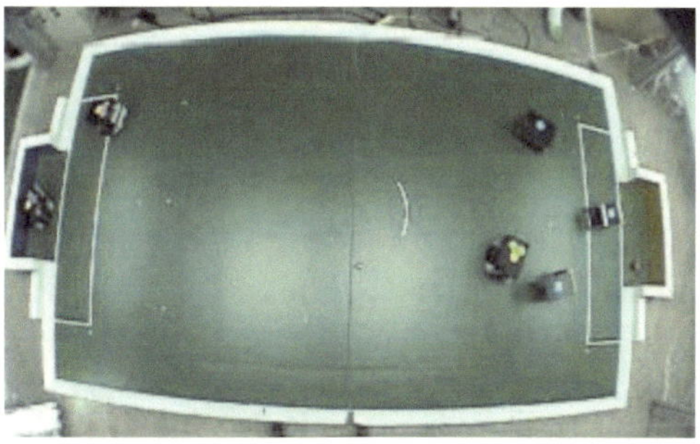

Fig. 18. (g) Game 7: ZSQ(λ) scored at the time 00:16, 02:18 and 03:07. The Green dribbled the ball toward the goal.

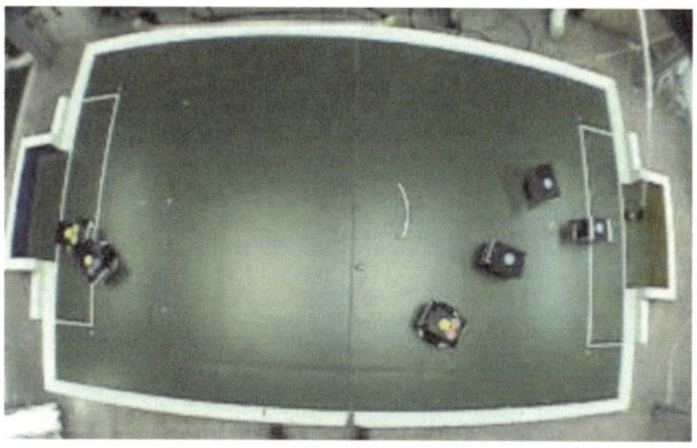

Fig. 18. (h) Game 8: ZSQ(λ) scored at the time 03:14. The Pink kicked the ball toward the goal.

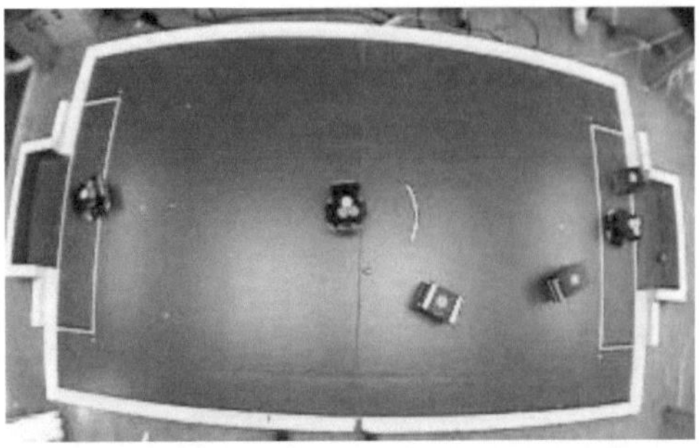

Fig. 18. (i) Game 9: ZSQ(λ) scored at the time 00:45, 01:37 and 03:39. The Green dribbled the ball toward the goal.

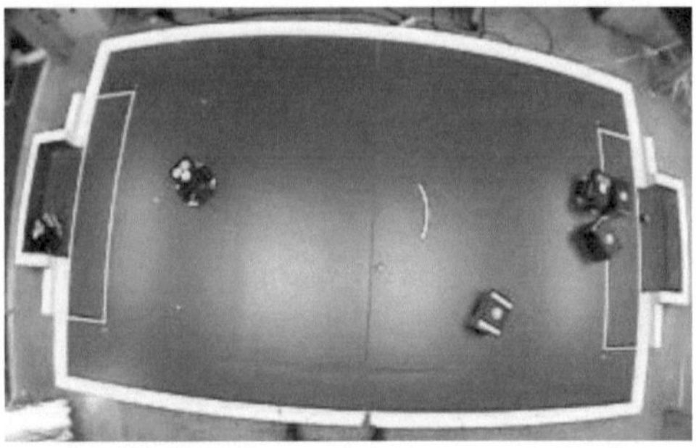

Fig. 18. (j) Game 10: ZSQ(λ) scored Expert scored at the time 03:43 and 04:23. The ball rebounded toward the goal by Green.

Fig. 18. The snapshots of the competitions between the ZSQ(λ) team that had trained and the Lingo team.

	Non-Trained Team		Trained Team	
Times	Scores	Time	Scores	Time
1	1:4	6′09″	1:1	4′50″
2	0:3	5′21″	5:0	4′38″
3	2:0	4′50″	2:1	3′28″
4	3:3	4′40″	3:0	3′51″
5	4:1	5″04″	3:0	5′51″
6	3:2	4′27″	2:0	5′08″
7	2:1	4′46″	3:2	4′12″
8	4:2	5′07″	1:1	3′41″
9	2:4	4′07″	3:0	3′50″
10	1:1	4′17″	2:0	4′29″
Total	22:21	48′54″	25:5	44′02″

Table 2. Results of experiments

6. Conclusions

Multi-robot at a game is an important application of MAS, which can be used to demonstrate the theories and mechanisms of MAS. In addition, because cooperation is central to the operations of MAS, we develop a mechanism to achieve cooperation and demonstrate it in a soccer game. According to the experimental results, using the task assignment mechanism with self-improving ability is a feasible and successful method for cooperation. Comparison with other methods also shows that the proposed method is better because the reinforcement learning in the mechanism gives it a good learning ability to adapt to varying environments. Moreover, the principle in game theories also successfully improves the action-selection ability of the reinforcement learning by considering opponent behaviors. Solving the cooperation problem of the MAS not only can be applied to soccer robot systems, but also can solve strategy determining problems of cooperation. Of course, this study is not the first to use the game theory to improve the reinforcement learning, but the idea of using the task-assignment mechanism with the reinforcement learning is novel. The tasks in the mechanism should be determined in advance, which means that determining appropriate tasks depends on questions, and is a problem for future research. However, the proposed method is still a new approach to cooperation problems, despite the fact that the task self-determining ability merely gives the mechanism more functions and reduces the work of preparation.

7. References

A Flache & M.W. Macy. (2002). The Power Law of Learning, *Journal of Conflict Resolution*, Vol. 46, no. 5, pp. 629-653.

A Gosavi. (2009). Reinforcement Learning: A Tutorial Survey and Recent Advances. *INFORMS Jounral on Computing*. Vol 21(2), pp. 178-192.

A. G. Barto, R. S. Sutton, & C. W. Anderson. (1983). Neuronlike Adaptive Elements That Can Solve Difficult Learning Control Problems, *IEEE Transactions on System, Man, and Cybernetics*, vol. 13, no. 5, pp. 834-846.

D. Xiao & A. H. Tan. (2007). Self-Organizing Neural Architectures and Cooperative Learning in a Multiagent Environment. *IEEE Transactions on systems, Man, and Cybernetics-Part B: Cybernetics*, vol. 37, no. 6, pp. 1567-1580.

Federation of International Robot-soccer Association (FIRA) ,2009. http://www.fira.com.

George A. Bekey. (2005). Autonomous Robots, *The MIT Press*.

H. Yanco & L.A. Stein. (1993). An adaptive communication protocol for cooperating mobile robots, *in J. A. Meyer, H. L. Roitblat, and S.W. Wilson, editors, Animals to Animats: Proceedings of the Second International Conference on the Simulation of Adaptive Behavior*, pages 478--485. MIT Press, Cambridge.

J. Hu & M. P. Wellman. (1998). Multiagent Reinforcement Learning: Theoretical Framework and an Algorithm, *Proceedings of 15th International Conference on Machine Learning*, Morgan Kaufmann, pp. 242-250.

J. R. Marden, G. Arslan, & J. S. Shamma. (2009). Cooperative Control and Potential Games. *IEEE transaction on Systems, man, and Cybernetics-Part B*, vol. 39, no. 6, pp. 1393-1407.

K. S. Hwang & J. S. Lin. (1998). Smoothing Trajectory Tracking of Three-Link Robot: A Self-Organizing CMAC Approach, *IEEE Transactions on System, Man, and Cybernetics-part B*, Vol. 28, no. 5, pp. 680-692, Oct. 1998.

K. S. Hwang. & Y. Z., Chen. (2007). Reinforcement Learning on Strategy Selection for Cooperative Robot System. *IEEE transaction on Systems, man, and Cybernetics-Part A*. Vol. 37, No. 6, pp. 1151-1157.

L. Busoniu, R. Babuska, & B. D. Schutter. (2008). A Comprehensive Survey of Multiagent Reinforcement Learning. *IEEE Transactions on systems, Man, and Cybernetics-Part C: Applications and Reviews*, vol. 38, no. 2, pp. 156-172.

L. P. Kaelbling, M. L. Littman, & A. W. Moore. (1996). Reinforcement Learning: A Survey, *Journal of Artificial Intelligence Research 4*, pp. 237-285.

Littman, M. L. (1994). Markov Games as a Framework for Multi-agent Reinforcement Learning, *in W. W. Cohen & H. Hirsh, eds., `Proceedings of the Eleventh International*

Conference on Machine Learning (ML-94)', Morgan Kauffman Publishers, Inc., New Brunswick, NJ, pp. 157-163.

M. Tan. (1993). Multi-agent reinforcement learning: independent vs. cooperative agents," Proceedings of the Tenth International Conference on Machine Learning, Amherst, Morgan Kaufmann, pp. 330-337.

S. P. Bingulac. (1994). On the compatibility of adaptive controllers, *in Proc. 4th Annu. Allerton Conf. Circuits and Systems Theory*, New York, pp. 8–16.

V. Vassiliades, A. Cleanthous, & C. Christodoulou. (2011). Multiagent Reinforcement Learning:Spiking and Nonspiking Agents in the Iterated Prisoner's Dilemma. *IEEE Transactions on Neural Network,* vol. 22, no. 4, pp. 639-653.

Part 4

Communication and Distance Measurement for Swarm Robots

14

Synchronous and Asynchronous Communication Modes for Swarm Robotic Search*

Songdong Xue[1], Jin Li[1], Jianchao Zeng[1], Xiaojuan He[2] and Guoyou Zhang[2]
[1]*Complex System and Computational Intelligence Laboratory, Taiyuan University of Science and Technology, 66 Waliu Road, Taiyuan 030024, Shanxi*
[2]*College of Electrical and Information Engineering, Lanzhou University of Technology, 85 Langongping, Lanzhou 730050, Gansu*
China

1. Introduction

Swarm robots are special multi-robots and usually considered being controlled with swarm intelligence-based method to complete some assigned complex tasks (Dorigo and Sahin, 2004). Similar to the biological counterparts in nature, swarm intelligence among such artificial system is emerged from local interactions between individual robots or individual robot and its environment (Beni, 2005; Şahin, 2005). It is obvious that interactions play a crucial role in emergence of swarm intelligence in swarm robotics (Schmickl and Crailsheim, 2008). In other words, communication mode taken in control process of swarm robotic search is important. How to control swarm robots with certain communication mode? We can borrow ideas from swarm intelligence-based optimization algorithms in general, and the particle swarm optimization (PSO) algorithm in particular, since the case of swarm robotic search can be mapped to the case of functions optimization with PSO. Later, this method is named as the extended particle swarm optimization (EPSO) method (Pugh and Martinoli, 2007). The particle swarm optimization algorithm is a global, stochastic search one, being derivative-free and population-based style (Schutte *et al.*, 2004). As one of tools of systemic modeling and cooperative control, it can be used to model swam robotic systems and control robots cooperatively. Bio-inspiringly, this algorithm works in parallel in nature. Learning from this, we can control swarm robotic search with special communication modes in similar way. As for the parallel algorithms, they can be classified by granularity (Xu and Zeng, 2005). Wang *et al.* (2007) present a parallel version of PSO based on parallel model with controller. Its communication cycle affects speedup of the algorithm. Huang and Fan (2006) propose parallel version of PSO by island population modeling. It partitions the group into several sub-groups and places them on different processors to evolve, communicating timely in the evolution procedures. Zhao *et al.* (2005) introduce an idea of migration into PSO, present a parallel version based on multi-groups evolving simultaneously. All sub-groups are collected to get the optima by comparison after several iterations. Then the particle having best

*Supported by the National Natural Science Foundation of China under Grand 60975074, Shanxi Natural Science Foundation under Grand 2009011017-1, Shanxi Research and Development Project of Colleges and Universities under Grand 20091130, and Startup Foundation for Doctors of Taiyuan University of Science and Technology under Grand 20102011.

fitness is transferred into all sub-groups as a migration. These mentioned algorithms are all attributed to coarse-grained parallelism. To the contrary, the fine-grained parallel algorithms are characteristic with majority advantages, i.e., maintaining diversity of groups, restraining pre-mature and holding the highest degree of parallelism. Therefore, Schutte *et al.* (2004) develop an approach to implement parallel PSO on multi-processors environment. Especially, to overcome the communication bottle-neck due to massively increasing in size of fine-grained PSO (Li *et al.*, 2006), Chang *et al.* (2005) design three types of communication strategies according to the degree of correlation between parameters. The above parallel versions of PSO are all synchronous paradigms. However, Koh *et al.* (2006) point out that asynchronous algorithm can increase efficiency in heterogeneous environment. Typically, the asynchronous PSO version proposed by Luo and Zhong (2005) makes each particle acted as an independent individual and search performed asynchronously. Aiming at the differences of heterogeneous computing environments and the cost of fitness evaluate, Venter and Sobieszczanki-Sobieksi (2006) introduce asynchronous pattern into PSO for speedup enhancement.

As mentioned above, swarm robots controlling inevitably involves parallel operation too (Henrich and Honiger, 1997). No doubt the extended PSO approach has to take parallelism into account in control algorithms designing. The individual robots distributing in search space makes cooperation control algorithms parallel in nature. Besides, differences in sampling frequency of sensors carried by robots and communication delays make it more realistic to control swarm robots in an asynchronous fashion. How to move nature-inspired algorithms to parallel, asynchronous and decentralized environment (Ridge *et al.*, 2005)? There has so far been fairly little research in this area. For this end, within the field of swarm robotic systems, one area that has received more attention is target search, where a group of robots work together to localize one or more targets. As a first step, a single target is considered here. Searching can be done massively in parallel, significantly decreasing the time-consuming to locate the targets and improving robustness against failure of single agents by redundancy as well as individual simplicity. Then, the problem of parallel asynchronous control swarm robots, i.e., swarm robotic search with different specific communication modes is proposed. As for this chapter, the remainder is organized as follows. Section 2 maps swarm robotic search to the particle swarm optimization algorithm. Then it models the swarm robots with EPSO method and describes the control following swarm intelligence principles. For taking the control mechanism effect, several definitions and assumptions are given for problem simplification without misunderstanding. In Section 3, the problem of communication modes in swarm robotic search is introduced. In order to describe the corresponding strategies developed and taken in swarm robotic search taken place in obstacle-free environment steadily, we set about this section from analyzing the properties of different versions of PSO because of the mapping relationship between swarm robotic search and PSO. Then the synchronous and asynchronous communication modes are discussed. Also, the corresponding control algorithm descriptions with specific communication modes are given here. Based on this, the authors explain the simulation settings, propose evaluate metrics such as searching efficiency and energy consuming, show the results from simulations, and discuss the implications through statistical evidence presentation in Section 4. Finally, we conclude this chapter in Section 5 by summarizing our current work and accompanying the future work.

2. Modeling and controlling

By extending PSO to model swarm robotic systems, Pugh and Martinoli (2007) firstly and Zeng and Xue (2010) subsequently investigate the problem of target search in an ideal environment. In PSO-type swarm robotic search, modeling and control mechanism are involved inevitably.

2.1 Mapping from swarm robotic search to PSO

Swarm intelligence is inspired from phenomena of individuals following simple rules only but emerging intelligence through local interactions in biological communities. Generally speaking, PSO is viewed as optimization tool, in which particles are guided by the best positions having optimal fitness to get the solution of given functions. Here, each particle has perfect knowledge about environment and its neighbor particles. While swarm robotic search works depending on individual robots' experience and social experience, for robots move according to their own behavioral decision making. The former comes from signals measurement by robot itself, and the latter from local communications within robot's communication neighborhood. Consider the similarities and differences between the two cases, we can map swarm robotic search to PSO, see Figure 1.

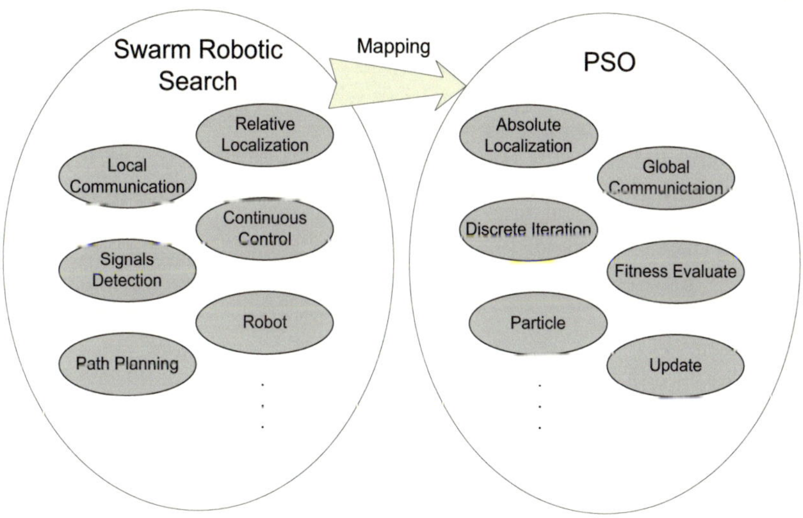

Fig. 1. Mapping swarm robotic search to the particle swarm optimization algorithm. Note that several one-to-one relationships of robot-to-particle, signals detection-to-fitness evaluate and local communication-to-global communication are the most important.

The PSO algorithm makes information about environment or potential solution shared among particles anywhere, as long as the particles are in the search space. Similarly, robots always have strict limitations on their maximum communication range due to the limited power consuming. In this case, we define robot's neighborhood structure as all others within some fixed geometrical distance from it. Since robots are constantly in motion, this means such structure is dynamic and time varying (Xue and Zeng, 2008a).

2.2 EPSO-based modeling

Based on the above mapping relationship between swarm robotic search and PSO, EPSO method can be taken to model the swarm robotic system (Pugh and Martinoli, 2007; Xue and Zeng, 2008b). In order to understand such method well, let us examine the particle swarm optimization algorithm at first. Particle swarm optimization is based on the sociological behaviors associated with bird flocking and other animals' moving (Zeng *et al.*, 2004). Each particle is capacitated to fly over the space with changeable velocity. And a series of positions in which particles are situated are viewed as potential solutions of problem. Then, the best position of particle itself and swarm respectively having the best fitness can be decided. Farther, the behavior of particle can be adjusted according to its inertia, individual experience (cognition) and social experience (learning). The velocity and position update equations of standard PSO at time $kt + 1$ are executed as follows:

$$\mathbf{v}_{k+1}^{i} = \omega_k \mathbf{v}_k^i + c_1 r_1 (\mathbf{p}_k^i - \mathbf{x}_k^i) + c_2 r_2 (\mathbf{p}_k^g - \mathbf{x}_k^i) \tag{1}$$

$$\mathbf{x}_{k+1}^{i} = \mathbf{x}_k^i + \mathbf{v}_{k+1}^i \tag{2}$$

where $\mathbf{x}_k^i$ is the position vector of particle i at time kt, and $\mathbf{v}_k^i$ the corresponding vector of velocity, subscript k the abbreviation of time increment kt. Note that the two vectors have the same dimensional variables. While $\mathbf{p}_k^i$ and $\mathbf{p}_k^g$ are the best-found positions of particle i itself and the swarm before time k respectively. The coefficient matrix ω_k is diagonal matrix whose diagonal elements are inertia coefficients with value range [0, 1] to slow down over time to prevent explosions of the swarm and ensure ultimate convergence. Similarly, r_1, r_2 are diagonal matrices whose diagonal elements are sampling of uniformly-distributed random variable in [0, 1]. And c_1, c_2 are diagonal matrices whose diagonal elements are cognition and social acceleration constants, respectively.
Then we can extend the standard version of PSO for swarm robots and farther model the swarm robotic system with the EPSO method, as is shown below:

$$\mathbf{v}_{k+1}^{i} = w_k \mathbf{v}_k^i + c_1 r_1 (\mathbf{p}_k^i - \mathbf{x}_k^i) + c_2 r_2 (\mathbf{p}_k^g - \mathbf{x}_k^i) \tag{3}$$

$$\mathbf{v}_{k+\Delta k}^{i} = \mathbf{v}_k^i + (\mathbf{v}_{k+1}^i - \mathbf{v}_k^i)\Delta k \tag{4}$$

$$\mathbf{x}_{k+\Delta k}^{i} = \mathbf{x}_k^i + \Delta k \mathbf{v}_{k+\Delta k}^i \tag{5}$$

where $\mathbf{x}_{k+1}^i$ is the expected velocity vector of robot i at time $k + 1$. Δk is a factor to decrease the step taken when robots move about in the search space. By the way, we add the Δk factor in order to make individual robots moved "smoothly" , and therefore a more refined search may be carried out. In addition, the parameter Δk is somehow different from the others, as it is not related to the physical nature of the problem. However, we can also understand it in this fashion: robot in real world has inertia due to its mass (Xue and Zeng, 2008b).

2.3 PSO-type controlling

As is shown in the above subsection, swarm robots can be modeled mathematically taking the form of PSO-type iteration equations. Therefore, the cooperative control over swarm robots can be carried out following swarm intelligence principles. Farther comparison of the two cases can be made. First, both work on the base of fitness evaluate, or signals detection. While the relative independency of individual robots demands fitness evaluate being complemented in their on-board processors rather than in processing center. And the limitations on hardware and power supply make it impossible that robots interact successfully beyond the maximum communication range of robots. Apparently, the swarm that each robot dwells in differs from others, since every robot selects itself and all other robots within some distance in the search space as its evolving swarm.

2.3.1 Time-varying character swarm

Each individual robot of the swarm system selects the close near neighbors as its temporary swarm members only because those neighbors within its maximum communication range are capable of interacting with it through communication. Accordingly, a concept of time-varying character swarm (TVCS) for computational evolution is presented naturally, see Figure 2. Take the position of robot i at time t as the center, the maximum communication range R of robot i as radius for a circle neighborhood constructing. The set of those robots covered by this neighborhood is named as TVCS of robot i at time instance. The size of TVCS depends on the maximum communication range of robot i and the relative position relationship of robots at time t. This implies the property of time-varying in character swarm of swarm robotic system, since a robot may be close to few or many other robots at different time instance.

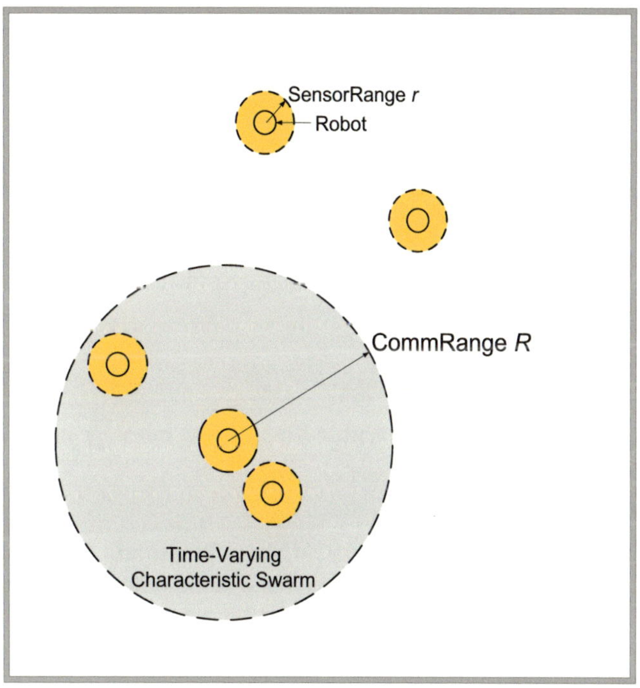

Fig. 2. Individual robot's time-varying character swarm constructing for swarm robots control following swarm intelligence principles according to the involved robot's maximum communication radius, which means this robot can interact with others within its TVCS only.

2.3.2 Signals detection

In PSO-type algorithms, motion control of individual robot depends on both its cognitive position and the best-found social position. While the two best-found experiential positions come from position evaluate. Each individual robot of the swarm is assumed to be equipped with one sensor to detect the intensity of signal emitted from potential target. This is of theoretical significance only. In fact, there are multiple types of signals in searching environment. To accomplish the search task, there is need to appropriately fuse the real-time heterogeneous signals with fusion algorithms to determine the decision sensor under different sensory conditions (Xue and Zeng, 2008c), following the working principle of swarm intelligence-based method. However, we still simplify the detection process as making measurement with a single sensor, for exploring the key effects caused by communication modes here. We simply this evaluate mechanism here by assuming each robot has a sensor to detect the intensity of the target signal within its maximum detection radius. This intensity $\mathrm{I}(d_i)$ is determined with model below:

$$\mathrm{I}(d_i) = \begin{cases} 0, & d_i > r \\ \frac{P}{d_i^2} + \eta(), & \text{otherwise} \end{cases} \tag{6}$$

where P is the target signal power, d_i the distance from robot R_i to target, r the radius of sensor detection and $\eta()$ a sampling of additive Gaussian noise.

2.3.3 Position evaluate and cognitive decision

As to individual robot R_i, its cognitive position at time t is determined following the rule:

$$p_i^*(t) = \begin{cases} x_i(t), & \text{if } \mathrm{I}(x_i(t) \geq \mathrm{I}(p_i^*(t-1)) \\ p_i^*(t-1), & \text{otherwise} \end{cases} \tag{7}$$

where $p_i^*(t)$ is the cognitive position of robot R_i at time t, $x_i(t)$ the current position, and $\mathrm{I}()$ the simplified evaluate function of measurement readings of target signals.

2.3.4 Best-found position in TVCS

Based on the definitions of TVCS and signals evaluate, the best-found position in swarm robots can be decided with the criterion:

$$p_{(i)}^*(t) = p_k^*(t), \arg_k \max\{\mathrm{I}(p_k^*(t)), k \in R_i\text{'s TVCS}(t)\} \tag{8}$$

where $p_{(i)}^*(t)$ be the best-found position within the TVCS of robot R_i at time t.

3. Communication modes

Now swarm robots can be controlled with EPSO-based method. But a problem should be considered. In PSO-type control, target signals have to be detected in parallel for position evaluate and such swarm system should be controlled in an asynchronous manner. Therefore, we present asynchronous communication mode in case of target search. Specifically, each robot independently detects signals emitted from target in a fine-grained parallel way and compares intensity of signals with the best in its TVCS. Then velocities and positions of individual robots are updated immediately. But the shared information within TVCS is updated asynchronously. As comparison, a synchronous mode is also given in this section.

We set about this problem at the beginning of analysis on the characteristics of PSO. The standard PSO algorithm has a key idea about velocity and position of particle (Eberhart and Shi, 2001; Kennedy and Eberhart, 1995; Zeng *et al.*, 2004), which is used to optimize nonlinear functions at the beginning of development and is extended to more applications gradually. The algorithm tries to find potential solutions of problem by imitating behaviors of social creature, e.g., birds flying over space. Taking fitness of given function as evaluate metrics, this algorithm adjusts the velocities and positions of particles representing solutions of problem to obtain optimum eventually. PSO is based on possessing many desirable properties that we would like to transfer to our PSO-type swarm robotic systems. One of them is that PSO operates in parallel and asynchronously (Ridge *et al.*, 2005), which is consistent with the biological significance of swarm algorithm. Thus we proceed with the analysis on characteristics of different versions of PSO.

3.1 Synchronous v.s. asynchronous

To explore characteristic of different versions of PSO, we can start in accordance with two issues, i.e., fitness evaluate in a serial or parallel way, synchronous or asynchronous communication mode of sensing and reacting to environment as well as velocity- and position-evolution. Therefore, we divide the different versions of PSO into four patterns.

3.1.1 Serial evaluate and synchronous update

Particle swarm optimization is traditionally considered to be implemented in serial and synchronous on single-processor computing environment. The execution procedure can be described with the following pseudo code (Koh *et al.*, 2006), see Algorithm 1. It can be seen that fitness evaluate of all particles is carried out one by one in optimization process through cost function computation. And the best positions both of particle itself (cognitive) and in its TVCS are determined by fitness comparison in the same way. Then the update of all velocities as well as positions occurs simultaneously at each iteration.

Algorithm 1 PSO with characteristics of serial evaluate and synchronous update.

initialize algorithm constants
initialize all particle velocities, positions
    For $k = 1$, number of iterations
        For $i = 1$, number of particles
            evaluate cost function
        End
    check convergence
    update $\mathbf{p}_k^i$, $\mathbf{p}_k^g$, $\mathbf{v}_{k+1}^i$,$\mathbf{x}_{k+1}^i$
    End
output results

3.1.2 Serial evaluate and asynchronous update

Immediately updates on velocity, position of certain particle as well as its history cognition and the best of swarm are carried out as soon as completing evaluate on the cost function of this particle. The procedure can be elaborated with the following pseudo code (Koh *et al.*, 2006), see Algorithm 2. It is clear that the evaluate and update process on different particles are not completed at the same time.

Algorithm 2 PSO with characteristics of serial evaluate and asynchronous update.

initialize algorithm constants
initialize all particle velocities, positions
    For $k = 1$, number of iterations;
        For $i = 1$, number of particles
            evaluate cost function
            check convergence
            update $\mathbf{p}_k^i$, $\mathbf{p}_k^g$, $\mathbf{v}_{k+1}^i$,$\mathbf{x}_{k+1}^i$
        End
    End
output results

3.1.3 Parallel evaluate and synchronous update

The most obvious PSO parallel implementation is to simplify fitness evaluate for particles at iteration in parallel, without changing the overall logic of the algorithm itself (Venter and Sobieszczanki-Sobieksi, 2006). And the property of synchronous refers to all particles being sent to parallel computing environment and moving from the current iteration to the next only if the fitness of all particles has been gotten (Schutte *et al.*, 2004). To demonstrate the internal relationship of logic better, we illustrate with flow chart rather than pseudo code, as showed in Figure 3 (Koh *et al.*, 2006). In this case, the existence of load imbalance in computing environment may significantly affect parallel performance. These factors are shown below:

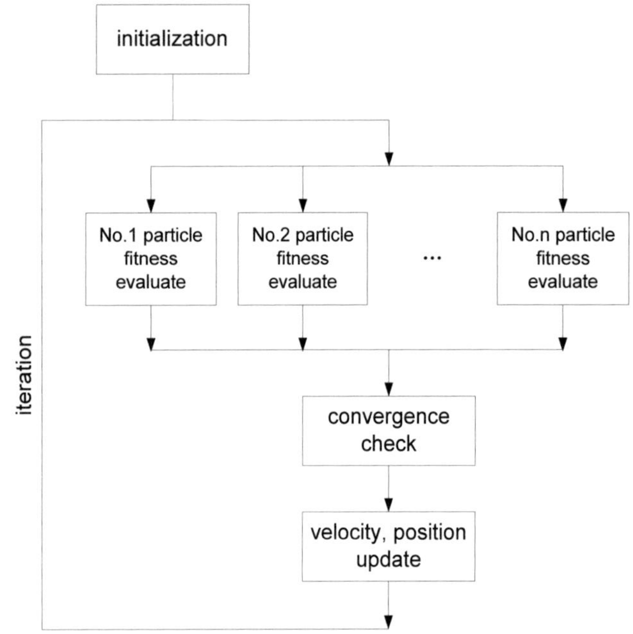

Fig. 3. Visual sketch for a version of PSO having characteristics of parallel evaluate on fitness and synchronous update for velocity and position.

- a heterogeneous distributed computing environment where processors with varying computational speed are combined into a parallel computing environment;
- time spent in fitness evaluate, i.e., using a numerical simulation to evaluate each particle, where the required simulation time depends on the particle being analyzed;
- the number of particles cannot be equally distributed among the processors in the computing environment, i.e., having a swarm size that is not an integer multiple of the number of processors (Koh *et al.*, 2006).

3.1.4 Parallel evaluate and asynchronous update

Parallel implementations being asynchronous in PSO can make the algorithmic computation efficiency enhanced (Venter and Sobieszczanki-Sobieksi, 2006). The asynchronous approach does not need a synchronous point to determine new velocities and positions, as showed in Figure 4 (Koh *et al.*, 2006). The optimization can proceed to the next iteration without waiting for the completion of all functions evaluate from the current iteration.

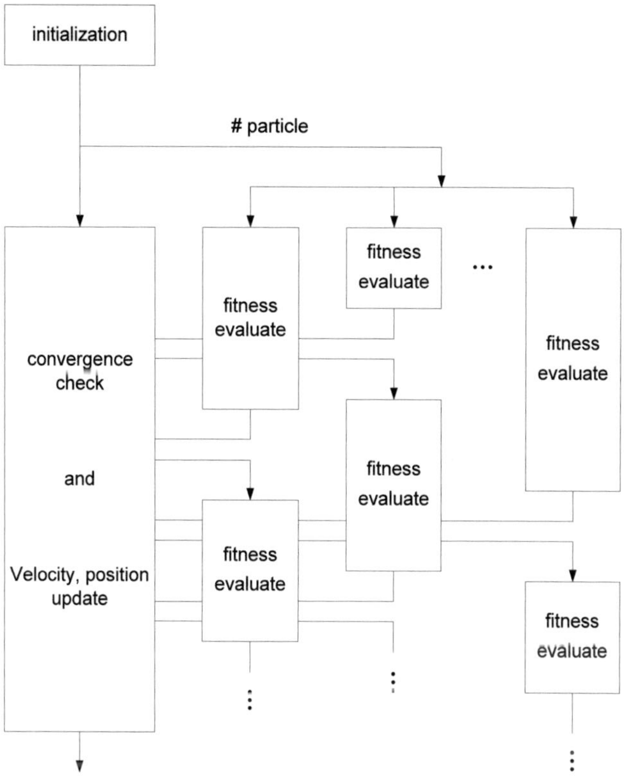

Fig. 4. Visual sketch for a version of PSO having characteristics of parallel evaluate on fitness and asynchronous update for velocity and position.

As stated above, different versions of PSO have different algorithmic properties when implementation. But the most desirable properties that we would like to transfer to our swarm robotic system may be controlled in parallel and asynchronously (Ridge *et al.*, 2005). Indeed, as one of nature-inspired algorithms, parallel and asynchronous version of PSO makes algorithm more efficient in execution.

3.2 Communication modes taken in swarm robotic search

Now, let us examine the case of swarm robotic search in a closed obstacle-free environment. According to the analysis above, controlling over robots should be done in a fine-grained way, as individual robots detect target signals independently at the same time to determine the best-found position by signals intensity comparison with their respective TVCS neighbors (Xue and Zeng, 2008b). Due to heterogeneous hardware caused by parameter distribution of sensors, difference of detection and evaluate time required among different positions because of signals diffusing, part of swarm robots completing signals detection early have to wait for synchronous update. The reason is that the update depends on the slowest robot (Schutte *et al.*, 2004). By this means, velocities and positions of all robots are updated at the same time after evaluate fully completing.
Here, asynchronous communication mode refers to that each robot compares at once with the optimal value of the swarm after iterating in the iteration process, if their detective signals are discovered stronger, updates immediately the optimal value of the swarm, thereby, other

robots can share the experience timely, without having to wait until given some synchronous moment, then realizes the non-synchronization of the robots in the search process.

3.2.1 Communication trigger

Clearly, the key to asynchronous implementation of control algorithm is to partition the individual from the group update behavior to take the different property into account. These update behaviors include updating the individual robot and the shared information within the swarm histories. Similarly, as for the asynchronous particle swarm optimization, the update action starts after fitness evaluate, while the update on swarm starts in the last at each iteration (Venter and Sobieszczanki-Sobieksi, 2006). For the target search with swarm robots, detection of target signals depends only on their respective on-board processors rather than processing center. In other words, such center does not exist in swarm robotic system. The processors work independently and in parallel between one another. The individual robot updates its velocity, position and history as soon as it completes target signals measurement and makes decision on the best-found by comparison with the best of its TVCS (Zhao *et al.*, 2005). But the update on the shared information should start in accordance with some special asynchronous control strategy. In fact, this is the decision on communication triggers. Differing from the ideal particles in PSO, robot possesses mass in real world that causes it to have inertia when moves about in the search environment. Therefore, as for same an evolution position of certain particle, it is unlimited to reach at any speed in PSO case, while robot may arrive at the same position in several sampling times due to constraint of kinematics and dynamics because the evolution position is only expected (Pugh and Martinoli, 2007). These factors should be taken consideration when we design the asynchronous interaction strategies. Based on this, some update strategies have been developed. One is communication cycle-based control principle. Here, communication cycle is named as evolution iterations. Similar to the coarse-grained parallel particle swarm optimization, we can make robot R_i communicate every n iterations to decide the best-found position within robot R_i's TVCS (Huang and Fan, 2006; Wang *et al.*, 2007; Zhao *et al.*, 2005). To improve systematic efficiency, a communication cycle can be assigned to several fixed times of sampling periods (Xue *et al.*, 2009). Besides, different robots can be allowed to have different sampling frequencies. On the other hand, the best-found fitness value and position of TVCS should be remembered in memory before the next iteration starts. Another update strategy is evolution position-based control principle (Xue and Zeng, 2009). According to this control principle, update of the shared information does not been carried out in the current iteration before the previous evolution position has not been reached. That is to say, robots communicate when they arrive at the decisive expected or desired evolution positions regardless of the iteration history and the next iteration required. No communication between two consecutive ideal evolution positions makes robot moving continuously, saving power and decreasing communication time-consuming.

3.2.2 Synchronous case in swarm robotic search

The difference between synchronous and asynchronous control lines in their update types and opportunities (Zhao *et al.*, 2005). As for the synchronous mode, update time points depend on the last particle completing fitness evaluate at each step. Thus the communication triggers do not need to consider in synchronous mode. As comparison, refers to the updates of all the robots being synchronous at same iterative procedure, they are aiming at the optimal value in the current iteration step, after all the updates are completed, all the robots proceed towards the goal synchronously, and accomplish the search task with common integral cognitive level.

3.3 Algorithm description

The corresponding algorithms with different communication modes can be listed in the following. Of them, the synchronous communication version is taken according to the characteristics of processing moments on signals detection, search completion judging as well as velocity and position updating, see Algorithm 3. Different from the synchronous communication mode, the moment that robot updates shared information of TVCS is more flexible in asynchronous communication mode. The details of corresponding algorithm can be found in Algorithm 4.

Algorithm 3 Controlling robot R_i involved in swarm robotic search with synchronous communication mode. For convenience, the default variables are for R_i unless those are marked explicitly for TVCS.

initialize
    $j \leftarrow 1$
    velocity $V(t=0)$, position $X(t=0)$
    target signals measurement $I(t=0)$
    $I_{best} \leftarrow I(t=0)$, $X_{best} \leftarrow X(t=0)$
    $I_{best} \leftarrow I(t=0)$, $X_{best} \leftarrow X(t=0)$ for TVCS
while target is not found out
    for $i = 1; i <= popsize; i++$
        calculate expected and real velocities
        calculate position
    end
    target signals measurement
    update I_{best} and X_{best}
    update I_{best} and X_{best} of TVCS
    calculate velocity
    move ahead one step
    $j \leftarrow j+1$
end

4. Simulations

The two control algorithms are performed and repeated for 10 runs respectively, for both are characteristic of random search in nature and the comparison can be made with statistics gotten from enough simulations.

4.1 Parameter settings

The parameters about working environment, individual robot and swarm robots affect the results directly. Thus the important parameters and their common configurations used in simulations are given in Table 1. The farther information of the symbols can be found in the third column of this table.

4.2 Performance metrics

In order to comparatively evaluate the running performance of control algorithms with different communication modes, some quantitative metrics need to be designed in advance. Here, two performance figures are considered and presented. Both are based on the definition of search success. We can define such term as the best position of swarm closes to target

Algorithm 4 Controlling robot R_i involved in swarm robotic search with asynchronous communication mode. Note that symbols share the same meanings as in Algorithm 3.

```
initialize
    set counter j ← 1
    velocity V(t = 0), position X(t = 0)
    target signals measurement I(t = 0)
    I_best ← I(t = 0), X_best ← X(t = 0)
    I_best ← I(t = 0), X_best ← X(t = 0) for TVCS
    calculate number of neighbors neighbor_number in TVCS
while target is not found out
    for i = 1; i <= neighbor_number; i++
        target signals measurement
        if best-found is gotten
            update I_best and X_best
        end
        calculate expected V_expec and real velocity V_real
        calculate X
    end
    target signals measurement
    update I_best and X_best
    update I_best and X_best for TVCS
    calculate velocity
    move ahead one step
    j ← j + 1
end
```

Symbol	Value	Meaning
$Space$	500×500	size of searching environment
$StartArea$	160×160	start area for robots at the beginning of simulations
$popsize$	$3, 5, 8, 10$	number of individual robots in swarm robotic system
R_{detec}	$250, 125$	maximum detection radius of sensors
R_{comm}	250	maximum communication radius of robot
V_{max}	5	maximum moving velocity of robot
P	1600	signals power emitted from target
T	70	constant of inertia element
Δt	0.8	contracted factor for a moving step of robots

Table 1. Parameter settings taken in simulations. Note that those parameters expressed in certain dimensions are assigned a corresponding proper one respectively.

enough so as that at least one robot can identify the target with some sensory ability (Kowadlo *et al.*, 2006). It means that if one or more robots approach and reach the area where target locates, the run is considered successful.

4.2.1 Search efficiency

Search efficiency is defined as reciprocal of mean steps required for one successful search. In fact, it concerns search speed by counting time steps for completion of a successful search, which indicates the elapsed time in a single simulation run indirectly. Because the sampling cycle in simulations has been determined in advance, a simple relation between time step and spent time can be established. The value equals to the reciprocal of average steps taken by all individual robots in a successful search. Clearly, the more the average time steps, the lower the search efficiency, and vice versa.

4.2.2 Energy consuming

The metric is distance principle-based one. It is expressed in form of the sum of passed distance of all the individual robots when the swarm robotic search task is completed. Since the energy consumption of robot is fixed per distance unit, the average energy consumption of individual robots can measure aspect of algorithm performance in economical efficiency. Compared with energy consuming, search efficiency seems more important in swarm robotic search control evaluate because we concern about higher algorithmic speed.

4.3 Results

Simulations with different setting configurations are conducted and repeated for 10 runs respectively, for reducing the effect caused by the inherent randomness from the swarm intelligence-based control algorithms. Then the results are shown in some figures and tables. Of them, the running screenshots of swarm robotic system with different sizes when simulated programs terminate at first, see Figure 5 for details. We can find that the robots start searching from the lower left corner of the working space at the beginning of each simulation, being limited to a area of 160×160. While the target position is initialized at the same time, being limited to the upper right corner of the searching environment. As for the statistics from simulations, they are shown in Table 2 – 5, and Figure 6 – 9.

comm. mode	3-rob swarm	5-rob swarm	8-rob swarm	10-rob swarm
synchronous mode	98.2 ±6.11	94.3 ±13.23	93.9 ±3.81	93.2 ±7.16
asynchronous mode	95.4 ±5.48	91.6 ±5.83	93.2 ±7.35	92.3 ±6.10

Table 2. Average control time steps spent in completion of target search under conditions of $R_{detec} = 250, R_{comm} = 250$.

comm. mode	3-rob swarm	5-rob swarm	8-rob swarm	10-rob swarm
synchronous mode	1165.3 ±74.52	1863.1 ±259.72	2965.3 ±119.51	3665.9 ±251.90
asynchronous mode	979.9 ±86.76	1681.2 ±144.11	2603.9 ±220.70	3315.4 ±186.62

Table 3. Average total distance spent by all robots for target search success under conditions of $R_{detec} = 250, R_{comm} = 250$.

4.4 Discussions

We can comparatively analyze and discuss the indications surrounding simulation results, trying to reveal effects of different communication modes on swarm robotic search.

- As for the same task and same parameter setting configurations, time steps decrease as swarm size increases regardless of which communication mode taken in control algorithm. It indicates that search efficiency enhances as swarm size expands.

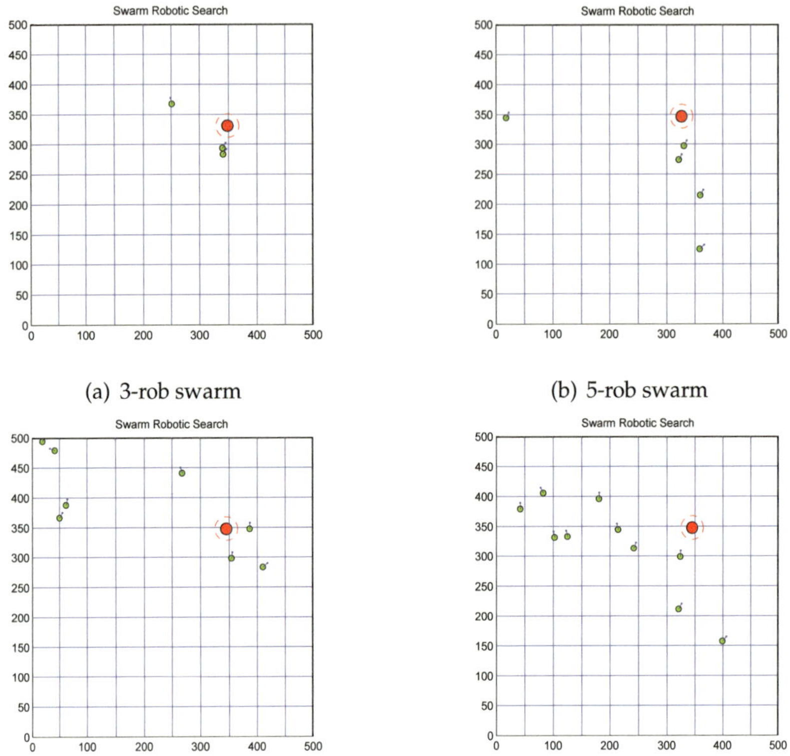

(a) 3-rob swarm

(b) 5-rob swarm

(c) 8-rob swarm

(d) 10-rob swarm

Fig. 5. Screenshots of swarm robotic search in cases of different size swarms. Note that at least one robot closes to the target enough when search succeeds.

comm. mode	3-rob swarm	5-rob swarm	8-rob swarm	10-rob swarm
synchronous mode	91.3 ±3.36	97.1 ±10.15	97.9 ±4.91	89.8 ±6.55
asynchronous mode	97.8 ±7.40	93.1 ±7.43	90.5 ±3.96	89.1 ±4.88

Table 4. Average control time steps spent in completion of target search under conditions of $R_{detec} = 125, R_{comm} = 250$.

comm. mode	3-rob swarm	5-rob swarm	8-rob swarm	10-rob swarm
synchronous mode	1081.0 ±40.95	1917.8 ±207.36	3117.9 ±154.86	3546.5 ±269.06
asynchronous mode	1017.6 ±82.28	1786.8 ±146.27	2628.1 ±244.16	3139.2 ±274.70

Table 5. Average total distance spent by all robots for target search success under conditions of $R_{detec} = 125, R_{comm} = 250$.

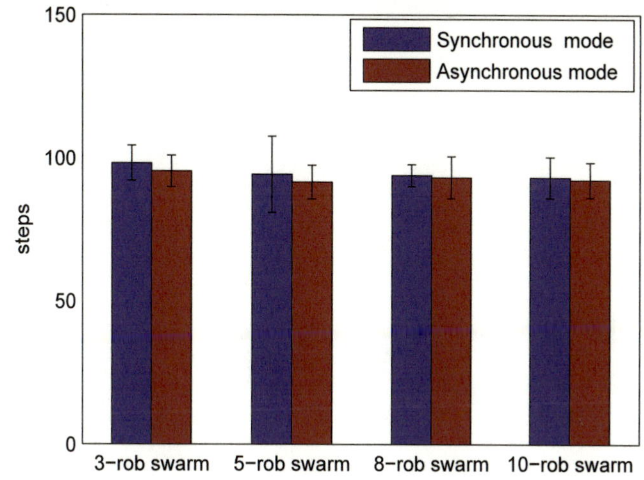

Fig. 6. Average time steps required to complete target search for 10 runs under conditions of $R_{detec} = 250, R_{comm} = 250$. Note that search efficiency is inversely proportional to time steps.

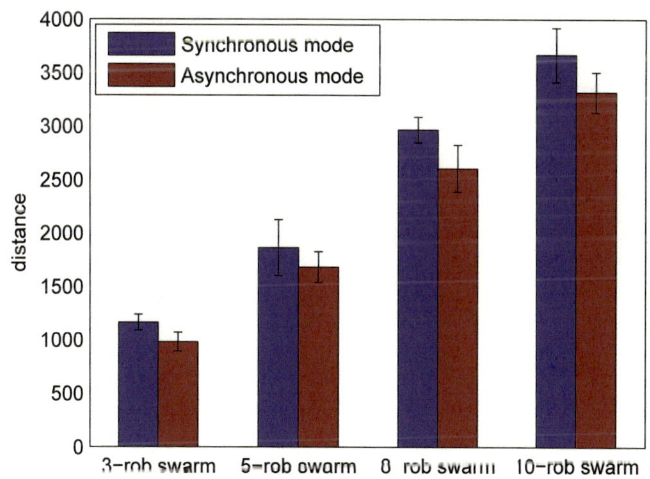

Fig. 7. Average total distance passed by all robots for 10 runs search success under conditions of $R_{detec} = 250, R_{comm} = 250$.

- As for the same parameter setting configurations, average total distance required for a success search varies in the same direction as swarm size increases regardless of which communication mode taken in control algorithm. It indicates that energy consuming largens as swarm size scales expansion. We can think that swarm robots carry out task of target search at compromise of time consuming and energy consuming.
- Swarm robotic search with asynchronous communication mode runs more efficiently than with synchronous communication mode, which seems to indicate that information and experience of certain dominant individual can be shared timely among its society for others

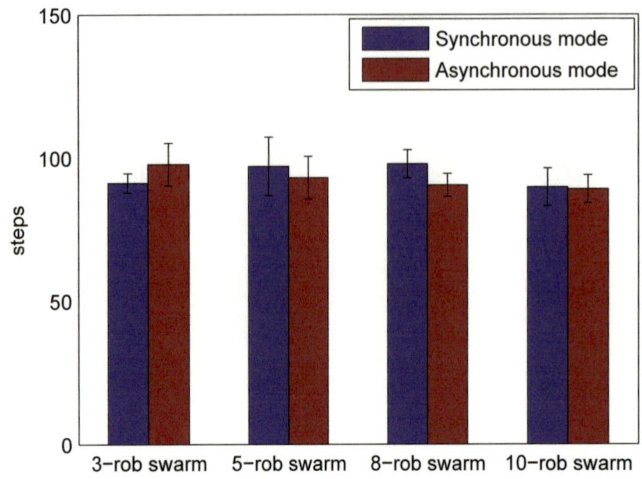

Fig. 8. Average time steps required to complete target search for 10 runs under conditions of $R_{detec} = 125, R_{comm} = 250$.

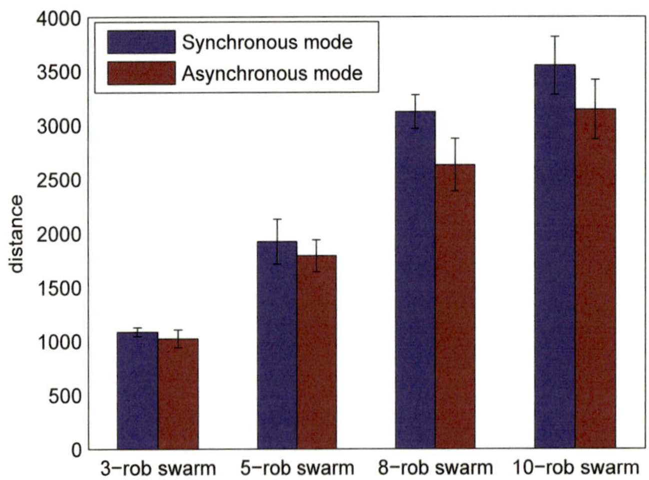

Fig. 9. Average total distance passed by all robots for 10 runs search success under conditions of $R_{detec} = 125, R_{comm} = 250$.

behavior decision making. The result seems that robots can adjust its own motion velocity and position and finally fasten the search process by learning from the optimal neighbors at different time steps.

- As to different parameter setting configurations, such as detection radius is set $R_{detec} = 250$ and $R_{detec} = 125$ respectively but the communication radius remains the same, search efficiencies and energy consuming do not vary obviously. The reason maybe that at the beginning of simulations, robots are far from the potential target so as not to be capable of detecting target signals either for case of $R_{detec} = 250$ or for $R_{detec} = 125$. Therefore, robots only move randomly without help of the social experience sharing.

5. Conclusions

By extending the particle swarm optimization algorithm, we model swarm robotic system and control it in PSO-type way for carrying out target search task. Because of the relationship between swarm robotic search and PSO, some ideal characteristics of PSO can be transferred to case of swarm robotic search. Inspired from asynchronous versions of PSO, we develop control strategy with asynchronous communication mode for search efficiency enhancement. To reveal the effect, we compare the algorithm with synchronous communication mode. From the statistics of simulation, a conclusion can be drawn that asynchronous communication mode is more efficient than synchronous mode under conditions of the same parameter settings in search efficiency. In our future work, we will explore the effect how to vary as controlled object and task change.

6. References

Beni, G. (2005) From swarm intelligence to swarm robotics, *Lecture Notes in Computer Science*, vol. 3342, pp. 1–9.

Şahin, E. (2005) Swarm robotics: From sources of inspiration to domains of application, *Lecture Notes in Computer Science*, vol. 3342, pp. 10–20.

Dorigo, M. and Sahin, E. (2004) Swarm robotics–special issue editorial, *Autonomous Robots*, vol. 17, nos. 2–3, pp. 111–113.

Schmickl, T. and Crailsheim, K. (2008) Trophallaxis within a robotic swarm: bio-inspired communication among robots in a swarm, *Autonomous Robots*, vol. 25, no. 1, pp. 171–188.

Chang, J.F., Chu, S.C., Roddick, J. F. and et al. (2005) A Parallel Particle Swarm Optimization Algorithm with Communication Strategies, *Journal of Information Science and Engineering*, no. 21, 809–818.

Eberhart, R.C. and Shi, Y. (2001) Particle swarm optimization: developments, applications and resources, in *Proceedings of the 2001 IEEE Congress on Evolutionary Computation*, Piscataway, NJ, USA, vol. 1, pp. 81–86.

Henrich, D. and Honiger, T. (1997) Parallel Processing Approaches in Robotics, in *Proceedings of IEEE International Symposium on Industrial Electronics*, Guimaraes, Portugal, July 7–11.

Huang, F. and Fan, X.P. (2006) Parallel Particle Swarm Optimization Algorithm with Island Population Model, *Journal of Control and Decision*, vol. 21, no. 2, pp. 175–179, 188. (in Chinese)

Kennedy, J. and Eberhart, R. (1995) Particle swarm optimization, in *Proceedings of IEEE International Conference on Neural Networks*, vol. 4.

Koh, B., George, A.D., Haftka, R.T. and et al. (2006) Parallel Asynchronous Particle Swarm Optimization, *International Journal of Numerical Methods in Engineering*, vol. 67, pp. 578–595.

Kolda, T.G. and Torczon, V.J. (2003) Understanding Asynchronous Parallel Pattern Search, *High Performance Algorithm and Software for Nonlinear Optimization*, pp. 316–335.

Kowadlo, G., Rawlinson, D., Russell, R.A. and et al. (2006) Bi-modal search using complementary sensing (olfaction/vision) for odour source localization, in *Proceedings of the IEEE International Conference on Robotics and Automation*, Orlando.

Li, J.M., Wan, D.L., Chi, Z.X. and et al. (2006) A Parallel Particle Swarm Optimization Algorithm Based on Fine-Grained Model with GPU-Accelerating, *Journal of Harbin Institute of Technology*, vol. 38, no. 12, pp. 2162–2166. (in Chinese)

Luo, J.H. and Zhong, Z.N. (2005) Research on the Parallel Simulation of Asynchronous Pattern of Particle Swarm Optimization, *Journal of Computer Simulation*, vol. 22, no. 6, pp. 68–70, 78. (in Chinese)

Pugh, J. and Martinoli, A. (2007) Inspiring and Modeling Multi-Robot Search with Particle Swarm Optimization, in *Proceedings of the 4th IEEE Swarm Intelligence Symposium*, Honolulu, Hawaii, USA, April, 1–5.

Zeng, J.C. and Xue, S.D. (2010) Modeling and Simulation Approaches to Swarm Robotic Systems, *Journal of System Simulation*, vol. 22, no. 6, pp. 1327–1330. (in Chinese)

Ridge, E., Kudenko, D., Kazakov, D. and et al. (2005) Moving Nature-Inspired Algorithms to Parallel, Asynchronous and Decentralised Environment, *Frontiers in Artificial Intelligence and Applications*, Vol. 135, pp. 35–49.

Schutte, J.F., Reinbol, J.A., Fregly, B.J. and et al. (2004) Parallel Global Optimization with the Particle Swarm Optimization, *International Journal of Numerical Methods in Engineering*, vol. 61, no. 13, pp. 2296–2315.

Venter, G. and Sobieszczanki-Sobieksi, J. (2006) A Parallel Particle Swarm Optimization Algorithm Accelerated by Asynchronous Evaluations, *Journal of Aerospace Computing, Information, and Communication*, no. 3, pp. 123–137.

Wang, Y.Y., Zeng, J.C. and Tan, Y. (2007) Parallel Particle Swarm Optimization Based on Parallel Model with Controller, *Journal of System Simulation*, vol. 19, no.10, pp. 2171–2176. (in Chinese)

Xu, Y.Z. and Zeng, W.H. (2005) The Development of Parallel Evolutionary Algorithms, *Pattern Recognition and Artificial Intelligence*, vol. 21, no. 2, pp. 183–192. (in Chinese)

Xue, S.D. and Zeng, J.C. (2008a) Using Relative Localization Observations for Swarm Robots Search, in *International Conference on Modelling, Identification and Control*, Shanghai, China, June 29–July 2.

Xue, S.D. and Zeng, J.C. (2008b) Control over Swarm Robots Search with Swarm Intelligence Principles, *Journal of System Simulation*, vol. 20, no. 13, pp. 3449–3454.

Xue, S.D. and Zeng, J.C. (2008c) Swarm Robots Search under Conditions of Heterogeneous Sensory Signals Fusion, in *Proceedings of the 2008 International Conference on Genetic and Evolutionary Methods*, Las Vegas, Nevada, USA, July 14–17.

Zeng, J.C., Jie, J. and Cui, Z.H. (2004) *Particle Swarm Optimization*, Beijing: Science Press. (in Chinese)

Zhao, Y., Yue, J.G., Li, G.Y. and et al. (2005) A Parallel Particle Swarm Optimization Algorithm Based on Multigroup for Solving Complex Functions Optimization, *Journal of Computer Engineering and Applications*, vol. 41, no. 16, pp. 58–60, 64. (in Chinese)

Xue, S.D., Zhang, J.H. and Zeng, J.C. (2009) Parallel asynchronous control strategy for target search with swarm robots, *International Journal of Bio-Inspired Computation*, vol. 1, no. 3, pp. 151–163.

Xue, S.D. and Zeng, J.C. (2009) Controlling swarm robots for target search in parallel and asynchronously, *International Journal of Modelling, Identification and Control*, vol. 8, no. 4, pp. 353–360.

15

Using a Meeting Channel and Relay Nodes to Interconnect Mobile Robots

Nassima Hadid, Alexandre Guitton
and Michel Misson
*LIMOS CNRS / Clermont Université, Campus des Cézeaux, 63 173 Aubiere Cedex
France*

1. Introduction

Mobile robots are increasingly used in industrial applications (generally in order to convey merchandises), in healthcare applications (for example, to transport patients or medicines) or in domotic applications (for instance, to sweep a floor). These robots are often mounted with several sensors and actuators, in order to carry out their tasks (*e.g.*, grabbing an object of interest, and bringing it a destination without colliding with obstacles). To reduce the complexity and weight of the wiring of all these sensors and actuators, these devices can be equipped with wireless capability. A central entity, called the central unit, is usually in charge of collecting the data from the sensors, integrating part of the data in a models, computing a behavioral response and operating the actuators accordingly. Another entity, called the coordinator, is in charge of managing the wireless entities, retrieving data from sensors and sending orders to actuators. The same device can be simultaneously the central unit and the coordinator. In the following, the wireless communications between sensors, actuators and coordinator are called intra-robot communications.

Real applications often require the collaboration of mobile robots in order to perform required tasks. For instance, when they are used in airports to transport luggage, mobile robots can cooperate to avoid colliding with each other. In search and rescue operations, mobile robots can inform their neighbors that an area has already been explored (Ferranti & Trigoni, 2008). Robots can also follow tracks on the floor (Hogg et al., 2002) or monitor an area collaboratively (Lambrou & Panayiotou, 2009). An allegorical example is shown on Fig. 1, where the waiter robot *Dro* has to collaborate with the waitress robot *Ide* in order to serve hot coffee to customers. These communications used for collaborations are called inter-robots communications.

The design of a network architecture that allows mobile robots to cooperate efficiently, without negatively impacting the performance of each intra-robot communication, is a challenging issue, and is the aim of this chapter.

The remainder of the chapter is the following. In Section 2, we introduce a multi-channel approach where intra-robot communications are performed on a robot-specific channel, and inter-robot communications are performed on a common meeting channel. In this way, mobile robots can communicate without impacting intra-robot communications. In Section 3, we discuss the protocols that allow mobile robots to communicate directly, when they are

Fig. 1. Mobile robots *Dro* (on the left) and *Ide* (on the right) have to collaborate to serve hot coffee to customers.

in range of each other. These protocols are based on several mechanisms: a data storage mechanism (which allows data to be stored until the intended destination mobile comes in range), a neighbor discovery mechanism (which detects mobiles in range), an association mechanism (which allows neighbors to identify each other), and a data exchange mechanism (which transfers the data). All these mechanisms are described in details. The concepts behind these protocols come from delay tolerant networks (DTN), as the mobility of robots is often difficult to integrate to applications, or even unpredictable in some applications. In Section 4, we discuss how to achieve indirect communications between mobile robots. We explain how road-side units can help mobile robots to communicate. These road-side units can also be required in some applications, where a message has to be disseminated in a specified geographical area. A generic addressing mechanism is useful in this case, when the ID of the destination mobiles cannot be known beforehand. We show that the overhead of implementing the generic addressing is limited. Finally, in Section 5, we conclude this chapter by summarizing the benefits of the meeting channel.

2. Interconnection of mobile robots using a meeting channel

Most applications using mobile robots require the mobile robots to communicate together in order to perform the required tasks. In the following, we consider a generic application where mobile robots follow tracks and carry items (such as robots carrying luggage in airports). The mobile robots communicate in order to avoid collisions, and to detect priority when crossing intersections. Each mobile is equipped with sensors and actuators that communicate with the central unit embedded on the mobile.

Although the communications between mobile robots can be performed on the same communication channels as the intra-mobile communications, this tends to reduce the

performance of the whole system. Indeed, if the communications between mobiles and the intra-mobile communications are on the same channel, the competition for the medium is high. Reducing the bandwidth of intra-mobile communications can have a disastrous impact on the robot behavior, for instance if the coordinator is not able to inform quickly the actuators (such as brakes) to be operated.
Thus, it is important to interconnect mobile robots using a multi-channel approach.

2.1 Multi-channel approaches

Multi-channel approaches have the main advantage to reduce the contention on each communication channel, and thus to improve the performance of the system. In the literature, several multi-channel approaches have been proposed (Bahl et al., 2004; Jovanovic & Djordjevic, 2007; Nasipuri et al., 1999; Pathmasuntharam et al., 2004). There are two types of multi-channel approaches: multi-channel approaches with several radio interfaces and multi-channel approaches with a single radio interface.

2.1.1 Multi-channel approaches with several radio interfaces

HMCP (Hybrid Multi-Channel Protocol) (Kyasanur & Vaidya, 2005) divides radio interfaces into two groups: static interfaces operating on static channels, and dynamic interfaces able to change the channel they use. Nodes periodically inform their neighbors of the channel of their static interfaces. Notice that neighboring nodes do not necessarily have their static interfaces on the same set of channels. When a sender has to transmit data to a neighbor, the sender switches one of its dynamic interface to a channel corresponding to one of the static interface of the neighbor. HMCP requires periodic exchange of control packets to maintain the neighbor tables, and is not suited for mobile nodes.
The DCA (Dynamic Channel Assignment) protocol (Wu et al., 2000) requires each node to have at least two radio interfaces. One of the interface is used for control frames, and the others for data frames. The control interface always communicates on the same channel, while the data interfaces can change channels dynamically. Neighboring nodes collaborate to reserve available channels for transmissions. The amount of messages exchanged in DCA makes it not suitable when contacts between mobile entities last for a short duration.
The DCSS (Dynamic Channel Selection with Snooping) protocol (Seo et al., 2008) also uses two radio interfaces: the control interface is always listening on a common given channel, while the data interface changes its channel depending on the reservation. Nodes perform reservations depending on their view of the channel states. Nodes overhear control packets in order to determine the set of channels locally used. They assess the state of all the channels using a round-robin algorithm while they are neither transmitting nor receiving a frame. Thus, they periodically update their view of the channel states. The main drawback of DCSS is that nodes spend a lot of energy in assessing the channels.
In all these protocols, nodes are required to have two radio interfaces, which consumes a lot of energy. Indeed, radio interfaces are the main source of energy consumption in low rate wireless solutions. Moreover, the control radio interface is often active all the time. Additionally, having several radio interfaces increases the cost of the hardware. Thus, having a multi-channel approach operating with several radio interfaces is not suited in our applications.

2.1.2 Multi-channel approaches with a single radio interface

The McMAC (Multi-channel Medium Access Control) (So & Walrand, 2007) protocol uses a channel reservation mechanism based on the periodic diffusion of beacons. The beacons allow nodes to be synchronized. The interval between successive beacons is divided into two periods: a control period, and a data period. During the control period, all the nodes communicate on the same channel in order to plan the data period. During the data period, node switch to and communicate on the channels reserved during the control period. The main drawback of McMAC in our applications is that it is difficult to maintain a synchronization between mobile entities. Also, nodes need to be identified before the control period in order to be able to reserve channels: if the delay to identify new mobile nodes is large, these new nodes might be unable to communicate during a long duration.

The TMMAC (TDMA-based Multi-channel Medium Access Control) (Zhang et al., 2007) protocol is similar to the McMAC protocol. It uses a channel allocation bitmap and a channel usage bitmap in order to reserve channels. The channel allocation bitmap is filled during the control period, and the channel usage bitmap is maintained by each node and transmitted to its one-hop neighborhood. The TMMAC protocol has the same drawbacks as the McMAC protocol.

The SMC MAC (Sensor Multi-Channel Medium Access Control) (Ramakrishnan & Vanaja Ranjan, 2009) protocol based on RTS (Request To Send) and CTS (Clear To Send) control frames. RTS and CTS frames are sent on the control channel. The RTS contains the list of available channels from the sender point of view. The CTS contains the chosen channel from the destination point of view, which is the first channel of the intersection of the available channels of the sender and the receiver. When transmitting and receiving a CTS, both the sender and the receiver switch to the chosen channel, and the other nodes mark this channel as busy. The main drawback of SMC MAC is that all the nodes in range share the same communication channels: the performance of intra-robot communications can be reduced by other intra-robot communications or by inter-robot communications.

2.2 Interconnection using a meeting channel

Our goal is to allow several mobile robots to communicate without requiring an expensive hardware. Thus, we plan to use a multi-channel approach with a single radio interface. Moreover, the performance of the intra-robot communications of a robot should not be reduced by the proximity of other robots in the neighborhood. Finally, the mechanism should be compatible with existing low-power wireless personal area networks standards.

To achieve these goals, we propose to consider two types of channels: the robot-specific channels and the meeting channel. Each robot is allocated its own robot-specific channel, in order to be able to perform intra-robot communications without suffering from severe and systematic interferences from other robots that are moving in the same area. In the case where there are more robots than available channels, it is possible that several robots share the same channel. However, it is important to allocate the same channel to robots having disjoint trajectories, in order to maintain a low level of interferences. Additionally, all the robots share the same meeting channel. This channel is used solely for communications between robots.

2.2.1 Energy-efficiency

Mobile robots are often equipped with a battery that powers the engine and the coordinator. Sensors and actuators are also powered by a battery (either by the main battery or by their

own). In order to maximize the lifetime of the robot, it is crucial that all embedded devices save energy, including network devices.
As the radio module of network entities consumes a significant amount of energy, most standards for low-power wireless personal area networks, such as IEEE 802.15.4 (IEEE 802.15, 2006), deactivates the radio module of nodes periodically. During this time, the nodes are sleeping and can neither receive nor transmit. They are activated at specific instants, usually during small fractions (typically less than 1%) of time.

2.2.2 Compatibility with standards

Our proposition is to allow some of the nodes to use the meeting channel for inter-mobile communications, while other nodes are sleeping. Note that most standards for low power wireless personal area network define an inactivity time of the standards (such as IEEE 802.15.4 standard for example). The nodes that remain awaken are called energy-rich (ER) nodes, and the nodes that go to sleep are called energy-limited (EL) nodes. In this way:

- inter-mobile communications are transparent to EL nodes,
- inter-mobile communications do not degrade intra-mobile communications,
- the activities on the robot-specific channel are compliant with the standard,
- ER nodes use the inactivity time, which is otherwise wasted.

Our mechanism does not require mobile robots to be synchronized so that their inactivity period occur at the same time. Such a synchronization among mobile entities would be very difficult to achieve and induce a significant amount of control traffic, especially as the number of robots in the fleet increases.
Our mechanism uses the fact that inactivity is typically larger than the activity time. Table 1 shows the expected time ratio during which all the n mobiles are simultaneously inactive, as the activity ratio varies. It can be seen that when the activity ratio is low, the simultaneous inactivity ratio becomes very high. For example, for an activity ratio of 1%, 10 mobile robots are simultaneously inactive (and thus potentially on the meeting channel) during more than 90% of the time.

Number of mobile robots n	Activity ratio	Expected simultaneous inactivity ratio
2	0.5	0.25
2	0.25	0.5625
2	0.01	0.9801
3	0.5	0.125
3	0.25	0.4219
3	0.01	0.9703
10	0.5	0.0010
10	0.25	0.0563
10	0.01	0.9044

Table 1. The ratio of time during which n mobile robots are simultaneously inactive is very high when the activity ratio of each mobile is low.

Allowing mobile robots to communicate during their inactivity period has several advantages. As inactivity periods are usually long, the probability that two different robots are simultaneously inactive is high. By activating a subset of network devices during the

inactivity period of each robot, it is possible to have these ER devices communicate with each other while the EL nodes sleep.
The remaining of this chapter describes how mobile robots can communicate using the meeting channel, either directly (when the destination robot becomes in range with the source robot) or indirectly (when the destination robot does not become in range with the source robot, or when the destination robot address is unknown by the source robot). Notice that intra-mobile communications can be performed according to the ZigBee and IEEE 802.15.4 standards for instance, and is not the focus of this chapter.

3. Direct communications between mobile robots

To allow direct communications between mobile robots, a DTN paradigm has to be used (*Delay-Tolerant Networking Architecture*, 2006). Indeed, communications between mobile robots are opportunistic in nature: they depend on the mobility pattern of each mobile robot and on the wireless propagation conditions, both of which are considered here as unpredictable. Moreover, the number of mobile robots in the area is small compared to the area size. Thus, mobile robots meet each other infrequently.
We propose to adapt the *store, carry and forward* principle from the DTN literature to our generic application. In DTNs, a new layer called the bundle layer is introduced between the application layer and the transport layer of the OSI model. The main role of this bundle layer is to mask to applications the effects of node mobility, and mainly the sporadicity of the network connectivity. To do so, data from the application layer is encapsulated in a composite structure, called a bundle. A bundle contains the application data, and additional data to deal with sporadic connectivity, such as timeouts and destinations. Using a bundle layer implies the need of a transport layer, as the bundle structure is often larger than the maximum frame size in a low-power wireless personal area network.
According to the DTN principle, messages are stored in bundle in specific nodes called custodians. When two mobiles meet, their custodians can exchange bundles. The bundles can circulate through the mobility of robots or through contacts between mobiles. The detection of a contact between mobiles is performed by specific nodes called gateways.
To summarize, we can consider the following types of nodes.

- EL nodes are only active on the robot-specific channel. They sleep during the robot inactivity period, and do not contribute to communications between mobile robots.
- There is one custodian node per robot. It is an ER node with larger storage capabilities than the other nodes. During the robot activity period, it works as the other nodes. During the robot inactivity period, it switches to the meeting channel, and manages bundles.
- There is at least one gateway node per robot. Gateways are ER nodes whose role is to detect neighboring robots on the meeting channel, during the robot inactivity period.
- There can be intermediate ER nodes per robot. Intermediate nodes are ER nodes that help interconnecting the custodian and the gateways on the meeting channel, during the robot inactivity period.

Fig. 2 describes the network stack of our architecture.
In this section, we first describe how the presence of ER nodes can improve the network performance of intra-robot communications. This feature is important as the mobile robots often have no neighbors in range. Second, we study the neighbor discovery mechanism and

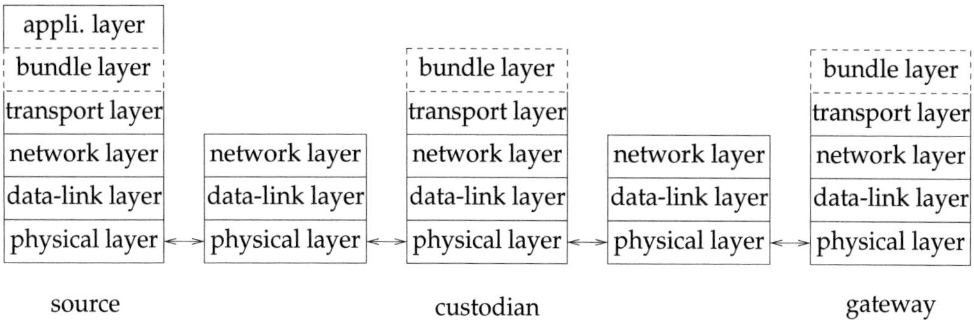

Fig. 2. Network stack and communication from a source to a gateway (on the same mobile), including the bundle layer.

we present the trade-off between the energy-efficiency of the neighbor discovery mechanism and the duration of the neighbor detection. Third, we show how communication links can be established between robots, and how the custodians of two neighbor robots can know each other. Fourth, we discuss direct communications between mobiles.

3.1 Optimized intra-robot communications

During the robot inactivity period, all ER nodes are active on the meeting channel. The gateways are trying to detect potential neighboring robots. However, most of the time, no robots are found in the vicinity. Instead of being idle while waiting for encounters, ER nodes can help relaying the intra-robot traffic.

3.1.1 Description

During the robot inactivity period, the ER nodes form a network called the ER network. Thus, ER nodes have two addresses: one during the robot activity period, and one during the robot inactivity period. EL nodes have a single address, for the robot activity period. We assume that there exists a routing function $\mathcal{R}$, such that $\mathcal{R}$ takes as input the destination addresses (in the robot network and/or in the ER network) and returns the next hop addresses (in the robot network and/or in the ER network). In the ZigBee standard, the function $\mathcal{R}$ is based on the hierarchical allocation of addresses. We also assume that there exists a neighbor address function $\mathcal{N}$ that takes as input the address of a next hop (in the robot network and/or in the ER network) and returns, if possible, the address of the next hop in both networks. It is not mandatory that functions $\mathcal{R}$ and $\mathcal{N}$ are always able to translate the address of a node from one network to another, but they can improve the global behavior of the mechanism.
To optimize intra-robot communications on the meeting channel, the MAC relaying scheme has to be modified in the following way. Let us assume that an ER node n has a frame for destination d. If the address of d in the ER network is known, this address can be used to compute the next hop h on the ER network, as $h = \mathcal{R}(d)$. n can send the frame to h. If the address of d in the ER network is not known, it is possible that the next hop h' of d computed on the robot network is also a neighbor of n in the ER network. If this is the case, $h = \mathcal{N}(h')$ is known: n can send the frame to h (in fact, both addresses h and h' are present in the frame). Otherwise, n is unable to send the frame during this period. Thus, n proceeds to the next frame in queue (without dropping the previous frame, as it will be sent during the next robot activity period).
To summarize, ER nodes can always relay frames to ER nodes.

3.1.2 Simulation results

To quantify the performance improvement when the meeting channel is used for intra-robot communications, we performed simulations using the network simulator NS2. We used the IEEE 802.15.4 physical layer and a custom IEEE 802.15.4 MAC layer that supports our meeting channel architecture and our relaying scheme. We used the ZigBee hierarchical routing protocol for both the robot network and the ER network. The propagation model used is the ITU indoor propagation model (*Propagation data and prediction method for the planning of indoor radio communication systems and local area networks in the frequency range of 900 MHz to 100 GHz*, 2005), with a path loss exponent of 3. The transmission power of nodes is set to -25 dBm. Frames have a size of 119 bytes at the MAC layer, which corresponds to 1000 bits at the physical layer. Queue lengths are set to 50 frames.

Simulations are performed on a network of 6 nodes with four pairs of source s and destination d, as depicted on Fig. 3. Solid nodes represent EL nodes, half-striped nodes represent ER nodes. The source is always three hops away from the destination.

In order to evaluate the advantages of communications on the meeting channel, we compare the scenario where all the nodes are ER nodes with a scenario where the two central nodes are ER nodes, and the source and destination both have a 50% probability of being ER nodes. Simulation results are averaged over 100 runs.

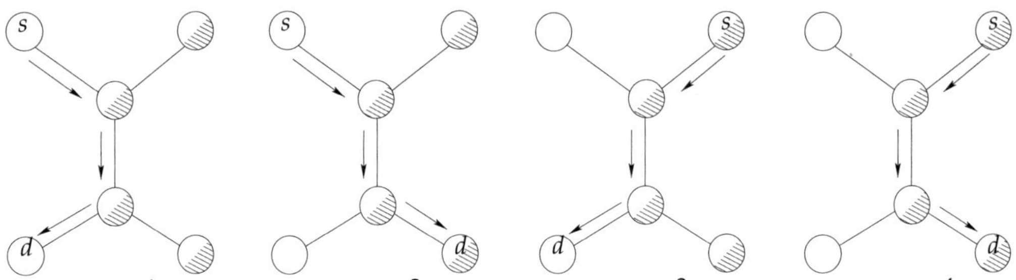

Fig. 3. Simulations are performed on a small topology, where sources and destinations both have a 50% probability to be ER nodes.

We considered two metrics: (1) the goodput, which is the number of frames correctly received by the destination node per second, and (2) the end-to-end delay.

Fig. 4 displays the goodput as a function of the frame generation rate, for several scenarios: one where the nodes are always active on the robot-specific channel, in dotted lines, one with only EL nodes and for various duty cycles on the robot-specific channel (from 12.5% to 50%), in solid lines, and one with both EL and ER nodes (as shown on Fig. 3) for various duty cycles on the robot specific channel (from 12.5% to 50%), in dashed lines. First, it can be noticed that the goodput increases with the frame generation rate, until it reaches a maximum. This behavior is typical of IEEE 802.15.4. As expected, the goodput with EL nodes only is strongly related to the duty cycle: it is high when the duty cycle is 100% (EL nodes are always active), and is reduced when the duty cycle goes from 50% down to 12.5%. When ER nodes are added, the goodput is significantly increased with respect to the scenarios with only EL nodes.

Fig. 5 displays the average end-to-end delay as a function of the frame generation rate, for the three same scenarios, and for various duty cycles. The average end-to-end delay increases with the frame generation rate, until it reaches a maximum. This maximum can be explained in the following way. The delay is computed based on the received frame only. Frames cannot

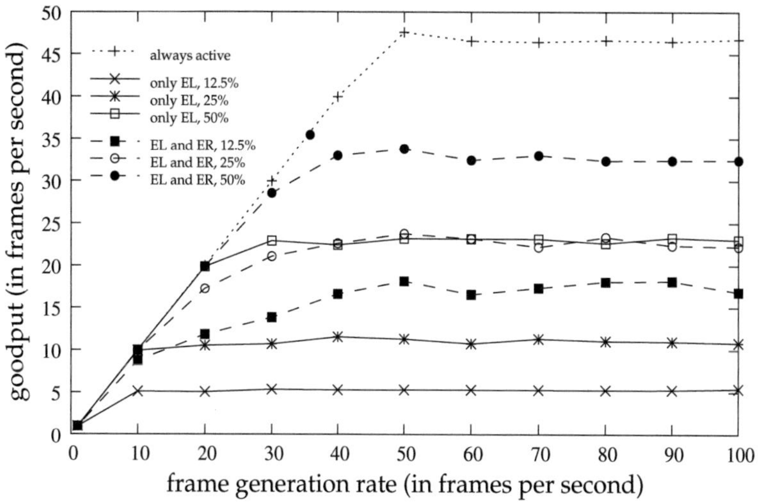

Fig. 4. The presence of ER nodes significantly improves the performance of the network, for all duty cycles.

wait for a long time in the frame queue, because of the limited size of the frame queue and because slotted CSMA/CA drops frames after a small number of attempts. When nodes are always active, the delay is very low. The delay increases when the network is saturated, which occurs at about 50 frames per second (which can be seen on Fig. 4). When there are only EL nodes, the end-to-end delay is large: indeed, EL nodes spend a large amount of time sleeping, which increases the delay. In the scenario with ER nodes, the end-to-end delay is improved compared to the scenario without ER nodes: indeed, some frames can be relayed during the inactivity period of EL nodes, which helps reducing the delay.

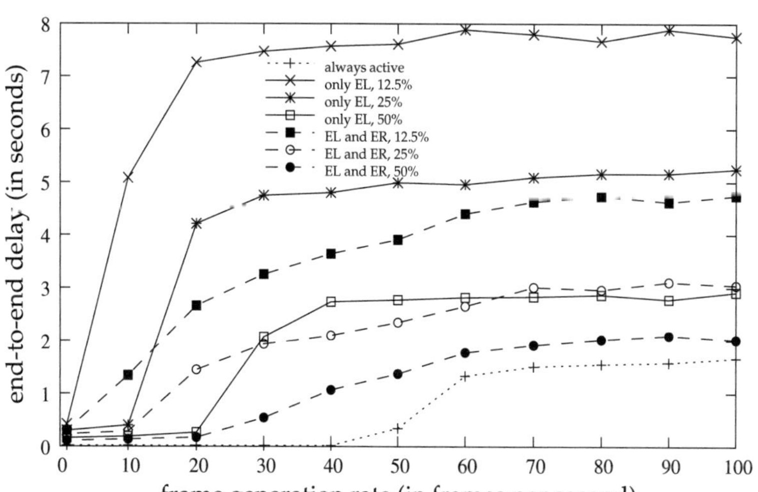

Fig. 5. The presence of ER nodes significantly reduce the end-to-end delay, for all duty cycles.

ER nodes consume more energy than EL nodes, because they are more often active. Thus, ER nodes allow to achieve a trade-off between energy consumption and performance.

3.2 Neighbor discovery

When a robot moves in range of another, the two robots need to discover each other before being able to communicate. The neighbor discovery mechanism Hadid et al. (2010) occurs during the inactivity period, and is performed by gateway nodes only.

Gateway nodes send periodic gateway advertisement (GW-ADV) frames. Each GW-ADV frame contains the identification of the gateway, and the identification of the robot that carries it. When receiving a GW-ADV frame, a gateway replies with a gateway reply (GW-REP) frame. Each GW-REP frame contains the identification of the local gateway, the identification of the distant gateway, and the identification of the robot that carries it. Upon receiving a GW-ADV or GW-REP frame, both gateways update their neighbor table. An entry in this table contains the identification of the neighbor mobile, the address of the distant gateway, and a time to live. It is possible for the table to contain two different entries for the same mobile, if the mobile is reachable through two gateways.

The neighbor discovery is not instantaneous, as gateways transmit GW-ADV periodically, and only when they are operating on the meeting channel. The neighbor discovery delay depends on the periodicity of the GW-ADV frames, and on the robot activity period. To study the impact of these two parameters, we carried out two sets of simulations. The first set of simulation concerns the neighbor discovery delay from the moment the two mobiles become in range. The second set of simulation concerns the total contact duration, between the moment the gateways identify each other, until the time the contact is broken due to robots mobility.

We used a simple mobility model[1]: two robots move on parallel lines, in opposite directions, with a constant speed of 2 meters per second. They start far away, at a position where they are not in range of each other. They eventually meet and discover each other, while continuing their movement. After some time, they are too far away to keep communicating, and the contact is broken. The minimum distance between robots is 3 meters, which is loss than the range of a node. The placement of gateways is shown on Fig. 6, where gateways are represented with nodes connected to a black rectangle.

We used a duty cycle of about 1 second[2]. Each simulation lasts for 100 second, and the results are averaged over 500 runs.

3.2.1 Neighbor discovery delay

The first simulation concerns the quantification of the neighbor discovery delay. It is computed as the difference between the time when the two mobiles become in range and the time when they are not in range anymore. To model the fact that mobiles are not synchronized, the activity (on the robot-specific channel) of mobile M_1 starts at t_0, and the activity (on the robot-specific channel) of mobile M_2 starts at $t_0 + \Delta$, where Δ is chosen randomly in the cycle. Moreover, we consider that the periodicity of GW-ADV might vary between mobiles, due to

[1] Our focus here is not to focus on the mobility model, but on the quality of the neighbor discovery procedure. That is why we prefer to use a simple mobility model rather than a more complex one. A discussion on the impact of the mobility model can be found in Bai et al. (2003).

[2] The duty cycle is in fact equal to 983.04 ms, which is the closest value for a duty cycle of 1 second for IEEE 802.15.4 solutions.

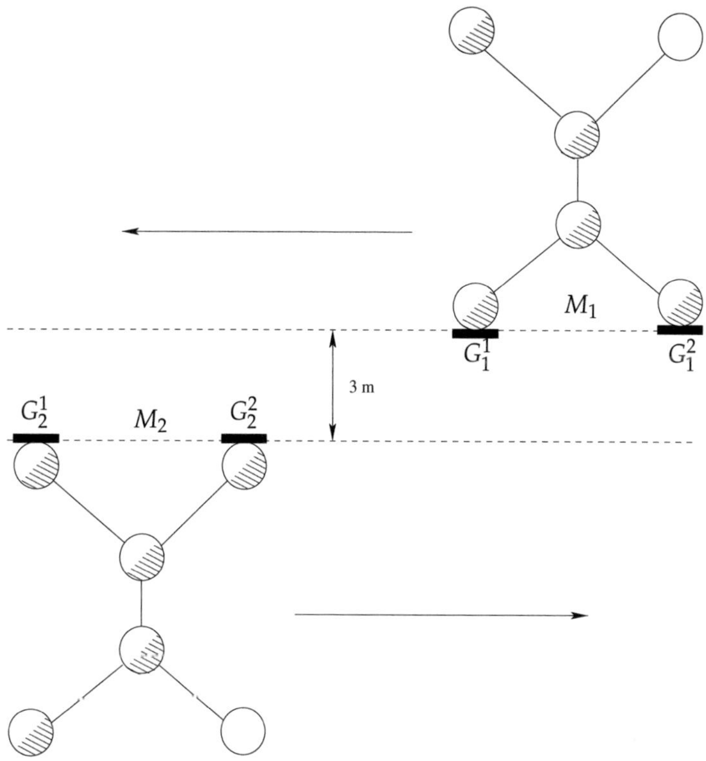

Fig. 6. Two mobiles M_1 and M_2 cross each other. The minimum distance between the gateways of M_1, G_1^1 and G_1^2, and the gateways of M_2, G_2^1 and G_2^2, is 3 meters.

clock drifts. The periodicity of GW-ADV for a mobile M_i is given by $T(1+\delta_i)$, where T is the basic periodicity, and δ_i is chosen randomly in $[-0.3; 0.3]$ (according to a uniform distribution). Fig. 7 shows the neighbor discovery delay as a function of the GW-ADV periodicity T, and as a function of the duration of the activity period on the robot-specific channel. The largest the periodicity, the longest the duration. However, the duration of the activity period has the most significant impact on the neighbor discovery delay: when the nodes spend 50% of their time on their robot-specific channel, the gateways of a mobile often miss the GW-ADV messages sent on the meeting channel by the gateways of the other mobile.

3.2.2 Contact duration

The second simulation concerns the quantification of the contact duration. It is computed as the difference between the time when the mobiles discover each other and the time when there is no gateway of one mobile in contact of one gateway of the other mobile. Notice that having several gateways per mobile might be a way to (slightly) improve the contact duration.
Fig. 8 shows the contact duration as a function of the GW-ADV periodicity T, and as a function of the duration of the activity period on the robot-specific channel. The largest the periodicity, the smallest the duration, as it takes some time for the nodes to discover each other. Again, the duration of the activity period has the most significant impact on the contact duration.

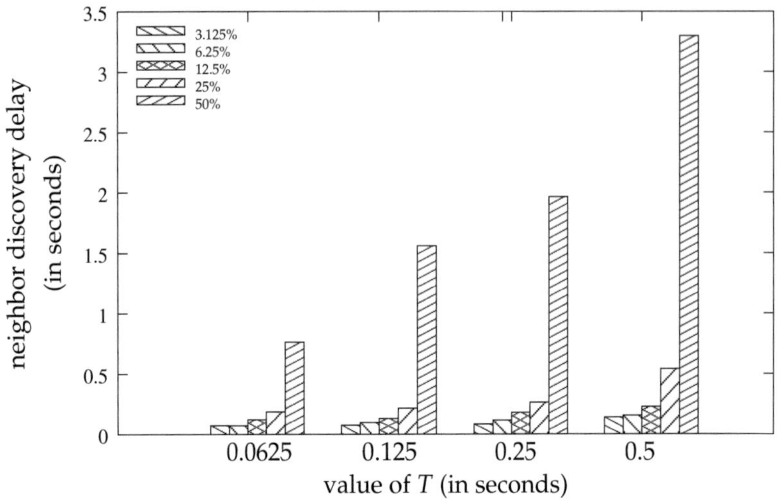

Fig. 7. The neighbor discovery delay is greatly impacted by the duration of the inactivity period.

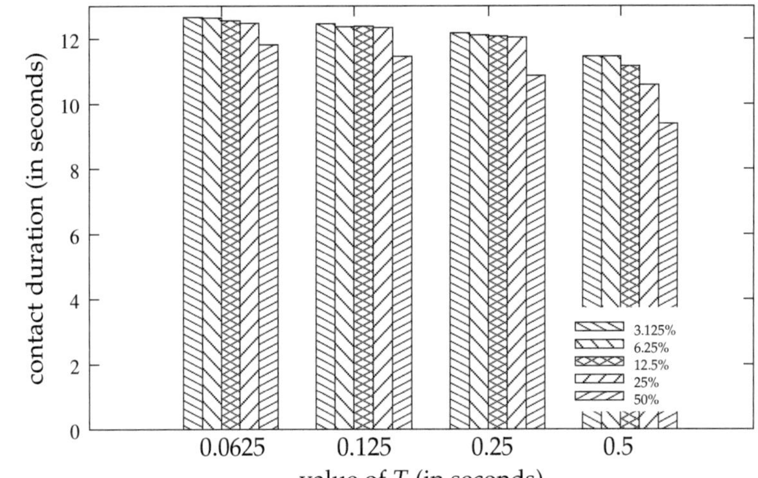

Fig. 8. The neighbor discovery delay is greatly impacted by the duration of the inactivity period.

3.3 Link establishment

Once a local gateway discovers a distant gateway, the local gateway notifies its custodian of its discovery using a notification (NOTIF) message. This is done either when the local gateway received a GW-REP, or received the GW-ADV from another gateway. The NOTIF message contains the identification of the distant mobile, the address of the distant gateway and the address of the local gateway. Upon receiving the NOTIF message, the custodian updates its contact table. An entry of the contact table contains the identification of the distant mobile, the address of the distant gateway and the address of the local gateway. There might be several local and distant gateways that lead to the same mobile.

If a gateway stops receiving GW-ADV messages during a period equal to the time to live of the neighbor entry, the gateway sends a message to its custodian, in order to remove the entry from the contact table. The mobile might still be reachable, through other local or distant gateways.

3.4 Direct data exchange between mobiles

The data exchange between a source node on a source mobile and a destination node on a destination mobile is performed according to the following steps:

- the source node sends the bundles to the custodian node of the source mobile,
- the custodian node of the source mobile sends the bundles to a gateway node of the source mobile, once a link between these two mobiles has been established,
- the gateway node of the source mobile sends the bundles to the gateway node of the destination mobile,
- the gateway node of the destination mobile sends the bundles to the destination node.

The transmission from the source node to the custodian node of the source mobile corresponds to an intra-robot communication, as both nodes are on the same mobile. This part of the communication can be achieved even if the two mobiles are not in range. The same goes for the transmission between the gateway of the destination mobile to the destination node.
Fig. 9 depicts this data exchange on a small example. Custodians are shown with a small database on the side, to indicate that they possess storage capabilities. As shown below the figure, several communication layers are involved in this process. The source node S generates data at the application layer, which is encapsulated within bundles at the bundle layer. The communication from S to C_1 is done at the bundle layer, and can require the transmission through intermediate nodes, such as X_1. Then, the bundles wait in C_1 for a notification concerning mobile M_2. Once this happens, C_1 communicates with its gateway G_1 at the bundle layer. G_1 and G_2 are able to communicate directly. Finally, G_2 sends the bundles to D, at the bundle layer, potentially involving intermediate nodes, such as X_2. Finally, D can extract the data at the application layer.
One bundle can correspond to several packets. The transport layer is in charge of splitting large bundles into several packets.

3.4.1 Description

The direct data exchange between mobiles raises four issues: (1) the choice of the source gateway by the source custodian, (2) the choice of the distant gateway by the source gateway, (3) the algorithm used by the custodian to send the bundles, and (4) the policy to retransmit or drop bundles.
When there are several gateways that can reach the destination mobile, the source custodian has to choose to which source gateway the bundles have to be sent. The custodian can apply different policies of load balancing, depending on the application, by choosing several different source gateways to allow parallel transmissions of the bundles. Thus, several bundles at a time are sent, one per gateway that discovered the destination mobile.
A given source gateway can be in range of several destination gateways. The source gateway has to choose to which destination gateway the bundles should be sent to. In this paper, we propose that source gateways choose destination gateways at random, to reduce

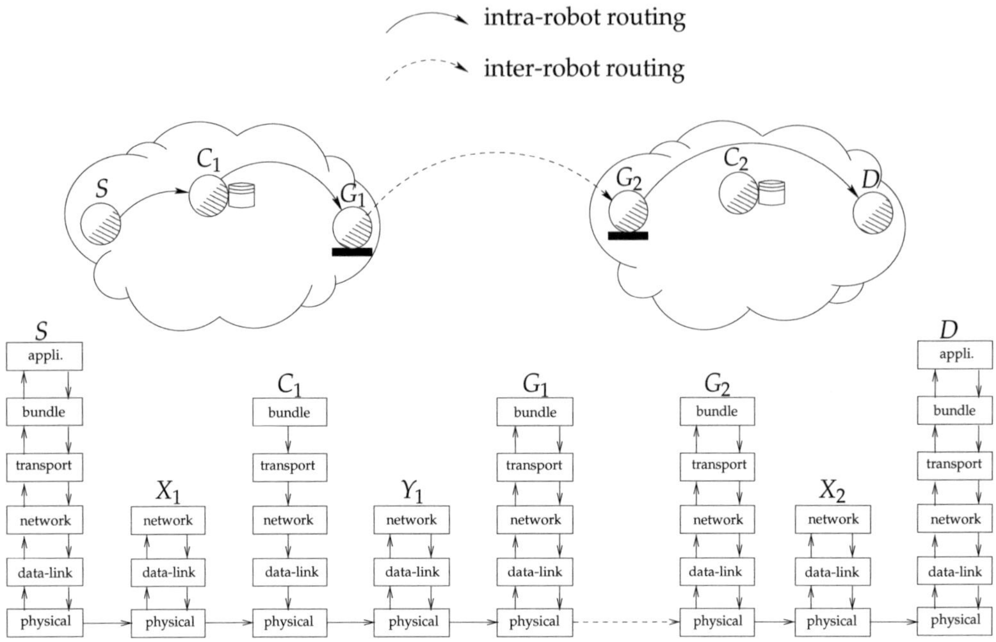

Fig. 9. The bundles sent by mobile M_1 to mobile M_2 first go from the source node S to the custodian C_1. Once M_1 and M_2 have established a link through their gateways, the bundles go from C_1 to G_1, from G_1 to G_2, and finally from G_2 to D.

the probability that two different source gateways choose the same destination gateway repeatedly.

The algorithm applied by the custodian is the following. When it receives a bundle, it checks whether there is an entry in its contact table corresponding to the destination mobile or not. If there is no entry, the custodian stores the bundle. Otherwise, the custodian computes the set of source gateways that can reach the destination mobile.

Bundles are not kept permanently in custodians. Indeed, they are associated to a time to live, which is decided at the application layer. Bundles are also removed by the source custodian when they are successfully transmitted. When a local gateway sends a bundle to a distant gateway, it waits for the acknowledgment of the transmission. Once the local gateway receives the acknowledgment, it sends a ROUTE-SUCCESS message to the custodian. Upon receiving this message, the custodian removes the bundle from its memory. If the local gateway does not receive acknowledgments from the distant gateway after a number of retries, the local gateway sends a ROUTE-FAILURE message to the custodian. The custodian can decide to transmit the bundle to another source gateway. The custodian does not delete bundles after ROUTE-FAILURE messages: it simply waits for another opportunity to transmit it, or for the time to live to expire. Notice that the ROUTE-SUCCESS and ROUTE-FAILURE messages only contain a bundle ID: these messages can fit in short frames.

3.4.2 Simulation results

We evaluated the performance of this data-exchange mechanism according to three metrics: (1) the bundle reception rate, (2) the average delay from the custodian of the source to the destination, and (3) the average path length for bundles.

Simulations are performed on a scenario close to the one depicted on Fig. 9. We assume that each mobile has two gateways, that the two gateways of the source mobile are one hop away from the custodian, and the destination is one hop away from the two gateways of the destination mobile. We assume that each bundle can fit in a single frame, and thus bundles do not require fragmentation and reassembly. The frame size for a bundle is set to 119 bytes at the MAC layer, so that a bundle is 1000 bits long at the physical layer. The duty cycle is set to about one second, and gateways send their GW-ADV frame with a periodicity of T =0.25 seconds (when they operate on the meeting channel). This value of T is a trade-off between the discovery delay and the overhead of the GW-ADV frames.
Fig. 10 shows the bundle reception rate as a function of the bundle generation rate. The bundle reception rate increases with the bundle generation rate, until it reaches a maximum (again, this behavior is typical of IEEE 802.15.4). When the duty cycle on the robot-specific channel is low (for instance, when the duty cycle is equal to 12.5%), the ER nodes spend most of their time on the meeting channel. Thus, the bundle reception rate is large. Contrarily, when nodes spend a significant part of their activity time on the robot-specific channel (for instance, when the duty cycle is 50%), the ER nodes cannot spend most time on the meeting channel, and the gateways of the different robots have less time to communicate (on average, the gateways of the two robots are on the same meeting channel during only 25% of the time), which negatively impacts the bundle reception rate.

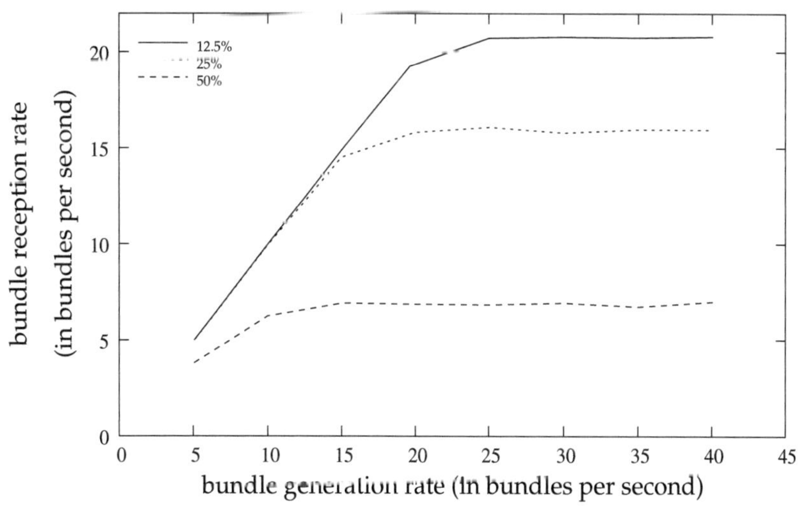

Fig. 10. The bundle reception rate (in bundles per second) increases as the duty cycle on the robot-specific channel decreases.

Fig. 11 shows the average delay to transmit a bundle from the custodian of the source mobile to the destination. Note that this delay does not take into account the time required to transmit a bundle from the source to the custodian. Indeed, most of the time, all the bundles are accumulated in the custodian rather than in the source node. As expected, the average delay increases with the bundle generation rate. When the bundle loss rate is too high, most bundles are lost, and the delay becomes stable (as it involves only bundles that are correctly received). The average path length for the bundles is computed as the average amount of transmission of a bundle. The minimum value for this metric is three in our topology, as the destination is

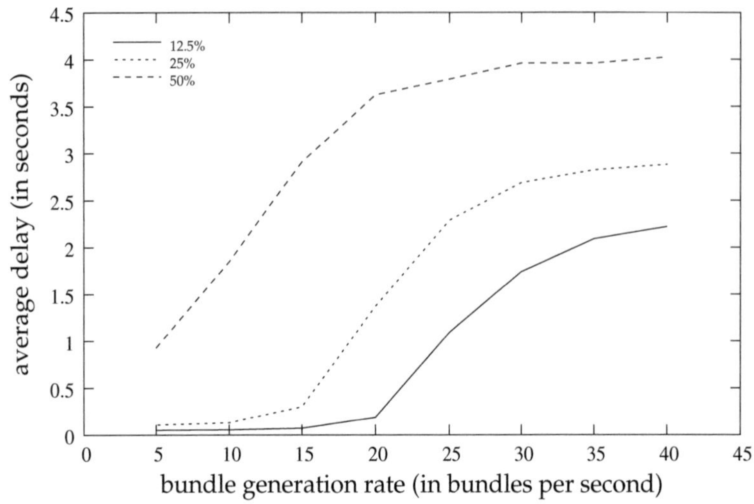

Fig. 11. The average delay from the source custodian to the destination node decreases when the duty cycle on the robot-specific channel decreases.

three hops away from the custodian of the source mobile. Path lengths larger than three occur in the following two cases.

- When one of the nodes on the path from the custodian to the destination has to repeat the transmission (usually due to the loss of a frame or of its acknowledgment), the path length increases accordingly.
- When the custodian C_1 sends the bundle to a local gateway G_1^1, it is possible that this local gateway is not connected anymore to the distant gateway. Indeed, this local gateway can be waiting for the time to live of the neighbor to expire in order to remove the entry in its neighbor table and to notify the custodian. After having tried several times to transmit the bundle, G_1^1 notifies the source custodian with a ROUTE-FAILURE message. The custodian can try to send the bundle to another local gateway G_1^2 which is still in contact with the destination mobile. In this case, the path length is composed of four hops: (C, G_1^1), (C, G_1^2), (G_2^2) and D. Note that while bundles are long frames, the ROUTE-FAILURE message is a simple notification. Thus, it is not counted in the path length.

Note that to simplify, we assume in our simulations that each bundle is sent in a single frame. Thus, the path length increases only by one each time a data frame is transmitted.
Fig. 12 shows the average path length for the bundles. When the duty cycle on the robot-specific channel is low (for instance, when it is only 12.5%), the average path length varies slightly around 3.2 hops. As the minimum path length is 3, this means that, on average, the probability for a bundle to be retransmitted (due to collision, loss of a frame or of its acknowledgment, or due to the mobility of mobiles) is only 6%. However, this percentage increases up to 26% when the duty cycle on the robot-specific channel is high.

4. Indirect communications between mobile robots

Direct communications occur when the following three conditions are met.

- A source robot has to communicate with a destination robot.

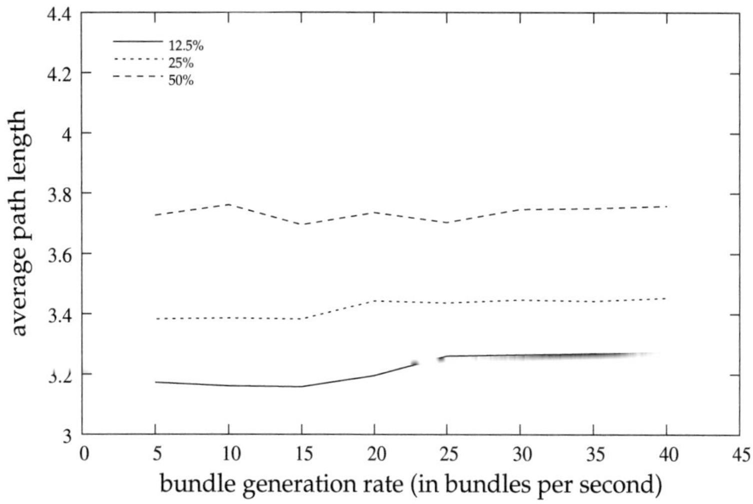

Fig. 12. The average path length for the bundles decreases when the duty cycle on the robot-specific channel decreases.

- The source robot knows the ID of the destination robot.
- The source robot and the destination robot meet directly.

In practice, the two last conditions are hard to fulfill. For example, when a robot has to perform an emergency brake, it has to inform the following robots. The ID of those robots might not be known by the source robot. Moreover, it is possible that a robot and the one following it on the same trajectory never meet: in this case, they are unable to communicate using direct communications.

Indirect communications help removing these constraints. In this section, we first consider how fixed relays can improve the overall connectivity of the network, and can enable indirect communications without requiring sophisticated routing protocols. Then, we consider how to identify mobile robots and nodes based on their role rather than based on their ID or address.

4.1 Communications through fixed-relay networks

Fixed-relay networks can be used to improve the communication opportunities between mobiles (Groenevelt et al., 2005; Zhao et al., 2006). When these relay networks are placed along the road, or at crossings, they can be used to store bundles from robots passing by. In turn, the bundles can be sent to other mobiles traveling in range of the relays.

4.1.1 Description

Fixed-relay can be used to improve the connectivity of the network. If they are deployed strategically at the intersection of mobile trajectories, they can store bundles temporarily, and retransmit them to mobiles passing by.

Each fixed-relay is composed of several nodes forming a network. There is a custodian node in each fixed-relay, and several gateways. The role of the custodian node and of the gateways is the same as for mobile robots.

The main difference between fixed-relays and mobile is that the nodes in fixed-relays are always working on the meeting channel. Thus, they are always available on this channel, and are thus easier to contact than mobiles.

4.1.2 Simulation results

To show the benefits of fixed-relays, and to evaluate the advantage of remaining on the meeting channel on the performance of the network, we performed numerous simulations. The simulation settings are mainly the same as in Section 4. Fixed-relays have a six-node topology, which is identical to the topology of the mobiles. They are static.

Fig. 13 shows the average delay for a mobile to discover a relay node, as a function of the periodicity T of the GW-ADV of both the mobile and the fixed-relay. The relay discovery delay increases as the periodicity T decreases and as the duty cycle on the robot-specific channel increases. The duty cycle on the robot-specific channel is the key parameter: its impact is the most significant. When comparing with Fig. 7, the advantage of fixed-relays on the discovery delay is obvious. In the case where the duty cycle on the robot-specific channel is 50%, and for a GW-ADV periodicity T of 0.5 seconds, the discovery delay of a mobile is 13.2 times larger than the discovery delay of a fixed-relay.

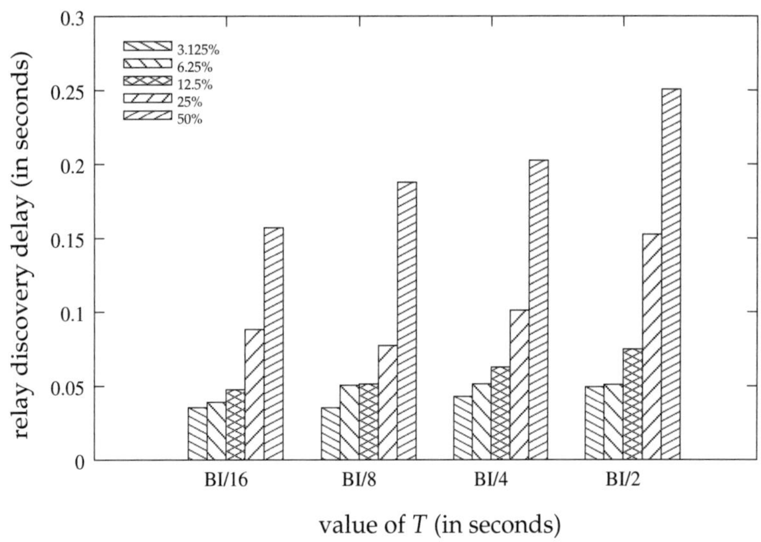

Fig. 13. The relay discovery delay for a mobile is very short, and depends mostly on the duty cycle on the robot-specific channel.

Fig. 14 shows the contact duration between a mobile and a fixed-relay. The contact duration decreases as the GW-ADV periodicity increases because it takes more time to discover the fixed-relay with a large periodicity. The contact duration also decreases as the duty cycle on the robot-specific channel increases, for the same reason. When comparing with Fig. 14, the contacts are about twice larger, as expected since only one robot is moving in this scenario with fixed-relays.

Fig. 15 displays the bundle reception rate as a function of the bundle generation rate, for several duty cycles on the robot-specific channel. The bundle reception rate increases with the bundle generation rate, until a maximum is reached. When comparing with Fig. 15, the results are significantly improved and show that using fixed-relays is highly beneficial. For instance, when the duty cycle is 50%, the maximum bundle reception rate is about 7 when two mobiles meet, and is about 14 when a mobile meets a fixed-relay (as the bundle reception rate is given in bundles per second, the increase of the contact duration in the scenario with

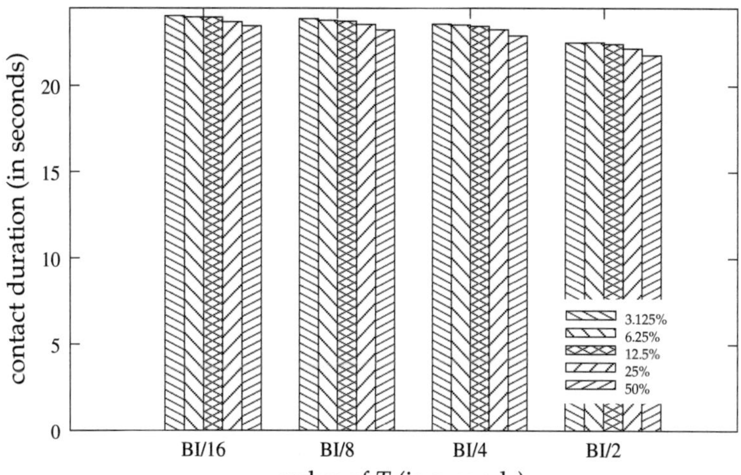

Fig. 14. The contact duration between a mobile and a fixed-relay depends mostly on the GW-ADV periodicity.

fixed-relays has little impact on this number). When the duty cycle is 12.5%, the maximum bundle reception rate is 21 when two mobiles meet, and 24 when a mobile meets a fixed-relay.

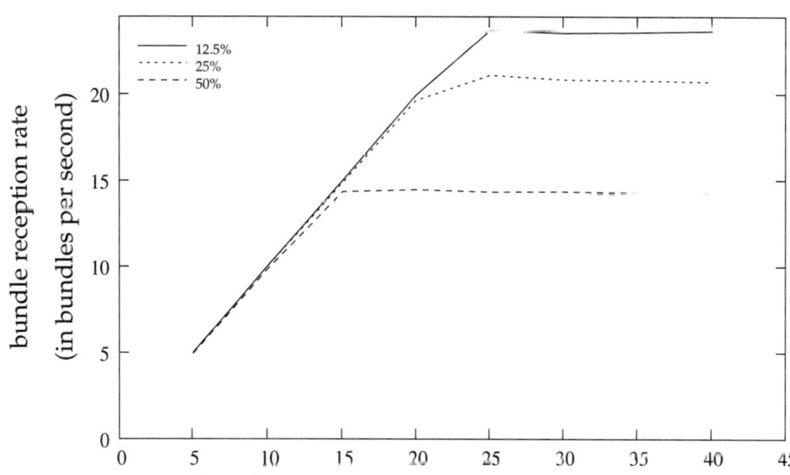

Fig. 15. The bundle reception rate increases when the duty cycle on the robot-specific channel decreases.

Fig. 16 shows the average delay as a function of the bundle generation rate, for several duty cycles on the robot-specific channel. The delay is the time difference between the first transmission of the bundle by the custodian of the source mobile, and the first reception of the bundle by the custodian of the fixed-relay. When comparing with Fig. 16, the results are similar. The gain in delay is not significant, except for the duty cycle on the robot-specific channel, where the delay is reduced by 25%.

Fig. 17 shows the average hop count as a function of the bundle generation rate, for several duty cycles on the robot specific channel. The y-axis is the same as in Fig. 17. It can be seen

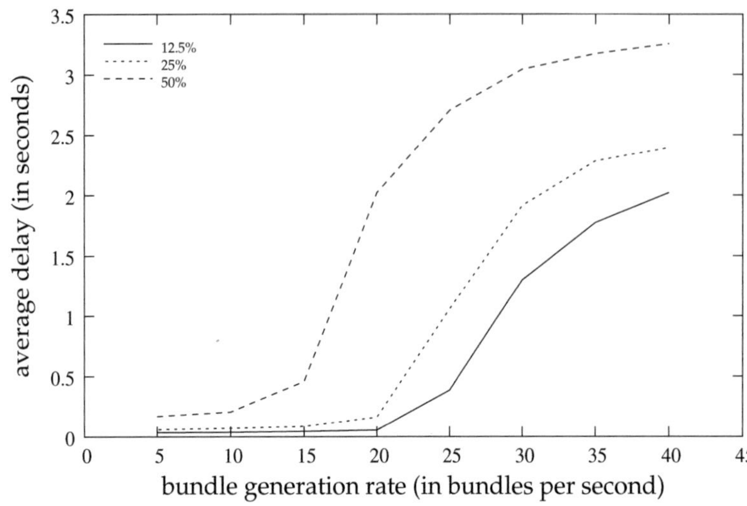

Fig. 16. The average delay from the custodian of the source to the custodian of the fixed-relay decreases when the duty cycle on the robot-specific channel decreases.

that the average hop count varies from 3 (which is the minimum value in our scenario) to 3.2, which constitutes an important improvement.

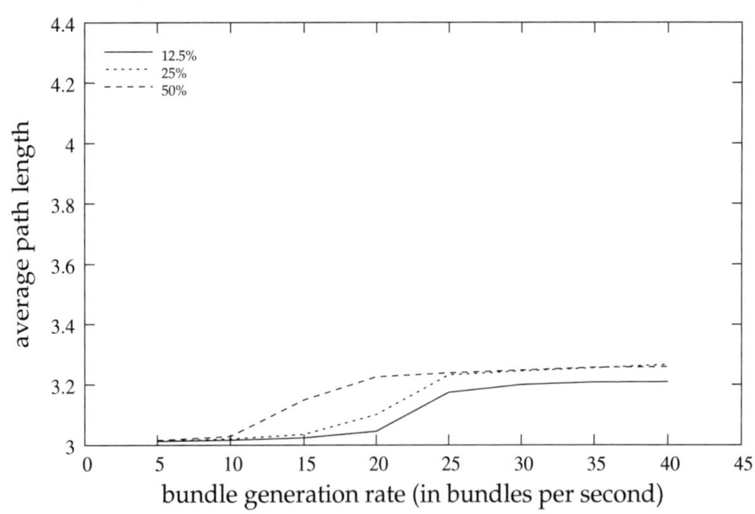

Fig. 17. The average length of the paths followed by bundles is close to the minimum value of 3 when a mobile meets a fixed-relay.

4.2 Generic addressing

The generic addressing consists in addressing network entities by their role rather than by their ID. There are two types of generic addressing: one for the mobile robots, and one for the nodes. These two types can be used simultaneously: it is possible to address a generic node on a generic mobile.

The address space has to be divided into two parts: the part for network addresses and the part for generic addresses. In the ZigBee protocol for instance, when the distributed address

allocation mechanism is used, the whole address space is not used for nodes. Among the 2^{16} available short addresses, only n can be allocated, where n is the maximum number of nodes in the hierarchical tree. n depends on the ZigBee network parameters R_m, C_m and L_m. It is possible to use the unused addresses to represent roles. The same is true for mobiles: it is unrealistic to consider that there are 2^{16} mobile robots in the network. Generally, the number of roles is much lower than the number of entities (either mobiles or nodes) in the network: it is not required to allocate a large number of generic addresses.
The definition of generic addresses is application-dependent, and it depends on the communication requirements. For mobiles, we identified the following requirements.

- Concerning unicast communications (that is, from one mobile to another mobile), two types of requirements can be considered.
  - Role *unicast-next*: the mobile has to communicate with the next mobile. The next time the source mobile meets another mobile, this new mobile becomes the destination of the bundles.
  - Role *unicast-next-geographical*: the mobile has to communicate with the next mobile entering a given area. This role makes use of the fixed-relay to transmit bundles. With this role, the source mobile sends the bundles to the first mobile or fixed-relay it meets. Once the bundles have reached a fixed-relay, it sends the bundles to the first mobile it meets.
- Concerning multicast communications (that is, from one mobile to several mobiles), three types of requirements can be considered. They are a generalization of the roles for unicast communications.
  - Role *multicast-next-all*: the mobile has to communicate with n next mobiles. The next time the source mobile meets another mobile, it decreases the value of n and sends all the bundles to this new mobile. If n is equal to zero, the bundles can be removed from the memory of the source mobile. Otherwise, the bundles are stored for the future encounters. Note that the source mobile has to maintain a list of mobiles it has already met: if a source mobile meets twice the same mobile, the source does not send the same bundles twice.
  - Role *multicast-next-chain*: the mobile has to communicate with a a chain of n next mobiles. This requirement is important when several mobiles are following each other. When the source mobile meets the first mobile of the chain, the source sends all the bundles, and does not keep a copy of them (once they are acknowledged by the gateway of the first mobile of the chain). Then, the first mobile becomes in charge of transmitting the bundles to the $n-1$ next mobiles. Additional care has to be taken in order to avoid loops: a mobile should not receive bundles it has previously received.
  - Role *multicast-next-geographical*: the mobile has to communicate with n next mobiles entering a given area. When a mobile has to transmit k times the bundles and meets a fixed-relay, the mobile transmits all its bundles to this fixed-relay. The fixed-relay becomes in charge of transmitting $k-1$ times the bundles. When a mobile has to transmit k times the bundles and meets another mobile, the former mobile become in charge of transmitting $\lfloor (k-1)/2 \rfloor$, and the latter mobile becomes in charge of transmitting $\lfloor k/2 \rfloor$. In this way, the bundles are disseminated rapidly in the area.

For nodes, several generic roles can be identified.

- the custodian of the mobile,
- one of the gateways of the mobile,
- all the gateways of the mobile,
- one of the sensors (for each type of sensor),
- all of the sensors (for each type of sensor),
- one of the actuators (for each type of actuator),
- all of the actuators (for each type of actuator).

As it can be seen, it is possible to represent several nodes through the same generic role. The duplication of messages from one generic message into several messages is performed in the node that translates the generic addresses, that is in the custodian node.
The generic addressing mechanism requires address translations (one for the generic mobiles, and one for the generic nodes). The translation for generic nodes is always performed at the custodian of the destination mobile, as only this custodian know the exact sensors and actuators deployed on its mobile. The translation for generic mobiles is performed by the custodian of a mobile that depends on the type of generic address.

- Role *unicast-next*. The translation is performed by the custodian of the source mobile.
- Role *unicast-next-geographical*. The translation is performed by the custodian of the source mobile, and potentially by the custodian of the fixed-relay.
- Role *multicast-next-all*. The translation is performed by the custodian of the source mobile.
- Role *multicast-next-chain*. The translation is performed by the custodian of each mobile of the chain, in turn.
- Role *multicast-next-geographical*. The translation is performed by all the custodian of the mobiles involved as intermediate mobiles, as well as by the custodian of all fixed-relays.

5. Conclusion

Several new applications require robots to move within an area, in order to monitor the environment while performing a task. The applications are likely to require that these mobile robots communicate, to avoid obstacles or to improve the monitoring for example. In this context, we have proposed a multi-channel architecture: each robot uses a robot-specific channel for intra-robot communications, and a common channel, called the meeting channel, is used to interconnect mobile robots that are in range of each other.
The use of a meeting channel allows mobiles to perform intra-mobile communications without interferences from neighbor robots, and it allows robots to communicate while some of their nodes save energy by sleeping. When there is no neighbor mobile in the vicinity, the meeting channel can be used to improve the intra-mobile communications. We conducted several simulations in order to quantify the benefits of the meeting channel, and we studied the performance of the network mainly in terms of throughput, delay and path length.
We proposed to use fixed-relays to improve the connectivity of the network, and to allow information to be stored at specific location in the environment. Fixed-relays are always on the meeting channel: this feature allows significant performance benefits.
Finally, we proposed the use of generic addresses in order to identify mobiles and nodes based on their role, rather than based on their address. This concept is especially important when a mobile robot has to communicate with the next mobiles.

6. References

Bahl, V., Chandra, R. & Dunagan, J. (2004). SSCH: Slotted seeded channel hopping for capacity improvement in IEEE 802.11 ad-hoc wireless networks, *ACM International Conference on Mobile computing and Networking*, pp. 216–230.

Bai, F., Sadagopan, N. & Helmy, A. (2003). IMPORTANT: a framework to systematically analyze the impact of mobility on performance of routing protocols for adhoc networks, *IEEE INFOCOM*, pp. 525–535.

Delay-Tolerant Networking Architecture (2006). *Request For Comments 4838*, IETF.

Ferranti, E. & Trigoni, N. (2008). Robot-assisted discovery of evacuation routes in emergency scenarios, *ICRA*, pp. 2824–2830.

Groenevelt, R., Nain, P. & Koole, G. (2005). The message delay in mobile ad hoc networks, *Performance Evaluation* 62(1-4): 210–228.

Hadid, N., Guitton, A. & Misson, M. (2010). Exploiting a meeting channel to interconnect 802.15.4-compliant mobile entities: discovery and association phases, *IEEE International Symposium on Computers and Communications*.

Hogg, R. W., Rankin, A. L., Roumeliotis, S. I., McHenry, M. C., Helmick, D. M., Bergh, C. F. & Matthies, L. (2002). Algorithms and sensors for small robot path following, *IEEE International Conference on Robotics and Automation*, pp. 11–15.

IEEE 802.15 (2006). Part 15.4: Wireless medium access control (MAC) and physical layer (PHY) specifications for low-rate wireless personal area networks (WPANs), *Standard 802.15.4 R2006*, ANSI/IEEE.

Jovanovic, M. D. & Djordjevic, G. L. (2007). TFMAC: Multi-channel MAC protocol for wireless sensor networks, *International Conference on Telecommunications in Modern Satellite, Cable and Broadcasting Services*, pp. 23–26.

Kyasanur, P. & Vaidya, N. H. (2005). Routing and interface assignment in multi-channel multi-interface wireless networks, *IEEE Wireless Communications and Networking Conference*, pp. 13–17.

Lambrou, T. P. & Panayiotou, C. G. (2009). Collaborative area monitoring using wireless sensor networks with stationary and mobile nodes, *EURASIP Journal on Advances in Signal Processing* .

Nasipuri, A., Zhuang, J. & Das, S.-R. (1999). A multi-channel CSMA MAC protocol for multi-hop wireless networks, *IEEE Wireless Communications and Networking Conference*, pp. 1402–1406.

Pathmasuntharam, J. S., Das, A. & Gupta, A. K. (2004). Primary channel assignment based MAC (PCAM) - a multi-channel MAC protocol for multi-hop wireless networks, *IEEE Wireless Communications and Networking Conference*, pp. 1110–1115.

Propagation data and prediction method for the planning of indoor radio communication systems and local area networks in the frequency range of 900 MHz to 100 GHz (2005). *Recommendation ITU-R P 1238-4*, ITU.

Ramakrishnan, M. & Vanaja Ranjan, P. (2009). Multi channel MAC for wireless sensor networks, *International Journal of Computer Networks and Communications (IJCNC)* 1(2): 47–54.

Seo, M., Kim, Y. & Ma, J. (2008). Multi-channel MAC protocol for multi-hop wireless networks: handling multi-channel hidden node problem using snooping, *IEEE Military Communications Conference*, pp. 1–7.

So, H. W. & Walrand, J. (2007). McMAC: A multi-channel MAC proposal for ad-hoc wireless networks, *IEEE Wireless Communications and Networking Conference*.

Wu, S.-L., Lin, C.-Y., Tseng, Y.-C. & Sheu, J.-P. (2000). A new multi-channel MAC protocol with on-demand channel assignment for multi-hop mobile ad hoc networks, *International Symposium on Parallel Architectures, Algorithms and Networks*, pp. 232–237.

Zhang, J., Zhou, G., Huang, C., Son, S. H. & Stankovic, J. A. (2007). TMMAC: an energy efficient multi-channel MAC protocol for ad hoc networks, *IEE ICC*, pp. 3554–3561.

Zhao, W., Chen, Y., Ammar, M., Corner, M., Levine, B. & Zegura, E. (2006). Capacity enhancement using throwboxes in DTNs, pp. 31–40.

16

Distance Measurement for Indoor Robotic Collectives

Mihai V. Micea, Andrei Stancovici and Sînziana Indreica
Politehnica University of Timisoara
Romania

1. Introduction

Location monitoring is a common problem for many mobile robotic applications covering various domains, such as industrial automation, manipulation in difficult areas, rescue operations, environment exploration and monitoring, smart environments and buildings, robotic home appliances, space exploration and probing.
A key aspect of localization is inter-robot distance measurement. In this chapter we consider the problem of autonomous, collaborative distance measurement in mobile robotic systems, under the following set of design and functional constraints:

a. indoor operation,
b. independence of fixed landmarks,
c. robustness and accuracy,
d. energy efficiency,
e. low cost and complexity.

This work significantly extends and updates the results previously published in (Micea et al., 2010). We present and discuss some of the most relevant state of the art techniques for robot distance estimation. Next, we introduce a framework for collaborative inter-robot distance measurement along with a procedure for accurate robotic alignment. The proposed alignment algorithm is based on evaluating and comparing the strength of ultrasonic signals at different angles, processing (filtering) the measured data and ensuring a good synchronization during the process. Further on, we present the CTOF (Combined Time-of-Flight) method for distance measurement, which brings significant improvements to the classical TOF technique, and we show how this new technique meets the above specified design constraints. Some of the most interesting test and evaluation results are presented and discussed. The experimental data show how the distance estimation accuracy can be increased by applying the Kalman filter algorithm on repetitive measurements. The final remarks and the reference list conclude this chapter.

2. Current techniques for robot distance estimation

The problem of inter-robot distance measurement and location monitoring is considered of key importance in the field and, consequently, a large number and variety of methods have been proposed and studied in the literature. For instance, the GPS system (Ohno et al., 2004; Reina et al., 2007) and landmark-based solutions such as the Cricket Indoor Location System (Cricket Project, 2005; Priyantha, 2005) are well established in the field. On the other hand, they do not comply with the constraints specified in the previous section (i.e. independence of fixed landmarks). In this section we discuss some of the most prominent techniques which can be used for indoor robotic collectives.

Time-Difference-of-Arrival (TDOA) measures the distance between two points by using two different types of signals (usually radio and acoustic) which cover the route connecting the two points with different speeds. To illustrate this technique, consider two points, A and B, located at distance *d* from each other. At a time instance, the transmitter from A sends simultaneously the signals S_1 and S_2, which cover the distance *d* at the speeds v_{S1} and v_{S2}, respectively. If, for example, $v_{S2} < v_{S1}$, then signal S_2 arrives at the receiver B after S_1, with a delay which depends on the distance *d*. This delay is measured at the destination point B and the value of *d* is consequently derived. Cricket Indoor Location System uses TDOA to measure the distance to the reference points. The system consists of several landmark transmission devices, depending on the size of the desired coverage area (at least three modules) and one or more mobile devices that play the role of receptors. In most cases, the transmission devices are attached to the upper part of the room so as to cover a large portion or the entire room. The reception devices are attached to robots, located on the floor. As shown in (Priyantha, 2005), the system relies on two types of signals to calculate distances: a RF (radio) signal and an ultrasonic signal. The radio signal is 10^6 times faster than the ultrasonic signal, and the distance is calculated by applying the principle of TDOA to the difference of the two propagation periods. The localization of mobile robots through this system is made at an accuracy of $1 \div 3$ cm. Similarly, the system presented in (Fayli & Kleeman, 2004) solves the localization problem based on four transmitters as fixed reference points and a wireless receiver.

Another well known technology used in robotics to calculate distance is based on infrared (IR) sensors. There are several types of IR sensors, each varying according to their parameters (e.g. maximum range and accuracy) and price. In comparison to ultrasonic devices, the IR sensors are cheaper and use light, which is much faster than the acoustic signal. They have nonlinear characteristics which depend on the surface reflectance of the objects. Based on measuring the intensity of light reflected by a target, the IR sensors can calculate the distance to it. This technique is presented and discussed in several works, including (Novotny & Ferrier, 1999; Ha & Kim, 2004; Mohammad, 2009). Hagisonic StarGazer (Hagisonic, 2009) is a location system for mobile robots, based on the analysis of infrared rays which are reflected by a passive landmark with a unique ID. The system works as follows:

1. The IR transmitter is located on the robot. It transmits infrared beams to the fixed landmark attached to the ceiling of the room.
2. The infrared rays are reflected from the landmark and reach the Stargazer, mounted on the robot.
3. Stargazer contains a CMOS camera able to estimate the angle of incidence of the reflected IR waves and the distance between the robot and the landmark.
4. Based on the angle of incidence and on the distance to the landmark, the position of the robot in the room can then be obtained through geometric techniques.

The advantage of such a system is its accuracy, which, according to (Hagisonic, 2009), can reach approximately 2 cm. The system can carry out 20 measurements per second. Its disadvantage is the high price and the reduced coverage area, which ranges from 2.5 to 5 m.

A set of radio-based methods use the power of the received signal to estimate the distance to the source. The mathematical model of the emitted signal power is given in (Fuicu et al., 2009) as:

$$\frac{P_t}{P_r} = \frac{4\pi d}{\lambda^2} = \frac{(4\pi f d)^2}{c^2} \tag{1}$$

where P_t and P_r are the signal power at the emitter and at the receiver, respectively, d is the propagation distance, λ and f are the carrier wavelength and frequency, respectively, and c, the speed of light. The accuracy of such systems though, is around 2-3 m. Another similar technique, based on modeling the signal power for the ZigBee (IEEE 802.15.4) protocol (ZigBee Standards Organization, 2007), is presented in (Grossmann, 2007).

An indoor GPS system, presented in (Kim et al., 2006), consists of two receivers as fixed reference points and a transmitter which uses both ultrasonic and RF signals. The receivers estimate the distance to the transmitter based on the delay between the received RF signal and the ultrasound waves. The two resulting distances are then used to calculate the location of the transmitter, through geometrical formulas. A Linear Kalman Filter is also used to minimize the errors and noise occurring in the measurements of the ultrasound signal.

GPS systems usually calculate the distance between a receiver and multiple transmitters based on the difference in the time-of-flight of the received signals (the TOF method). Through the TOF method, more precise results can be obtained. However, TOF is influenced by the synchronization accuracy of system, environment temperature or other factors which could yield calculation errors. As a result, filtering of measurement values is frequently used. A common solution is the Linear Kalman Filter, as presented in (Kim et al., 2006; Ko et al., 2008; Welch & Bishop, 2006). Other approaches use Bayesian filters (Fox et al., 2003), which estimate the state of a probabilistic dynamic system from observations drowned in noise. Using statistic techniques, they operate in a deterministic manner and are suitable for systems with multiple sensors with different characteristics.

The Building Positioning System (Reynolds, 1999), determines the position of a mobile device by receiving radio signals from fixed devices. These are designed to transmit radio signals in a manner which is similar to the operation of the Cricket or GPS modules. Compared with the GPS, this system uses a much lower frequency, making the radio waves propagate with a relatively low attenuation. The system requires only 4 fixed transmission antennas attached to four different corners of the building. The accuracy of such a system is about 5 cm.

Image processing methods are also widely used, many of them using passive cameras. A microcontroller drives a motor to focus a target image located at a certain distance from the sensor. Based on several parameters, like motor position or lens properties, the distance to the target object can be determined. Some systems are based, for instance, on visual information from a 360 degree camera (Tamimi et al., 2006). Such systems must be trained before being used, by capturing representative images from the environment and associating them with the corresponding locations.

3. Example framework for collaborative inter-robot distance measurement

The proposed distance measurement method and inter-robot alignment algorithm have a common set of requirements for the target robotic system. Such a framework, which has been used to implement and test the proposed techniques, is the CORE-TX platform (Cioarga et al., 2006).

CORE-TX (COllaborative Robotic Environment - the Timisoara eXperiment) is designed to provide theoretical and applicative support for the study of intelligent sensor networks and robotic collectives. Its architecture is structured on three main layers (see Fig. 1):

1. Perception and Operation Layer, consisting mainly of autonomous microsystems with embedded intelligence, called WITs (Wireless Intelligent Terminals),

2. Collaborative Communication Layer, based on ad-hoc wireless data communication techniques (currently, the basic support is provided by the ZigBee protocol), and
3. Background Control and Supervision Layer, with a central node called BRAIN (Background Robotic Activity Induction Node).

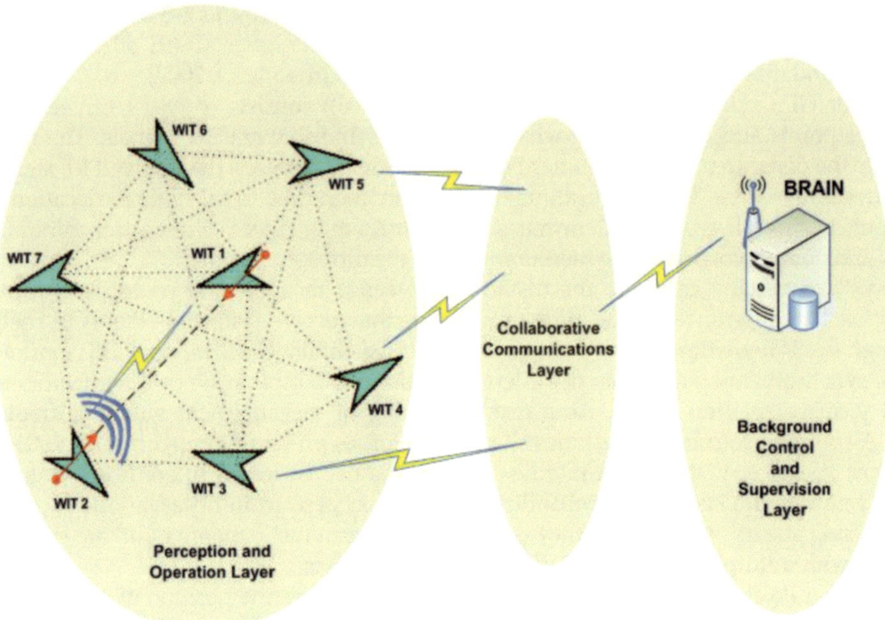

Fig. 1. General architecture of the CORE-TX system

3.1 General architecture of the robotic elements

The WIT elements may have perception functions (intelligent sensors), operating functions (autonomous mini-robots), or combined. They have been designed using a modular approach (Fig. 2) which specifies a motherboard (the Base Processing Module), interconnected through a system bus to a set of specialized daughter boards. Such daughter boards are the Power Management Module, the Perception Module and the Communication Module. The additional Support and Operation Module transforms the WIT, from a static intelligent sensor, into an autonomous mini-robot.
Currently, the WIT communication board uses the XBee wireless module (Digi International, 2009), which is based on the Zigbee protocol. This module provides a unique 64-bit ID. Other features are: size of 2 cm x 3 cm, operating range of up to 30 m indoor and up to 90 m outdoor, with a maximum consumption of 50 mA at a voltage of 3.3 V. Communication uses the 2.4 GHz radio frequency band with 16 channels.

3.2 Design of the perception module

The schematic design of the Perception Module is shown in Fig. 3. The main processor is the ARM7-based LPC2294 (NXP Semiconductors, 2008) which runs the Hard Real-Time Operating Kernel, HARETICK (Micea et al., 2006), for predictable operation. Another important part of the module is the coprocessor ATxmega128A1 (Atmel Corporation, 2010), used for fast, periodic data acquisition and processing operations. It was also chosen for its high ADC performance and multiple resources, such as 16 ADC channels, 4 DMA channels, 8 timers, 24 PWM channels, 4 SPI interfaces and a large set of I/O ports.

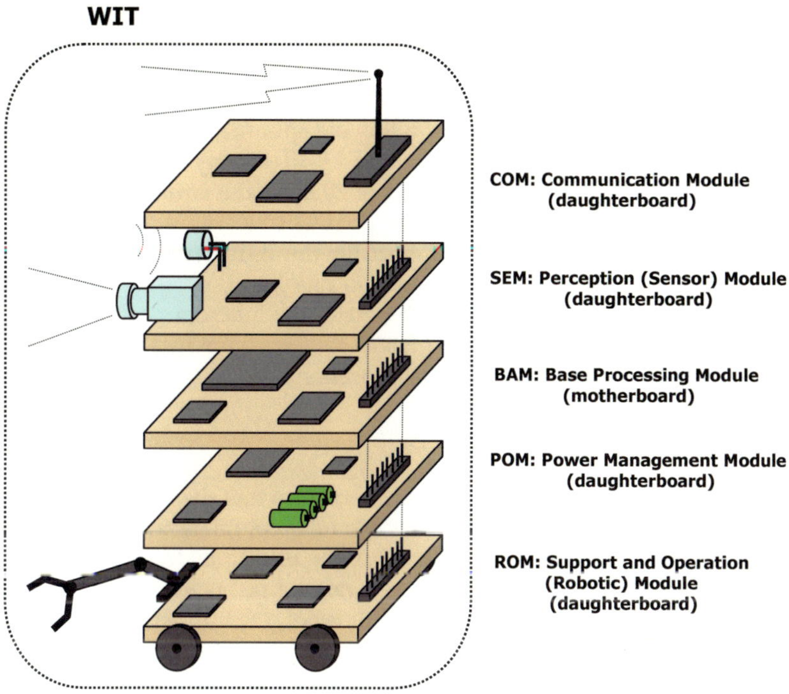

Fig. 2. Diagram of the Wireless Intelligent Terminal

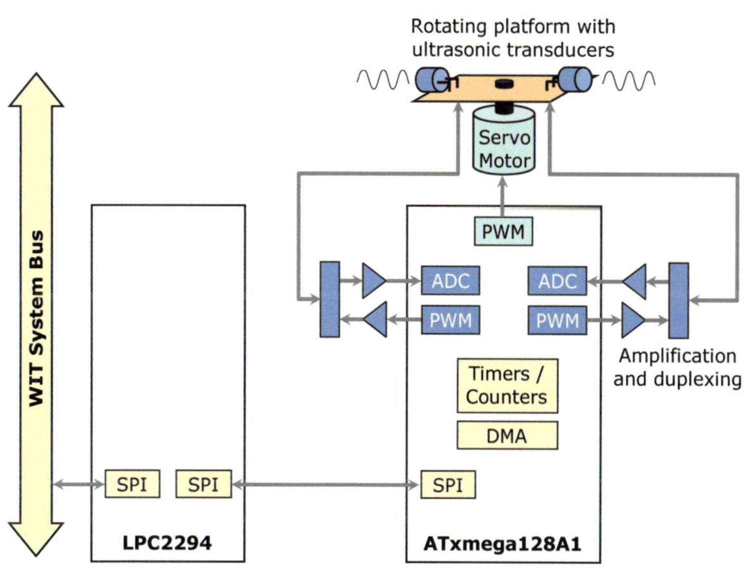

Fig. 3. Schematic of the WIT Perception Module

Two similar transducers are used both for transmitting and for receiving ultrasonic signals. The BPU-1640IOAH12 device (Bestar Electronics, 2006) has been selected, due to its convenient features, which include low cost, bidirectional operation, nominal frequency of 40 kHz, and maximum input voltage of 120 Vpp. Signal duplexing at the transducer level (bidirectional operation) has been implemented using SI4808DY MOSFET circuits.
To perform a fast alignment process, the two ultrasonic transducers are mounted back to back at 180 degrees on a rotating platform, which is driven by a servo motor. The motor is a TowerPro SG-50 (Tower Pro, 2008) with the following specifications: weight 5 g, dimensions 21.5 × 11. 7 × 25.1 mm, speed 0.1 s/60 degrees (at 4.8 V), supply voltage 4 ÷ 6 V. The servo motor is driven by a PWM signal with a period of 20 ms and a variable duty cycle. Rotation is between 0 (minimum pulse duration) and 180 degrees (maximum pulse duration).
Choosing the design based on a turret support for the ultrasonic transducers has several advantages when compared to other designs. On one hand, avoiding the rotation of the entire robots during the alignment process eliminates the inherent positioning errors, while also lowers the power consumption of the system. Other advantages of this solution are the increase of the alignment accuracy at a higher process speed.

4. Inter-robot alignment algorithm

To obtain correct results, the proposed techniques require that the pair of robots performing the distance measurement procedure must successfully complete the alignment algorithm. Correct alignment means the sensing devices (i.e. the ultrasonic transducers) of the robots are facing each other, as close as possible to the straight line between them (see Fig. 4). This procedure also provides key angle values of each robot position and orientation related to the local reference system (Fig. 5):

- a_{12} is the angle between the *orientation axis* (Ox_1) of the first robot (W_1) and the direct line between the two robots.
- φ_1 is the angle defined by the Ox_1 axis and the ultrasonic sensor axis of W_1.
- a_{21} is the angle between the *orientation axis* (Ox_2) of the second robot (W_2) and the direct line between W_1 and W_2.
- φ_2 is the angle defined by the Ox_2 axis and the ultrasonic sensor axis of W_2.

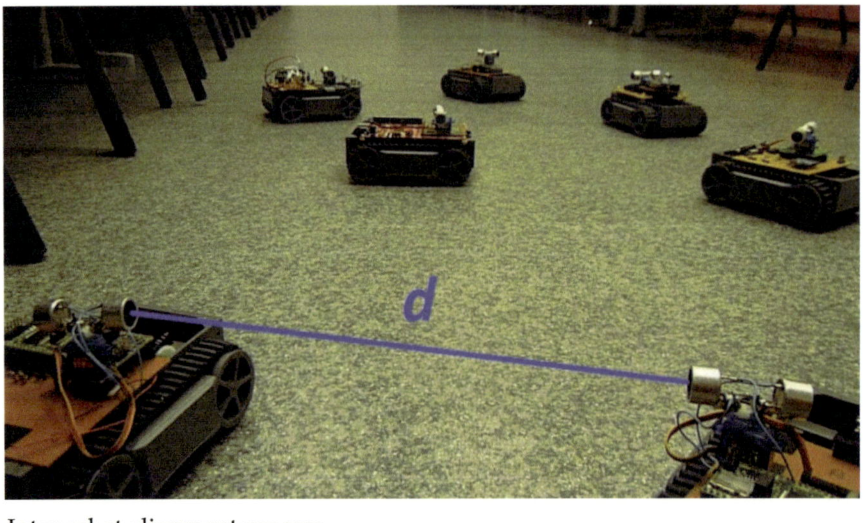

Fig. 4. Inter-robot alignment process

Thus, the ideal alignment situation is when $a_{12} = \varphi_1$ and $a_{21} = \varphi_2$. In this case, the alignment error is 0.

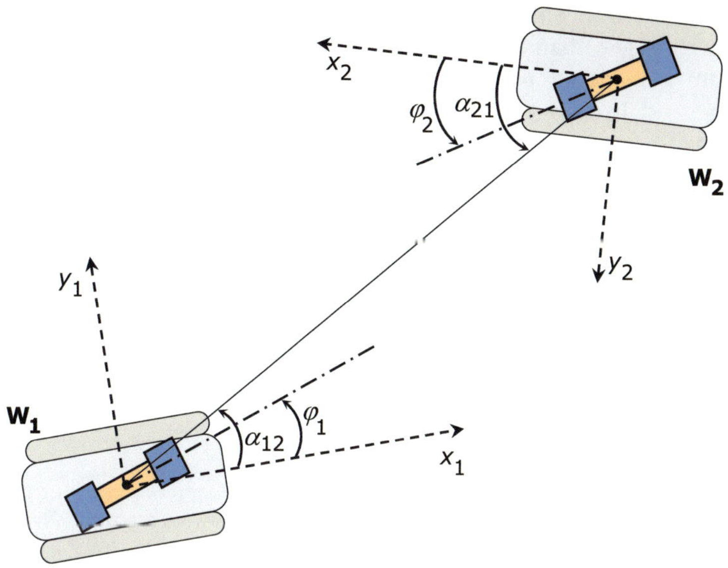

Fig. 5. Key angles on the alignment process

4.1 Description of the algorithm

The alignment procedure uses the wireless communication interfaces of the robots to enable the two corresponding peers exchange the required commands and messages, and is based on the continuous measurement of the Sonar acoustic intensity. It is initiated and conducted by one of the robots, which acts as the master (W_M), while the other robot, the slave (W_S), executes the commands received from the master through the wireless link. The master will operate in the Sonar receive mode and the slave in Sonar transmit mode.

The procedure is based on the high directivity of ultrasonic waves used by the Sonar. As the two robot turrets rotate, the master calculates the average strength of the ultrasonic signal received from the slave, at each rotation step of 1 degree. If W_M senses this average signal strength has increased from the previous rotation step, it continues the procedure until a decrease is encountered. Then, the two robots will change the rotation directions of their turrets to return to the previously detected maximum.

Fig. 6 shows all steps of the alignment algorithm. The process starts with the reading of the ADC results. The two decision blocks labeled "done" refer to extracting a predetermined number of samples, respectively, re-reading the values stored in memory. Next, the measured values are optimized by finding the peak amplitude of each period, applying a Kalman filter and comparing the results to each other to determine a global maximum. This global maximum is further used as the amplitude of the signal in the next steps.

The block labeled "pre_align==0?" determines whether it is the last alignment phase of the process. If not, the algorithm determines the trend of the signal: if two consecutive increases are detected, "flag_inc" is set and if there has been a previous decrease, the current stage of the algorithm is finished. On the other hand, if there are two consecutive decreases, the

direction of rotation changes and, if "flag_inc" is set, "flag_dec" also will be set. This phase also determines and updates the highest value of the signal reached so far.
In the last phase of the alignment, the algorithm tries to trace back the position with the highest measured values. Thus, if it detects an increase of the measured signal and the current value is larger than or equal to the stored peak, the process ends and the transducers are considered aligned. If, instead, a decrease is detected, the rotation direction is changed. In all cases, if the process doesn't end, the flow of the algorithm goes back to ADC readings.

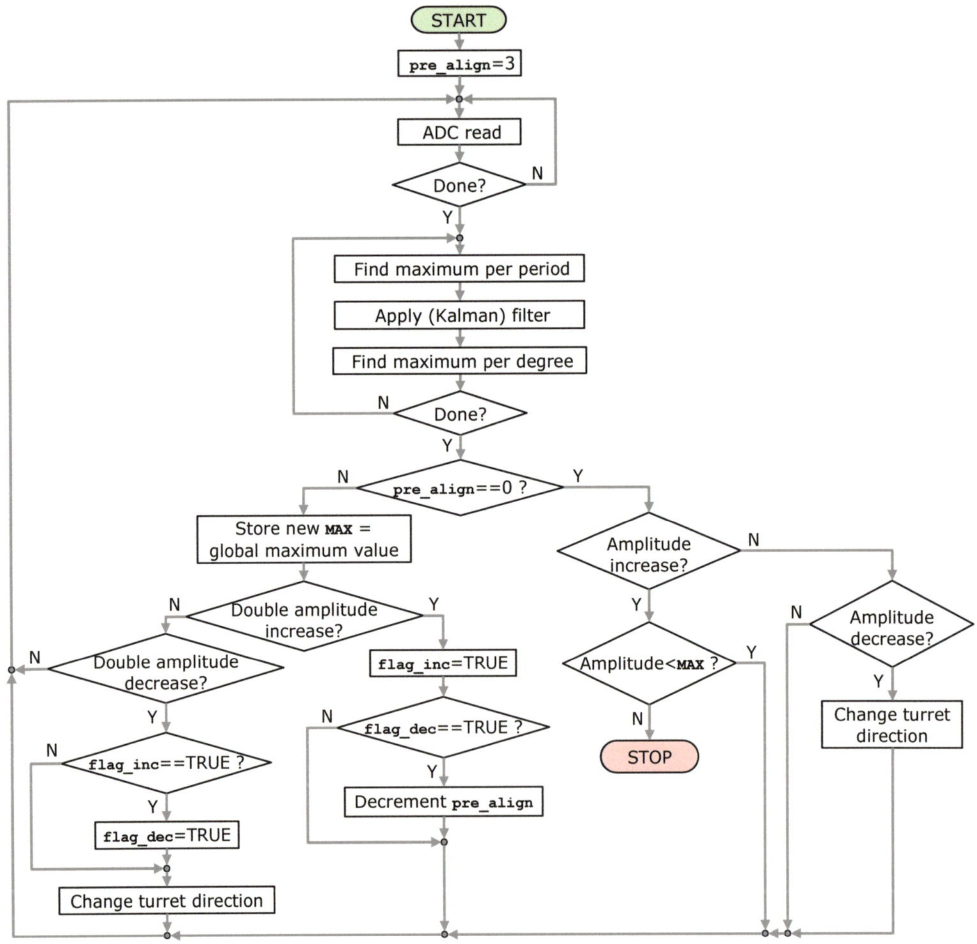

Fig. 6. Robotic alignment algorithm

4.2 Performance evaluation of the alignment algorithm

To evaluate the robotic alignment procedure, a custom simulation application has been developed using the WPF (Windows Presentation Foundation). Two cases have been considered:

1. the ultrasonic transducers are fixed on the robot case and thus the robots rotate themselves (using the wheels) during the alignment process, and

2. the current design in which the robots are equipped with turrets, rotated at 180 degrees by a servo motor.

The data collected was used to improve the alignment algorithm.

To simulate the ultrasonic signal transmission we considered an idealized representation - an isosceles triangle which represents the longitudinal cross-section of the propagation cone. This triangle will represent the "active space" of the robot, and it has an opening of 60 degrees. In this case, the power of the signal (its amplitude) varies with the distance from the bisector of the considered angle. In other words, a perfect alignment takes place when the bisectors of the receiver and transmitter concur (Fig. 7).

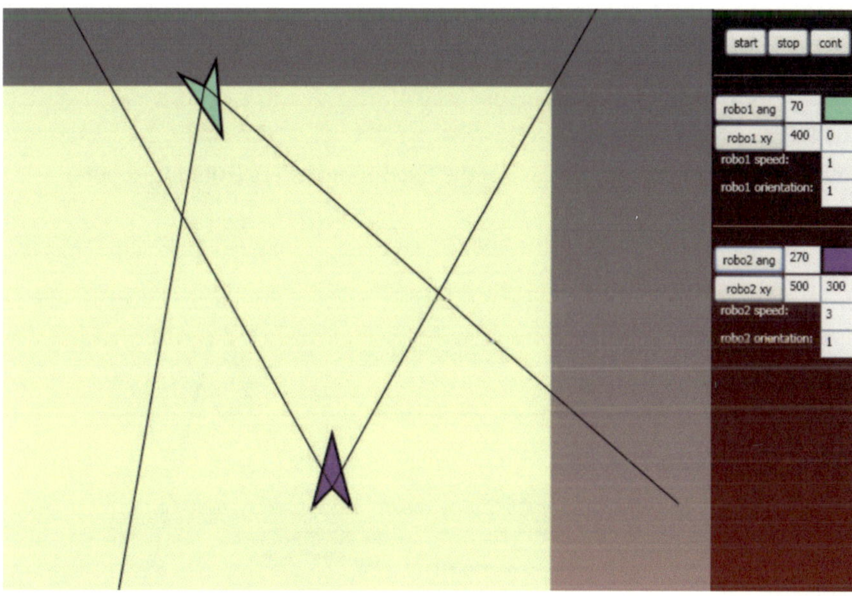

Fig. 7. Robotic alignment simulation tool

Because the distance between the robots is constant throughout the alignment process, its importance in the real case lies in the fact that the receiver may not "read" anything from the transmitter if it is too far away. Also, when further apart, the visibility angle broadens and more precise alignments may occur. In the application, we consider the distance a constant parameter and simulate the most usual scenarios, i.e. where the robots are not too far apart for the angle to broaden.

To be able to determine the position of the robots (of the sensors) relative to each other, we have to calculate the signal strength from both directions and then multiply the results. For the correction of the final alignment, the maximum value of the signal is determined. With the simulator, this maximum value is calculated from the initial conditions. In the real case, the best value must be searched for the signal currently measured, because the distance between robots is unknown.

Fig. 8 and Fig. 9 show the elapsed time for the alignment, versus the initial angles of the robots, at various ratios of the rotation speed (calculated as [rad/(s·10³)]). The thick lines mark the best two series, which minimize the alignment duration, at a corresponding rotation speed ratio.

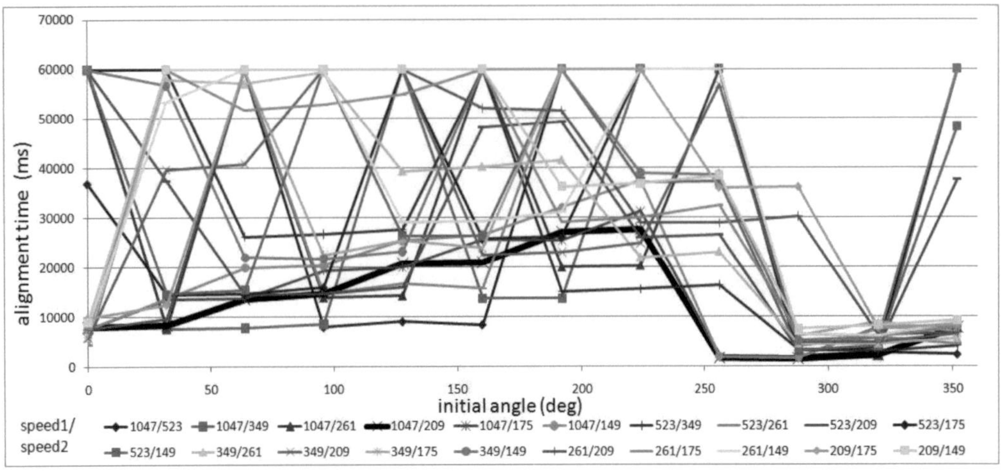

Fig. 8. Inter-robot alignment in the case of the fixed transducers (rotating robots)

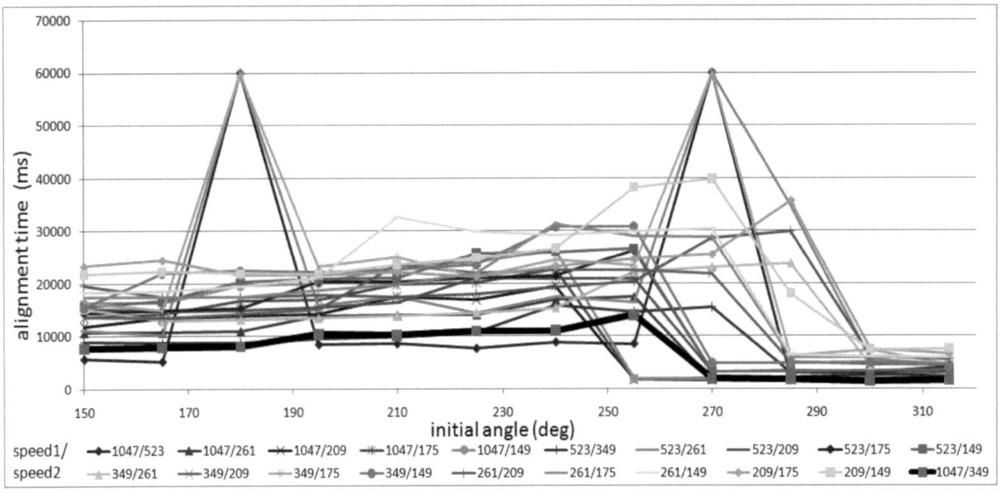

Fig. 9. Inter-robot alignment in the case of the transducers mounted on the turret

For the design without turrets, most of the rotation speed ratios have cases in which the alignment time exceeds the limit imposed (60 s). They are shown on the graph as being equal to the limit. For the turret case, the sensors have to scan only half of the circle and, consequently, they find each other more quickly.

In many cases of small rotation ratios, the sensors cannot find each other, because they are "following" each other closely behind. On the other hand, for increased values of the rotation ratios, the entire process is becoming slow. Considering these facts, the simulations found an optimal rotation ratio of 1047/209 (i.e. 5/1) for the case in Fig. 8 and 1047/349 (i.e. 3/1) for the rotating turret (Fig. 9). Taking also into consideration the alignment time, its maximum value is about 27 s in the first case and 14 s for the second. These results prove that by using the turret design, the alignment time can be almost reduced to half.

5. Distance measurement with the CTOF method

CTOF, Combined Time-of-Flight, is based on the TOF technique and involves two robots. Although a little more complicated, CTOF has several advantages over the MTDOA method, proposed in (Micea et al., 2010). Thus, the CTOF procedure does not require an additional robot (the third one) to coordinate the distance calculation. It also does not depend on the delays implied by the wireless communication interfaces of the robots.

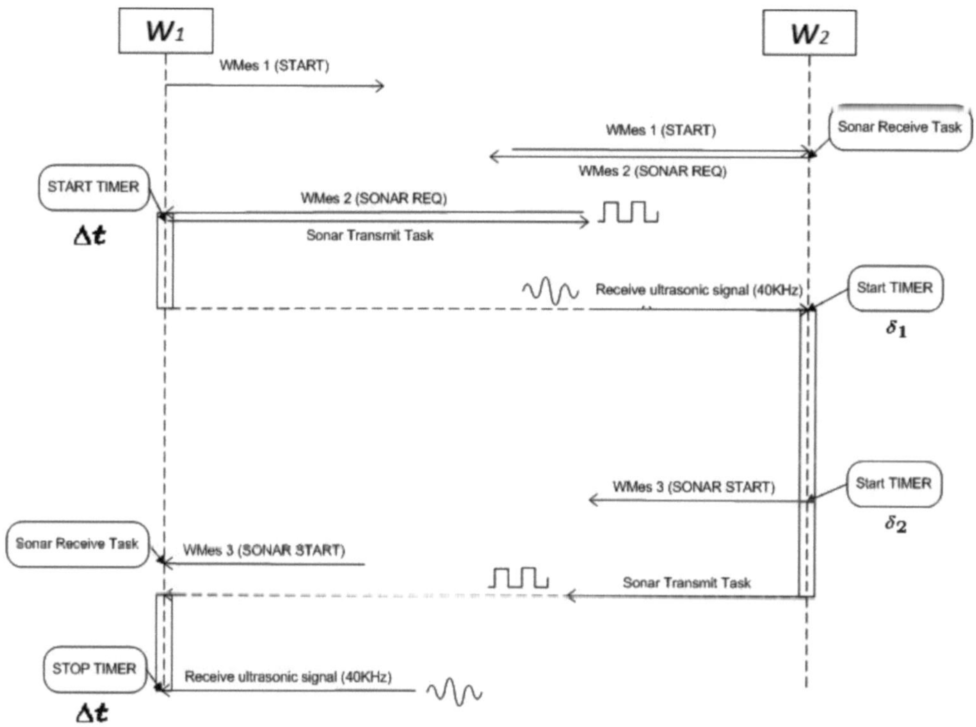

Fig. 10. Distance measurement with the CTOF method

Fig. 10 depicts the CTOF technique. Robot W_1 initiates the procedure by sending a "START" wireless message (abbreviated "WMes" in Fig. 10) to its peer, W_2. The latter acknowledges the start of its part of the procedure with the "SONAR REQ" message, while simultaneously launching its own Sonar Receive Task. As a response to the second message, W_1 starts the Sonar Transmit Task and activates the timer which will count the elapsed time of the entire procedure, Δt. Upon receiving the ultrasound signal, W_2 activates a delay timer with a predefined value, δ_U, which is empirically determined to cover the total duration of the ultrasonic transmission from W_1. After the δ_U delay, W_2 sends a "SONAR START" message to W_1 and starts a second timer, with a value δ_W empirically established to cover the maximum communication delay over the wireless link and the corresponding interfaces. When W_1 receives the "SONAR START" message, it launches its Sonar Receive Task. After the δ_W timer expires, W_2 starts its Sonar Transmit Task and sends the corresponding ultrasonic signal towards W_1. Finally, when W_1 receives the signal, it stops the timer to produce the Δt period. As a result, the Δt period contains the two predefined delays, δ_U and δ_W, and twice the propagation delay of the ultrasound signal, from W_1 to W_2 and backwards. Based on this ultrasound propagation delay, the distance between the two robots can be derived:

$$d = \frac{c_{\text{air}}\left(\Delta t - \delta_U - \delta_W\right)}{2} \tag{2}$$

where c_{air} = 343.4 m/s is the velocity of acoustic waves in air at room temperature and at normal pressure. When considering the threshold-based detection method of the received ultrasonic bursts and the fact that the ultrasonic measurements are not perfectly linear, an additional calibration offset is needed for the distance formula in (2):

$$d = \frac{c_{\text{air}}\left(\Delta t - \delta_U - \delta_W - \theta_{UC}\right)}{2} \tag{3}$$

where θ_{UC} is the ultrasonic signal calibration offset and has an experimentally determined value (in our case studies, θ_{UC} = 290 μs).

6. Experimental results

An extensive set of experiments have been conducted in the DSPLabs using the robotic system of the CORE-TX platform. The experimental setup consisted in several mobile robots, out of which two of them were randomly chosen to perform the alignment and the distance calculation procedures for each experiment. The robots have been placed at a distance ranging from 100 mm to 3000 mm and, for each 100 mm in this range, a set of over 50 pairs of measurements have been performed. Before each measurement, the robots have been positioned in random directions with respect to each other.

Since the proposed techniques are based on Sonar and are specifically designed for indoor measurements, the experiments, evaluations and results consider normal room values for the air parameters (such as temperature, humidity, pressure, etc.). These parameters could otherwise influence the speed of ultrasonic waves used in equations (2) and (3).

Fig. 11 shows several periods for the raw received ultrasonic signal, at the output of the LM6134 amplification circuit. Further on, the values of the signal peaks (which occur every 25 μs) are extracted and interpreted as the received Sonar signal.

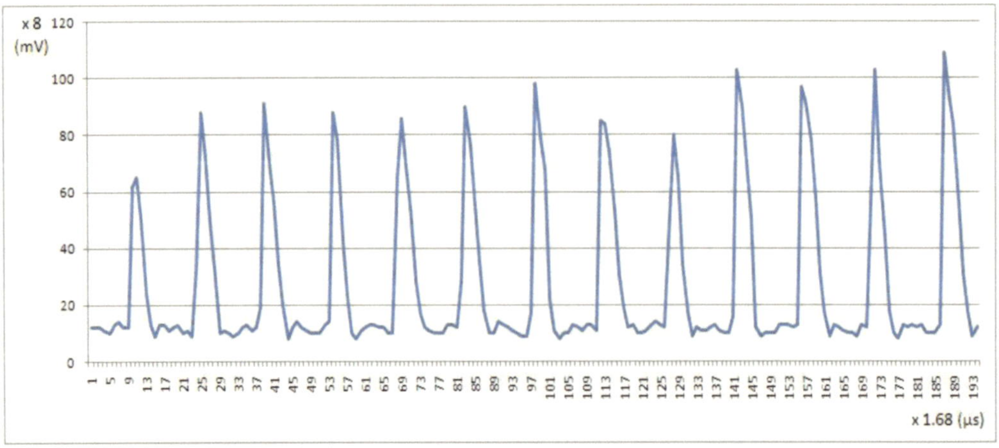

Fig. 11. Amplitude of the received ultrasonic signal, after amplification

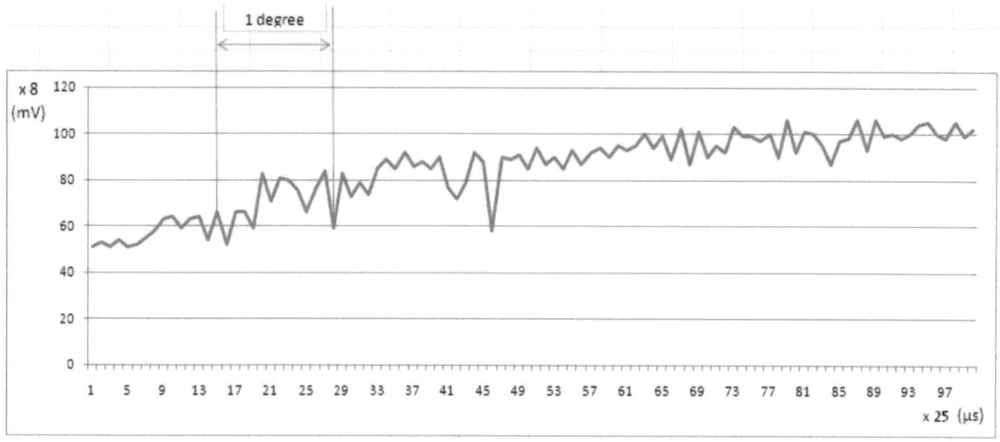

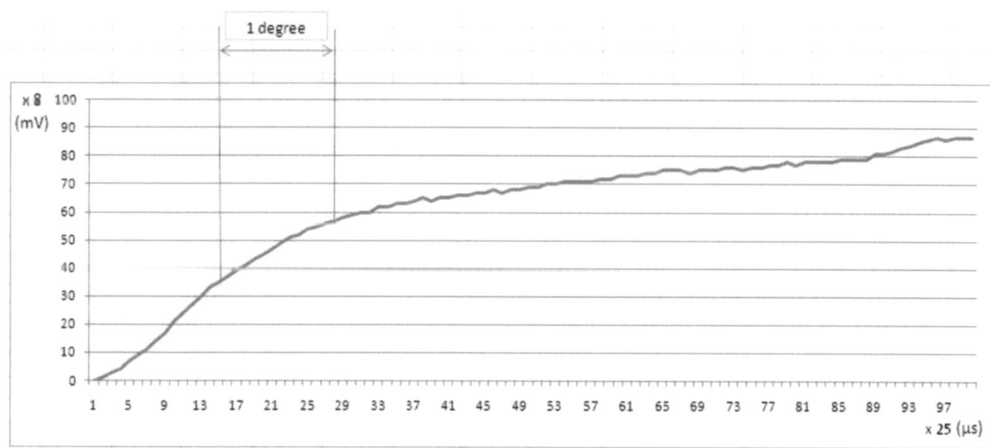

Fig. 12. Received signal without filtering (up) and after the Kalman filtering (down)

As depicted in the upper diagram of Fig. 12, the received Sonar signal contains a relatively significant amount of noise, which can be reduced by applying a Kalman filter. As shown in the lower diagram of Fig. 12, the filter significantly reduces the random variations of the signal, while retaining its trend. Even with the filtering process, accidental variations of the signal can occur. Therefore, empirical thresholds have been specified to establish when the signal amplitude changes. Setting the threshold values is a key calibration step of the alignment procedure.

The results for a full scan of the ultrasonic transducer, i.e. a rotation of 180 degrees, are presented in Fig. 13, both with and without applying the Kalman filter. The target robot is detected at around 93 degrees, with a maximum of the Sonar signal occurring at approximately 116 degrees.

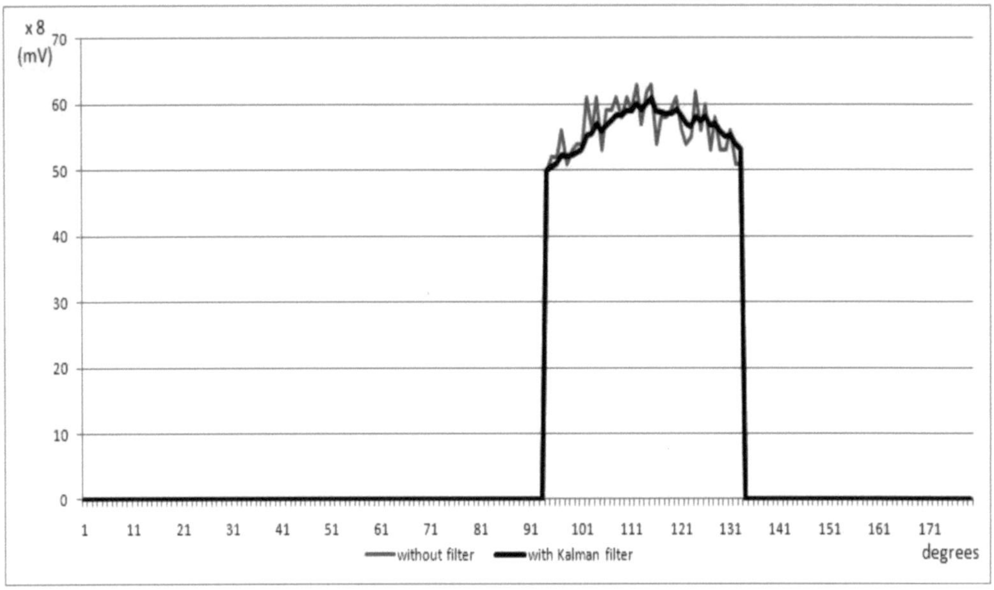

Fig. 13. Reading of the Sonar signal for a 180 degree rotation (scan) of the transducer

Some of the most interesting experimental results for the alignment process are depicted in Fig. 14. The alignment accuracy varies even for the same distance between the robots, due to the frequent changes of the turrets rotation and to the different time instances the transmitted ultrasonic signal is acquired. With few exceptions, the alignment correction angle remains lower than 10 degrees. The results presented in Fig. 14 also show the improvement of the alignment accuracy with the increase of the distance between the two robots.

Table 1 presents the distance measurement results for the proposed CTOF method. The maximum absolute error is 4.8 cm, when the robots are 3000 mm apart. The result and error analysis of the CTOF procedure show that, after the necessary calibrations, the measurement characteristics are linear and follow very closely the real distance. It proves also to be independent from the random delays introduced by the wireless modules.

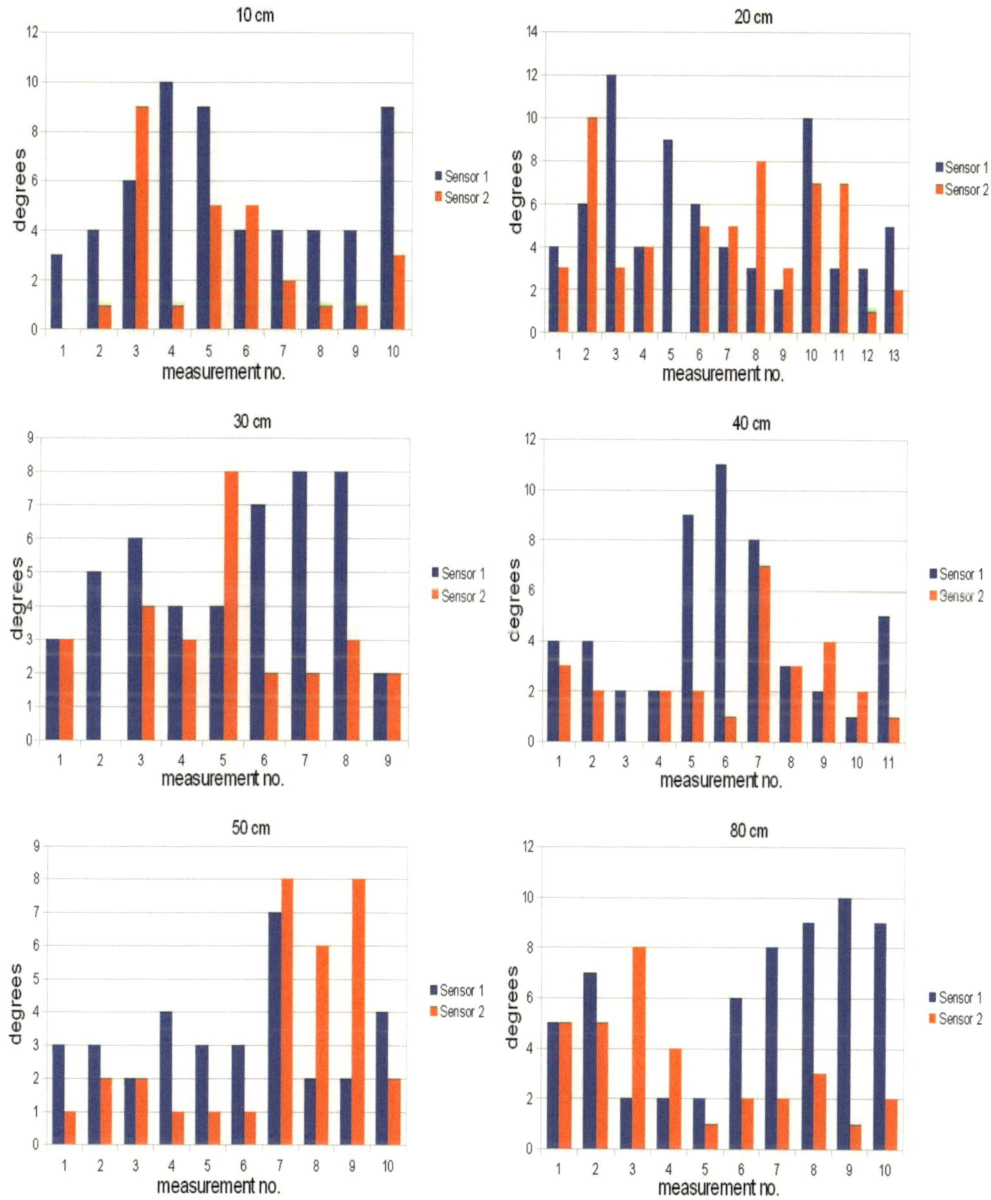

Fig. 14. Maximum variation of the correction angle for experimental alignment procedures

Table 2 presents distance measurements with the CTOF method, after applying the Kalman filter to the results. As we can see, the results are very good in terms of accuracy. The disadvantage is the time of measurement, which increases for repetitive measurements. Some comparative distance evaluation results with and without filtering the data are depicted in detail in Fig. 15. From the experiments we conclude that the system provide optimal results for approximately 10 repetitive measurements.

Real Distance [mm]	CTOF Measured Distance [mm]			Procedure Duration [μs]
	Min	Average	Max	
100	92	96	99	20849
200	199	201	207	21461
300	298	300	303	22037
400	401	404	410	22643
500	504	508	515	23249
600	604	607	612	23825
700	700	706	710	24402
800	803	807	813	24990
900	906	911	916	25596
1000	1013	1019	1026	26225
2000	2024	2033	2043	32130
3000	3018	3031	3047	37943

Table 1. Distance measurement results for the CTOF method

Real Distance [mm]	CTOF Measured Distance with Filtering [mm]			Procedure Duration [ms]	Error reduction with Filtering [%]
	Min	Average	Max		
100	98	100	101	208÷688	66
200	198	200	205	215÷708	29
300	298	300	301	220÷727	62
400	397	399	401	226÷747	50
500	498	500	508	232÷767	63
600	598	599	601	238÷786	68
700	697	699	703	244÷805	19
800	796	800	802	250÷825	56
900	897	899	901	256÷845	56
1000	997	1001	1004	262÷865	49
2000	1997	2002	2006	321÷1060	40
3000	2993	3000	3006	379÷1252	57

Table 2. Distance measurement results for the CTOF method with Kalman filtering

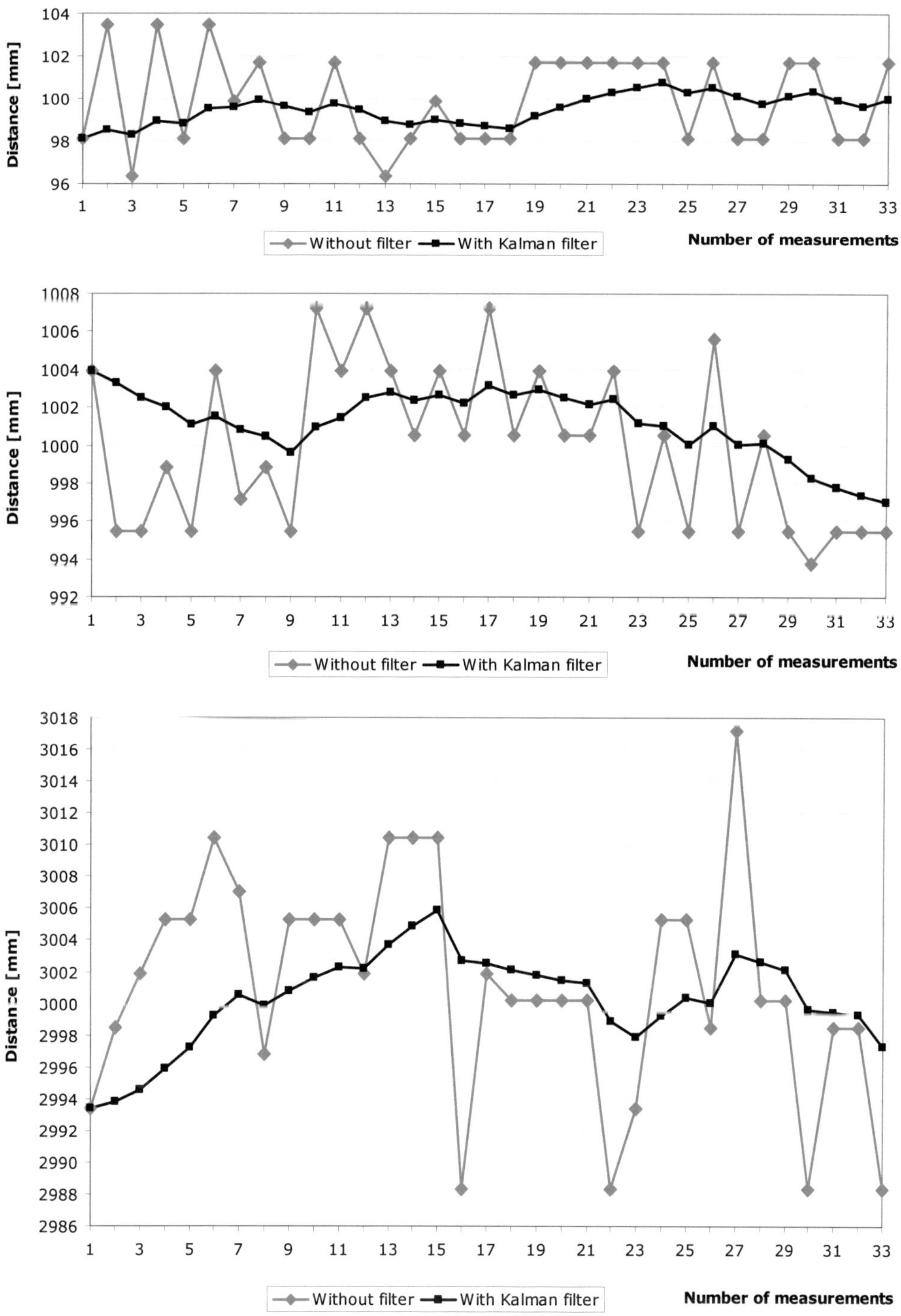

Fig. 15. Repetitive CTOF distance measurements with and without Kalman filtering, for various distances between robots (100 mm - top, 1000 mm - middle, and 3000 mm - bottom)

7. Conclusion

This work extends the discussion on the CTOF method of inter-robot distance measurement, introduced in (Micea et al., 2010). An extended discussion has also been made on the prerequisite sensor (robot) alignment procedure. The custom designed software simulation application provided the optimal ratio between the rotation speeds of the robots or their turrets with the ultrasonic transducers.

After the alignment process, the distance between two robots can be measured with the CTOF method. It has been shown that CTOF is independent of the communication propagation errors. We have also shown how the CTOF method meets the requirements of indoor, low-cost, energy-efficient robotic applications, reaching an accuracy of 4.8 cm for distances of 3 m. Furthermore, by applying the Kalman filter to repetitive distance measurements, an accuracy of 1 cm has been achieved for distances of 3 m, without the need of fixed landmarks.

The proposed distance measurement method and inter-robot alignment algorithm rely on inter-robot collaborative procedures and, therefore, these techniques are independent of fixed landmarks. Nevertheless, if the system further requires accurate localization of the mobile robots, at least the initial position of one of the robots must be known prior to the start of the system operation.

8. References

Atmel Corporation. (2010). *ATxmega64A1/128A1/192A1/256A1/384A1 Preliminary Data Sheet*, (Rev. M), Atmel Corporation, September 2010

Bestar Electronics. (2006). *BPU-1640IOAH12 Ultrasonic Sensor Datasheet*, Bestar Electronics Industry Co., Ltd., China, October 2006, Available from http://www.be-star.com/

Cioarga, R.D.; Micea, M.V.; Ciubotaru, B.; Chiuciudean, D.; Stanescu, D. (2006). CORE-TX: Collective Robotic Environment - the Timisoara Experiment, *Proceedings of the 3rd Romanian-Hungarian Joint Symposium on Applied Computational Intelligence, SACI 2006*, pp. 495–506, ISBN 963-7154-46-9, Timisoara, Romania, May 25-26, 2006

Cricket Project (2005). *Cricket v2 User Manual*, MIT Computer Science and Artificial Intelligence Lab, Cambridge, USA, Jan 2005

Digi International. (2009). *XBee/XBee-PRO RF Modules: Product Manual*, Digi International, Inc., USA, 2009, Available from http://www.digi.com/

Fayli, S.; Kleeman, L. (2004). A Real Time Advanced Sonar Ring with Simultaneous Firing, *Proceedings of 2004 IEEE-RSJ International Conference on Intelligent Robots and Systems*, Sendai, Japan, September - October 2, 2004

Fox, V.; Hightower, J.; Lin, L.; Schulz, D.; Borriello, G. (2003). Bayesian Filtering for Location Estimation. *IEEE Pervasive Computing*, Vol.2, No.3, (September 2003), pp. 24 - 33.

Fuicu, S.; Marcu, M.; Stratulat, B.; Gîrban, A. (2009). Effectiveness and Accuracy of Wireless Positioning Systems. *WSEAS Transactions on Computers*, Vol.8, No.9, (September 2009), pp. 1471-1483, ISSN 1109-2750

Grossmann, R.; Blumenthal, J.; Golatowski, F.; Timmermann, D. (2007). Localization in Zigbee-based Sensor Networks, *Proceedings of the 1st European ZigBee Developer's Conference, EuZDC 2007*, Munchen-Dornach, Germany, 2007

Ha, Y.S.; Kim, H.H. (2004). Environmental Map Building for a Mobile Robot Using Infrared Range-Finder Sensors. *Advanced Robotics*, Vol.18, No.4, (2004), pp. 437-450

Hagisonic. (2009) *Localization System StarGazer for Intelligent Robots: User's Guide*, Hagisonic Co., Ltd., Korea, 2009, Available from http://www.hagisonic.com/

Kim, D.-E.; Hwang, K.-H.; Lee, D.-H.; Kuc, T.-Y. (2006). A Simple Ultrasonic GPS System for Indoor Mobile Robot System using Kalman Filtering, *Proceedings of the SICE-ICASE International Joint Conference*, pp. 2915-2918, Busan, Korea, 2006

Ko, S.-I.; Choi, J.-S.; Kim, B.-H. (2008). Indoor Mobile Localization System and Stabilization of Localization Performance Using Pre-Filtering. *International Journal of Control, Automation, and Systems*, Vol.6, No.2, (April 2008), pp. 204-213

Micea, M.V.; Cretu, V.I.; Groza, V. (2006). Maximum Predictability in Signal Interactions With HARETICK Kernel. *IEEE Transactions on Instrumentation and Measurement*, Vol.55, No.4, (August 2006), ISSN 0018-9456

Micea, M.V.; Stancovici, A.; Chiciudean, D.; Filote, C. (2010). Indoor Inter-Robot Distance Measurement in Collaborative Systems. *Advances in Electrical and Computer Engineering*, Vol.10, No.3, (August 2010), pp. 21-26, ISSN 1582-7445

Mohammad, T. (2009). Using Ultrasonic and Infrared Sensors for Distance Measurement. *World Academy of Science, Engineering and Technology*, Vol.51, (2009)

Novotny, P. M.; Ferrier, N. J. (1999). Using Infrared Sensor and the Phong Illumination Model to Measure Distances, *Proceedings of the 1999 IEEE International Conference on Robotics and Automation*, pp. 1644-1649, Detroit, USA, May 1999

NXP Semiconductors. (April 2008). *UM10114: LPC21xx and LPC22xx User Manual* (Rev. 03), NXP Semiconductors N. V.

Ohno, K.; Tsubouchi, T.; Shigematsu, B.; Yuta, S. (2004). Differential GPS and Odometry-Based Outdoor Navigation of a Mobile Robot. *Advanced Robotics*, Vol.18, No.6, (2004), pp. 611-635

Priyantha, N.B. (2005). *The Cricket Indoor Location System*, Ph.D. Thesis, Department of Electrical Engineering and Computer Science, Massachusetts Institute of Technology, USA, Jun. 2005

Reina, G.; Vargas, A.; Nagatani, K.; Yoshida, K. (2007). Adaptive Kalman Filtering for GPS-based Mobile Robot Localization, *Proceedings of IEEE International Workshop on Safety, Security and Rescue Robotics, SSRR 2007*, pp. 1-6, 2007

Reynolds, M.S. (1999). *A Phase Measurement Radio Positioning System for Indoor Use*, Master's Dissertation, Department of Electrical Engineering and Computer Science, Massachusetts Institute of Technology, USA, Feb 1999.

Tamimi, H.; Andreasson, H.; Treptow, A.; Duckett, T.; Zell, A. (2006). Localization of Mobile Robots with Omnidirectional Vision Using Particle Filter and Iterative SIFT. *Robotic and Autonomous Systems*, Vol.54, 2006, pp. 758-765

Tower Pro. (2008). *Servomotor Information*, Specification Note, Tower Pro, January 2008, Available from http://www.towerpro.com.tw/driver/drivers/Towerpro%20 servo%20spec.pdf

Welch, G.; Bishop, G. (2006). *An Introduction to the Kalman Filter*, Technical Report, TR 95-041, University of North Carolina at Chapel Hill, USA, July 24, 2006

ZigBee Standards Organization. (2007). *ZigBee Specification*, ZigBee Alliance, Inc., California, SUA, Available from http://www.zigbee.org/

Part 5

Mobile Robot Operator Training

17

Improvement of RISE Mobile Robot Operator Training Tool

Janusz Będkowski and Andrzej Masłowski
Institute of Mathematical Machines
Poland

1. Introduction

In the chapter the framework for RISE (Risky Intervention and Environmental Surveillance) mobile robot operator training design is presented. The basic idea of the framework is to provide advanced software tools to improve the performance of the mobile robot simulation design that are applicable for operator training. The simulator is the end product of the framework. To achieve realistic rigid body simulation and realistic rendering the NVIDIA PhysX engine and the OGRE (Open Graphic Rendering Engine) are used. The compatibility between the framework software and CAD modeling tools such as SolidWorks was essential, therefore virtual models are exchangeable using COLLADA format. The simulator can be integrated with control panel of the mobile robot. It is possible to simulate the behavior of several types of robots such as caterpillar or wheeled.

The improvement presented in this chapter is related to the problem of automatic environment model generation based on autonomous mobile robot observations. The approach related to semantic mapping is used and in consequence semantic simulation engine is implemented. Semantic simulation engine is composed by data registration module, semantic entities identification module and environment model generation module. Presented result is a new approach related to mobile robots applications and potentially can improve the process of advanced mobile robot application design and professional training of the operators.

Training of the mobile robot operators is very difficult task not only because of the complexity of the mobile robot but also because of several factors related to different task execution. The study of the human-robot interactions (HRI) during a real rescue is presented in Casper & Murphy (2003) Bedkowski, Kowalski & Piszczek (2009). Several robotic systems are developed for rescue tasks Wei et al. (2009) Zhang et al. (2009). The research on simulation and training systems for mobile robots is shown in Xuewen et al. (2006). In the field of robotics applied for medical purposes such as surgery the simulation and training environments are developed and described in Schlaefer et al. (2008). The problem of learning via Web is common application used in several cases and as remote experiment system for distance training is discussed in Hong et al. (2007). Virtual labs have been meaningful and very popular in distance education in engineering oriented fields of study, since they allow to perform interesting experiments with the equipment in the labs remotely available through Internet Masar et al. (2004). An advance computer techniques such augmented reality approach applied in training robotics is shown in Kowalski et al. (2008)Bravo & Alberto (2009). The simulation environment used as a testbed for simulation based approach for unmanned system command and control is demonstrated in Drewes (2006). The research related to the training and the hypotheses through interactions with unmanned systems using computer mediated gesture recognition is shown in Varcholik et al. (2008), where the presented methodology employs the Nintendo Wii Remote Controller (Wiimote) to retrieve and classify one- and two-handed gestures that are mapped to an unmanned system command set. The Modeling and simulation for the mobile robot operator training tool was presented in Bedkowski, Piszczek, Kowalski & Maslowski (2009).

Semantic information Asada & Shirai (1989) extracted from 3D laser data Nüchter et al. (2005) is recent research topic of modern mobile robotics. In Nüchter & Hertzberg (2008) a semantic map for a mobile robot was described as a map that contains, in addition to spatial information about the environment, assignments of mapped features to entities of known classes. In Grau (1997) a model of an indoor scene is implemented as a semantic net. This approach is used in Nüchter, Surmann, Lingemann & Hertzberg (2003) where robot extracts semantic information from 3D models built from a laser scanner. In Cantzler et al. (2002) the location of features is extracted by using a probabilistic technique (RANSAC RANdom SAmple Consensus) Fischler & Bolles (1980). Also the region growing approach Eich et al. (2010) extended from Vaskevicius et al. (2007) by efficiently integrating k-nearest neighbor (KNN) search is able to process unorganized point clouds. The improvement of plane extraction from 3D Data by fusing laser data and vision is shown in Andreasson et al. (2005). The automatic model refinement of 3D scene is introduced in Nüchter, Surmann & Hertzberg (2003) where the idea of feature extraction (planes) is done also with RANSAC. The semantic map building is related to SLAM problem Oberlander et al. (2008) Se & Little (2002) Davison et al. (2004). Most of recent SLAM techniques use camera Castle et al. (2010) Williams & Reid (2010) Andreasson (2008), laser measurement system Pedraza et al. (2007) Thrun et al. (2000) or even registered 3D laser data Magnusson, Andreasson, Nüchter & Lilienthal (2009). Concerning the registration of 3D scans described in Magnusson et al. (2007) Andreasson & Lilienthal (2007) we can find several techniques solving this important issue. The authors of Besl & Mckay (1992) briefly describe ICP algorithm and in Hähnel & Burgard (2002) the probabilistic matching technique is proposed. In Magnusson, Nüchter, Lörken, Lilienthal & Hertzberg (2009) the comparison of ICP and NDT (Normal Distributions Transform) algorithm is shown. In Rusu et al. (2008) the mapping system that acquires 3D object models of man-made indoor environments such as kitchens is shown. The system segments and geometrically reconstructs cabinets with doors, tables, drawers, and shelves, objects that are important for robots retrieving and manipulating objects in these environments.

In this paper the application of framework for RISE mobile robot operator training design is presented. The framework is developed to improve the training development using advanced software tools. The semantic simulation engine is proposed to automatic environment model (composed by walls, door, stairs) generation and task execution with mobile robot supervision. As a result, it is demonstrated the task execution of training example of RISE robot, also examples of automatically generated scenes are shown. It should be noticed that the automatic generation of robot environment is still open problem and needs further developments.

2. Framework overview

The platform is going to support specialized tools for designing physical models, spatial models and others. The standard implemented and used in the platform accepts Collada and PhysX files for designing physical models. For spatial models 3ds format is considered as the best choice. Physical models can be developed using commercially available tools such as Solid Works with Collada plug-in, or Maya with ColladaMaya plug-in. It is also possible to use freeware tool under GNU/GPL license such a Blender with Collada plug-in. Similarly for designing spatial models one can use Autodesk 3ds Max as a non-free product which native file standard is 3ds or free competitor on the market Blender which can handle 3ds file as well. Utilization of presented tools ensures long term support for the designed platform. The framework consists of several software tools for training components development. Figure 1 shows the scheme of the expected end product of the framework - RISE robot operator training.

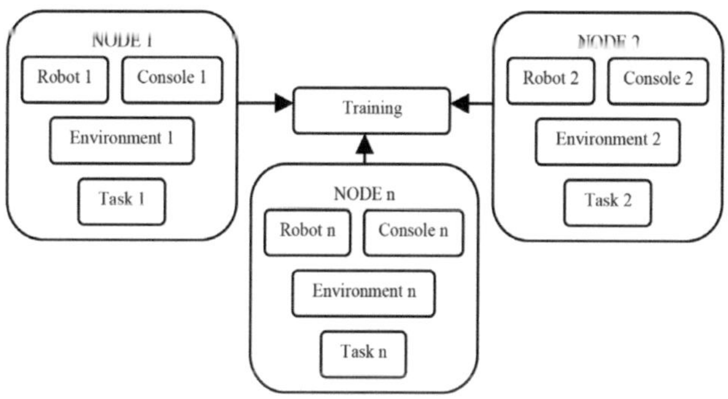

Fig. 1. The scheme of the expected end product of the framework - RISE robot operator training.

The training is composed from n nodes. Each node consist of robot model, console model, environment model, and task model. Framework provides software tools for model design and integration such as: Robot Builder, Environment Builder, Console Builder, Task Builder, Training Builder. The main concept and the achievement is the limitation of the programmer effort during training design (all activities can be done using the framework advanced GUI applications), therefore the process of models design and integration is improved and the time and potential costs are decreased. The training can be performed simultaneously for several operators. It is possible to construct training classroom for multiple operator training. To demonstrate this, the single training architecture has to be discussed. Figure 2 shows the software architecture of single training.

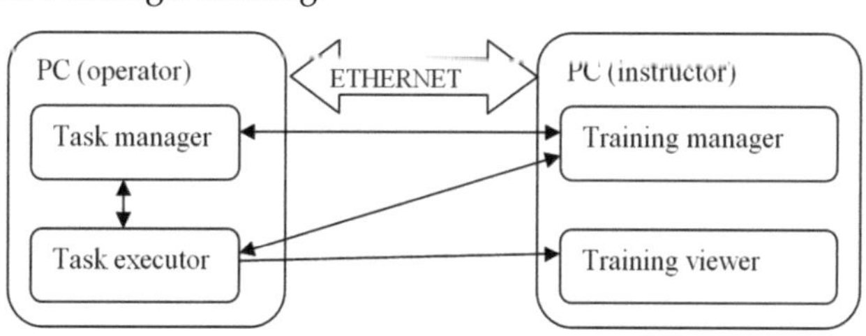

Fig. 2. Software architecture of single training.

To perform the single training 2 PCs with framework software has to be taken into account. PC (operator) consists of: task manager and task executor programs. These programs are responsible for proper task execution and sending the result to the PC (instructor) where the training process is supervised with training manager program. Instructor can observe the operator behavior using training viewer software that visualizes the simulated scene. For the multiple training PC (instructor) can use separate training manager and training viewer programs to perform single training on several operator PCs simultaneously.

2.1 Model of a robot

Virtual model of robot is the basic element of a simulation. The figure 3 presents main elements of the robot model.

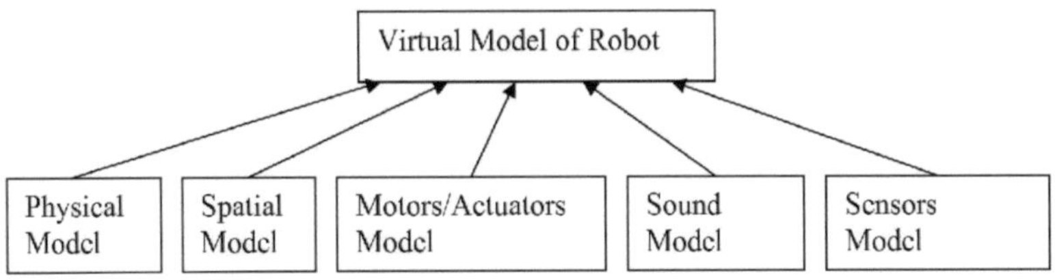

Fig. 3. Scheme of components of virtual model of robot.

Virtual model of robot is composed of the following models:
a) physical (represents mechanical properties of a designed robot),
b) spatial (represents what is visualized),
c) motors/actuators (represents all driven parts of a robot),
d) sound (represents a set of sounds connected to associated to the model of motors/actuators),
e) sensors (represents a set of sensors mounted onto a robot).

Each listed model is described in a single file, therefore full virtual model of robot requires five files and the information about their interaction. Framework offers a tool called Robot Builder for designing a robot. The window of the application is presented in the Figure 4.

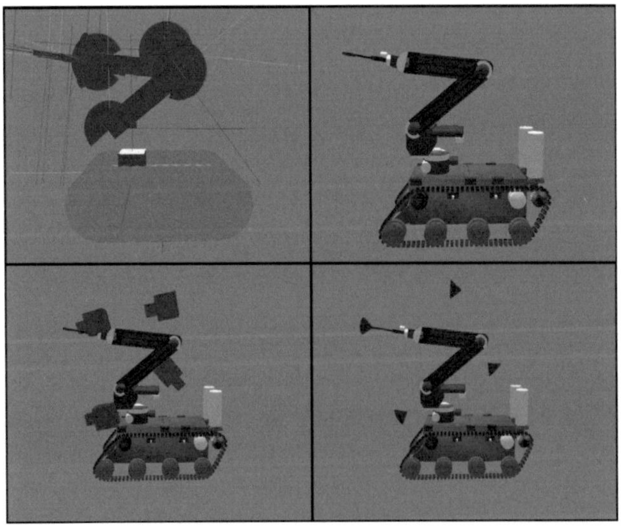

Fig. 4. Robot Builder: up left - physic model, up right - spatial model, bottom left - sensor model (cameras), bottom right - sensor model (lights).

2.2 Model of an environment

Virtual model of an environment similarly as a models of control panel and robot consists the following elements: physical model, spatial model and sound model. The scheme shown in figure 5 presents main elements of an environment.

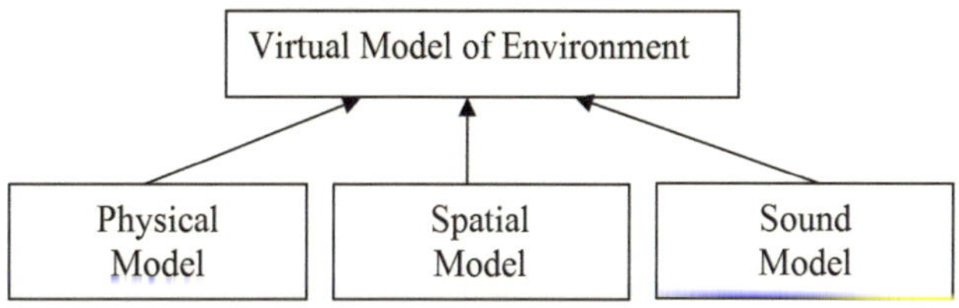

Fig. 5. Composition of a virtual model of an environment.

The environment in the simulator is represented by Virtual Model of Environment. It is accepted that this model will be designed in similar manner as virtual model of robot described previously. The platform supplies such a tool which can support a designer in the process of creating the environment. It will be possible to utilize components from the Components Base. This Base will consists a set of sample models such as buildings, furniture (chairs, closets, tables, etc.) , vehicles, trees, different kinds of basis (i.e. soil, road, grass) (see figure 6). All objects from the Base will have physical and spatial properties. Apart from the physical properties, there will be a spatial model attached and visualized. Also special effect such as particle system, fog, animation and water can be integrated into the scene.

Fig. 6. Environment of the robot.

2.3 Model of a control panel

Virtual model of a control panel is a second basic element of a simulation. The figure 7 presents the main elements of a control panel. It is assumed that control panel consists of joysticks, buttons, switches and displays. The control panel communicates with a robot via communicating module or directly dependently whether a real or a virtual panel is in use. Control panel can be virtual or real thus the trainee can accustom to real device controlling the virtual robot using real control panel. Described technology is presented in the figure 8.

2.4 Task

In order to design Task, the following models are required: robot model, control panel model and environmental model. A training designer has the ability to set the beginning position of

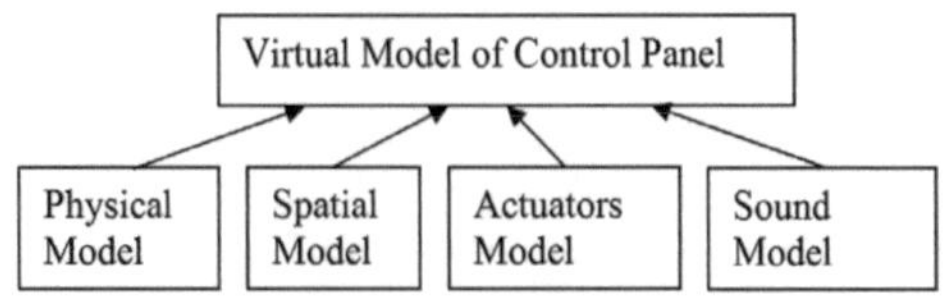

Fig. 7. Composition of necessary elements of a control panel.

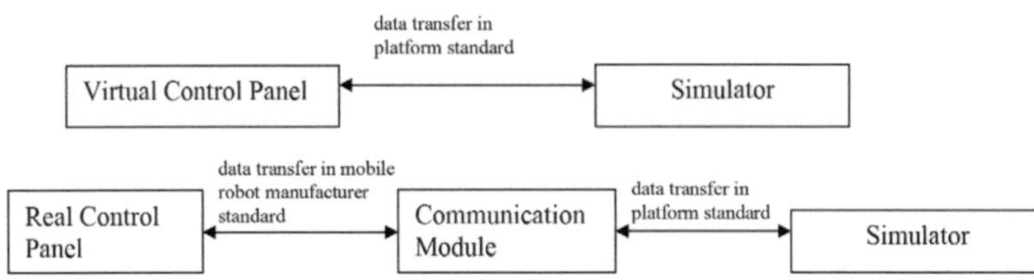

Fig. 8. Composition of necessary elements of a control panel and robot communication.

the robot and after that define all possible mission events he concerns as important. Next, he can define time thresholds for the mission. Such e-Task is then exported and can be used to prepare the full multilevel training. The task designing steps are shown in figure 9. For this purpose, framework supplies e-Tasks Generator. The tool supplies the base of events. The Task designer will be able to move about the scene with the robot and environment read in then chose one actor or group of actors and attach an event with specific parameters. The possible events defined for the single Task are: a) robot path to chosen localization, b) time exceeded, c) move an object to chosen localization, d) touch an object, e) damage/disable robot, f) neutralize/destroy an object (e.g. shoot with water gun, put an explosives).

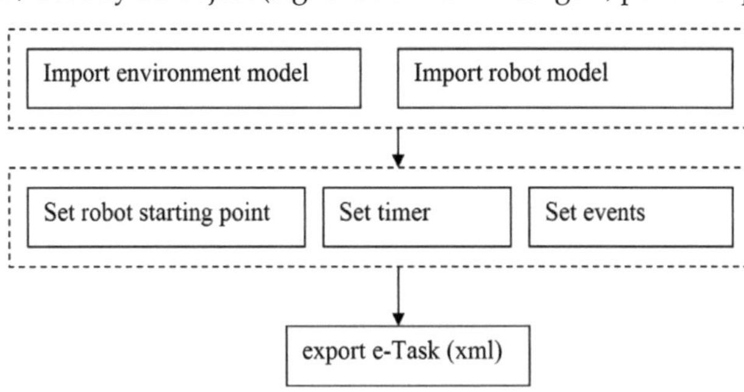

Fig. 9. Task designing steps.

2.5 Training

When a designer get through the previous steps the last what is required is to prepare the training. Using previously defined Tasks one can prepare a graph of the training. The sample graph is presented in figure 10. The training starts with the Mission 1 and leads through two or three subsequent missions. In the first case when the condition C_3 is satisfied it means that Mission 2 is non requisite otherwise if the condition C_2 is satisfied the training path will lead through the Mission 2. Under conditions C_1 C_4 C_5 C_8 appropriate missions will be repeated. At the end of each training the summary is presented on the screen and saved to file. The file is imported to the training manager. Figure 11 shows the Training manager GUI. Training manager GUI informs the user about currently task report, also the training report is available for further tasks execution. Training manager can visualize the simulated scene for analyses purposes. Figure 12 shows executed task taskMETRO. The main objective was to transport hazardous object into defined place of neutralization.

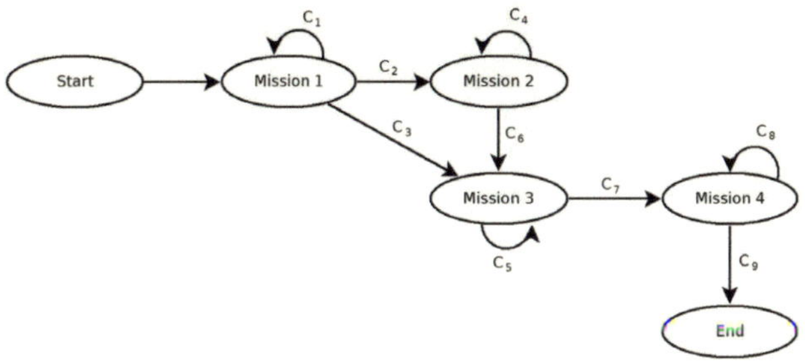

Fig. 10. Scheme of the training graph. Each node must be one of three: start, end or mission; transition to the next mission is represented by arrow and is triggered under specific condition denoted as C_1 ... C_9

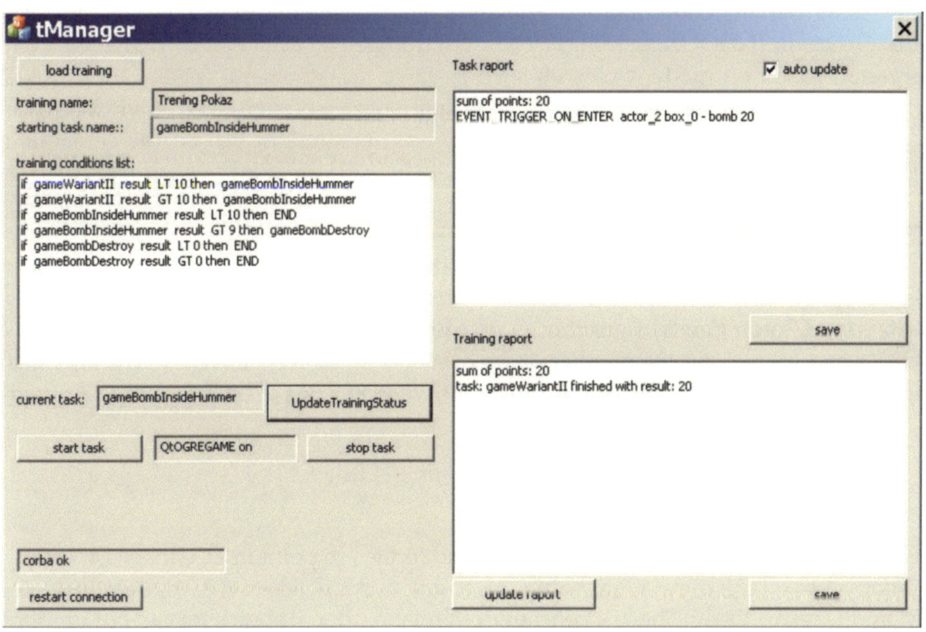

Fig. 11. Training manager GUI.

3. Framework improvement

Framework for robot operator training design is improved by semantic simulation engine. Semantic simulation engine J. Bedkowski (2011b) is a project that main idea is to improve State of The Art in the area of semantic robotics Asada & Shirai (1989) Nüchter & Hertzberg (2008) Grau (1997) with the focus on practical applications such as robot supervision and control, semantic map building, robot reasoning and robot operator training using augmented reality techniques Kowalski et al. (2011) J. Bedkowski (2011a). It combines semantic map with rigid body simulation to perform supervision of its entities such as robots moving in INDOOR or OUTDOOR environments composed by floor, ceiling, walls, door, tree, etc.

Fig. 12. Task taskMETRO execution. The goal is to transport hazardous object to appropriate location.

3.1 Semantic simulation engine

Semantic simulation engine is composed of data registration modules, semantic entities identification (data segmentation) modules and semantic simulation module. It provides tools to implement mobile robot simulation based on real data delivered by robot and processed on-line using parallel computation. Semantic entities identification modules can classify several objects in INDOOR and OUTDOOR environments. Data can be delivered by robot observation based on modern sensors such as laser measurement system 3D and RGB-D cameras. Real objects are associated with virtual entities of simulated environment. The concept of semantic simulation is a new idea, and its strength lies on the semantic map integration with mobile robot simulator.

3.2 Data registration

Data registration module provides accurate alignment of robot observations. Aligning two-view range images with respect to the reference coordinate system is performed by the ICP (Iterative Closest Points) algorithm. Range images are defined as a model set M and data set D, N_m and N_d denotes the number of the elements in the respective set. The alignment of these two data sets is solved by minimization with respect to **R**,**t** of the following cost function:

$$E(\mathbf{R},\mathbf{t}) = \sum_{i=1}^{N_m}\sum_{j=1}^{N_d} w_{ij} \left\| \mathbf{m}_i - \left(\mathbf{R}\mathbf{d}_j + \mathbf{t}\right) \right\|^2 \tag{1}$$

w_{ij} is assigned 1 if the i-th point of M correspond to the j-th point in D. Otherwise w_{ij}=0. **R** is a rotation matrix, **t** is a translation matrix. $\mathbf{m}_i$ and $\mathbf{d}_i$ corresponds to the i-th point from model set M and D respectively. The key concept of the standard ICP algorithm can be summarized in two steps Segal et al. (2009):

1) compute correspondences between the two scans (Nearest Neighbor Search).

2) compute a transformation which minimizes distance between corresponding points.

Iteratively repeating these two steps should results in convergence to the desired transformation. Because we are violating the assumption of full overlap, we are forced to add a maximum matching threshold d_{max}. This threshold accounts for the fact that some points will not have any correspondence in the second scan. In most implementations of ICP, the choice of d_{max} represents a trade off between convergence and accuracy. A low value results in bad convergence, a large value causes incorrect correspondences to pull the final alignment away from the correct value. Classic ICP is listed as algorithm 1 and the ICP algorithm using CUDA parallel programming is listed as algorithm 2. Parallel programming applied for ICP computation results the average time 300ms for 30 iterations of ICP. The proposed solution is efficient since it performs nearest neighbor search using a bucket data structure (sorted using CUDA primitive) and computes the correlation matrix using parallel CUDA all-prefix-sum instruction.

We can consider data registration module as a component executing 6D SLAM - Simultaneous Localization and Mapping, where 6D is related to vector describing robot position in 3D (x,y,z,yaw,pitch,roll). The 6D SLAM algorithm executes three steps: a) ICP alignment of 2 robot observations, b) loop closing, c) map relaxation. The example result of data registration module is shown in figure 13

Algorithm 1 Classic ICP

INPUT: Two point clouds A = $\{a_i\}$, B= $\{b_i\}$, an initial transformation T_0
OUTPUT: The correct transformation T, which aligns A and B
$T \leftarrow T_0$
for $iter \leftarrow 0$ to $maxIterations$ **do**
  for $i \leftarrow 0$ to N **do**
    $m_i \leftarrow$ FindClosestPointInA($T \cdot b_i$)
    if $\|m_i - T \cdot b_i\| \leq d_{max}$ **then**
      $w_i \leftarrow 1$
    else
      $w_i \leftarrow 0$
    end if
  end for
  $T \leftarrow \underset{T}{argmin} \left\{ \sum_i w_i \|T \cdot b_i - m_i\|^2 \right\}$
end for

Algorithm 2 ICP - parallel computing approach

INPUT: Two point clouds M = $\{m_i\}$, D= $\{d_i\}$, an initial transformation T_0
OUTPUT: The correct transformation T, which aligns M and D
$M_{device} \leftarrow M$
$D_{device} \leftarrow D$
$T_{device} \leftarrow T_0$
for $iter \leftarrow 0$ to $maxIterations$ **do**
  for $i \leftarrow 0$ to N {in parallel} **do**
    $m_i \leftarrow$ FindClosestPointInM ($T_{device} \cdot d_i$) {using regular grid decomposition}
    if $foundClosestPointInNeighboringBuckets$ **then**
      $w_i \leftarrow 1$
    else
      $w_i \leftarrow 0$
    end if
  end for
  $T_{device} \leftarrow \underset{T_{device}}{argmin} \left\{ \sum_i w_i \|T \cdot d_i - m_i\|^2 \right\}$ {calculation T←R,t with SVD}
end for
$M \leftarrow M_{device}$
$D \leftarrow D_{device}$
$T \leftarrow T_{device}$

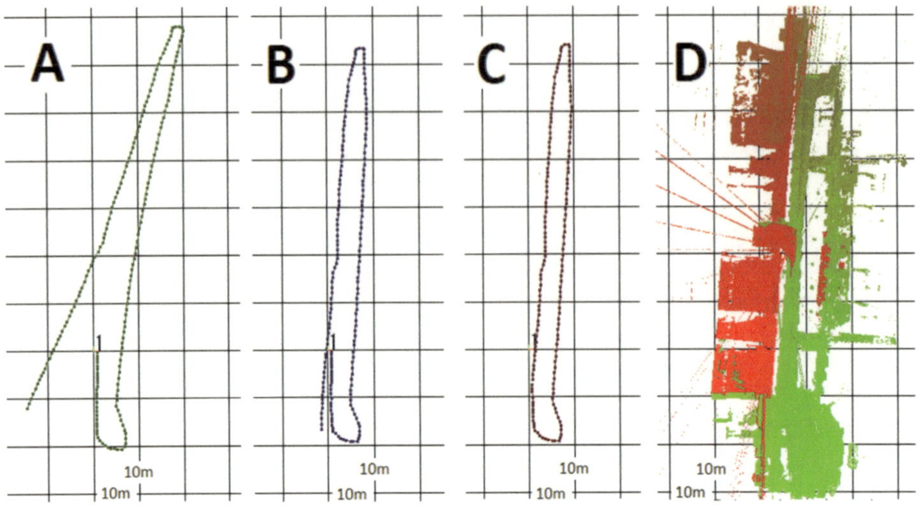

Fig. 13. Example of a result of data registration module. A - odometry with gyroscope correction, B - ICP, C - loop closed, D - final 3D map. Data was acquired in Royal Military Academy building (Brussels, Belgium) using PIONEER 3AT robot equipped with 3DLSN unit (rotated LMS200).

3.3 Semantic entities identification

The procedure of prerequisites generation using image processing methods is used. The set of lines obtained by Hough transform from projected $3D$ points onto $2D$ OXY plane is used to obtain segmentation of cloud of points, where different walls will have different labels. For each line segment the orthogonal $plane_{orth}$ to $plane_{OXY}$ is computed. The intersection between this two planes is the same line segment. All $3D$ points which satisfy the condition of distance to $plane_{orth}$ have the same label. In the first step all prerequisites of walls were checked separately - it is data segmentation. To perform the scene interpretation semantic net is used (figure 14). The feature detection algorithm is composed of cubes generation method, where each cube should contain measured 3D point after segmentation (see figure 15). In the second step of the algorithm wall candidates are chosen. From this set of candidates, based on relationships between them, proper labels are assigned and output model is generated (see figure 16 left). The image processing methods are also used for stairs prerequisites generation. The set of parallel lines (obtained by projected single 3D scan onto OXY plane) in the same short distance between each other is prerequisite of stairs. Possible labels of the nodes are L = {stair}. The relationships between the entities are R = {parallel, above, under}. Figure 16 right shows resulting model of stairs generated from 3D cloud of points. In this spatial model each stair (except first and last one obviously) is in relation r=above&parallel with the previous one and in relation r=under&parallel with next one.

3.4 Environment model generation

Model of the environment is automatically generated using semantic net from set of identified semantic entities. The engine basic elements for INDOOR environment are: semantic map nodes(entities) L_{sm}={Wall, Wall above door, Floor, Ceiling, Door, Free space for door, Stairs...}, the L_{sm} set can be extended by another objects, what is dependent on robust and accurate 3D scene analysis, robot simulator nodes(entities) L_{rs}={robot, rigid body object, soft body

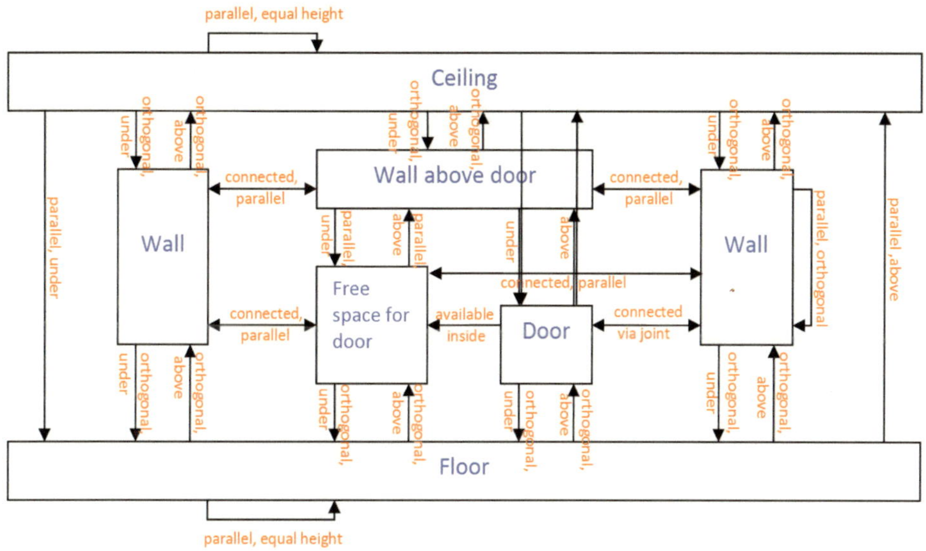

Fig. 14. Semantic net defined for semantic entities identification.

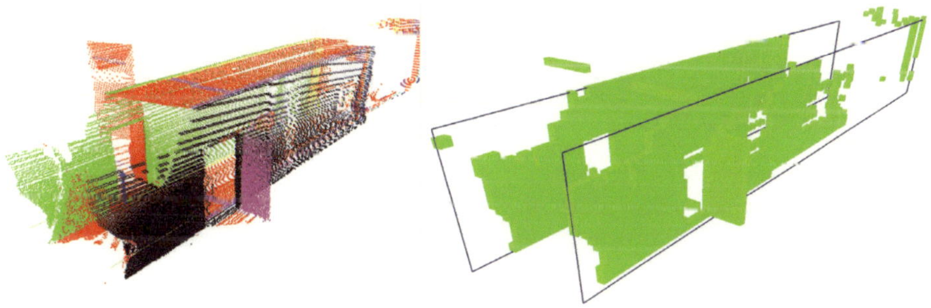

Fig. 15. Left - segmentation of 3D cloud of points, right - boxes that contain measured points.

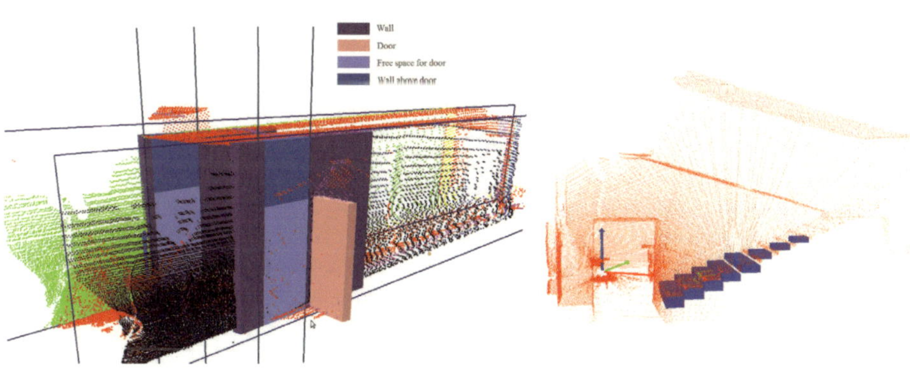

Fig. 16. Scene interpretation left - door, walls, right - stairs.

object...}, semantic map relationships between the entities R_{sm} = {parallel, orthogonal, above, under, equal height, available inside, connected via joint...}, robot simulator relationships between the entities R_{rs} = {connected via joint, position...}, semantic map events E_{sm} = robot simulator events E_{rs} = {movement, collision between two entities started, collision between two entities stopped, collision between two entities continued, broken joint...}. Robot simulator is implemented using NVIDIA PhysX library. The entities from semantic map correspond to actors in PhysX. L_{sm} is transformed into L_{rs} based on spatial model generated based on registered 3D scans. R_{sm} is transformed into R_{rs}. All entities/relations R_{sm} has the same initial location in R_{rs}, obviously the location of each actor/entity may change during simulation, therefore accurate object tracking system is needed. The transformation from E_{sm} to E_{rs} effects that events related to entities from semantic map correspond to the events related to actors representing proper entities. Following events can be noticed during simulation: robot can touch each entity (see figure 17), open/close the door and enter empty space of the door (see figure 18), climb the stairs (see figure 19), damage itself -broken joint between actors in robot arm (see figure 20), brake joint that connects door to the wall.

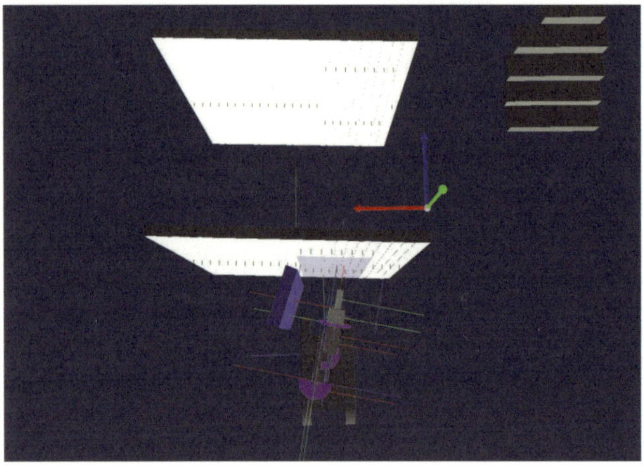

Fig. 17. Simulated robot touches entity.

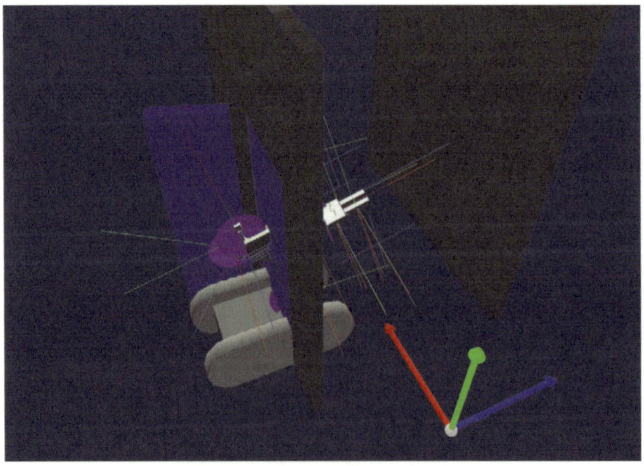

Fig. 18. Simulated robot enters empty space of the door.

Fig. 19. Simulated robot climbs the stairs.

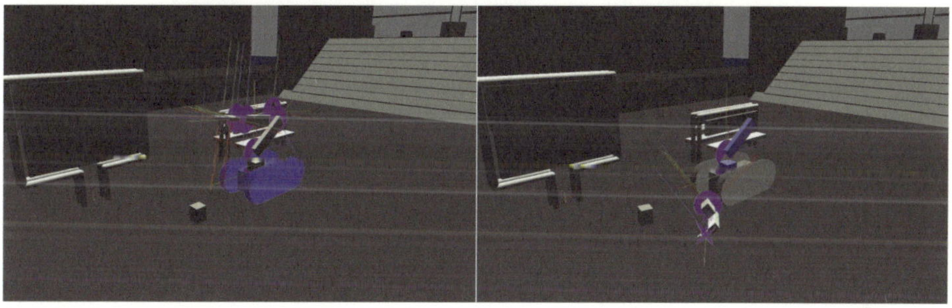

Fig. 20. Simulated robot damages itself.

4. Conclusion and future work

The chapter presents the framework for RISE mobile robot operator training design with an improvement based on semantic simulation engine. The framework is composed by advanced software tools to improve the performance of the mobile robot simulation design that are applicable for operator training. The simulator is the end product of the framework. The effort to build the simulation software is decreased by developed programs that allow to build several model such as robot model, and environment model. The framework can help in RISE robot training classroom building for several task execution in simultaneous mode, therefore multiple training can be performed. We believe that developed software can help also in multi robotic system design and development by providing advanced techniques for simulation process, therefore the framework can be applicable in several mobile robotics applications.
New concept of semantic simulation engine, composed of data registration modules, semantic entities identification modules and semantic simulation module is demonstrated. The implementation of parallel computing applied for 6D SLAM, especially for data registration based on regular grid decomposition is shown. Semantic simulation engine provides tools to implement mobile robot simulation based on real data delivered by robot and processed on-line using parallel computation. Semantic entities identification modules can classify door, walls, floor, ceiling, stairs in indoor environment. Semantic simulation uses NVIDIA PhysX for rigid body simulation. Future work will be related to AI techniques applied for semantic entities identification (furnitures, victims, cars, etc...), localization and tracking methods.

Concerning 6D SLAM, semantic loop closing techniques will be taken into the consideration as promising methods delivering conceptual reasoning.

5. References

Andreasson, H. (2008). *Local Visual Feature based Localisation and Mapping by Mobile Robots*, Doctoral thesis, Orebro University, School of Science and Technology.

Andreasson, H. & Lilienthal, A. J. (2007). Vision aided 3d laser based registration, *Proceedings of the European Conference on Mobile Robots (ECMR)*, pp. 192–197.

Andreasson, H., Triebel, R. & Burgard, W. (2005). Improving plane extraction from 3d data by fusing laser data and vision, *Proceedings of the IEEE/RSJ International Conference on Intelligent Robots and Systems (IROS)*, pp. 2656–2661.

Asada, M. & Shirai, Y. (1989). Building a world model for a mobile robot using dynamic semantic constraints, *Proc. 11 th International Joint Conference on Artificial Intelligence*, pp. 1629–1634.

Bedkowski, J., Kowalski, P. & Piszczek, J. (2009). Hmi for multi robotic inspection system for risky intervention and environmental surveillance, *Mobile Robotics: Solutions and Challenges, Proceedings of the Twelfth International Conference on Climbing and Walking Robots and the Support Technologies for Mobile Machines*.

Bedkowski, J., Piszczek, J., Kowalski, P. & Maslowski, A. (2009). Modeling and simulation for the mobile robot operator training tool, *The 5th International Conference on Information and Communication Technology and Systems (ICTS), Surabaya, Indonesia*, pp. 229–236.

Besl, P. J. & Mckay, H. D. (1992). A method for registration of 3-d shapes, *Pattern Analysis and Machine Intelligence, IEEE Transactions on* 14(2): 239–256.

Bravo, J. & Alberto, C. (2009). An augmented reality interface for training robotics through the web, *Proceedings of the 40th International Symposium on Robotics. Barcelona : AER-ATP, ISBN 978-84-920933-8-0*, pp. 189–194.

Cantzler, H., Fisher, R. B. & Devy, M. (2002). Quality enhancement of reconstructed 3d models using coplanarity and constraints, *Proceedings of the 24th DAGM Symposium on Pattern Recognition*, Springer-Verlag, London, UK, pp. 34–41.

Casper, J. & Murphy, R. R. (2003). Human-robot interactions during the robot-assisted urban search and rescue response at the World Trade Center., *IEEE transactions on systems, man, and cybernetics. Part B, Cybernetics : a publication of the IEEE Systems, Man, and Cybernetics Society* 33(3): 367–385.
URL: *http://dx.doi.org/10.1109/TSMCB.2003.811794*

Castle, R. O., Klein, G. & Murray, D. W. (2010). Combining monoslam with object recognition for scene augmentation using a wearable camera, 28(11): 1548 – 1556.

Davison, A., Cid, Y. G. & Kita, N. (2004). Real-time 3D SLAM with wide-angle vision, *Proc. IFAC Symposium on Intelligent Autonomous Vehicles, Lisbon.*

Drewes, P. (2006). Simulation based approach for unmanned system command and control, *Conference on Recent Advances in Robotics, FCRAR Miami, Florida*, pp. 189–194.

Eich, M., Dabrowska, M. & Kirchner, F. (2010). Semantic labeling: Classification of 3d entities based on spatial feature descriptors, *IEEE International Conference on Robotics and Automation (ICRA2010) in Anchorage, Alaska, May 3.*

Fischler, M. A. & Bolles, R. (1980). Random sample consensus. a paradigm for model fitting with apphcahons to image analysm and automated cartography, *Proc. 1980 Image Understandtng Workshop (College Park, Md., Apr i980) L. S. Baurnann, Ed, Scmnce Apphcatlons, McLean, Va.*, pp. 71–88.

Grau, O. (1997). A scene analysis system for the generation of 3-d models, *NRC '97: Proceedings of the International Conference on Recent Advances in 3-D Digital Imaging and Modeling*, IEEE Computer Society, Washington, DC, USA, p. 221.

Hähnel, D. & Burgard, W. (2002). Probabilistic matching for 3D scan registration, *In.: Proc. of the VDI - Conference Robotik 2002 (Robotik)*.

Hong, S. H., Park, J. H., Kwon, K. H. & Jeon, J. W. (2007). A distance learning system for robotics, *Lecture Notes in Computer Science, Computational Science ICCS 2007* 4489/2007: 523–530.

J. Bedkowski, A. M. (2011a). Methodology of control and supervision of web connected mobile robots with cuda technology application, *Journal of Automation, Mobile Robotics and Intelligent Systems* 5(2): 3 11.

J. Bedkowski, A. M. (2011b). Semantic simulation engine for mobile robotic applications, *Pomiary, Automatyka, Robotyka 2/2011, Automation 2011 6-8 April, Warsaw* pp. 333–343.

Kowalski, G., Bedkowski, J., Kowalski, P. & Maslowski, A. (2011). Computer training with ground teleoperated robots for de-mining, *Using robots in hazardous environments, Landmine detection, de-mining and other applications* pp. 397–418.

Kowalski, G., Bedkowski, J. & Maslowski, A. (2008). Virtual and augmented reality as a mobile robots inspections operators training aid, *Bulletin of the Transilvania University of Brasov. Proceedings of the international conference Robotics08, special issue vol. 15(50) series A, no. 1, vol. 2, Brasov, Romania*, pp. 571–576.

Magnusson, M., Andreasson, H., Nüchter, A. & Lilienthal, A. J. (2009). Automatic appearance-based loop detection from 3D laser data using the normal distributions transform, *Journal of Field Robotics* 26(11–12): 892–914.

Magnusson, M., Duckett, T. & Lilienthal, A. J. (2007). 3d scan registration for autonomous mining vehicles, *Journal of Field Robotics* 24(10): 803–827.

Magnusson, M., Nüchter, A., Lörken, C., Lilienthal, A. J. & Hertzberg, J. (2009). Evaluation of 3d registration reliability and speed âĂŞ- a comparison of icp and ndt, *Proc. IEEE Int. Conf. on Robotics and Automation*, pp. 3907–3912.

Masar, I., Bischoff, A. & Gerke, M. (2004). Remote experimentation in distance education for control engineers, *Proceedings of Virtual University, Slovakia*.

Nüchter, A. & Hertzberg, J. (2008). Towards semantic maps for mobile robots, *Robot. Auton. Syst.* 56(11): 915–926.

Nüchter, A., Surmann, H. & Hertzberg, J. (2003). Automatic model refinement for 3D reconstruction with mobile robots, *Fourth International Conference on 3-D Digital Imaging and Modeling 3DIM 03*, p. 394.

Nüchter, A., Surmann, H., Lingemann, K. & Hertzberg, J. (2003). Semantic scene analysis of scanned 3d indoor environments, *in: Proceedings of the Eighth International Fall Workshop on Vision, Modeling, and Visualization (VMV 03)*.

Nüchter, A., Wulf, O., Lingemann, K., Hertzberg, J., Wagner, B. & Surmann, H. (2005). 3d mapping with semantic knowledge, *IN ROBOCUP INTERNATIONAL SYMPOSIUM*, pp. 335–346.

Oberlander, J., Uhl, K., Zollner, J. M. & Dillmann, R. (2008). A region-based slam algorithm capturing metric, topological, and semantic properties, *ICRA'08*, pp. 1886–1891.

Pedraza, L., Dissanayake, G., Miro, J. V., Rodriguez-Losada, D. & Matia, F. (2007). Bs-slam: Shaping the world, *Proceedings of Robotics: Science and Systems*, Atlanta, GA, USA.

Rusu, R. B., Marton, Z. C., Blodow, N., Dolha, M. & Beetz, M. (2008). Towards 3d point cloud based object maps for household environments, *Robot. Auton. Syst.* 56(11): 927–941.

Schlaefer, A., Gill, J. & Schweikard, A. (2008). A simulation and training environment for robotic radiosurgery, *International Journal of Computer Assisted Radiology and Surgery* 3(3-4): 267–274.

Se, S. & Little, D. L. J. (2002). Mobile robot localization and mapping with uncertainty using scale-invariant visual landmarks, *The International Journal of Robotics Research* 21(8): 735–758.

Segal, A., Haehnel, D. & Thrun, S. (2009). Generalized-icp, *Proceedings of Robotics: Science and Systems*, Seattle, USA.

Thrun, S., Burgard, W. & Fo, D. (2000). A real-time algorithm for mobile robot mapping with applications to multi-robot and 3d mapping, *ICRA*, pp. 321–328.

Varcholik, P., Barber, D. & Nicholson, D. (2008). Interactions and training with unmanned systems and the nintendo wiimote, *Interservice/Industry Training, Simulation, and Education Conference (I/ITSEC), Paper No. 8255*, pp. 189–194.

Vaskevicius, N., Birk, A., Pathak, K. & Poppinga, J. (2007). Fast detection of polygons in 3d point clouds from noise-prone range sensors, *IEEE International Workshop on Safety, Security and Rescue Robotics, SSRR*, IEEE, Rome, pp. 1–6.

Wei, B., Gao, J., Zhu, J. & Li, K. (2009). A single-hand and binocular visual system for eod robot, *Second International Conference on Intelligent Computation Technology and Automation, ICICTA '09*, Vol. 3, pp. 403 – 406.

Williams, B. & Reid, I. (2010). On combining visual slam and visual odometry, *Proc. International Conference on Robotics and Automation.*

Xuewen, L., Cai, M., Jianhong, L. & Tianmiao, W. (2006). Research on simulation and training system for eod robots, *IEEE International Conference on Industrial Informatics*, pp. 810 – 814.

Zhang, W., Yuan, J., Li, J. & Tang, Z. (2009). The optimization scheme for eod robot based on supervising control architecture, *Proceedings of the 2008 IEEE International Conference on Robotics and Biomimetics*, pp. 1421 – 1426.